Mineral Deposit Evaluation

Mineral Deposit Evaluation

A practical approach

Alwyn E. Annels
Department of Geology, University of Wales, Cardiff

CHAPMAN & HALL
London · New York · Tokyo · Melbourne · Madras

UK	Chapman & Hall, 2–6 Boundary Row, London SE1 8HN
USA	Van Nostrand Reinhold, 115 5th Avenue, New York NY10003
JAPAN	Chapman & Hall Japan, Thomson Publishing Japan, Hirakawacho Nemoto Building, 7F, 1-7-11 Hirakawa-cho, Chiyoda-ku, Tokyo 102
AUSTRALIA	Chapman & Hall Australia, Thomas Nelson Australia, 102 Dodds Street, South Melbourne, Victoria 3205
INDIA	Chapman & Hall India, R. Seshadri, 32 Second Main Road, CIT East, Madras 600 035

First edition 1991

Typeset in 10/12pt Bembo by Acorn Bookwork, Salisbury
Printed in Great Britain by Clays Ltd., St. Ives plc.

ISBN 0 412 35290 7 0 442 31305 5 (USA)

British Library Cataloguing in Publication Data

Annels, Alwyn E.
Mineral deposit evaluation: A practical approach.
I. Title
553

ISBN 0-412-35290-7

Library of Congress Cataloging-in-Publication Data

Annels, Alwyn E.
Mineral deposit evaluation : a practical approach/Alwyn E. Annels. — 1st ed.
p. cm.
Includes bibliographical references and index.
ISBN 0-442-31305-5
1. Mine valuation. I. Title.
TN272.A56 1991 90-2454
622'.1—dc20 CIP

To my wife

Anita

For her patience and tolerance
during the writing of this book.

Contents

Preface

Although aspects of mineral deposit evaluation are covered in such texts as McKinstry (1948), Peters (1978), Reedman (1979) and Barnes (1980), no widely available in-depth treatment of the subject has been presented. It is thus the intention of the present book to produce a text which is suitable for both undergraduate and postgraduate students of mining geology and mining engineering and which, at the same time, is of use to those already following a professional career in the mining industry. An attempt has been made to present the material in such a way as to be intelligible to the average geologist, or engineer, who is perhaps daunted by the more mathematical approach to the subject of ore-reserves found in more specialist books and papers. Although most of the theory in this book is written using metric units, individual case histories are described using the units employed at each mine at the time of writing.

The following chapters will thus examine the role of the mining geologist in the sampling of mineral deposits and in the calculation of mineral inventories and mineable reserves by both 'classical' and geostatistical methods. The techniques available for this purpose will be examined and actual case-history examples of their use presented. It is essential that the geologist in the mining industry has a full understanding of the advantages and disadvantages of each technique so that a judgement can be made as to their applicability to a particular deposit and the mining method proposed or used. Too often, a lack of this expertise results in the ore-reserve calculation being undertaken at head-office or, indeed, by the survey department on the mine, and being treated as a 'number crunching' or geometric exercise divorced from geology. It is essential that mine ore-reserves are calculated at the mine by those geologists who are most closely associated with the local geology and who are thus best able to influence and/or constrain the calculation. Where the reserves are determined by computerized techniques it is particularly important that they understand the algorithm used and not be kept in the dark as to exactly how each block of ground is valued. Geologists are not there to be merely keyboard operators for they must be in a position to assess whether the results being produced are meaningful and also be able to modify various user specifications to produce the desired result.

Other fields in which mining geologists will play an important, if not dominant and essential role, are in (a) metallurgical test sampling, where they must ensure that the samples sent for analysis and pilot-plant testing are representative of the ore (plus expected dilution) to be mined in different areas of the deposit; (b) in grade control, where they must ensure that mining is confined to the ore-zone and that dilution is kept within acceptable limits; (c) in the representation of mine geological and assay data, and (d) in the assessment of the economic viability of a deposit and proposed mining method during mine feasibility and design studies. During pre-production and production phases they will be heavily involved in assessing the economic impact of the hydrogeology of the mine catchment area and will thus be involved with aspects of mine and aquifer drainage. They will also be heavily involved with geotechnical surveys and rock mechanics and strata-control problems for these will have considerable influence on mine viability. Environmental impact studies and the location of plant sites, tailings dams, etc. will be an additional call on the expertise of the mining geologist.

In order to face the demands outlined above, mining geologists need to have a high level of technical competence, especially in the fields of ore-deposit geology, computing and geostatistics and also need to possess a working knowledge of mining methods applicable to different types of orebody/mineral deposit. Over and above these technical attributes, a wide range of abilities are required which are listed below.

(1) The ability to communicate with, and transmit ideas to, both professional and production personnel on the mine.

(2) Although a strong educational base in geology and mining science is essential, they must have a good deal of of initiative and the ability to think logically. Having made an interpretation or decision on geological grounds, they must be able to assess how this will effect the day-to-day operation of the mine and whether the proposed course of action is feasible, practical and cost effective.

(3) They must be decisive and not afraid of making mistakes. When these are made, however, it is essential that an assessment is made of what went wrong for future reference. Too often geologists are criticized for 'sitting on the fence' and presenting alternative explanations or possibilities. Hard-nosed miners are not interested in philosophical analyses of the situation, but require clear guidance as to what their next move should be. New mine geologists face a particular dilemma for they may be asked to predict what is happening in areas to which there is no access or for which there is little or no geological information. Very often their understanding of the nature, origin and geological controls of the deposit are limited and they feel that they are being asked to 'crystal-ball gaze'. The longer they have worked on a particular mine the greater is the 'feeling' they have for it and the more likely it is that an educated guess will prove correct. There is no substitute for experience.

(4) They must be able to balance the return against the cost of collecting data. For example, the maximum account of information must be gleaned from drilling programmes which should include not only grade and thickness data, but also information pertinent to the hydrogeology of the mine; geotechnical data which may assist in assessing the amount of ore dilution and whether ground control problems may occur; structural and lithological information which may help in the interpretation of the factors controlling mineralization; and finally, mineralogical data which may have relevance in the fields of ore genesis and mineral processing.

(5) They must show a high degree of patience with mine personnel and with the operation itself for, although it may appear obvious what should be done to improve existing procedures, techniques, etc., they must accept that, by their very nature, mining operations are slow to react. A reasoned case should be presented to justify the changes proposed without being excessively forceful. Eventually, if the suggestions are practicable and involve a significant cost saving, they will be accepted and implemented.

(6) They must learn to deal with miners, mine captains and mining engineers to gain their respect and confidence and to understand their

needs/requirements and the limitations placed on them by the mining method, ground conditions and by the equipment available to them. This is essential for a good working relationship and in this way an active geological department can ensure the success of a mining operation. Failures have occurred in the past because miners have not accepted the need for close geological control or because the geologists have failed to establish a practical working relationship with them.

(7) They must accept that there is much to learn from experienced miners and other members of the geology and engineering departments. In this way they will become invaluable members of the team.

(8) The ability to develop a sense of self preservation in the work environment, not just in personal relationships, but also in awareness of personal safety and the safety of others. An awareness of the inherent dangers in the mine is essential and of the need to recognize the tell-tale signs of impending danger. It is important that they ensure that their assistants are working in a safe manner and in a safe working environment.

(9) New geologists must realize that, at an early stage, they will have to prove their worth, as credibility is established with the mine management through experience. This will take time and they should not become disheartened if initially there is a reluctance to respond to suggestions.

(10) The ability to write concise technical reports with the minimum of geological jargon. These should be tailored to be intelligible to the person(s) who will have to act on the information and thus they must be unambiguous.

Those geologists who takes on board the above suggestions and who develop a working knowledge of the techniques described in this book will become an invaluable asset to their employers.

Cardiff, Wales — Alwyn E. Annels

REFERENCES

Barnes, M. P. (1980) *Computer-assisted Mineral Appraisal and Feasibility*, Society of Mining Engineers of AIMM and Petroleum Engineers Inc., 167 pp.

McKinstry, H. E. (1948) *Mining Geology*, Prentice-Hall, New Jersey, 680 pp.

Peters, W. C. (1978) *Exploration and Mining Geology*, John Wiley, New York, 696 pp.

Reedman, J. H. (1979) *Techniques in Mineral Exploration*, Applied Science, London.

Acknowledgements

The author would like to acknowledge the help given by his colleagues in the mining industry who have willingly provided case history material or who have given practical advice and moral support during the writing of this book. In particular he would like to thank the following people and Companies.

P.C. Atherley, Horseshoe Gold Mine Project, Barrack Mine Management Pty Ltd.
J.H. Ashton, Tara Mines Ltd.
J.C. Balla, ASARCO Incorporated, Northwest Exploration Division.
W.L. Barrett, Tarmac Quarry Products Ltd.
R.A. Birch, Hepworth Minerals and Chemicals Ltd.
R. Bird, Mole Engineering Pty Ltd.
E.B. Boakye, Dunkwa Goldfields Ltd.
D. Brame, Newmont Australia Ltd.
P. Brewer, Tarmac Roadstone Ltd.
R. Corben, Surpac Mining Systems Pty Ltd.
S. Czehura, Montana Resources Ltd.
J. Davis, University of Wales, Cardiff.
J. Forkes, RTZ Technical Services Ltd.
F. Foster, Golden Sunlight Mines Inc.
P. Fox, Fox Geological Consultants Ltd.
R.A. Fox, RMC Group Plc.
R. Haldane, ZCCM Ltd.
T.S. Hayes, US Geological Survey.
E.G. Hellewell, University of Wales, Cardiff.
R. Holmes, Cleveland Potash Ltd.
D. Hopkins, Tarmac Quarry Products Ltd.
J.T. Hunt, Cleveland Potash Ltd.
R.H. Jones, Blue Circle Cement Technical Services Division.
S.A. Lambert, Horseshoe Gold Mine Project, Barrack Mine Management Pty Ltd.
L.T. Lynott, Scitec Corporation.
J. Luchini, ASARCO Incorporated, Northwestern Mining Department.
R. MacCallum, Opencast Executive, British Coal Corporation.
J.F. McOuat, Watts, Griffis and McOuat Ltd.
C.J. Morrissey, Riofinex North Ltd.
P.W. Morse, Tarmac Roadstone Ltd.
A.E. Mullan, Datamine International.
R. Naish, ZCCM Ltd.
J.P. Odgers, The Charles Machine Works, Inc. (Ditch Witch).
A. Peacock, ICI Tracero, ICI Chemicals and Polymers Ltd.
J.H. Reedman, J.H. Reedman and Associates Ltd.
D.I. Roberts, ARC Ltd.
A.G. Royle, University of Leeds.
P.F. Saxton, Mascot Gold Mines Ltd.
R.W. Seasor, Copper Range Company Ltd.
S. Schenk, Pegasus Gold Corporation.
G.J. Sharp, Riofinex North Ltd.

W.M. Snoddy, Montana Tunnels Mining Inc.
G.M. Steed, University of Wales, Cardiff.
J. Tweedie, GeoMEM Software.
R. Whittle, Whittle Programming Pty Ltd.
R.W. Vian, Stillwater Mining Company.

The author gratefully acknowledges the help given by Liesbeth Diaz in the word-processing of this book. Some diagrams in Chapter 4 were also produced by Margaret Millen. Both are colleagues in the Geology Department at the University of Wales, Cardiff. In particular the author would like to apologize to his academic colleagues and to his wife for being overly preoccupied during the gestation period of this book. Material for this book was collected whilst on a study tour of North America which was partly financed by an award from the Institution of Mining and Metallurgy, London (G. Vernon Hobson Bequest).

1

Representation of Mine Data

1.1 INTRODUCTION

Before embarking on a discussion of the various ways in which mineral deposits can be sampled and evaluated, it is necessary to review the methods available for the subdivision of these deposits and for representation of the geological and assay information gained during their exploration and exploitation. The methods employed must also allow the geological information to be related to mine development (e.g. drives, haulages, cross-cuts, etc.) in an underground mine, or to benches in an open-pit operation. This requirement in turn necessitates a brief review of the nature of this mine development and of nomenclature used in a mining operation so that terms used in later chapters will be clearly understood. It is not my intention in this book to become involved in a discussion of methods of working ore-deposits and thus the interested reader is referred to the available literature on this subject, e.g. Woodruff (1966), volumes 1 and 2 of the *SME Mining Engineering Handbook* (Society of Mining Engineers, 1973) or Thomas (1973).

1.2 MINE NOMENCLATURE

1.2.1 The orebody

Where a mineral deposit is tabular or podiform, as is the case for stratiform and stratabound base-metal deposits, iron ores, veins and shear zone deposits, it is possible to define hangingwall formations. These are the host rocks lying structurally (and possibly stratigraphically) above the orebody. Similarly, those beneath are the footwall formations. The orebody itself is defined on the basis of a cut-off grade, which is the lowest grade of mineralization which can be incorporated into a potentially economic intersection (section 3.3.1). We can thus delineate the assay hangingwall and assay footwall. The use of these terms is illustrated in Figure 1.1.

1.2.2 Access to the ore-body

A summary of the mine development associated with an orebody is presented in Figure 1.2.

Shaft

Shafts are normally vertical, or subvertical, accesses to underground workings which contain cages for man-transport, skips (large bucket-like containers) for hoisting ore or waste, and a services compartment (power, compressed air, water column, communications, etc.). In the case of tabular, inclined orebodies, the shafts are located in the footwall formations and are separated from the orebody by a shaft-pillar, in which no mining is allowed.

Some shafts may be specially dedicated and thus we have service shafts and ventilation shafts.

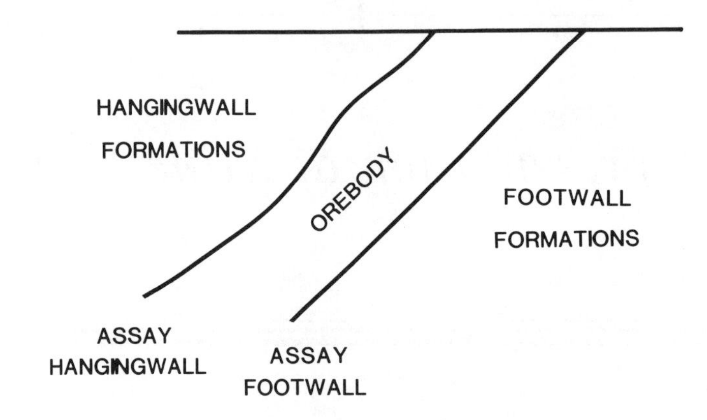

Fig. 1.1 An orebody and its host rocks – definitions.

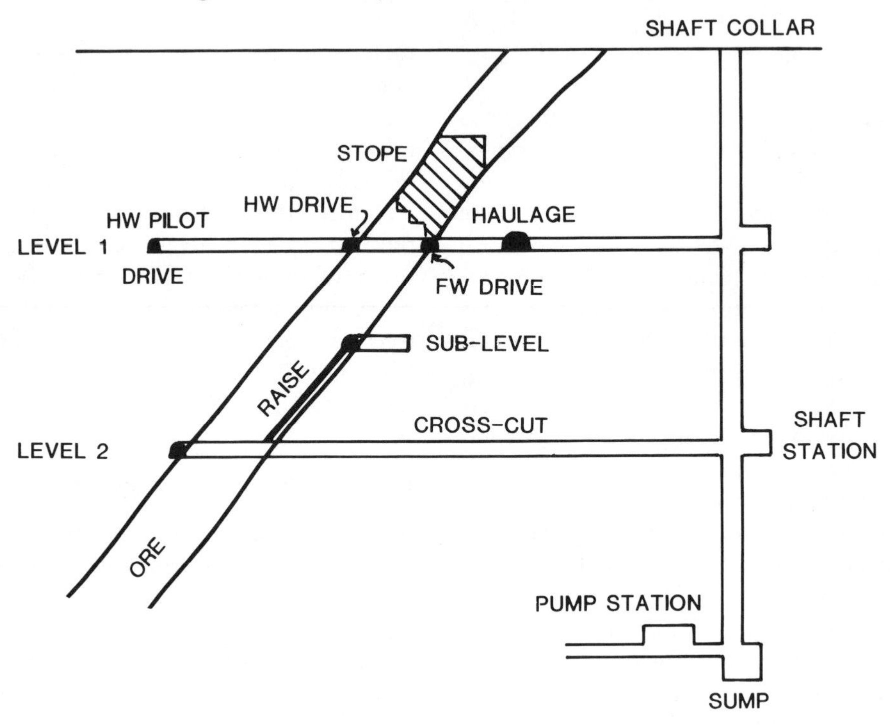

Fig. 1.2 Generalized layout of mine development.

At regular intervals down the shaft, we have shaft-stations, enlarged areas around the shaft to facilitate loading and unloading of materials and men, and the movement of tram cars, etc. Each of these stations corresponds to a mine level which is numbered on the basis of depth below the shaft-collar. At the base of the shaft is the shaft-sump, into which water circulating through mine workings eventually drains. This is linked with the main pump-station which pumps the water to the surface via the shaft-column.

In some mines, an internal shaft may exist which is collared underground and which allows access to a deeper section of ore. Internal tramming is thus necessary between this shaft and the main shaft for the ore or waste to reach the surface. In other cases, such an internal shaft may also be an inclined shaft which may be driven obliquely to the strike of the orebody to reduce the gradient and allow a strike offset to be gained from the main shaft. Where a shaft has been driven down the dip of the orebody, to minimize the cost of mining waste, it is referred to as a winze.

An adit is a horizontal access tunnel from a hillside and will be superseded by a shaft once mining penetrates below valley floor. A decline is an inclined adit (gradient perhaps 1 : 8 or 1 : 10) which allows direct access to deeper levels by wheeled vehicles, thus obviating the need for shafts, other than ventilation shafts. In this instance, ore may be brought to the surface by truck or conveyor belt. Drift coal mines utilize this method of access.

Development

At each shaft-station, an access cross-cut links the shaft with the main haulage which is a large diameter tunnel driven parallel to the strike of the orebody on the mine level. It has a slight gradient (e.g. 1 : 200 to 1 : 400) to allow free drainage towards the access cross-cut. The haulage is kept at a minimum distance from the footwall of the orebody to maintain a safety pillar, or sill-pillar, perhaps 25–30 m thick. This haulage is the main access to the orebody on the level and carries all the main services. It allows tramming (via rail cars), or haulage (via Load-Haul-Dump trucks – LHDs), of ore to the shaft, to ore-passes, or to underground primary crusher stations. The haulage may, in larger mines, be accompanied at a lower level by a main return airway (MRA). The two are connected at regular intervals by raise-borer, or smaller diameter holes, to allow drainage of the main haulage. As the name implies, however, the main role of the MRA is for ventilation purposes.

Further cross-cuts from the haulages allow direct access, at regular intervals, to the orebody itself where a footwall drive may act as a gathering drive, allowing transport of the ore to a central ore-pass. In some cases, the drive may be a lode drive, i.e. it is driven in ore with the hangingwall contact just exposed in the top corner of the drive. Those mining methods designed specifically for thick ore-deposits may also require the development of a hangingwall drive, which is connected to the footwall drive by an orebody cross-cut. Additional development may also exist, further into the hangingwall, to allow exploration drilling to greater depths or drainage of aquifers. Thus we have hangingwall exploration cross-cuts or drainage cross-cuts. These may in turn link up with a hangingwall pilot drive which is a smaller diameter drive whose purpose is to allow fan drilling at regular strike intervals to prove extensions of ore or to penetrate, and drain, aquifers further in the hangingwall. A small cross-cut into the side of the drive to allow the setting up of a diamond drill is referred to as a drill-cubby.

Development on levels, intermediate to the main levels, may also exist hence we can have sub-level cross-cuts and sub-level drives. These are generally linked with the main levels via raises which are driven upwards on ore to the overlying levels and which are generally directly related to the mining of the ore. Access to intermediate levels may also be gained by ramps (inclines or declines), which are merely inclined cross-cuts, or by spiral ramps which allow deep access for trackless mining.

The walls of a drive or cross-cut are referred to as sidewalls and the wall on the right hand side, as you penetrate further into the mine away from the mine portal or shaft station, is the right sidewall, while the opposite wall is the left sidewall. The roof of the tunnel is often referred to as the back.

1.2.3 Stoping

A stope is the hole underground from which ore has been, or is being, exploited. Hence the verb to stope and the term stoping-limits. In the case of a flat-lying stope, the roof will be referred to as the stope hangingwall or back. Thus a 'heavy back' refers to a roof that is somewhat unstable and liable to spall or collapse. In this instance, stope-pillars will be left to support the back or artificial pillars, or packs, will be constructed for this purpose. Adjacent stopes along strike will be separated across the dip by rib-pillars, while a crown-pillar separates the stope from surface and weak weathered ground. En-echelon benches may be worked in a stope to minimize the impact of sudden stress build up in one area. The ore is then exposed in the stope-face. When the dip is too flat to allow gravity feed of the broken ore to lower levels in the stope, it may be dragged, using a slusher or scraper and a winch, to a centre-gulley which is, in reality, a flat-lying raise. Gravity feed of broken ore from the stope to a gathering drive is via an ore-chute or boxhole and the ore may then be transmitted to a lower level in the mine via an ore-pass fitted with a grizzley. The latter is a grill which prevents large lumps of ore falling into the ore-pass and causing a hang-up, i.e. a blockage which may have to be removed by secondary blasting.

In a steep dipping orebody being mined by cut and fill methods, the mined-out portions of the stope may be backfilled with waste rock or tailings (fine sand from which the ore minerals have been removed). Each slice of rock mined by this method is referred to as the stope-lift and the stope-breast is the horizontally advancing face of the current lift which may be 3 m high and greater than 1.2 m wide. The latter is the minimum stoping width (section 3.3.1). Each stope will be accessed by a central man-way, a shuttered column through the fill, fitted with offset ladders and platforms, and an associated ore-pass or mill-hole, perhaps constructed of circular steel sections each able to slot into the one beneath.

The broken rock at an active face, after the blasting of a round of explosives, is referred to as the muck-pile. Its removal is referred to as mucking or lashing. The widening of a stope or drive to win additional ore is referred to as slyping or slashing.

Finally, the target set by the mine for production over a given period is referred to as the mine-call.

1.2.4 Open-pits

Fewer terms exist which are specific for open-pit or open-cast operations. A bench is a horizontal slice of ore which will be mined as a unit. Each bench becomes progressively smaller in area as the pit deepens and thus pit walls are developed with an angle or batter which is considered safe given the condition of the rock and the depth of mining. The width of each bench remaining on the pit wall is termed the berm. A bench may be subdivided, for the purpose of selective mining, into flitches. These are subslices of the bench perhaps 2.5 m high. A ramp or haul road is developed on the berms of one pit face to allow access to the pit bottom and trucking of ore and waste to the mill or dumps respectively, perhaps aided by 'trolley-assist' where diesel electric motors on the trucks are given extra power via a pantograph.

1.3 SUBDIVISION OF OREBODIES

1.3.1 Underground mines

Where a mine is working a steeply dipping tabular or podiform deposit underground, it is usual practice to project this orebody horizontally on to a vertical plane (a Vertical Longitudinal Projection (VLP) – section 1.6) and then to subdivide this projection into mining blocks. Each mining block is bounded by adjacent mine levels and by mine structure sections (section 1.4), thus creating a rectangular block whose dimensions may be 30–80 m vertically and 50–100 m horizontally, depending on the chosen section spacing. This block thus represents the horizontal projection of an inclined section of ore with a dip length equal to $L/\sin\theta$, where L is the level spacing and θ the orebody dip. Each block is then assigned a unique number related to the level on which it is based and on the distance

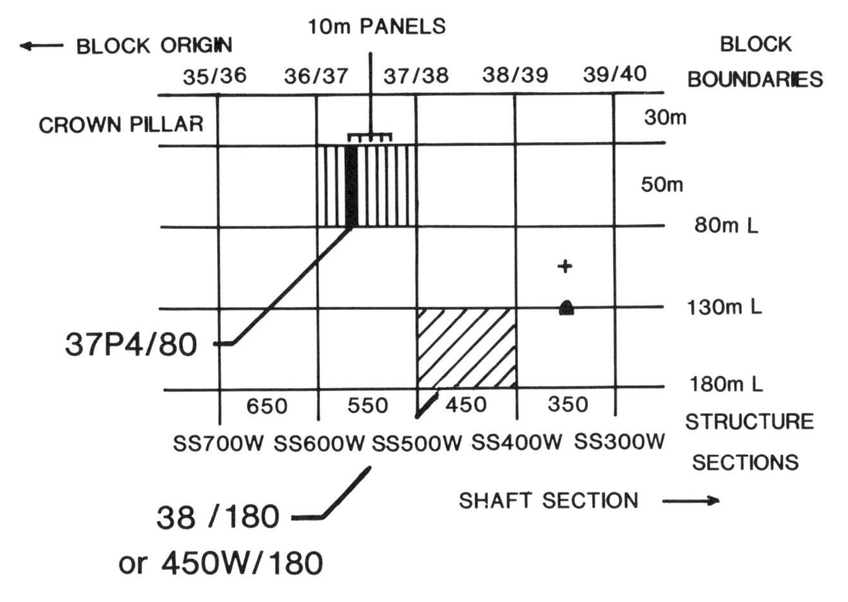

Fig. 1.3 Subdivision of orebodies into mining blocks and panels.

from the mine origin or from the shaft section, SS0. Figure 1.3 shows two ways in which mining blocks can be numbered. The first is based on a left to right progression from a mine origin at the furthest extremity of the orebody to the left, while the second relates to distance from a central shaft section, SS0, so that SS300W is 300 m west of this shaft section. The shaded block in Figure 1.3 is thus referred to as the 38/180 or 450W/180L mining block.

Further subdivision of a mining block is possible to create panels, whose width allows an exact number to fit into the mining block. These vertical panels of ore may thus be 10 m wide in a 100 m wide block. In Figure 1.3, therefore, the marked panel would be referred to as 37P4/80L. The boundary between the P5 and P6 panels may thus be described as the 37 mid. In some cases, the panel width may be related to pillar width in a stope. This method of orebody subdivision not only allows blocks of ore to be defined, but also allows the exact location of a cross-cut or sample to be determined. Thus the cross-cut illustrated in Figure 1.3 would be referred to as the 39 mid/130L X/C or 350W/130L X/C, while the drill-hole intersection in the centre of the overlying block would be located at 350W/105 SL (SL = sub-level) or 39 Mid/105 SL.

In some mines a local mine coordinate system may be used so that the term mine-north indicates the dip direction of the orebody. South is thus towards the footwall and east and west are the strike directions. Thus, Loop N 36P3/180L represents a loop which cross-cuts the orebody into the hangingwall formations in the third panel of the 36 mining block based on the 180 m level.

The above discussion presents two different methods of defining ore-blocks in a mine but there are many variations on a theme. However, the general principle of the method is the same.

1.3.2 Open-pits

The situation in an open-pit operation is generally simpler. Here, the basic subdivision of the orebody is by benches, whose thickness may vary from 2 m to 15 m depending on the nature of the ore and the equipment used. Each of these benches is then subdivided into a series of rectangular blocks representing the production ore-reserve blocks. These may be centred on geological/drill section lines or bounded by adjacent lines. The benches are usually numbered on the basis of the elevation above mean sea-level of the base, or toe, of the bench, while each block is denoted by the X and Y coordinates of its central point. At a later stage, these blocks may be subdivided into smaller blocks on the basis of in-pit drilling during production. This drilling may be by reverse-circulation or rotary percussive techniques or it may be blast-hole drilling. In all cases, it produces assay information for the current bench, and even some underlying benches, to allow a more detailed evaluation of the grade of ore-blocks and of the lithological or metallurgical type for each.

1.4 MINE SECTIONS

1.4.1 Production of mine sections

The first step in the production of a series of mine geological or structure sections is the design of a mine coordinate system. This is followed by the drawing of a section layout plan. Figure 1.4 shows the outline of an orebody at suboutcrop, i.e. at the rock-head beneath the overburden. A baseline is drawn in the footwall of the orebody so that its strike is parallel to the average strike of the orebody. It is often convenient if this line could pass through the collar of a centrally positioned shaft as shown. A series of structure section lines can now be established at uniform intervals along this line, e.g. 50 m. Where a central shaft exists, the first line can be positioned so that it passes through it, and at right angles to the baseline. All other lines are then positioned outward from this line (SS0) and numbered on this basis (e.g. SS1W and SS1E or SS50W and SS50E, depending on whether a system based on consecutive numbers is used or one based on line spacings). The mine coordinate system is completed by drawing in a series of lines parallel to the baseline and at spacings of say 100 m in the dip-direction of the orebody (Figure 1.4).

Sections can now be drawn for each section line, using information from surface, from boreholes and from underground development, if available. In the case of the boreholes, it may be necessary to project the information along strike on to the plane of the section. The scale used for the sections depends on the magnitude of the

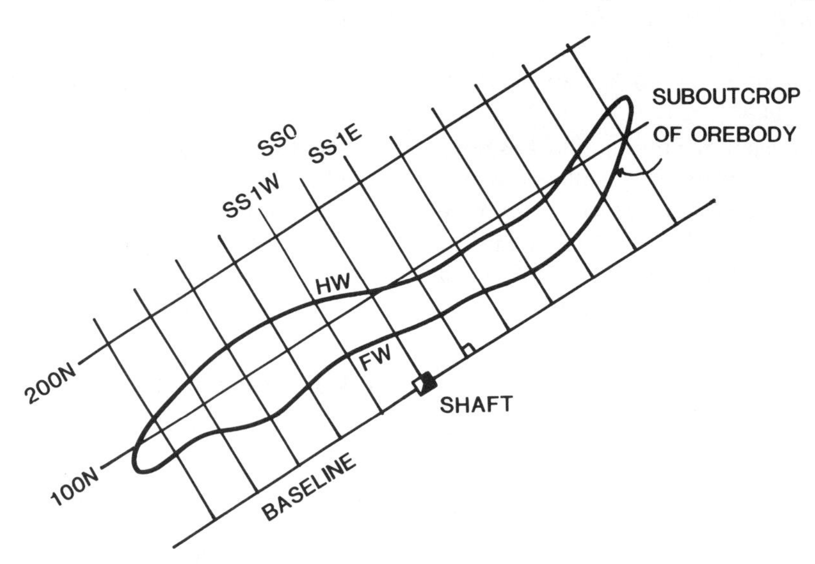

Fig. 1.4 Layout of mine sections and base-line (plan).

orebody itself and on the detail required, but typical values would range between 1 : 500 and 1 : 2000. In no circumstances should a vertical exaggeration be introduced so that actual dips can be plotted, or represented, without distortion. The convention that is used for drawing the sections is based on the 'Look North–Look West' rule. Thus, section lines which are closer to northerly than westerly, as in Figure 1.4, would have sections drawn from a viewpoint to the E, so that you are looking W. As a result, the northern end of the section is to the right of the drawing and the suboutcrop and the baseline are to the left. Where a section line is closer to E–W, as in Figure 1.5, the section is drawn looking N so that W is on the left, as one might expect on a map.

Figure 1.5 shows the main elements required for a mine section. These are also summarized below:

(1) Unrolled or projected traces of boreholes on line of section (LOS) showing orebody intersections and end-of-hole positions (EOH).

(2) Location of mine levels, numbered on the basis of depth below shaft-collar (the mine origin).

(3) Horizontal lines indicating elevation above mean sea-level.

(4) Vertical lines indicating distance (and direction) from the mine baseline.

(5) Section orientation (e.g. 275° T).

(6) Location of drives and haulages crossing the section and any cross-cuts on LOS.

(7) Geological units intersected in boreholes and underground workings showing interpreted structure, together with proposed or proven faults.

(8) Lease or claim limits if applicable. The mineral rights for different portions of the orebody may be owned by different people/companies and may be worked by different operators.

(9) A plan strip on which the topographic features, surface geology, drill-hole collars and surface coordinate systems are plotted in a zone centred on the geological section line. This indicates the deviation of holes from the LOS in plan view and the distances individual holes have been projected on to section.

(10) A title box (not shown on Figure 1.5) in which are recorded:

(a) section number;
(b) mining company/mine name;
(c) scale (e.g. 1 : 1250) or scale bar if diagram is likely to be reduced;
(d) date drawn and by whom;
(e) by whom checked/approved;
(f) dates listed for modifications;
(g) drawing number.

1.4.2 Use of mine sections

In a mining operation, sections can be used for any combination of the following:

(1) To show mineralized intervals in boreholes, and also those above the stipulated cut-off grade (i.e. economic intervals). These can be differentiated by using open and shaded boxes respectively, as in Figure 1.5(a), or by using colours on the borehole trace to indicate various grade categories (waste, low grade, run of mine grade, high grade, etc.). Alternatively, histograms of metal grade can be plotted alongside the mineralized interval, or assay summary boxes inserted adjacent to the hole. These contain the true thickness of ore, its weighted grade and the acid-soluble or oxide metal grade, if applicable (Figure 1.5(b)).

(2) Existing and planned surface drill-holes.

(3) Detailed drill sections of parts of an orebody for use in stope-layout (Figure 1.6).

(4) Existing and planned development.

(5) Location of stoped-out areas on the section and any associated cave-line induced by stoping.

(6) Location of aquifers and of the water-table within them at a specified date.

(7) Fringes of mineralization, oxidized near-surface zones and the limits of ore-reserve calculations.

(8) Location of sections of ore sterilized

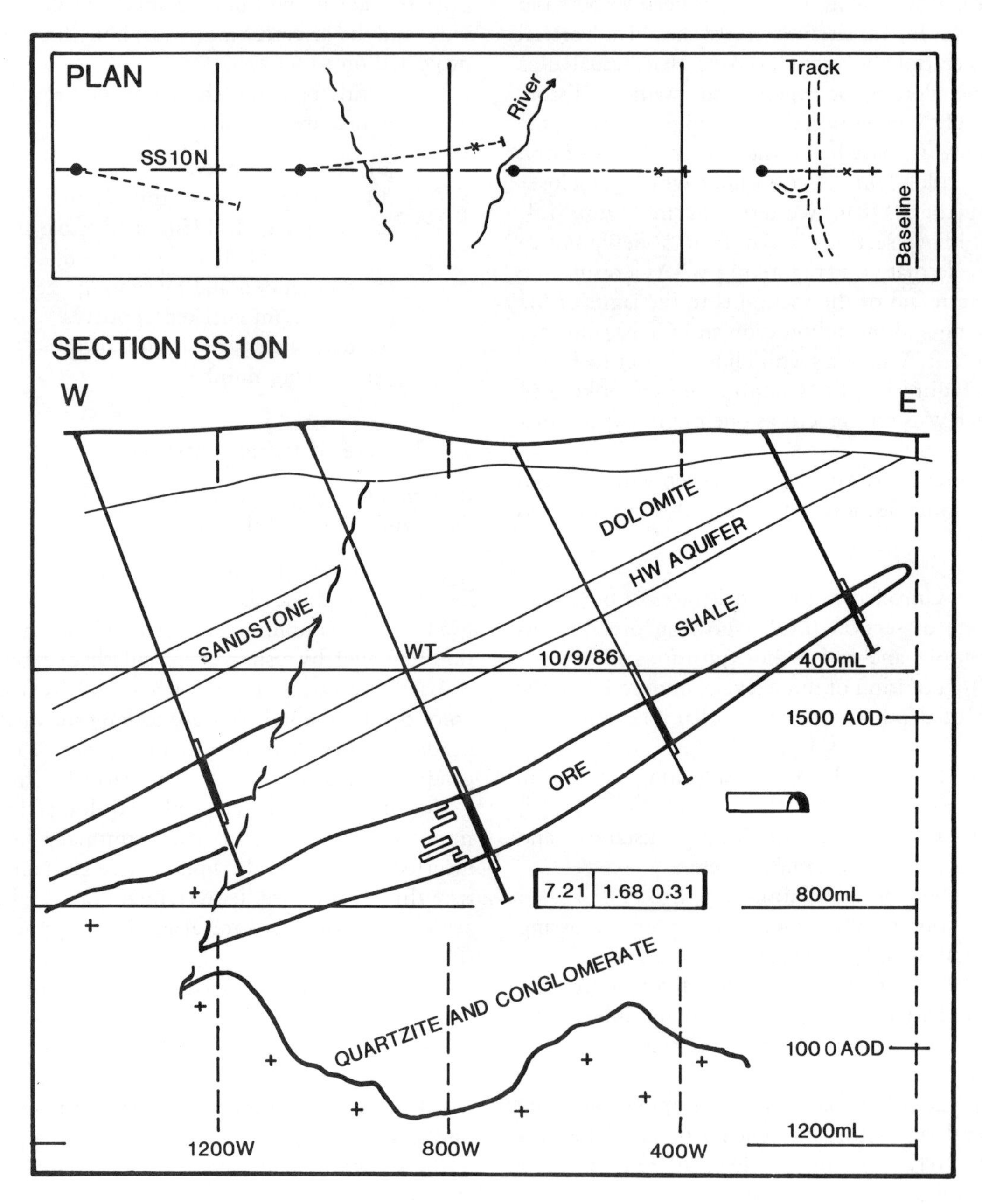

Fig. 1.5 (a) Main elements incorporated in a mine section.

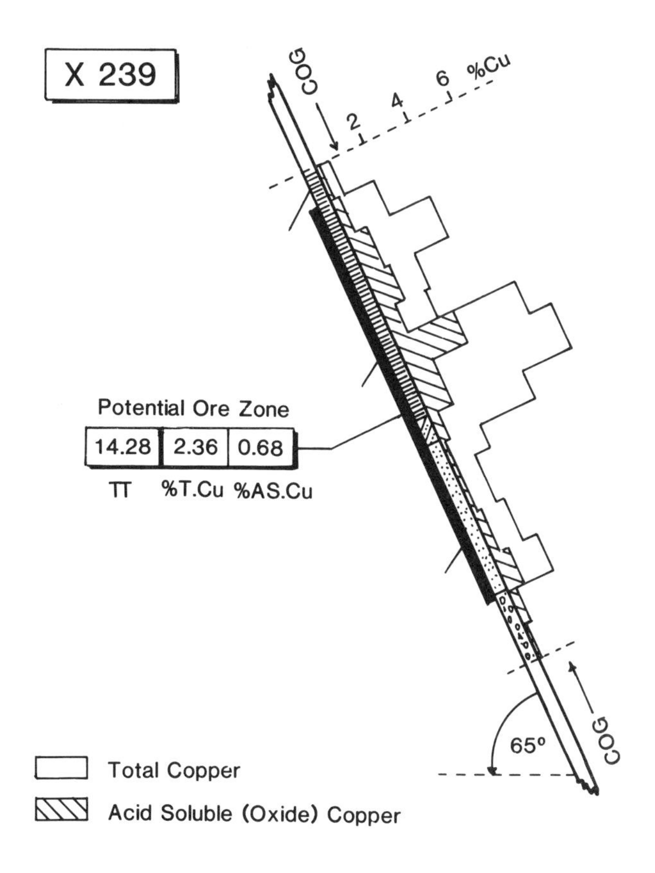

Fig. 1.5 (b) Representation of assay data from a potentially economic intersection of a mineralized zone.

because of shaft, crown-pillars or fault-pillars, or because of the need to prevent surface caving beneath buildings, etc.

1.4.3 Longitudinal sections

The descriptions in sections 1.4.1 and 1.4.2 refer specifically to transverse sections which are applicable in the case of inclined tabular or lenticular bodies. Where the mineralization occurs as a large irregular body with no preferred orientation/dip, it may be necessary to produce a series of longitudinal sections at right angles to the transverse sections and parallel to the mine baseline. These will allow a better understanding of the overall shape of the deposit. They are particularly important when considering the design of an open-pit operation and the slope and location of the pit walls. Obviously it is important to capture as much ore-grade material in the pit outline as possible.

1.5 MINE PLANS

It is not felt necessary here to discuss in detail the method of producing mine plans for these follow naturally from the production of mine sections. What follows is thus a review of the various types that could be drawn.

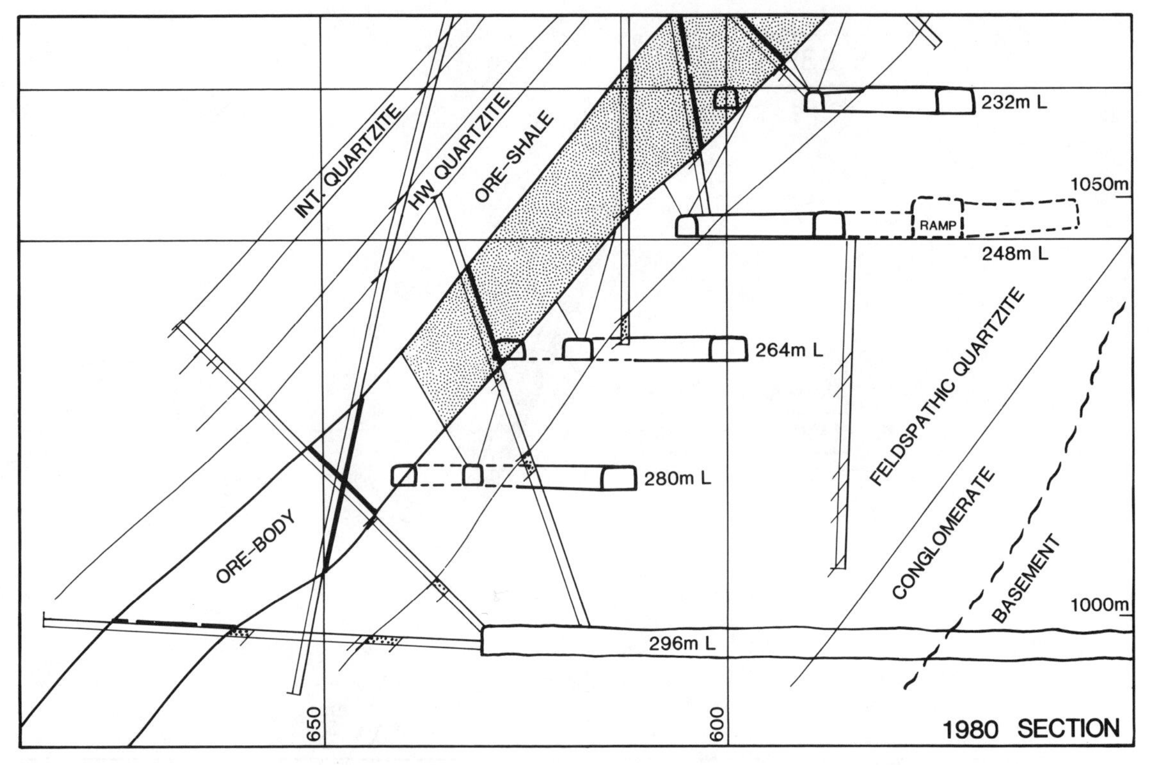

Fig. 1.6 Underground drilling at Chambishi mine, Zambia, showing stoped-out area with drawpoints.

1.5.1 Surface exploration plans

These are basically drill-hole progress plans which were produced during the various phases of exploration, evaluation and mining of the deposit concerned. The main elements of these are summarized in Figure 1.7. As well as drill-hole collars, drill-hole traces, mid-orebody intersection points and a tabulation of assay results, they might also show in plan projection:

(1) Present limits of stoping;
(2) Fringes of the mineral deposit;
(3) Limits of ore;
(4) Mining lease limits;
(5) Surface subsidence – net annual or total;
(6) Proposed drilling;
(7) Suboutcrops of mineralized zones;
(8) Location of geochemical and geophysical traverses;
(9) Outcrops;
(10) Topographical/geographical features, particularly streams, etc.

1.5.2 Level plans

Plans will be produced for each mine level showing all existing and proposed development, together with drill-holes collared on the level. These holes, together with sidewall exposures, will then allow stratigraphic/lithological correlations to be made for the country rock and also the location of orebody contacts and faults. The plan will also show all geological structure section lines, together with the mine coordinate system. Figure 1.8 shows an example from Chambishi mine in Zambia. In addition, more specialized plans may be produced to show:

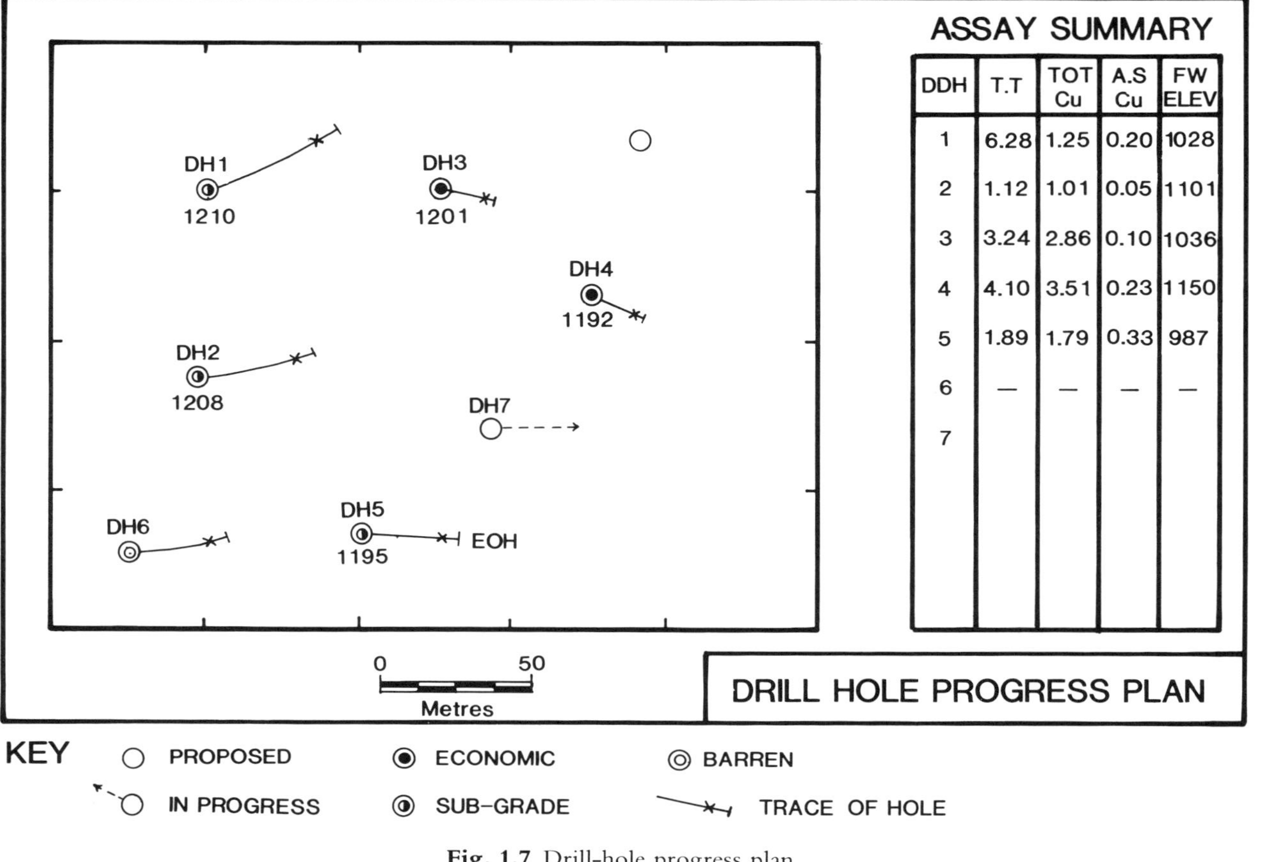

DDH	T.T	TOT Cu	A.S Cu	FW ELEV
1	6.28	1.25	0.20	1028
2	1.12	1.01	0.05	1101
3	3.24	2.86	0.10	1036
4	4.10	3.51	0.23	1150
5	1.89	1.79	0.33	987
6	—	—	—	—
7				

Fig. 1.7 Drill-hole progress plan.

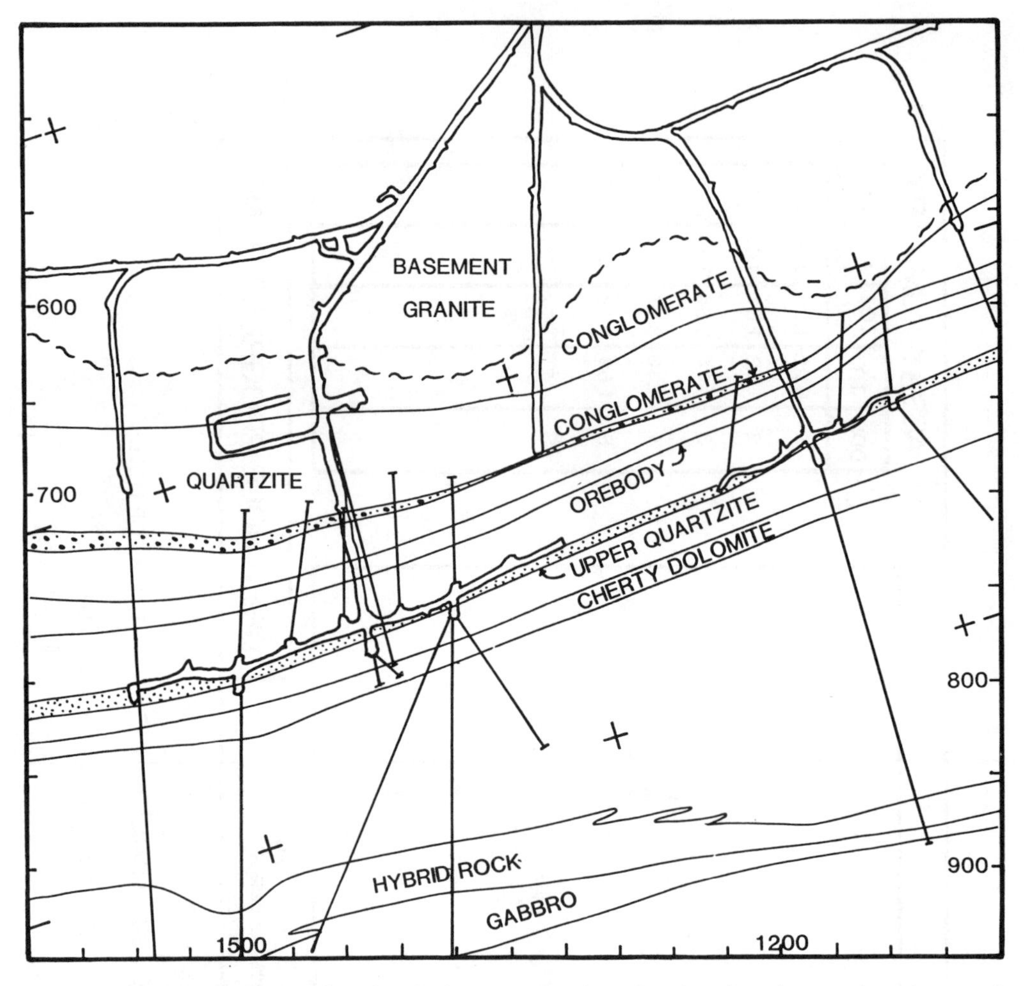

Fig. 1.8 Portion of the 400 mL plan for Chambishi mine, Zambia, showing the mine stratigraphy together with drainage development and drilling.

(1) All development and drilling undertaken for mine/aquifer drainage (Figure 1.8). This will record the locations of all 'V-notch' weirs and flow or pressure gauges and current flow rates. All aquifers and aquicludes will also be defined as accurately as possible and may also be annoted with details of porosity, permeability, etc.

(2) All assay information for the level, including blast sampling and sidewall and face sampling, together with drill-hole assays.

(3) Ore-blocks, tonnages and grades, where level plans are used for ore-reserve purposes.

(4) All geotechnical information.

(5) Development associated with stoping, pillars, blast-hole section lines, raises, manways, etc.

In addition to the level plans described above, the geologist may also be asked to produce sub-level or stope-lift plans by interpolation between level plans. The stope-lift plans will be used for grade control purposes and will show the location of all stope samples and their grades, the stope limits in relation to the geologist's orebody

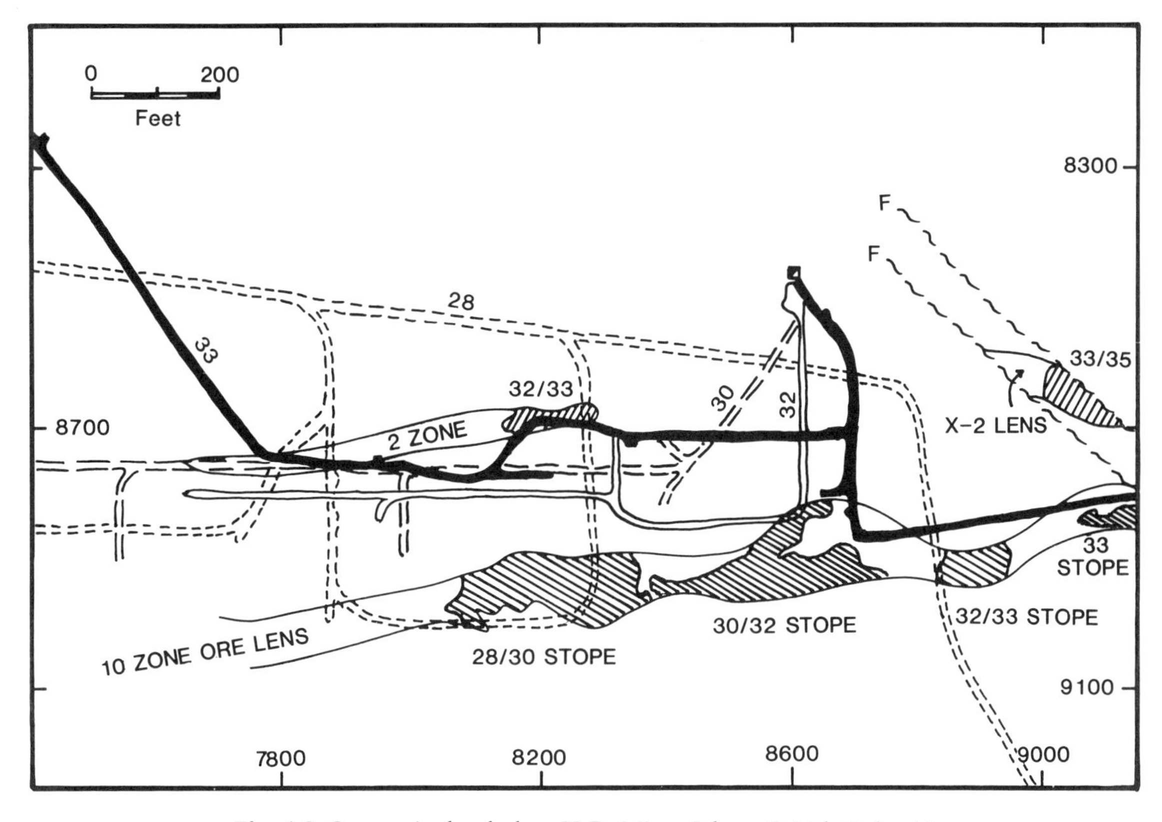

Fig. 1.9 Composite level plan, H.B. Mine, Salmo, British Columbia.

limits and any other relevant factors, such as faults and post-ore dykes. Composite level plans may also be required, but on a less detailed scale than single level plans, and may thus be drawn on a scale of 1 : 5000 to show the outlines of the ore-zone at each level, projected vertically on to the same plan. Each level could then be coloured, or shaded, differently to distinguish between them and a simplified development layout included for each level. In general, the plan will only show the main shaft(s), access cross-cuts and haulages. Its intention is to demonstrate the changes in morphology and attitude of the ore-body with depth and along strike and it thus gives a better 3D image of the deposit. Figure 1.9 shows a portion of the composite level plan from the H.B. Mine, Salmo, British Columbia. This mine is now closed, but was exploiting lead-zinc lenses in Lower Cambrian limestones. The plan shows the location of stopes serviced from different mine levels and the outline of the ore-zones as correlated from level to level.

1.5.3 Inclined plans

These are drawn in the plane of the lode so that bounding levels and raises can be included on the same plan. An example is shown in Chapter 2 (Figure 2.15) which is taken from Wheal Jane Tin Mine in Cornwall. Such plans are mainly used, as in this case, for the representation of blast advance samples in both raises and levels or for channel/chip sample assays. They allow contouring of true thickness data (isopachyte maps) and can thus be used for ore-reserve purposes. They

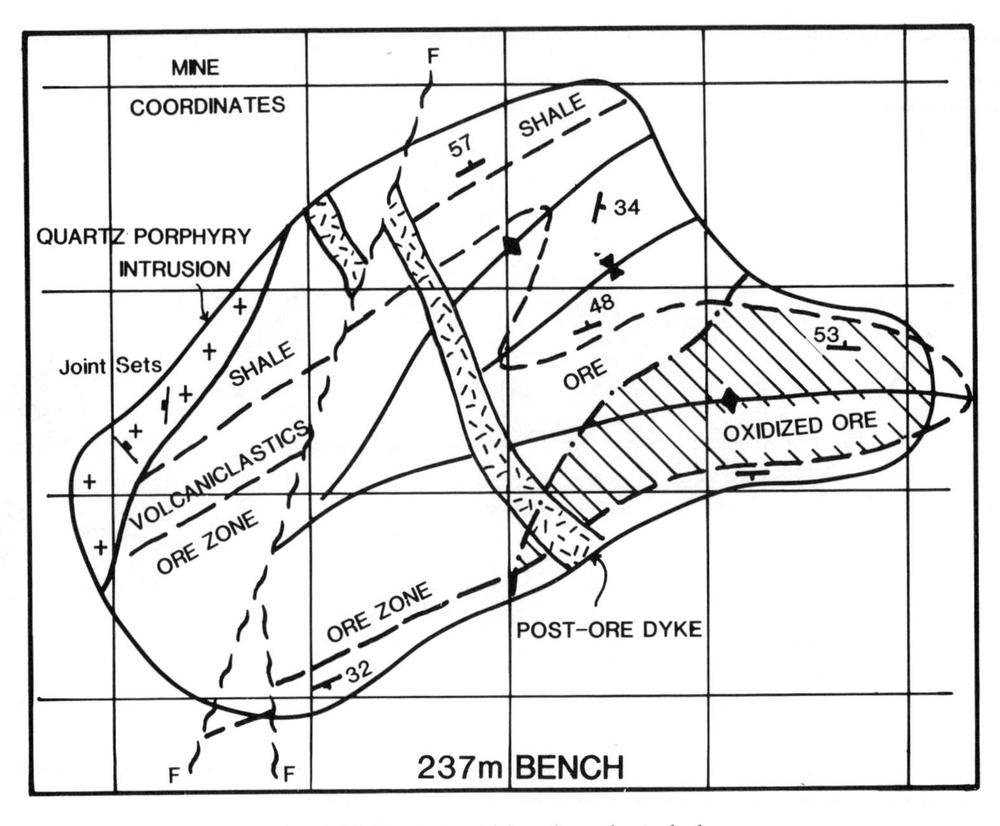

Fig. 1.10 Typical mid-bench geological plan.

are generally detailed plans of a small section of the mine.

1.5.4 Bench plans

Figure 1.10 summarizes some of the information that can be included on a bench geological plan for an open-pit operation. Such plans are frequently drawn at the mid-bench position and show the most significant elements, from the point of view of mining. Similar plans could also be produced showing:

(1) Pre-production ore-reserve blocks and bench intersection points of drill-holes. These blocks may be rectangular or polygonal.
(2) In-pit drill-holes with bench composite grades, perhaps for several benches.
(3) Blast-hole layouts with bench composite grades from cuttings.
(4) Kriged block grades from 1, 2 or 3 produced by geostatistical methods described in Chapter 4.
(5) Metallurgical grade zones or blocks for selective mining.
(6) Areas rippable, as opposed to areas requiring fragmentation by blasting.
(7) Geotechnical information pertaining to the stability of slopes or ramps.

1.6 VERTICAL LONGITUDINAL PROJECTIONS (VLPs)

These projections are one of the most useful methods of representing mine development and geological and assay data pertaining to steeply dipping ore-deposits.

1.6.1 Production of VLPs

To construct a VLP, various details of the ore-body have to be projected horizontally on to a vertical plane located in the footwall of the deposit. The method is most suited to those deposits which are tabular or podiform and which are steeply inclined. Where the dip is significantly less than 45°, there is unacceptable compression of the information and the VLP is of little use to miner or geologist.

Ideally, the VLP should be located along the baseline passing through the shaft, as in Figure 1.4, so that its strike is parallel to the overall average for the orebody. The diagram is drawn so that the final result is what would be seen looking in the dip direction through the VLP plane (i.e. towards the orebody from the footwall side). A plan is thus produced, on which section lines run vertically from top to bottom. Mine levels and elevations, above mean sea-level, are shown as horizontal lines (Figure 1.11). The levels are numbered on the basis of depth below shaft-collar and the sections are numbered as described in section 1.4.1. The topographic surface will be represented by an irregular line whose elevation varies relative to the shaft-collar.

The section produced in Figure 1.12 shows some of the information that can be projected horizontally on to the VLP plane. Thus for each section on the VLP, a series of points can be plotted representing each of these features. This is repeated for all the mine sections.

1.6.2 Use of VLPs

Assay information

If every orebody mid-point, from each drill-hole intersecting the orebody, is projected on to the VLP, then a series of VLPs can be produced on which grade, horizontal thickness and horizontal metal accumulation (grade × horizontal thickness) values are plotted and then contoured. On the same diagrams the fringes of the deposit can also be defined (Figure 1.13). Such diagrams are useful for depicting changes in these parameters,

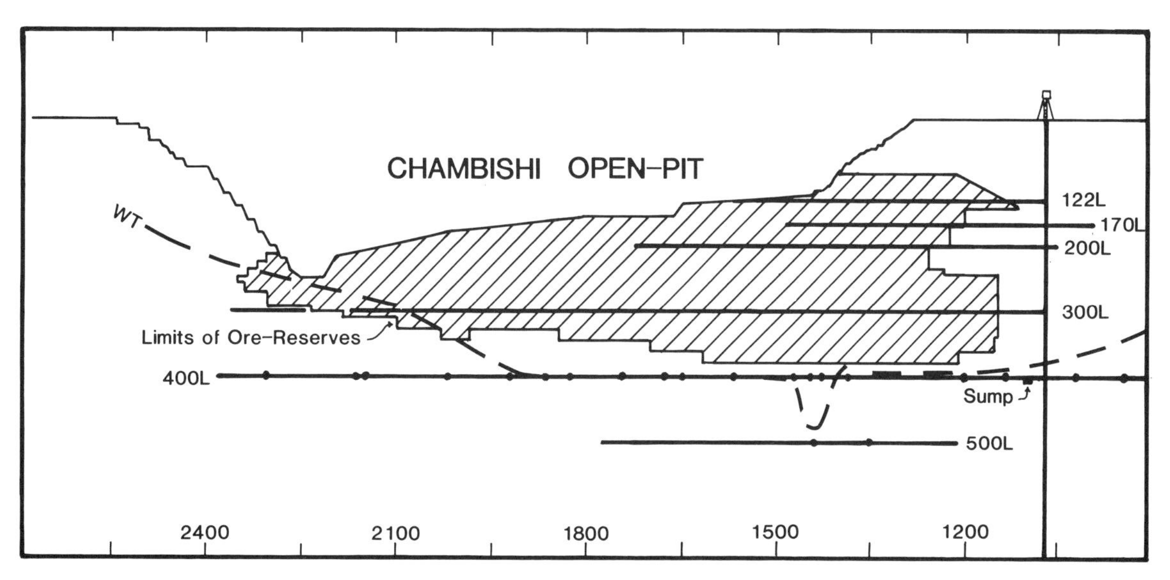

Fig. 1.11 Vertical longitudinal projection showing the limits of ore reserves beneath the Chambishi open-pit and the cone of dewatering for the Upper Roan based on drainage drilling on the 400 mL.

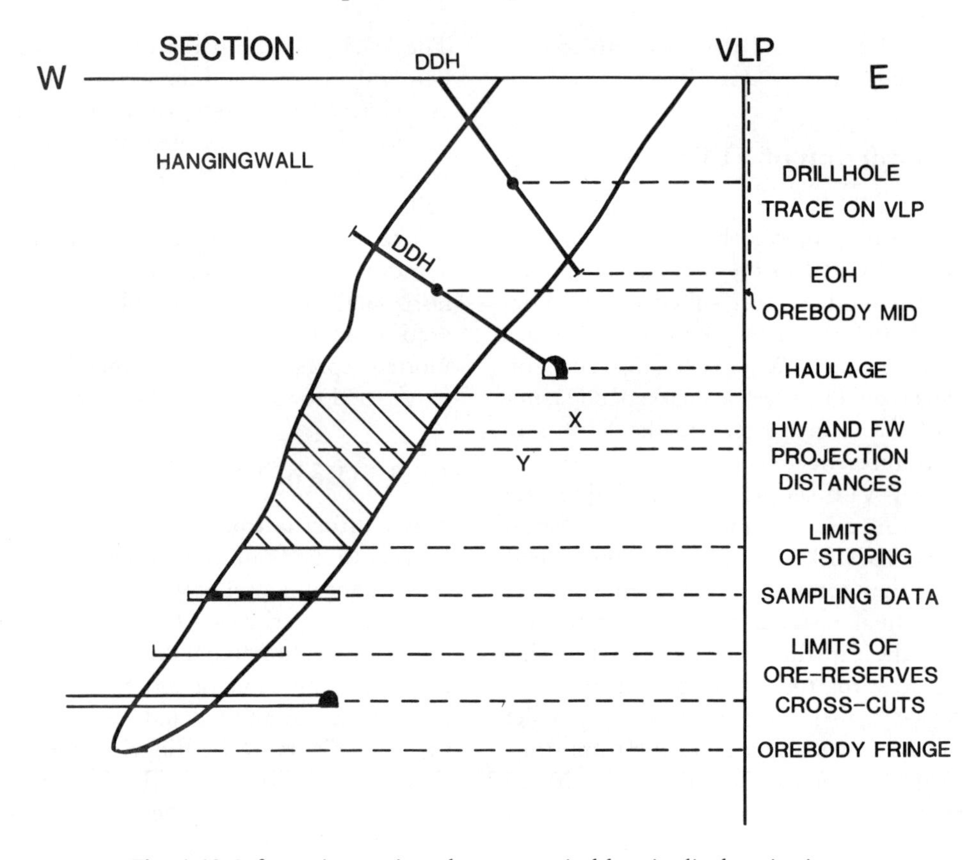

Fig. 1.12 Information projected on to vertical longitudinal projections.

in guiding future exploration drilling and in calculating ore-reserves. To assist the mill and concentrator, VLPs can also be produced showing the distribution and grade of deleterious elements, which incur a smelter penalty, or pose an environmental threat, if in significant concentrations.

Mine development

All haulages, shafts, cross-cuts and stoped-out areas can be depicted together with proposed future development. Figure 1.14 is such a VLP for H.B. Mine near Salmo, British Columbia. The area shown relates directly with that depicted in Figure 1.9. Figure 1.15 shows the stoped-out areas on a silver vein at Galena Mine, Wallace, Idaho.

Drilling

All drill-hole traces can be projected, and the collar, end-of-hole and orebody mid-point positions shown.

Ore reserves

The limits of the different classifications of ore-reserves can be defined (Figure 1.11).

Mine drainage

The current position of the water-table (cone of exhaustion) can be drawn in for each aquifer, either on separate plans or combined using different colours for each (Figure 1.11).

Faults/geological structure

Wherever faults, or the axial planes of folds, intersect the orebody, these can be projected

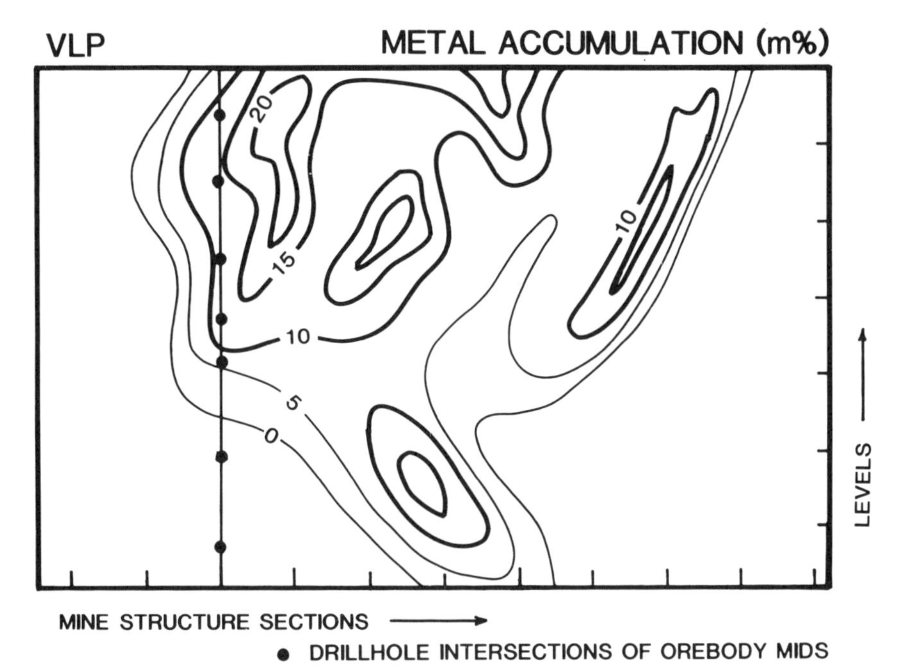

Fig. 1.13 Production of a mine assay VLP.

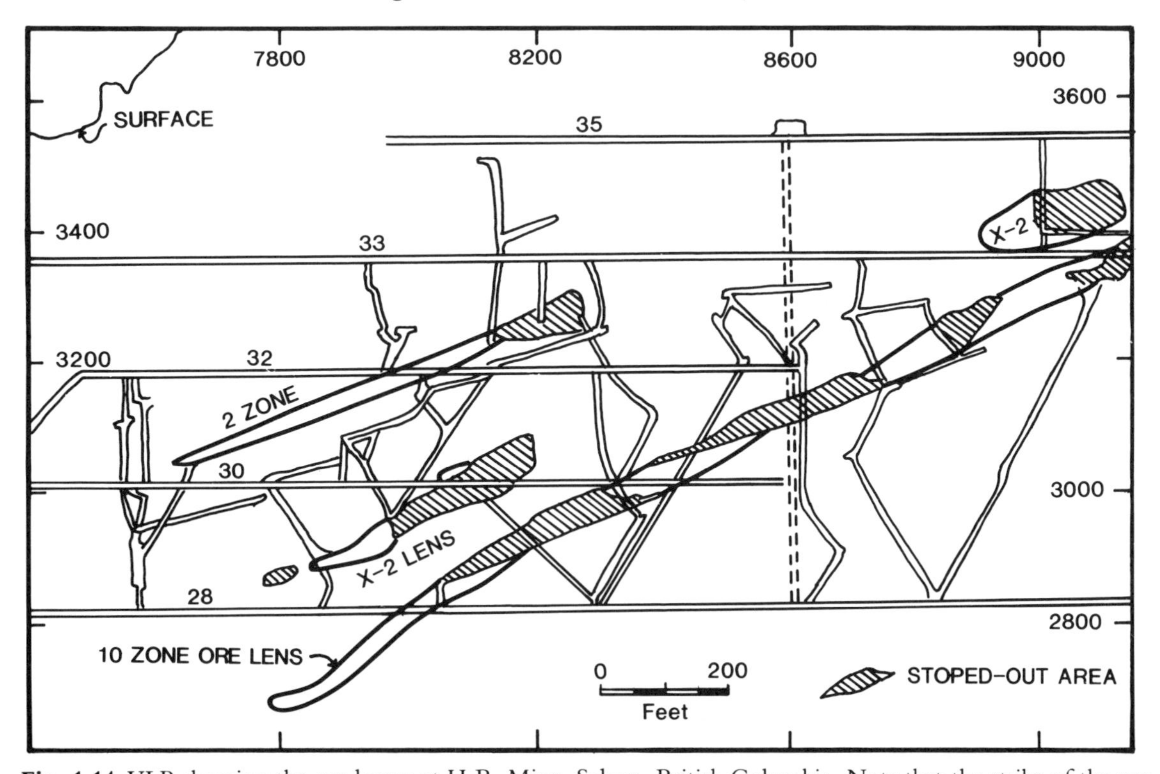

Fig. 1.14 VLP showing the ore lenses at H.B. Mine, Salmo, British Columbia. Note that the strike of the ore lenses (flat or subvertical) is not parallel to the VLP plane.

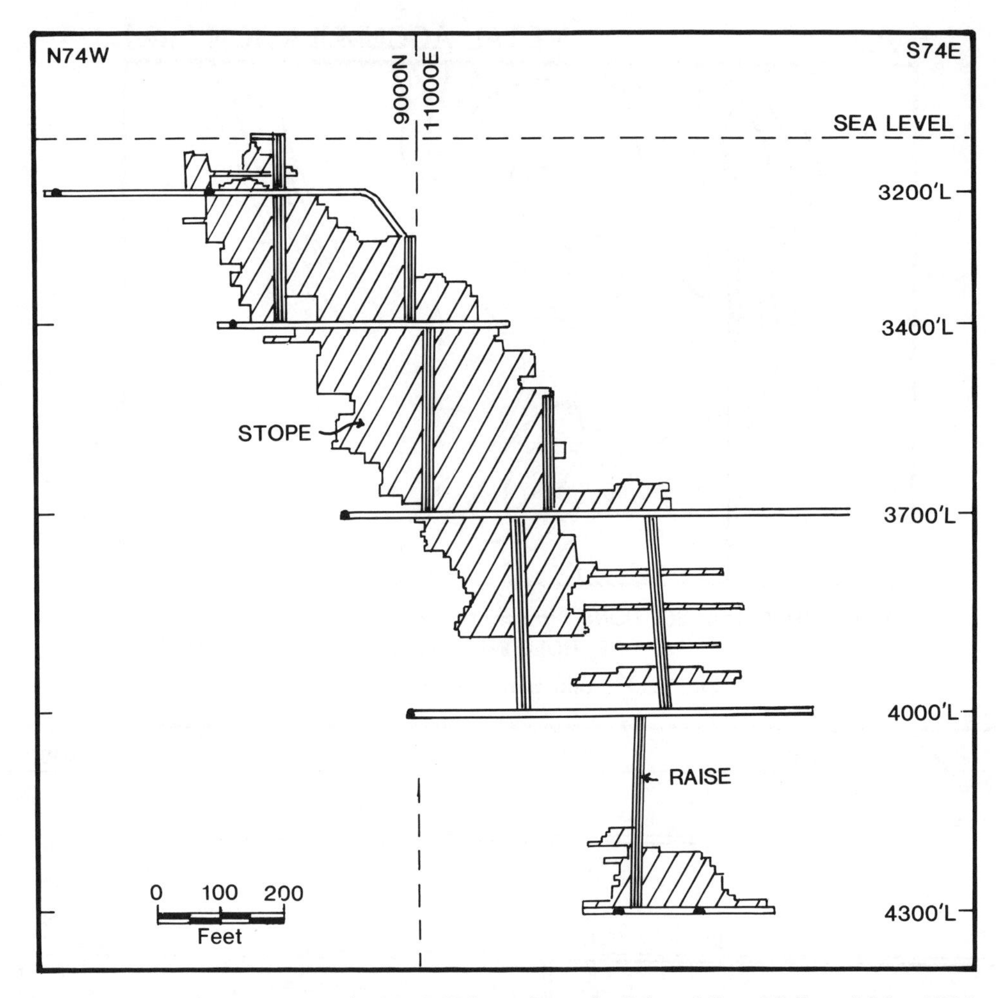

Fig. 1.15 Stoped-out areas on the No 7 Vein at Asarco's Galena Mine, Wallace, Idaho, USA.

on to the VLP and the points linked up to give an overview of the tectonic disturbance of the deposit.

Mineralization

Changes in the assemblage of ore minerals can be represented, thus highlighting any systematic zonation that may exist (Figure 1.16) and its relationship, if any, to other geological features similarly projected on to the VLP. Typical fringe conditions, perhaps indicated by the occurrence of chalcopyrite, pyrite and minor pyrrhotite, may also be recognized. Changes from processable ore to refractory mineralization, e.g. sulphides to chrysocolla, malachite and cupriferous vermiculite, can also be depicted on a VLP. Depths of penetration of leaching, oxidation and supergene enrichment can be drawn on a VLP, together with any zones of deep oxidation related to groundwater movements in fault or shear zones and in particularly porous or permeable or pervious horizons.

Sedimentological features

These include palaeo-current directions, palaeo-channels, facies changes and palaeo-topographic

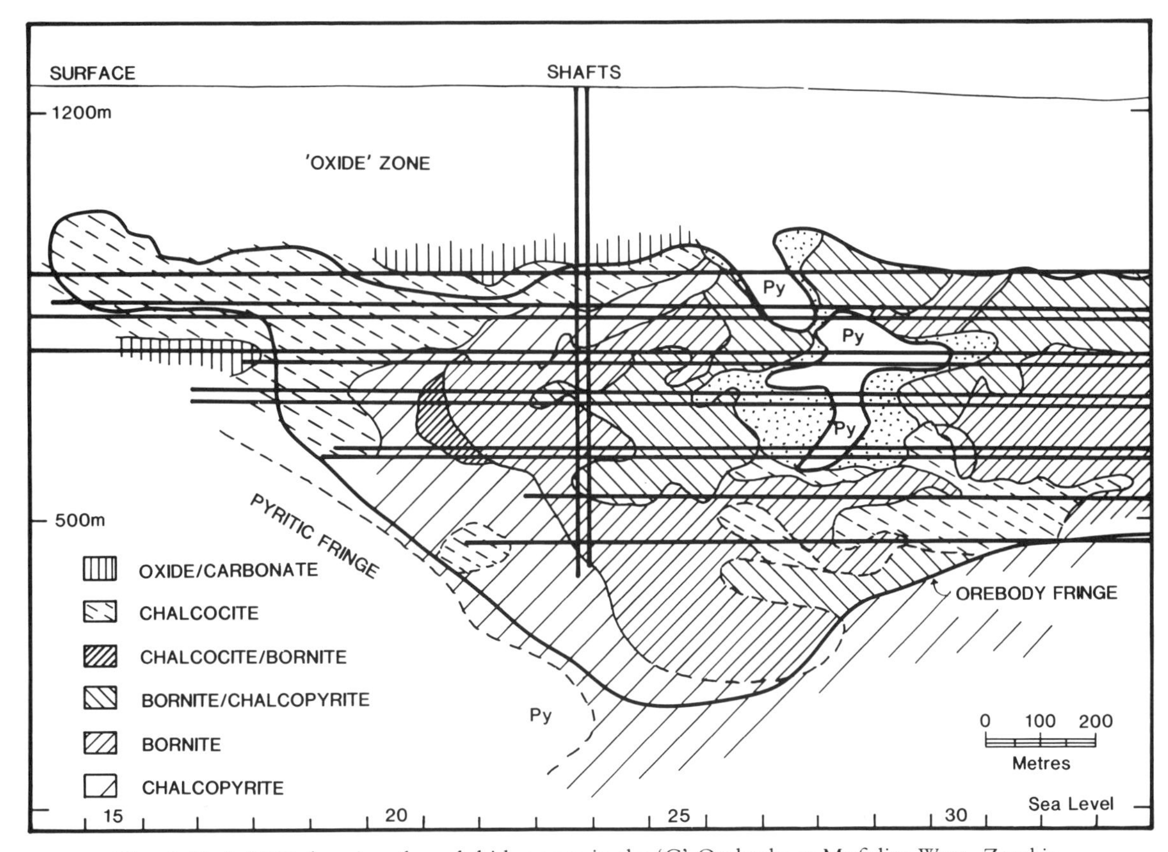

Fig. 1.16 A VLP showing the sulphide zones in the 'C' Orebody at Mufulira West, Zambia.

relief of an underlying basement complex. The latter aspect can be depicted by plotting isopachs of formations beneath a marker horizon, such as the footwall of the orebody in the case of a stratiform ore-deposit. Isopachyte VLPs can also be produced for specific geological formations or lithofacies types. Sedimentological parameters such as grain size, degree of rounding and sorting, maximum pebble dimensions, etc., can also be plotted and directly related to the grade VLPs for the metal(s) concerned.

1.7 STRUCTURE CONTOUR PLANS

1.7.1 Projection on to horizontal plan

Intersections of an orebody footwall in diamond drill-holes and in underground exposures allow the production of a structure contour map on which the elevation of this contact above mean sea-level, or above, or below, a mine datum (e.g. shaft-collar elevation), can be plotted. These are then contoured at a suitable interval. Figure 1.17 shows such an example in which the structure contours of the orebody are truncated by a fault. The fault plane is also represented by structure contours and the trace of the orebody cut-out is produced by joining the points where contours of the same altitude intersect. Similar maps can be produced to represent fold structures and the intersection of dykes, or other intrusive contacts, with the orebody or marker horizon. Such plans are important in a mining operation as they allow the prediction of orebody cut-outs in areas of the mine not yet penetrated by development. The magnitude of fault displacements can also be

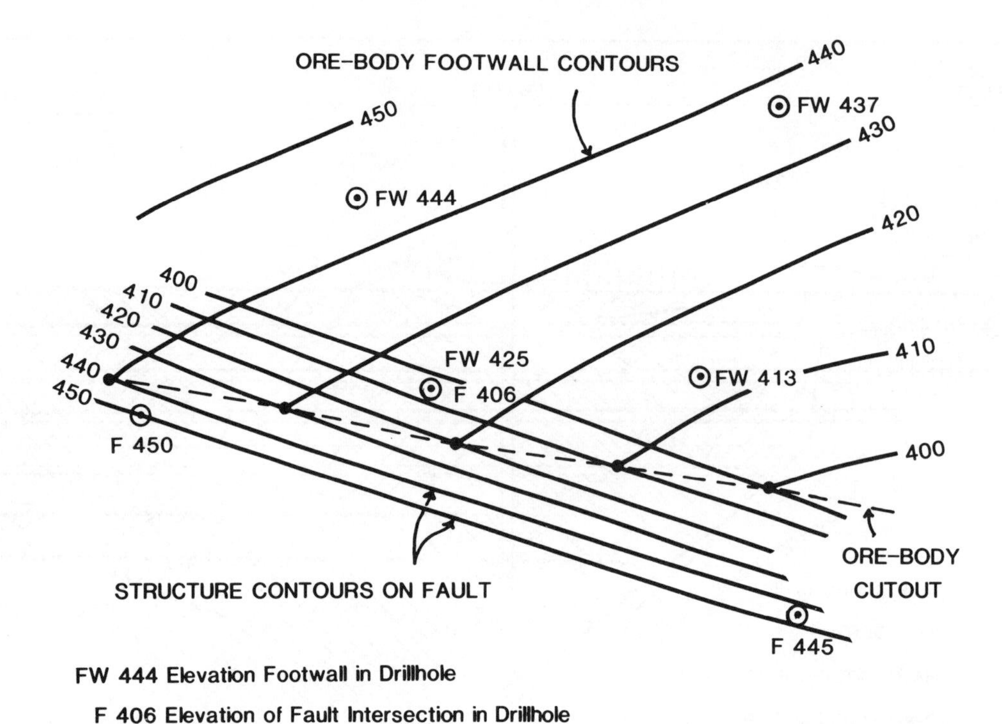

Fig. 1.17 Structure contours on an orebody footwall and on a cross-cutting fault.

determined and thus the location of the down-faulted section of ore. Where a post-ore dyke has been located in the upper levels of a mine (or open-pit), its location within a deeper mining block, or bench, can be estimated by extrapolation of structure contours. This allows its impact on block or bench reserves to be assessed.

1.7.2 Projection on to VLP

Figure 1.18 represents a section of an orebody showing the location of the VLP plane and also the horizontal projection distances (D1, etc.) of the hangingwall to this plane. These distances are measured at regular depth intervals and then

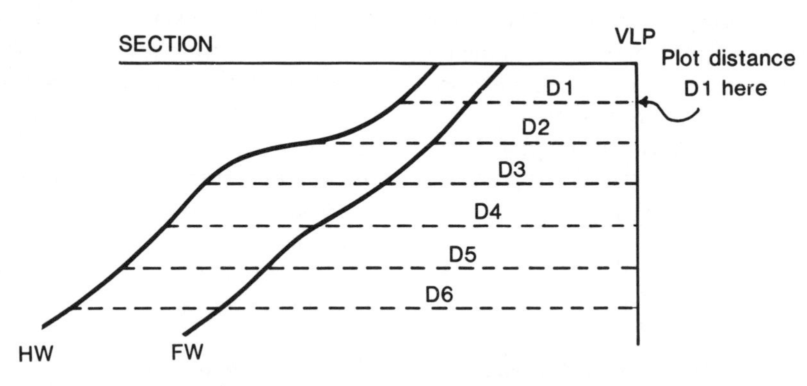

Fig. 1.18 Structure contours on VLP.

plotted on the relevant section line on a VLP. This exercise is repeated for all the mine sections and then the results are contoured. A uniformly dipping tabular deposit would thus be represented by a series of parallel, equally spaced, contours whose value increases with depth down the VLP. Deviations of the hangingwall would then be reflected by small deflections in these contours. A similar VLP can also be produced for the footwall of the orebody.

1.8 CONNOLLY DIAGRAMS

Morphological changes in an orebody with depth can be better depicted by use of the Connolly diagram, which allows the plotting, on a VLP, of the horizontal projection distance of either the hangingwall, or the footwall, to an inclined plane. This overcomes the problem incurred in the use of structure contours as described in section 1.7.2, in which small deviations in attitude or shape are swamped by the impact of increasing distance to the VLP plane at lower levels.

1.8.1 Production of the Connolly diagram

The first stage of this exercise is to measure the average dip of the orebody on each section-line and then calculate the overall mine average dip. A reference plane with this dip is then drawn in the orebody footwall on each section-line. It could also be located so as to intersect the mine VLP at shaft-collar elevation. The horizontal projection distance, from either the hangingwall or footwall, to this reference plane can now be measured at regular depth intervals (X1, etc. on Figure 1.19) and this value plotted on the VLP. The results for each section line are thus contoured to give a much clearer picture of the shape of these assay contacts. What we have produced is roughly equivalent to a residual geochemical anomaly once the inclined geochemical background has been removed. Figure 1.20 shows an example of a Connolly diagram produced for the hangingwall of an orebody. This indicates the existence of easterly plunging fold structures.

1.8.2 Use of the Connolly diagram

Other than the representation of orebody morphology, the Connolly diagram can be used to investigate the structural controls of veins hosted by faults. Sudden changes in dip of the host fissure, in response to changes in rock type and competancy, may be associated with a pinch-out or widening of the payable zone. Thus a comparison of vein grade, and/or thickness, with contours on a Connolly diagram may allow the exact nature of any structural control to be assessed. For example, in Figure 1.21(a) we see the

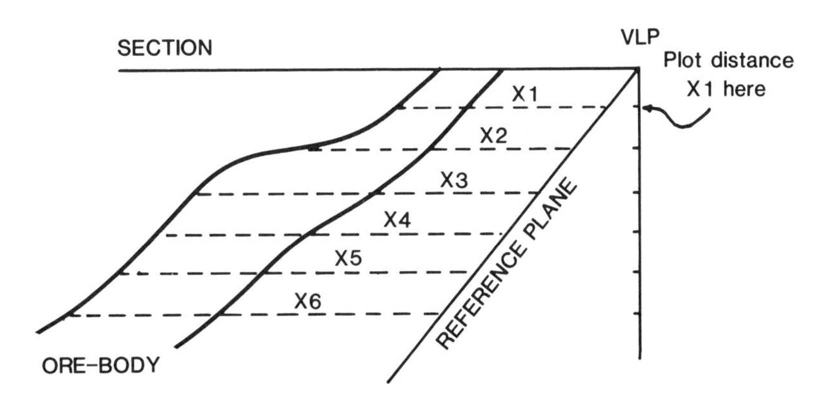

Fig. 1.19 The Connolly diagram reference plane.

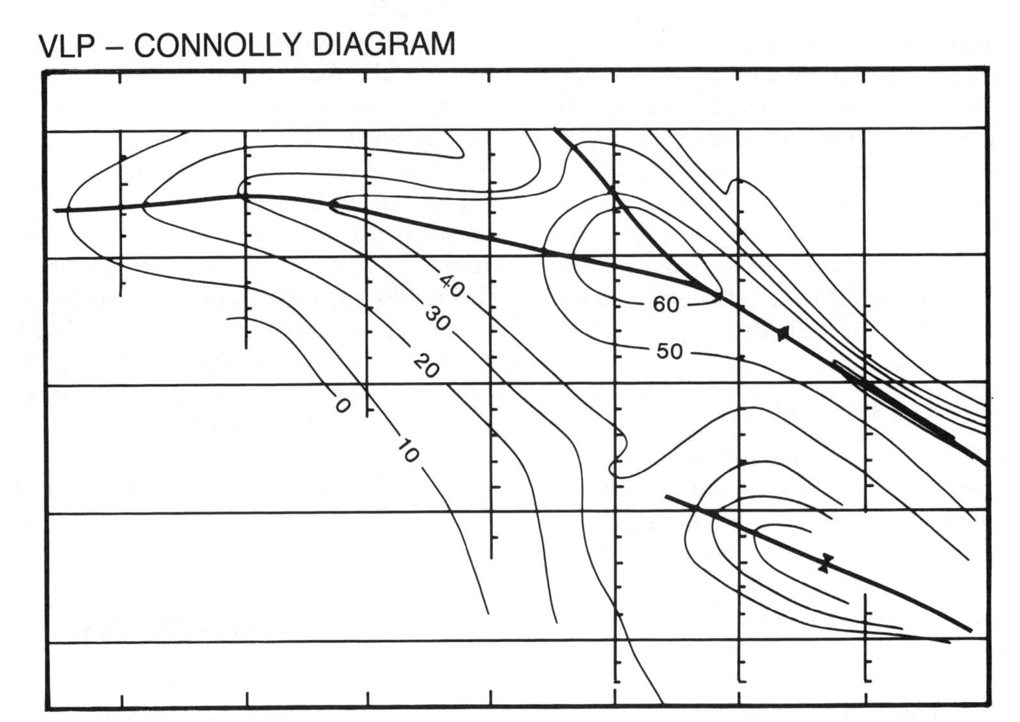

Fig. 1.20 A Connolly diagram showing fold structures affecting the hangingwall of an orebody. Contours represent distances in metres from the reference plane whose dip = 52°.

impact of normal faulting on vein thickness. Here, the thicker section of the lode corresponds to an area in which there will be a rapid change in the projection distance to the reference plane. On the VLP, therefore, there will be a bunching of structure contours which will coincide with stoped areas and areas of payable vein. On the other hand, in Figure 1.21(b), the host fault is a reverse fault and in this instance the area of rapid change in projection distance corresponds to the tight section of the fault with the dilationary zones corresponding to the flatter sections of the fault which are roughly parallel to the reference plane. Thus on the VLP, stoped areas will correspond to those with a wide structure contour spacing and avoid areas of bunched contours.

We can thus see that the Connolly diagram will allow us to assess the nature of the host fault and also allow us to predict where other pay zones may exist. Such zones may correlate with the intersection of a particular lithology with the fault plane.

1.9 DIP CONTOUR MAPS

On each mine section line, the dip of the orebody between adjacent levels can be measured (or calculated) at the orebody median plane position. These values are then plotted on a VLP midway between levels and contoured in the usual way, as in Figure 1.22. This map can then be compared with the orebody thickness or grade VLPs, as for the Connolly diagram, to assess whether the attitude of the deposit plays any role in determining whether it is payable or not. These diagrams are also useful for planning mining methods to be employed in different areas of the mine and to assess whether gravity feed will be possible from stopes, or whether it will have to be induced in some way.

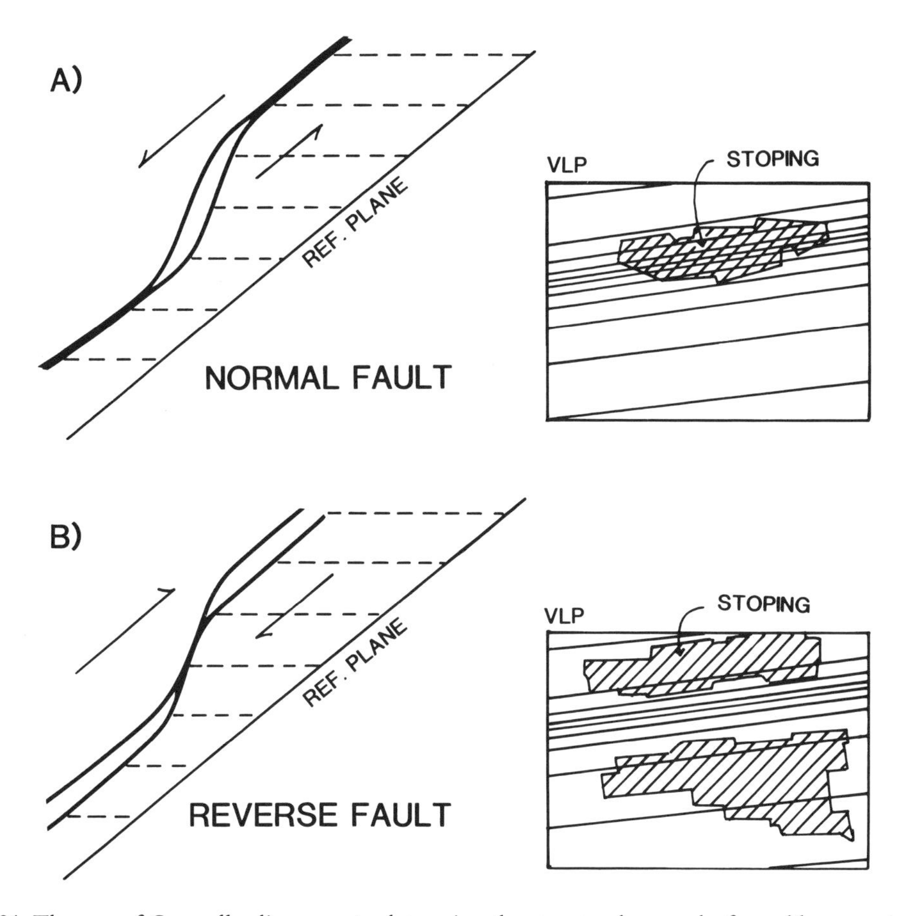

Fig. 1.21 The use of Connolly diagrams to determine the structural control of payable zones in a vein.

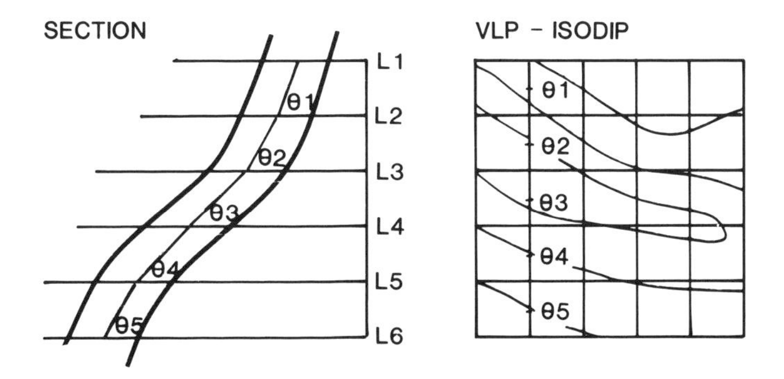

Fig. 1.22 A dip contour map on VLP.

1.10 STRUCTURAL UNROLLING – PALINSPASTIC MAPS

When the orebody under consideration is concordant with the bedding of the host strata (stratiform, stratabound and sedimentary ore deposits), then the effects of folding and faulting can be removed by the process of structural unrolling.

1.10.1 Production of palinspastic maps

The first step is to select a marker horizon which could be:

(1) The assay hangingwall;
(2) The assay footwall;
(3) The median plane;
(4) A thin distinctive and ubiquitous marker horizon in the orebody or in the immediate footwall formations, e.g. a purple shale bed or a green tuff bed.

Having selected the marker horizon to be unrolled, we must now select the axis, or axes, about which to unroll the orebody.

Single axis
Where only one period of folding has affected the deposit, we can locate the axis of unrolling at a point where a mine level intersects the marker horizon centrally down the dip-length of the orebody. This will thus reduce the amount of distortion. Alternatively, we can position the axis where the first mine level intersects the marker horizon. In Figure 1.23 we have taken the latter option, although it will increase distortion at the lower levels of the orebody. The median plane has also been taken as the marker horizon to the unrolled.

Unrolling proceeds by measuring the distance from the axis along the median plane to:

(1) Axial planes of folds;
(2) Drill-hole intersections;
(3) Mine levels;
(4) Faults;
(5) Base of the zone of oxidation;
(6) The topographic surface;
(7) The orebody fringe.

The distance of each, measured around the folds, is then used to plot its location, as shown in Figure 1.23. This is repeated for all sections using the same unrolling axis. Where the marker horizon is cut by a fault, its effect is removed by continuing measurement on the other side of the fault. Fold axes, mine levels, orebody fringes and the points at which a fault was located, can now be joined up between all the mine sections to indicate where they lie relative to the unfolded orebody. We have now produced a palinspastic base-map.

Two fold axes
Where two fold axes can be shown to exist, then the procedure is considerably more involved and tedious. The first stage is to produce a structure contour map of the median plane of the orebody using the elevations of this plane above a selected datum. On the basis of this map, the interpreted positions of the two unrolling axes are drawn in, corresponding to the original fold axes (Figure 1.24). The locations of all drill-holes intersecting this plane can also be plotted. Next, a series of section-lines are drawn parallel to the first axis (Axis 1 in Figure 1.24) so as to pass through as many drill-hole intersection points as possible. Each section is then unrolled along the median plane in a direction away from Axis 2. Drill-hole DDH1 thus moves out to position A. The exercise is repeated with section lines running parallel to Axis 2, so that drill-hole DDH1, for example, moves out to position B. We thus have the displacements in the two directions from which the resultant displacement position C can be determined. Thus, for each drill-hole we can produce the new unrolled coordinates of the intersection point.

1.10.2 Use of palinspastic maps

Their main use is for geological reconstructions of orebodies prior to tectonism in order to

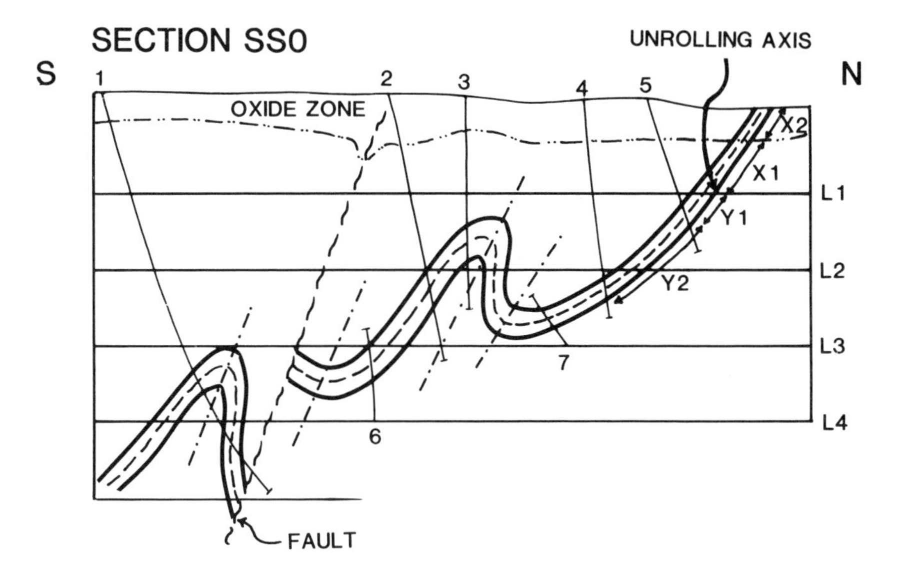

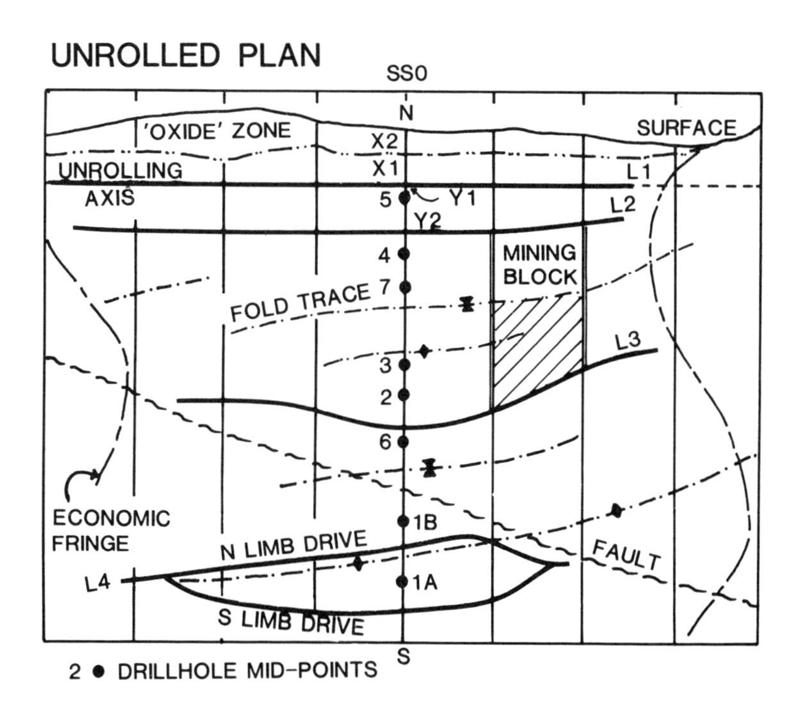

Fig. 1.23 Production of an unrolled plan of a folded and faulted orebody.

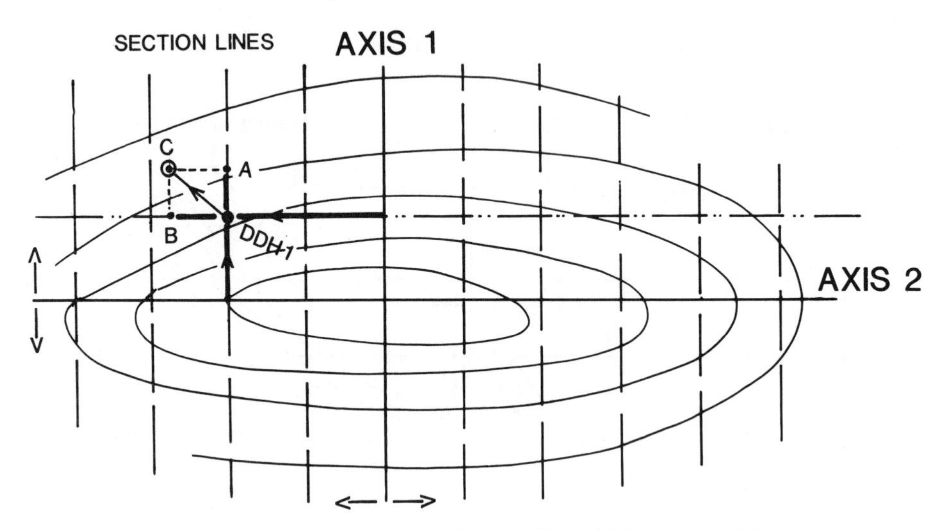

Fig. 1.24 Use of the structure contour map for unrolling folds generated by two fold axes.

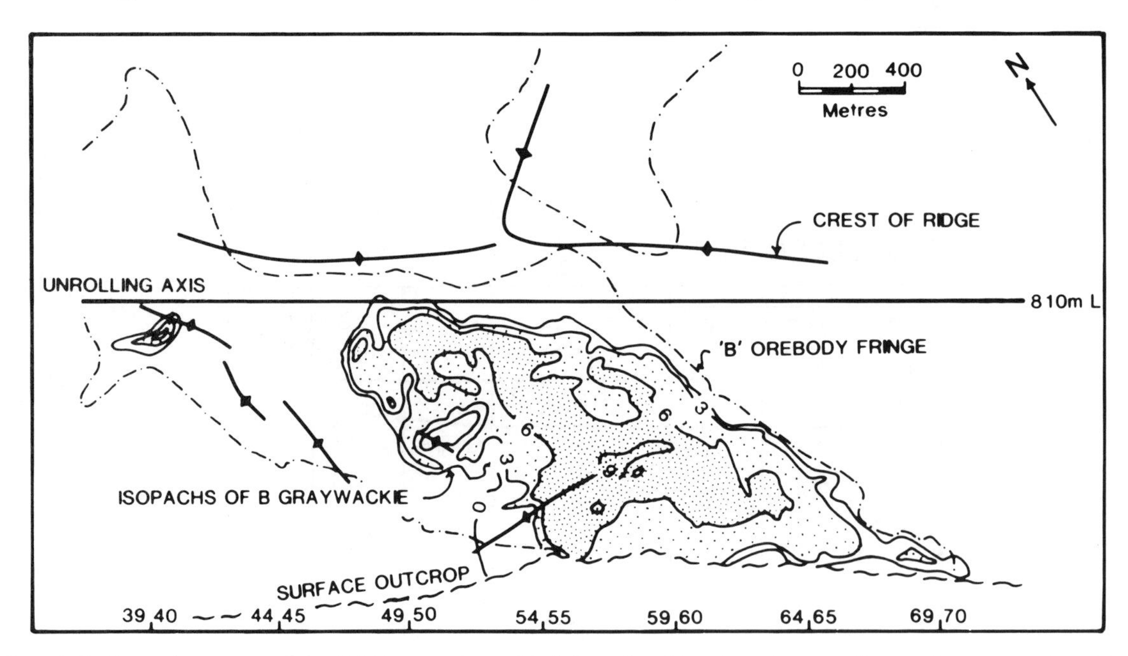

Fig. 1.25 Isopachyte map of the 'B' Graywacke at Mufulira, Zambia, plotted on a palinspastic map produced by unrolling about the 810 mL intersection with this horizon.

determine the true nature, and mode of origin, of the deposits. We can thus plot any of the following:

(1) Grades;
(2) Ore-mineral types and zones;
(3) Alteration zones;
(4) Lithofacies;
(5) Palaeo-environmental indicators (e.g. current directions)
(6) Palaeo-shorelines;
(7) Grain-size variations;
(8) True thickness of horizons or mineralized zones (Figure 1.25).

From these maps, the relationship of later structural features to any of these features can be assessed to see if structure played any role during sedimentation or during deposition of the ore minerals.

Palinspastic maps are also used for ore-reserve purposes where folding and fault disturbances make other methods difficult (Chapter 3). In this instance, true thickness values are plotted and contoured to produce an isopachyte map. A similar map is produced for metal accumulations. Ore-blocks can now be superimposed on these maps, in order to allow the calculation of the reserves. These blocks may be rectangular and will thus only allow the production of a global reserve (total orebody) estimate, or they may be defined by mine section lines and curved mine level lines on the palinspastic maps (Figure 1.23). In the latter case, mining block reserves can be produced. Unfortunately, such maps are not readily accepted by mining engineers because of the distortion introduced.

Isopachyte maps, based on structural unrolling, can also be used to detect palaeotopographical highs, and lows, on a depositional surface (perhaps unconformity) by using multiple linear regression techniques. The isopachyte map is based on the true thickness between a footwall marker horizon and the surface under investigation. This thickness, and the unrolled X- and Y-coordinates of the intersection point, are then compared using a multiple linear regression equation:

$$\text{Thickness} = A_0 + A_1(\text{X-coord.}) + A_2(\text{Y-coord.})$$

where A_0–A_2 are the regression coefficients.

Once the values of the coefficients are determined from a large data set, the predicted thickness value at each point can be calculated by the computer and compared with the observed value. Positive and negative residuals can thus be plotted on the palinspastic map which theoretically should correspond to palaeotopographical highs and lows respectively. The multiple regression technique is merely a method for fitting a regional trend surface to the isopach data from which anomalies can then be pin-pointed. Such a technique has been applied in South African gold mines in an attempt to locate auriferous palaeochannels.

1.11 2D AND 3D BLOCK MODELS

One of the most useful methods of representing mine geological information in the case of open-pit operations, is the production of block models of the orebody, either bench by bench (2D model), or for the entire deposit (3D model). The area is subdivided into a matrix of blocks whose dimensions are governed by the mining method as well as the geology. To each block, a grade can be assigned by a variety of methods (inverse distance weighting (section 3.11) or block kriging (Chapter 4)). Each block is then assigned a geological code depending on the dominant rock type within the block, a metallurgical code indicating the category of ore (e.g. oxide or sulphide) and, in some cases, a geotechnical code which indicates the competence or strength of the rock. From the 3D model, block sections and block plans can be produced which can be modified by the geologist to more accurately reflect local geological conditions. From the modified model, the geologist and engineer can then produce an optimum design for the pit which will allow the mining of the maximum amount of metal/mineral at a minimum cost.

2D block models can also be produced for underground mines exploiting tabular deposits. In this case, blocks are drawn on VLPs to which grades, horizontal thicknesses, rock type codes, etc., are assigned.

1.12 3D OREBODY PROJECTIONS

The advent of computer CAD software, such as AUTOCAD, has considerably speeded up the production of 3D projections of ore deposits and furthermore has allowed the user to change his viewpoint at will and output the result to a flat-bed or drum plotter (e.g. Roland DPX3300 or the Calcomp 1023, respectively). However, there are two main alternatives for the manual production of 3D diagrams, viz. the isometric projection and the true perspective diagram. Both are briefly discussed below.

1.12.1 Isometric projection

Figure 1.26 illustrates the method of production of the isometric projection. Three axes, 1–3, are drawn at angles of 120° to one another and Axis 2 is subdivided at equal intervals, representing the mine section spacing. A series of parallelogram sections is then constructed, each showing the mine levels and the outline of the ore deposit. The diagram can then be left in this form, or the various construction lines removed and the cross-sectional areas of the deposit linked to produce a 'solid' 3D representation of the orebody.

1.12.2 True perspective diagram

The method for the production of perspective drawings of mine workings and orebodies was first described by McPherson (1938) and the interested reader is referred to this paper for details of the method. The diagram produced is equivalent to that obtained by taking a photograph from a selected viewpoint at a given elevation in a specific direction. Convergence thus exists towards the perspective point on the horizon. Such diagrams are difficult to draw and do not allow the measurement of distances between points on the diagram. Computer software is now available to produce perspective diagrams using user-specified viewpoints and horizons (e.g. Surfer, section 1.16.1).

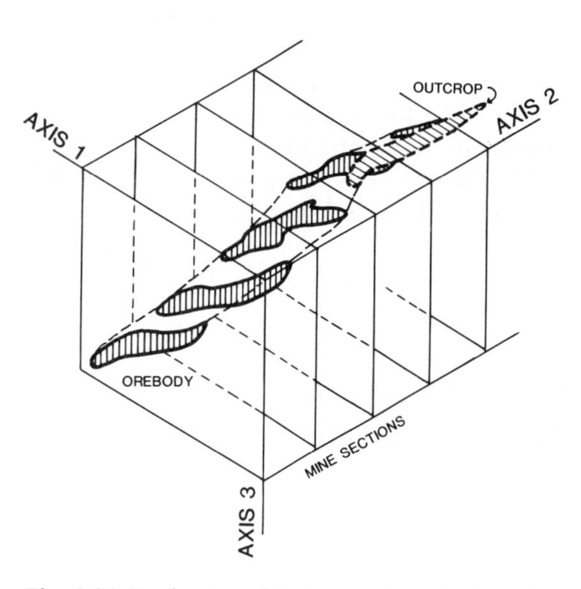

Fig. 1.26 Production of the isometric projection using mine sections.

1.13 HISTOGRAMS AND CUMULATIVE FREQUENCY PLOTS

Numerical data from a mine can often be best summarized by use of histograms for these can be plotted on sections or plans. Examples are:

(1) Histograms of metal grade against drill-hole traces on section.
(2) Histograms of chip sample grades in cross-cuts.
(3) Histograms of geotechnical information (e.g. number of discontinuities per metre) in cross-cuts and drives.
(4) Histograms of total mine assay data.
(5) Histograms of monthly discharge rates from aquifers under drainage.

Many other examples could no doubt be listed. Mine-ore reserves can also be depicted in this way showing, in histogram form, the percentage of the mine tonnage in each grade category. However, this information is more often represented in a cumulative frequency plot from which the percentage of the total tonnage above a given cut-off grade can be quickly read off. The production of these plots is discussed further in section 4.14 whilst their use for global grade estimation is covered in section 3.5.

1.14 ROSE DIAGRAMS

These diagrams plot the azimuth or bearing of directional information and at the same time give a general impression of their relative frequency of occurrence. Examples of the type of information that can be represented in this way are, the strike of subvertical joints, faults, shear zones, palaeo-current directions and palaeo-wind directions.

A set of measurements of one of these is subdivided into groups, on the basis of their orientation. For example, we may divide the data into 36 groups each spanning 10° of azimuth. The frequency of occurrence of readings in each group is then calculated as a percentage of the total population. The next step is to draw a circle whose radius is set equivalent to a percentage just greater than the largest frequency of occurrence. For example, if this were 26%, then the scale length of the radius would be made equivalent to 30%. The perimeter is then divided into 10° intervals and a point is plotted along the median lines of each sector so produced, whose distance from the centre represents the frequency of occurrence in this particular sector. Each of these points is then connected to adjacent points, as in Figure 1.27, to produce a rose diagram. This gives a visual impression of the dominant directions that may exist in the data set.

1.15 STEREOGRAPHIC PROJECTIONS

Space does not permit a description of stereographic projections and stereonets, and thus the reader is referred to standard texts on the subject,

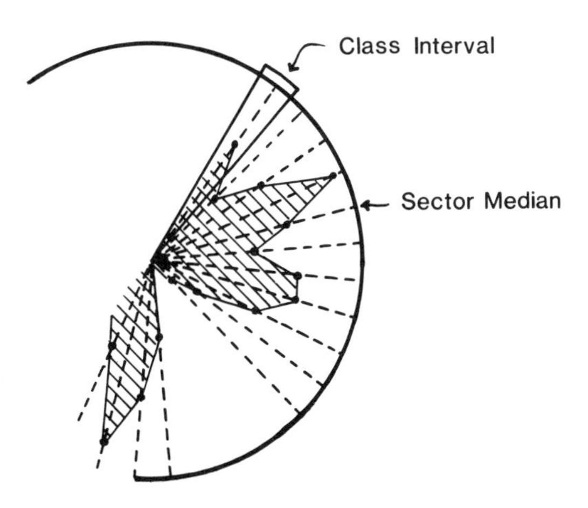

Fig. 1.27 Construction of a rose diagram.

such as Phillips (1960) or Priest (1985). Their role in the representation of mine data is three-fold:

(1) To allow the determination of true dips from apparent dip readings taken in different directions (i.e. in adjacent cross-cuts and drives), or to determine apparent dips in specific directions given the true dip.
(2) To allow the plotting of foreset orientations in current bedded units so that a correction can be made for tilt of the host formations.
(3) To allow the compilation of large amounts of geotechnical or structural information where both dip and dip direction must be recorded.

In the latter case, poles are plotted for each plane measured which allow the discrimination of sets of structures with a common orientation, e.g. joint sets. Such stereonets could be produced for each lithological unit in the mine, whereas others could represent data in specific areas of the mine. For example, a stereonet could be plotted for each mine level, or each mining block, to illustrate changes in the relative importance of different discontinuity sets with depth or laterally along strike.

1.16 COMPUTER SOFTWARE

A wide range of software is now available to the mining geologist or engineer for use on IBM-compatible machines which will allow the production of high quality graphics. The examples quoted below are but a selection of this range to illustrate their use in the mining environment.

1.16.1 Graphics programs

Electronic spreadsheets

Spreadsheet software, such as Lotus 1-2-3, Supercalc 3, Symphony and the Lotus clone, 'AsEasyAs', are invaluable for they not only allow the production, vetting and updating of drill-hole data and assay files (section 3.4.2), but they can be used to undertake a basic statistical analysis of this data. From the worksheet, we can produce bar charts showing the frequency

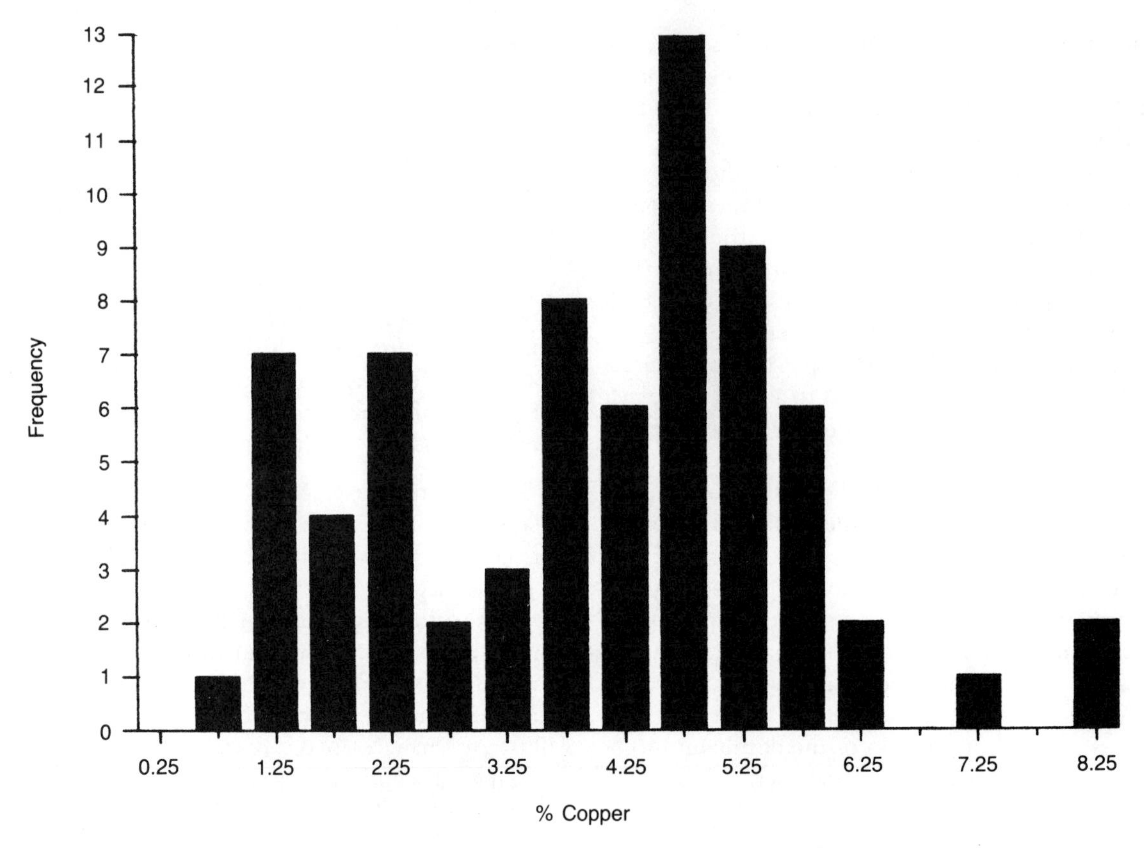

Fig. 1.28 Bimodal histogram of ore-zone copper grades from the Kalulushi East Prospect, Zambia, produced using Lotus 1-2-3.

distribution of the data and a wide range of line graphs (e.g. cumulative frequency curves) and XY regression plots. Figure 1.28 is a histogram of copper grades for the Kalulushi East copper prospect in the Zambian Copperbelt, which was produced by Lotus's 'PrintGraph' software. Figure 1.29 is a cumulative frequency plot of log-transformed copper grades from the 3170 ft bench in the Ingerbelle Pit, British Columbia, which was exploiting a porphyry copper deposit, while Figure 1.30 is a grade–thickness X–Y plot for the Kalulushi East prospect.

Contouring and surface modelling packages
Perhaps one of the most useful packages for the production of contour maps from irregularly spaced data points, is Golden Software's SURFER program. This program will allow the import of ASCII data-files or Lotus (WK1) worksheets, which consist of the X and Y coordinates of each data point together with a measured variable, e.g. footwall or surface elevation or orebody vertical thickness. Alternatively, the 'Grid' routine from this package has its own worksheet facility from which an ASCII file can be produced. This worksheet is useful in that an imported ASCII file can be vetted and modified to delete unwanted data. The 'Grid' routine can then be employed to create a regular data matrix, for the variable concerned, from the irregular data set within a user-specified area and with a user-specified grid dimension. This grid is generated by inverse distance weighting (Chapter 3) or by kriging (Chapter 4) methods. Search areas

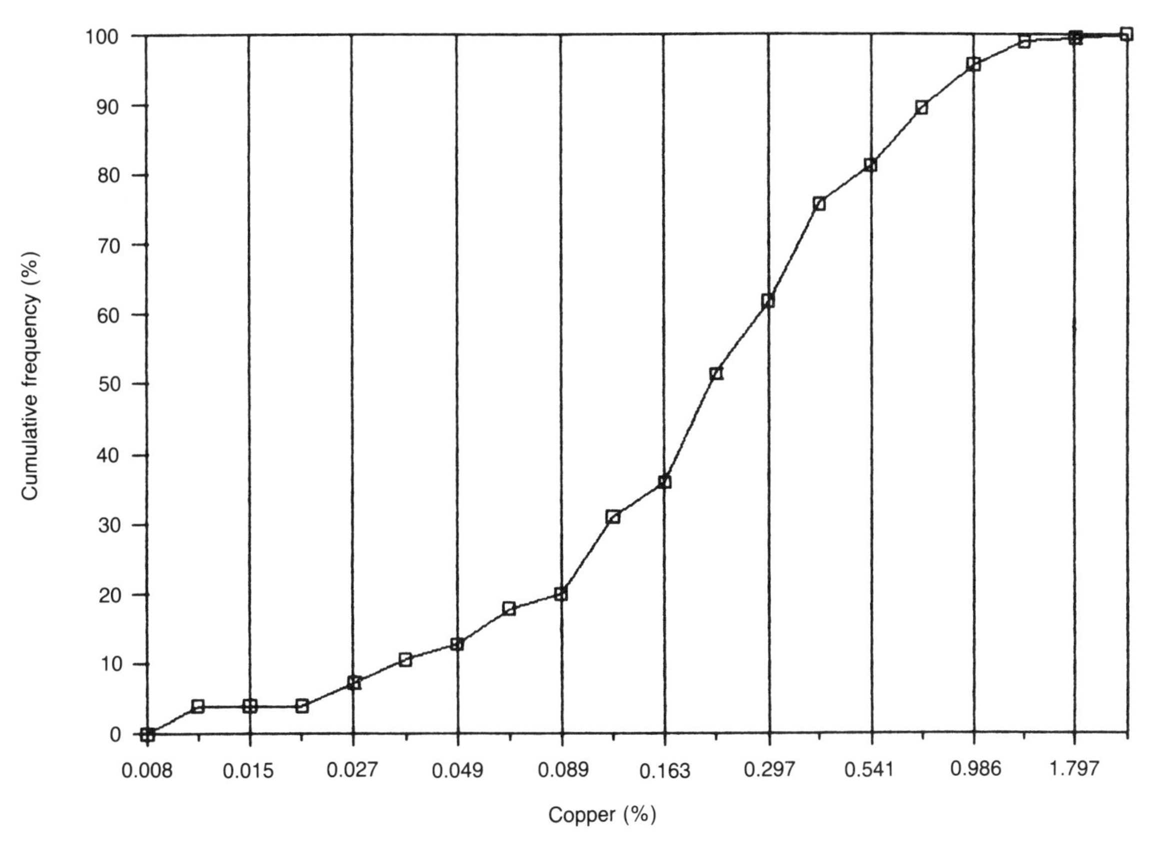

Fig. 1.29 Cumulative frequency plot of log-transformed blast-hole copper grades from the 3170 ft bench in the Ingerbelle Pit, British Columbia.

around each block centre to be valued are also defined by the user (circular, quadrant and octant), together with the maximum number of points to be used in each search area (e.g. the nearest 10 points). Additional smoothing of the data can be achieved by a cubic spline interpolation technique which increases the number of grid elements in the interior of a previously created grid and then connects sets of points with a smooth curve. Once data matrices have been created using 'Grid' they can be combined in various ways using mathematical functions. Grid to grid subtraction of, for example, hangingwall and footwall elevation matrices will produce a vertical thickness isopach map.

The output file from 'Grid' (*.GRD) can then be imported into the 'Topo' routine which generates the contour map. Again the user can modify the contour interval and the labelling of the final product at will. The map file (*.PLT) can then be directly printed (or plotted) or later imported for direct printing via the 'Plot' routine. Figures 1.31–1.33 show the impact of changing the gridding parameters on a contour map of vertical metre per cent total copper in the footwall arenite (RL7) orebody at Kalulushi East, Zambia. All use inverse square distance weighting with a circular search area but the first uses the default setting of 1215.4 m search radius and a ten-point search with no smoothing. Note the angular nature of the contours and the extrapolation of contours beyond the posted data points (*). Figure 1.32 was produced by reducing the search radius to 200 m. The contours are now

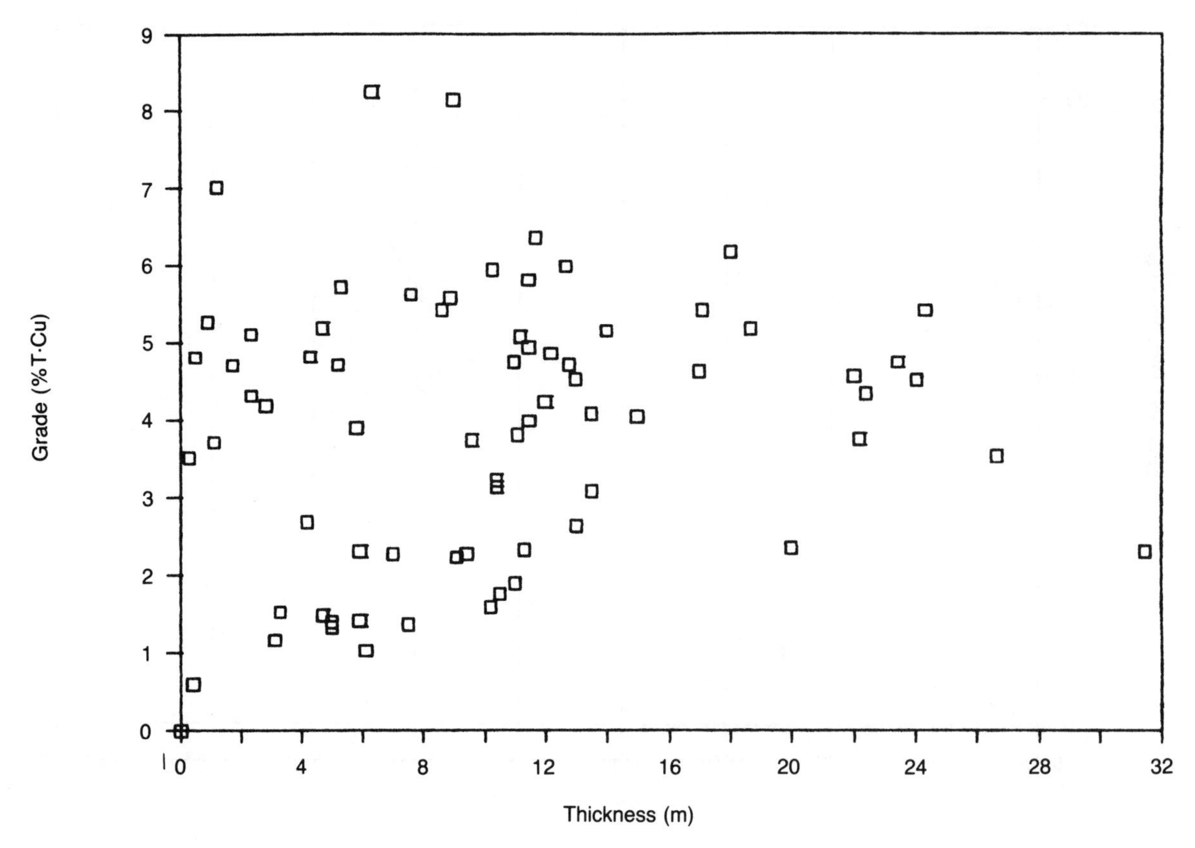

Fig. 1.30 X–Y plot of ore-zone thickness versus copper grade from Kalulushi East, Zambia.

largely limited to the area covered by drilling. In the case of Figure 1.33, the search radius was further reduced to 100 m, with an eight-point search, and with the smoothing facility employed. The result is more aesthetically pleasing and the contours now closely match those obtained by manual contouring. Additional lines can be superimposed on the maps if, for example, the suboutcrop trace of the orebody was required. A blanking facility also exists to remove areas of the map not required. Ore-evaluation blocks can be superimposed as the program allows a user-defined grid to be printed on the map. Further enhancement of the final map can be achieved by transferring the map to Harvard Graphics, a program which allows the production of high quality graphics suitable for reports and publications.

The 'Surf' routine enables the user to produce either orthographic or perspective diagrams of topographic, geochemical or geophysical surfaces directly from the 'Grid' files. Figure 1.34 is a contour map of the elevation of the basement complex beneath the Kalulushi East prospect while Figure 1.35 is an orthographic view of this palaeotopographical surface, viewed from the SE at an angle of 45°. No vertical exaggeration has been introduced but this is available, if required. Modelling of the surface can also be made using contours, rather than form lines parallel to the X and Y axes of the 3D model, as in this case.

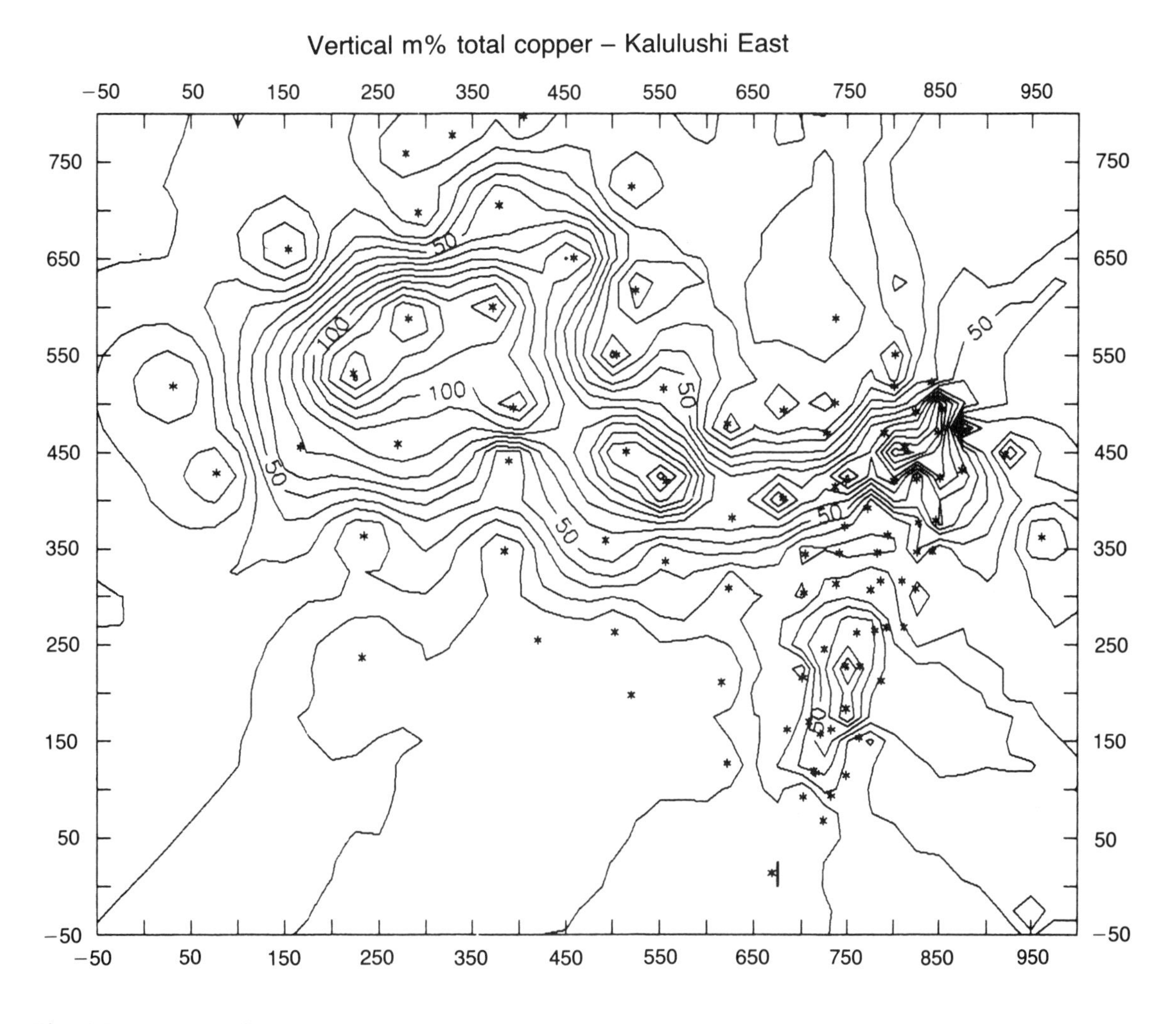

Fig. 1.31 Contour plot of vertical metal accumulation (%) for Kalulushi East produced using 'Surfer'. (Inverse distance weighting, 10-point search). Coordinate value in m.

Borehole software

Software for the production of graphic borehole logs, geological sections and plans is now widely available in the industry. Drillog 5 (GeoMEM Software) is a typical example which allows the production of lithological logs of boreholes. These can be descriptive or symbolic, as illustrated in Figure 1.36. The format of the graphic log can be specified by the user and descriptive text can be printed in one or two columns. Additional columns are available for the plotting of shaded or unshaded histograms and line graphs (both on normal or log scales) of numerical data imported from ASCII data files produced by word processors/text editors, spreadsheets or specialist software. Such data could include assay results, as in Figure 1.36, geophysical data, as in Figure 1.37 (gamma, density, neutron, caliper,

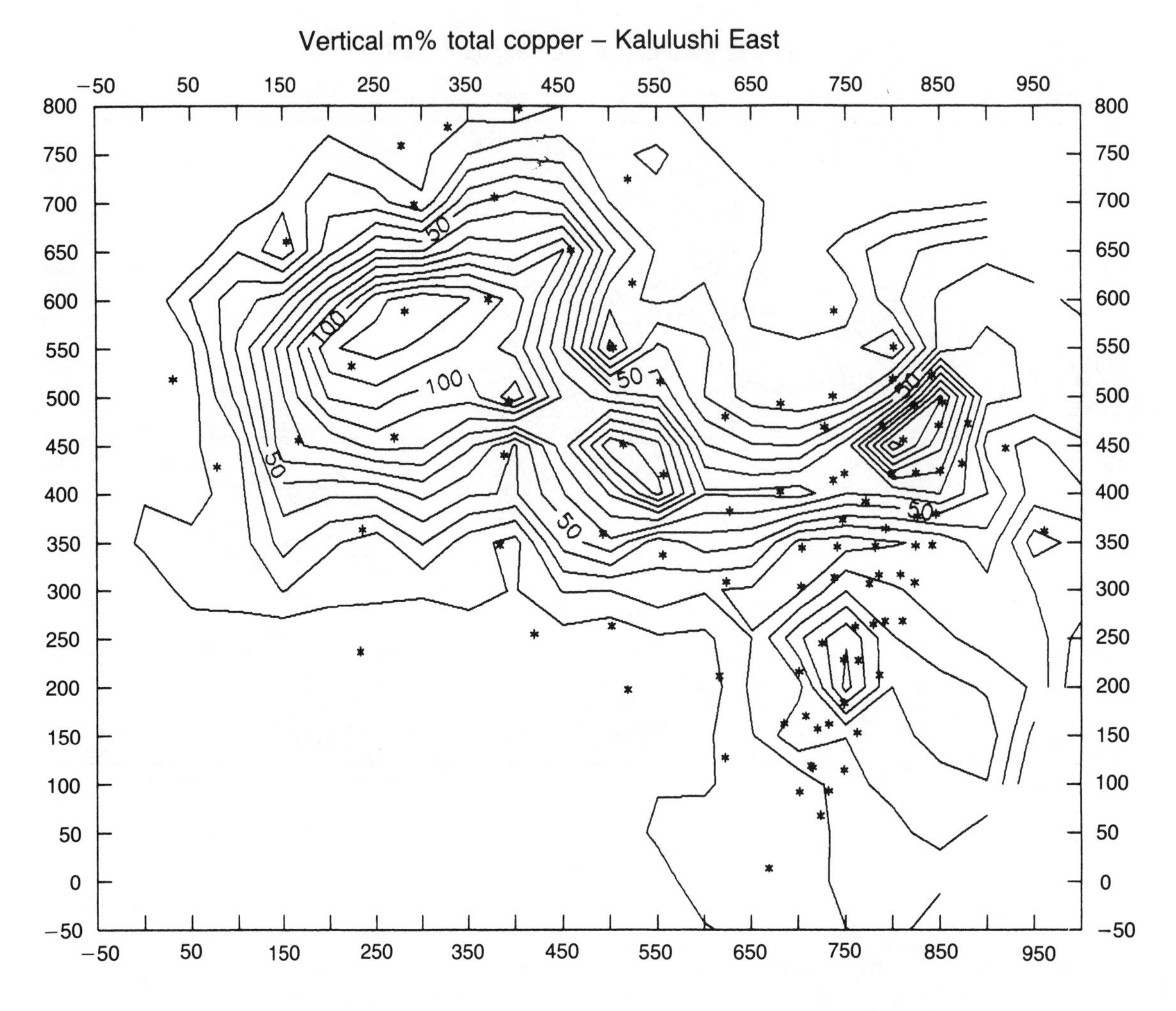

Fig. 1.32 The impact of a reduction in the search radius to 200 m (compare with Fig. 1.31). Coordinate value in m.

resistivity, etc.), aggregate grading data, penetration rates, RQD, porosity, moisture content, etc. Depth scales are also user specified. A coordinate transfer utility program (DLXYZ) is also available for extracting the coordinates of specified down-the-hole points (e.g. orebody/seam footwalls) and then transferring the results to ASCII XYZ files for use in other contouring or specialist programs.

Another useful software package is Borsurv (J. H. Reedman & Associates Ltd) which allows the production of borehole geological sections (Figure 1.38), plan projections of boreholes (Figure 1.39), longitudinal sections and level/bench plans. Lithological units are represented symbolically while assays can be printed out in numerical form or as bar graphs alongside the hole trace. Weighted averages of assayed intervals can also be plotted together with geophysical data profiles. Text can be added as keys or as drill-hole annotations. The facility also exists to output the print files to DXF files for further processing

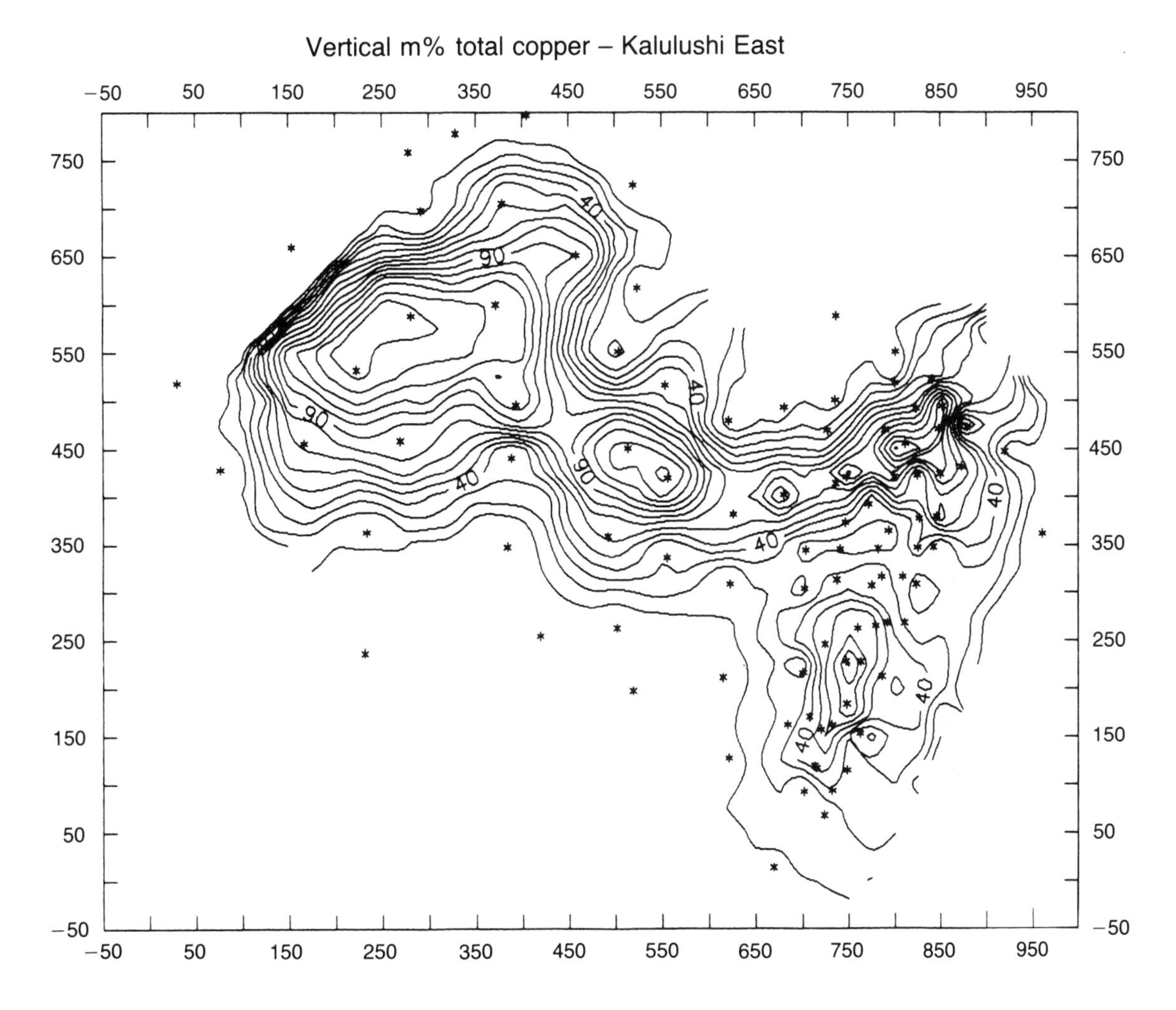

Fig. 1.33 A further reduction in the search radius to 100 m combined with an eight-point search and the use of the smoothing facility (compare with Fig. 1.31 and 1.32). Coordinate value in m.

with Autocad or similar CAD programs (see below).

CAD software

Internationally, Autocad (e.g. Autocad Release 10, ADE3) is accepted as being the industry standard for computer-aided drafting and design, although there are other cheaper alternatives such as Generic Cadd 3.0. These can be operated with IBM-compatible desk-top personal computers with at least 640K ram and a maths coprocessor, an A3 or larger graphics tablet (e.g. Cherry Mk II) and a mouse (e.g. Microsoft). Autocad is menu driven, using either a cross-hair propelled by the mouse or directly from the keyboard. The range of facilities offered by the system is too large to discuss here, but these include the automatic drawing of standard geometric forms such as user-defined arcs, circles, ellipses, lines, points and polygons. The angles, and distances,

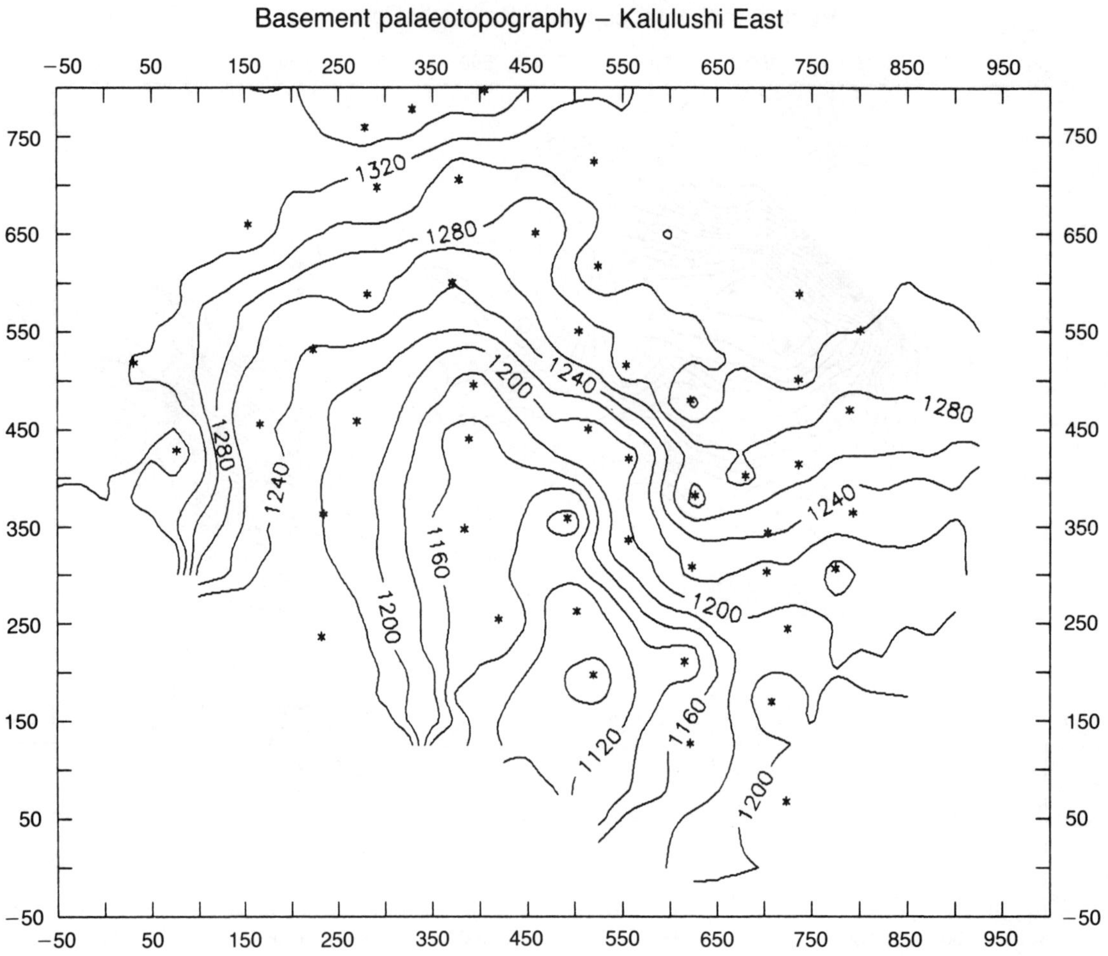

Fig. 1.34 A contour map of the basement palaeotopography at Kalulushi East which shows that a deep valley underlies this copper deposit. Coordinate value in m.

between points can be determined, as can areas and the length of lines (or perimeters of areas). A useful facility is the digitization of points on the screen (e.g. drill-hole collars) and the ability to receive data from a digitizing table (e.g. Calcomp 91360 A1 or Benson 6301 AO digitizers) and to output to a range of hardcopy devices, such as dot-matrix printers, colour printers or plotters (Roland DPX3300 flatbed or Benson 1600 series drum plotters). Drawings can be panned across, rotated, zoomed on to, or can be broken down into segments and later reconstructed again. Text can be inserted and a variety of ornaments can be generated using different styles, thickness and colours of lines. Scales can also be changed at will.

CAD systems are thus used to create 2D drawings such as geological sections, mine level plans,

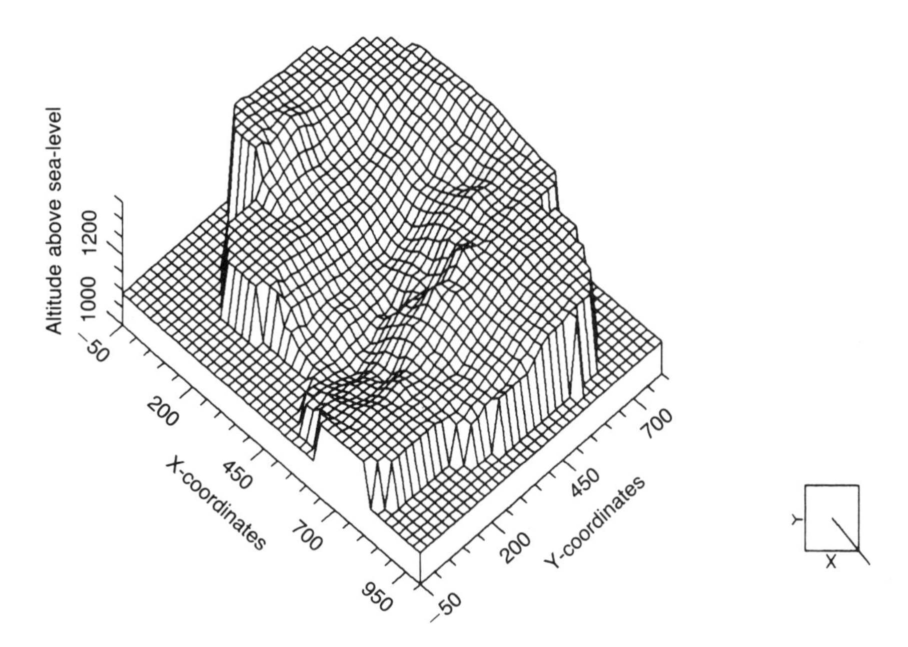

Fig. 1.35 An orthographic projection produced by 'Surfer' showing an 3D representation of the basement surface at Kalulushi East.

vertical longitudinal projections and bench plans. Plans and sections of mine development can be produced with, or without, geological contacts, faults, dykes, orebody outlines, etc., and portions of these can be extracted and blown up to any required scale. Bench and stope plans can also be produced showing ore-blocks, current, or past, face positions, blast-hole collars, sampling lines, etc. 3D models of orebodies, or mine development, from any specified viewpoint, can be created while level or bench plans can be combined to produce 3D models of the geology. Stope and drilling layouts can be designed and modified directly on the screen and the results viewed from different directions. Advance planning of mine development, in relation to the ore zone, can be undertaken.

A major advantage of CAD is the ability to update, or modify mine plans and produce multicolour plots in a very short time. Such plots can be of a very high quality and can be used directly in reports, etc. Also, as mentioned in the previous section, DXF files created by other software can be imported and modified as necessary before transmission to a plotter.

1.16.2 Specialist mine software

Two main packages are beginning to emerge as industry standards; these being DATAMINE (Datamine International Ltd) and a more recent arrival on to the mining scene, SURPAC (Surpac Mining Systems Ltd). Both attempt to provide the user with a comprehensive system which can be used at all stages in the life of an operation. They can thus be used at the early exploration stage, during feasibility and evaluation studies, for mine design (both open-pit and underground) and, finally, in the operational phases. As such, they combine the facilities provided by spreadsheets, graphics and statistical packages, ore-reserve programs and mine design software.

Both programs have the facility to input drillhole survey, assay and geological information,

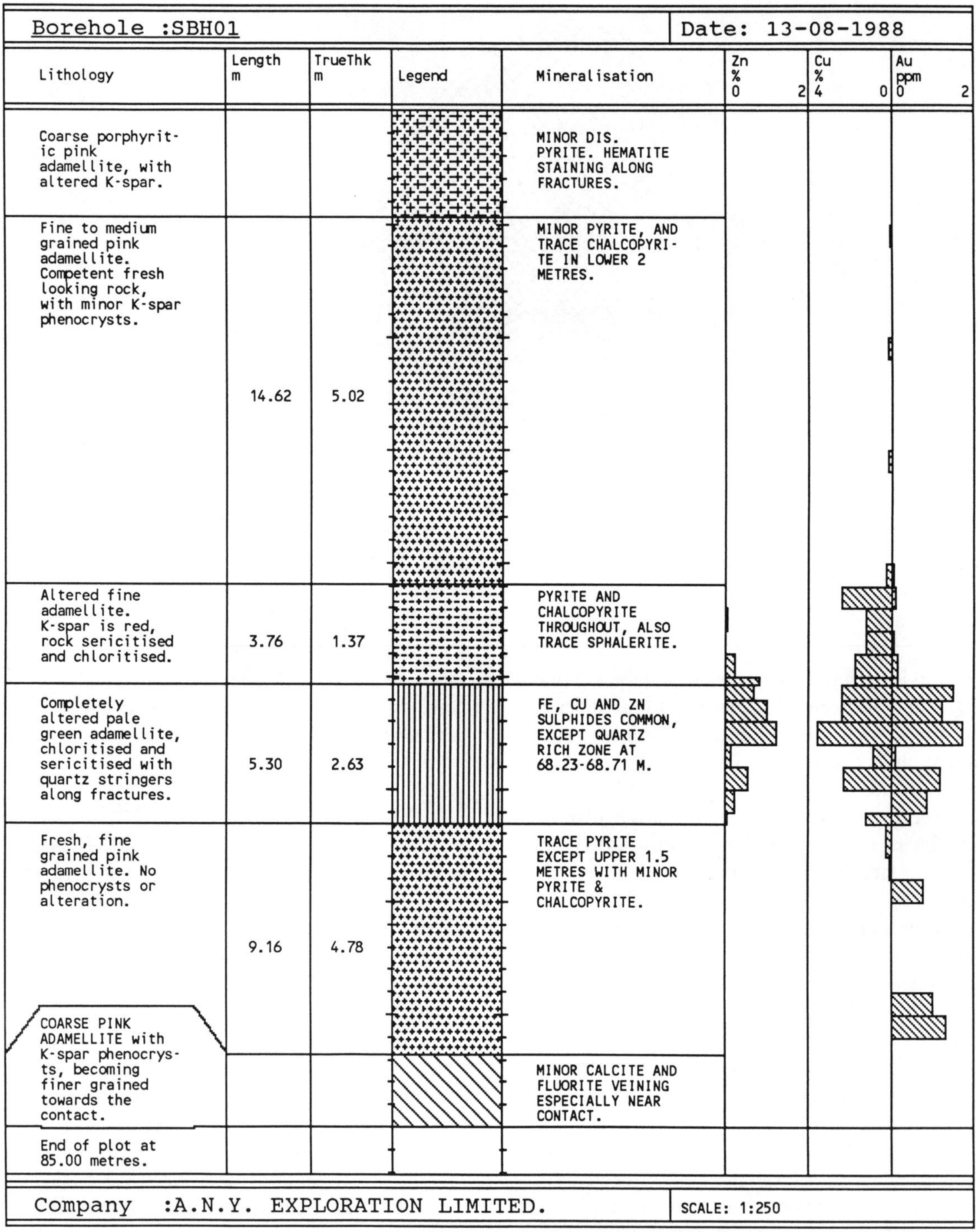

Borehole :SBH01					Date: 13-08-1988		
Lithology	Length m	TrueThk m	Legend	Mineralisation	Zn % 0 2	Cu % 4 0	Au ppm 0 2
Coarse porphyrit- ic pink adamellite, with altered K-spar.				MINOR DIS. PYRITE. HEMATITE STAINING ALONG FRACTURES.			
Fine to medium grained pink adamellite. Competent fresh looking rock, with minor K-spar phenocrysts.	14.62	5.02		MINOR PYRITE, AND TRACE CHALCOPYRI- TE IN LOWER 2 METRES.			
Altered fine adamellite. K-spar is red, rock sericitised and chloritised.	3.76	1.37		PYRITE AND CHALCOPYRITE THROUGHOUT, ALSO TRACE SPHALERITE.			
Completely altered pale green adamellite, chloritised and sericitised with quartz stringers along fractures.	5.30	2.63		FE, CU AND ZN SULPHIDES COMMON, EXCEPT QUARTZ RICH ZONE AT 68.23-68.71 M.			
Fresh, fine grained pink adamellite. No phenocrysts or alteration.	9.16	4.78		TRACE PYRITE EXCEPT UPPER 1.5 METRES WITH MINOR PYRITE & CHALCOPYRITE.			
COARSE PINK ADAMELLITE with K-spar phenocrys- ts, becoming finer grained towards the contact.				MINOR CALCITE AND FLUORITE VEINING ESPECIALLY NEAR CONTACT.			
End of plot at 85.00 metres.							
Company :A.N.Y. EXPLORATION LIMITED.					SCALE: 1:250		

Fig. 1.36 A graphic borehole log produced by 'Drillog 5' showing a mineralized intersection in a borehole. (Plot courtesy of GeoMEM Software Ltd.)

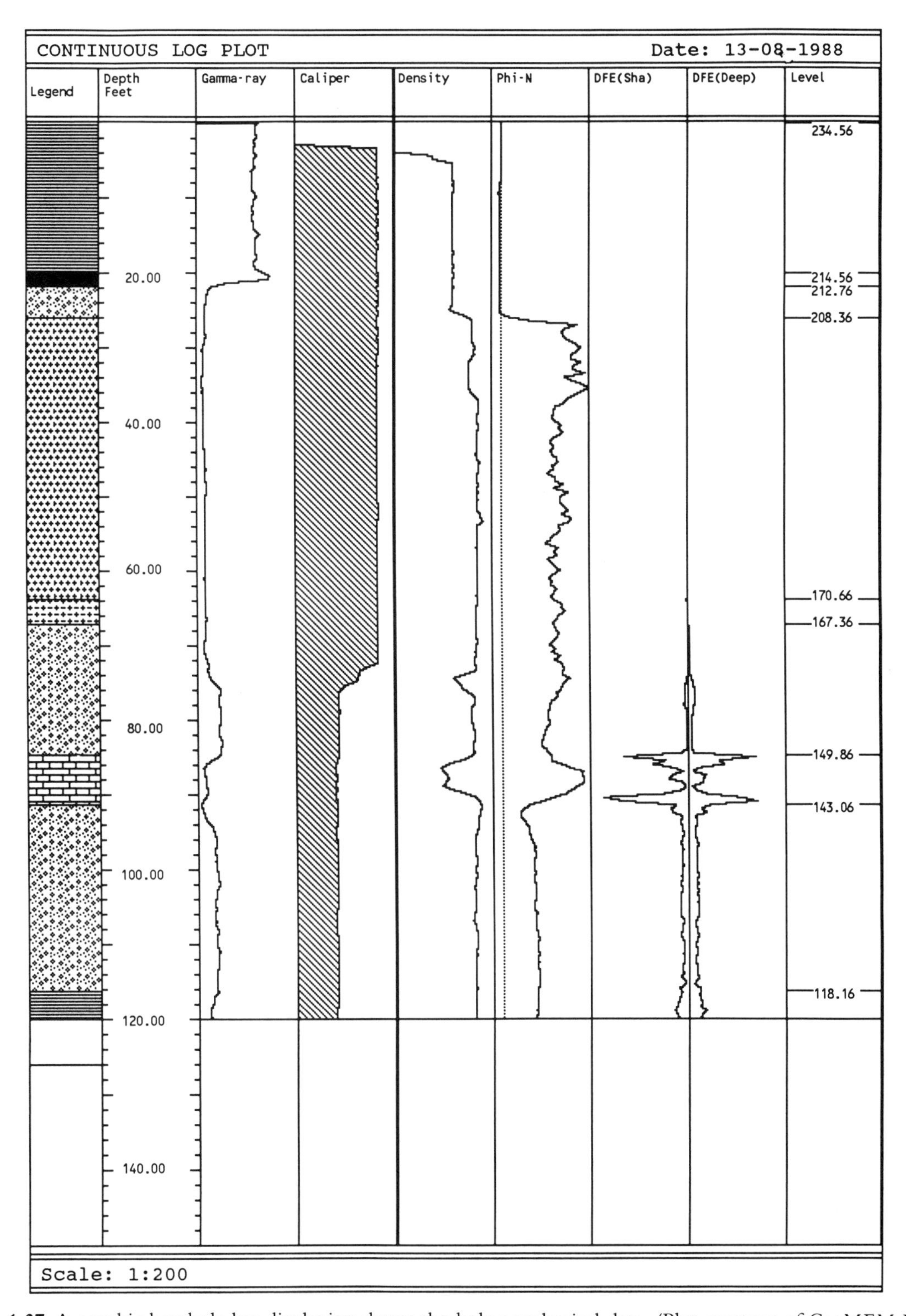

Fig. 1.37 A graphic borehole log displaying down-the-hole geophysical data. (Plot courtesy of GeoMEM Ltd.)

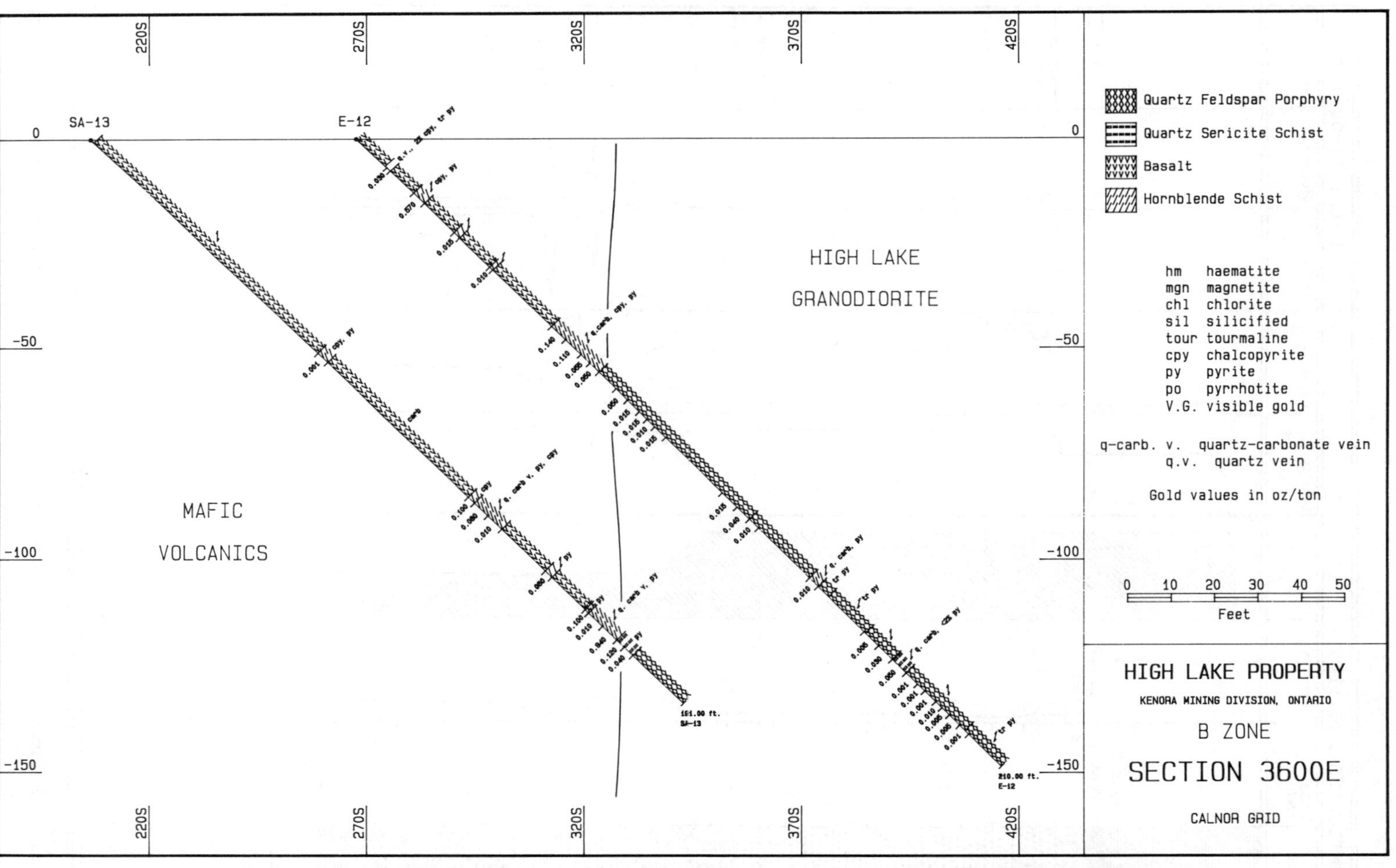

Fig. 1.38 A geological and assay section produced by 'Borsurv'. (Courtesy of J.H. Reedman.)

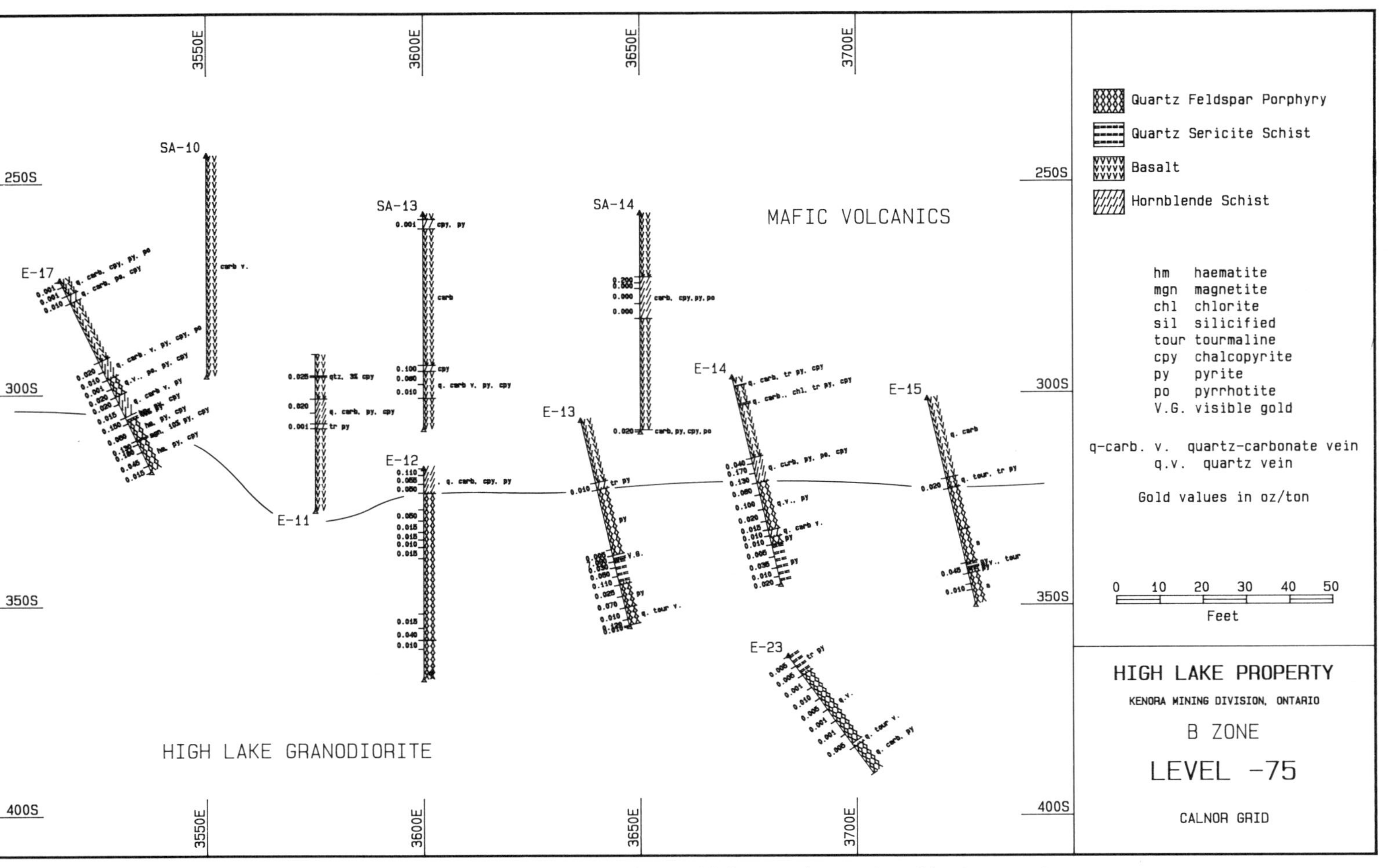

Fig. 1.39 Plan projection of drill holes in a Canadian gold deposit also produced by 'Borsurv'. (Courtesy of J.H. Reedman.)

from which 3D sample coordinates can be calculated. The drill-hole collar coordinates can be combined with digitized surface contour plans to produce drill-hole progress plans on which the plan-projected traces of drill-holes can be inserted. Compositing of sample assays is also possible over specified intervals which may relate to orebody/seam widths, to equal depth intervals down the holes, to bench thicknesses, and finally to rock-type zones.

Raw and log-transformed assay data can be analysed by basic statistical techniques and the results tabulated or presented as scatter plots, bar charts, line plots, contour plans, etc. 2D or 3D geostatistical modelling of directional semivariograms can also be undertaken (sections 4.5–4.7).

Drill-hole logs can be produced with lithological columns, geochemical, geophysical or geotechnical profiles and brief geological descriptions. Drill-hole sections may also be drawn to any standard layout showing the topographical profile, all the unrolled drill-hole traces, each with histograms of metal grades and lithological units/codes (Figure 5.22). On these sections, the geologist is able to outline the orebody limits, correlate stratigraphic units and model the structure. More information on these programs is presented in Chapter 5, where their use in orebody modelling, reserve estimation and open-pit design, is described. A brief case history is also provided in Chapter 8, section 8.8.

REFERENCES

McPherson, G. (1938) The preparation of true perspective views from mine plans by simple methods either graphical or numerical. *Trans. IMM*, **XLVII**, 1937–8, 347–379.

Phillips, F. C. (1960) *The Use of Stereographic Projection in Structural Geology*, Edward Arnold, London, 86 pp.

Priest, S. D. (1985) *Hemispherical Projection Methods in Rock Mechanics*, George Allen and Unwin, London, 124 pp.

Society of Mining Engineers, (1973) *SME Mining Engineering Handbook*, vols 1, 2 (ed. I. A. Given, AIME and Petroleum Engineers Inc.), New York.

Thomas, L. J. (1973) *An Introduction to Mining*, Hicks Smith and Sons, Sydney, 436 pp.

Woodruff, S. D. (1966) *Method of Working Coal and Metal Mines*, vol. 3, Pergamon, Oxford, 571 pp.

2

Mine Sampling

2.1 INTRODUCTION

The sampling of metalliferous and industrial mineral deposits is undertaken for a variety of reasons and at various stages in their evaluation and exploitation. During the exploration phase, the sampling is largely confined to the analysis of drill cuttings, or cores, and is aimed at the evaluation of individual, often well spaced, intersections of the deposit. It thus gives the *in situ* grade and thickness of an intersection but provides little evidence of the continuity of potentially economic mineralization and generally takes little account of mining constraints.

During the exploitation phase, sampling is used to define assay hangingwalls and footwalls, together with the grade over mineable thicknesses, taking into account not only the mineralized zone, but also its potential dilution by low grade or barren material. Sampling is much more intense in this situation and is undertaken to allow the assignment of overall weighted grades to individual ore-blocks or stopes. It is used to define internal zones of low grade or waste material or zones of differing metallurgical type, i.e. refractory or oxide ore, as opposed to sulphide ore amenable to flotation. It may also be used to delineate areas containing certain deleterious elements (Hg,As,Bi, etc.) or valuable by-product metals (Au,Ag, etc.). Also at this stage, sampling will be used to extend existing reserves and attempt to prove new ore-zones accessible from existing underground development. Perhaps one of the most important applications of sampling during the exploitation phase is in grade control, i.e. the regular monitoring of face grades in active stopes, or bench grades in an open-pit, or of broken rock being trammed or trucked as waste or mill feed. It is thus used to assess when a stope is exhausted, when a drive on lode has penetrated beyond the limits of ore-grade/mineable material and whether the material being sent to the mill is still of economic grade. Sampling will also aid blending of ores from different stopes, or portions of a pit bench, to ensure constancy of mill feed. Grade control, as such, will be considered in more detail in Chapter 7.

Barnes (1980) defines a sample as '. . . a representative part or a single item from a larger whole, being drawn for the purpose of inspection or shown as evidence of quality' and that it is '. . . part of a statistical population whose properties are studied to gain information about the whole'. The type of sample, and number collected, depends on a range of factors which include:

(1) The type of mineral deposit and the distribution and grain size of the valuable phase.
(2) The stage of the evaluation procedure.
(3) Whether direct accessibility exists to the mineralization.
(4) The ease of collection, which is related to the nature and condition of the host rock.
(5) The cost of collection, funds available and the value of the ore. Obviously, the cost of intense sampling of a low grade, or low value, deposit, may be prohibitive.

It is the intention of this chapter to examine these factors and to describe the various methods used to collect samples, illustrating each with examples taken from different deposits around the world.

2.2 CHARACTERIZATION OF MINERAL DEPOSITS FOR SAMPLING PURPOSES

The mode of occurrence and morphology of a mineral deposit has considerable impact on the type and density of sampling, and on the amount of material required. This aspect can be illustrated by considering some of the main groupings of mineral deposit in turn.

2.2.1 Veins

The following are the main features of vein type deposits from the point of view of their sampling:

(1) The valuable mineral/metal components are unevenly distributed across, and within, the plane of the vein.
(2) The ore mineral may be coarsely crystalline requiring a large sample volume for it to be representative.
(3) Many veins are narrow (compared to the stoping width) and hence are susceptible to heavy dilution.
(4) Most veins are associated with faults, fissures and shear zones and the condition of the wall-rock may affect dilution and thus the way in which they are sampled.
(5) The assay cut-offs are usually sharply defined and correspond to the wall-rock/vein contacts, however, wall-rock impregnations and subsidiary stringer vein systems may require additional sampling beyond the vein margins.
(6) Although fluctuations in thickness are fairly predictable, and lie within limited ranges, those of grade are highly erratic and unpredictable, thus requiring closer spaced sampling.
(7) They are frequently difficult to sample due to the hardness or brittle nature of the vein constituents and it is thus very difficult to avoid biased sampling because of variable quantities per unit length (with the exception of whole drill core sampling).
(8) Because mine development is concentrated on lode, sampling tends to be concentrated on faces or muckpiles requiring regular visits to the face, a time-consuming commitment.
(9) Advance sampling is sometimes limited by lack of remote development from which underground drilling can be accomplished.

2.2.2 Stratiform deposits

These include stratiform base-metal deposits, such as those of the Zambian Copperbelt, and stratiform gold deposits, such as those at Hemlo, Ontario and the Carling type deposits. These are thus deposits which lie within a particular lithological unit or lithofacies and whose contained ore minerals are largely controlled by bedding and other sedimentary features. Their significant features, from the point of view of sampling, are:

(1) They tend to be thick (i.e. up to 20 m).
(2) They have a large areal extent.
(3) They are commonly either gently inclined, or highly folded and tectonized, thus presenting problems for sampling and evaluation.
(4) Grades tend to be uniform and predictable, except when affected by late stage remobilization, metamorphism or veining.

(5) Gradual and systematic lateral changes in grade allow a wider sampling interval.

(6) Many deposits contain fine grained mineralization thus requiring a smaller sample volume.

(7) Being hosted by meta-sediments, they tend to be easier to sample and less prone to bias from variable sample size, rock hardness or high nugget effect (high grade patches).

(8) Many have a moderately high grade and thus sampling errors are less significant.

(9) Assay cut-offs may be gradational, especially at the hangingwall, and difficult to pinpoint underground, so that more samples are needed to ensure that the complete ore zone is covered.

2.2.3 Sedimentary deposits

These include coal, ironstones, potash, gypsum and salt deposits and are characterized by:

(1) Clear-cut contacts against country rocks.

(2) Gradual fluctuations in quality indicators, e.g. % ash, % sulphur and % chlorine in coal.

(3) Sampling is often controlled by the presence of interbedded layers of 'dirt', i.e. shale partings.

(4) Gradual and predictable changes in seam thickness and in the proportion of in-seam 'dirt' bands occur, except where washouts, low angle thrusts and faults interrupt continuity.

(5) Vertical, bed by bed, fluctuations in grade may occur which should be recognized by the sampling strategy.

2.2.4 Porphyry copper – molybdenum deposits

These are characterized by:

(1) Their large dimensions, which require that sampling be via drilling (diamond or percussion) on large grids, e.g. 100–150 m.

(2) Their generally non-tabular form.

(3) Their low and erratic grades which require large sample volumes, sometimes obtained from trial winzes, exploration adits or trenches.

(4) Their variable intensity and style of mineralization, from patchy disseminated sulphides, to vein- or fissure-type sulphides.

(5) The existence of superficial leached or oxidized zones, supergene-enriched zones and hypogene zones which should be recognized by the sampling.

(6) Higher grade hypogene mineralization is often concentrated along fracture systems and thus the orientation of drill-holes must be carefully considered to avoid drilling subparallel to their dip.

(7) Their internal zonation of sulphides, metals and associated wall-rock alteration which should be taken into account during sampling.

(8) Changes in host rock lithology from quartz-monzonite to diorite, from latite to andesite, from intrusive plutons to lavas and pyroclastics, from silicified rocks to rocks showing potassic, phyllic, argillic or propylitic alteration, all of which will result in changes in rock competence and rippability and in the spacing and size of blast-holes needed for their fragmentation.

2.2.5 Shear zone and epithermal gold deposits

These present a particular problem for sampling as:

(1) They have poorly defined and irregular geometries.

(2) The envelope containing ore-grade mineralization is difficult to define.

(3) They are internally complex and zones of low grade or barren host rock result in heavy dilution.

(4) High grade zones are narrow and associated with thin stringer veins and their immediate host rocks (impregnations).

(5) Continuity of individual veins is not great due to their podiform to en-echelon nature.

(6) They are difficult to sample because of wide ranges in competency of the rocks due to alteration and/or tectonism.

(7) Their grades are low and bulk sampling is really necessary to obtain a reliable grade.

(8) The valuable metal is not always visible and sampling must be controlled by indirect methods such as abundance of pyrite or fuchsite or carbonization or tourmalinization, etc.

(9) Grade distributions are strongly skewed and outliers exist (anomalously high values outside the main population) which result in grade bias.

2.2.6 Carbonate hosted stratabound/ Mississippi Valley deposits

These are characterized by:

(1) The abundance of sphalerite and galena so that grades are often in excess of 10% combined metals.

(2) Sulphides which are often coarse grained or massive, and which are thus clearly visible and easy to sample.

(3) Sharp cut-offs against country rock so that sampling is easily controlled and limited to potentially economic mineralization.

(4) A form which may vary from tabular (e.g. Navan in Eire), reflecting their diagenetic emplacement into bedded carbonates; lenticular, due to their emplacement into porous/permeable reef complexes (e.g. Pine Point, NWT, Canada); to pipe like due to their presence as cements to karstic solution collapse breccias (e.g. Tri-State Pb-Zn deposits, USA).

(5) In many cases they possess a gentle inclination (due to the deposition of the host carbonates in a tectonically stable environment) which requires the use of mining methods such as 'room and pillar' and thus allows direct access to stopes for sampling, and which makes surface drilling an effective sampling technique.

2.2.7 Volcanogenic massive sulphides/ exhalites

These are characterized by:

(1) Changes in the style of mineralization from feeder-type stringer ores, which are low grade and associated with ramifying networks of mineralized quartz veins within highly silicified, chloritized or sericitized rocks, to massive ores or sulphidites consisting largely of pyrite and/or pyrrhotite with variable concentrations of chalcopyrite, galena and sphalerite.

(2) Vertical variations in metal type and grade which must be taken into account during sampling (e.g. Cu to Pb to Zn).

(3) Morphologies which may be tabular but which are usually lenticular/podiform/discoid.

(4) Considerable widths, high grades and often steeply inclined attitudes, which are amenable to mining by block caving methods and which thus restrict direct access for sampling and require remote sampling by diamond drilling.

2.2.8 Classification of ore deposits

Carras (1987) has proposed a classification which takes into account the geometry, the grade distribution and the coefficient of variation C (s.d. × 100/mean) of the assay population which is summarized below:

Type A – low coefficient of variation

Type A1 – simple geometry and simple grade distribution.

Examples: Coal, iron, bauxite, lateritic nickel and stratabound (stratiform?) copper.

Type A2 – simple geometry and complex grade distribution.

Examples: disseminated copper, gold stockworks, Witwatersrand gold.

Only minor dilution is likely to take place during mining. Both classical and geostatistical ore-reserve methods give similar results although geostatistics may be more suited to type A2.

Type B – complex geometry and simple grade distribution with a low coefficient of variation

Examples: Basemetal deposits, e.g. skarn copper deposits (Craigmont, BC).

These deposits tend to have irregular margins, leading to high edge dilution, and their complex geometry makes it more difficult to apply geostatistics.

Type C – complex geometry and complex grade distribution with a high coefficient of variation
Examples: Archaean gold (e.g. Kalgoorlie and Canada).

These have internal and edge dilution due to the patchy nature of the mineralization and its ill-defined or irregular envelope. Carras states that classical ore-reserve methods are best but are susceptible to high margins of error.

2.3 GRADE ELEVATION

This is an arbitrary datum plane along which all geological mapping and horizontal sampling is done, or on to which all non-horizontal sampling assay data are projected. It is at approximately waist height along the sidewalls, or at the face, and is thus approximately 1 m above the floor. When a cap-lamp is hung around the neck, the level at which the light beam touches the wall can also be taken as grade elevation. The actual elevation of this plane can be determined from the nearest mine survey peg in the back. Figure 2.1 illustrates the procedure. A weighted chain or string of length L is hung from the survey peg whose elevation Z is known. The plumb-bob is thus at an elevation of $(Z - L)$. The distance X between grade elevation and the plumb-bob is measured, thus giving the required elevation as $(Z - L) + X$. The offset from the survey peg to the sampled sidewall can be measured (e.g. 1.85 m to right sidewall) together with the distance Y to the commencement of the first sample at grade elevation.

2.4 POSSIBLE LOCATIONS FOR UNDERGROUND SAMPLING

Figure 2.2 summarizes some of the locations underground from which, or in which, sampling can be undertaken either by chip or channel sampling or by diamond drilling. To this can be added, (a) sampling in raises driven on lode, and (b) face and muckpile sampling in stopes where access is possible (e.g. cut and fill stopes). In the case of drill-holes, the sample values will be projected down-dip or up-dip on to the nearest level or sub-level plan, or will be projected horizontally on to the mine VLP.

2.4.1 Chip/channel sampling locations

Sidewalls
Here samples can be taken horizontally at grade elevation in cross-cuts through the mineralization or at right angles to tabular inclined bodies. In the latter case, the results are projected down-dip on to grade elevation as shown in Figure 2.3(a). If both walls of a cross-cut are sampled in this way, then the orientation of the assay hangingwall and footwall can be determined (Figure 2.3(b)).

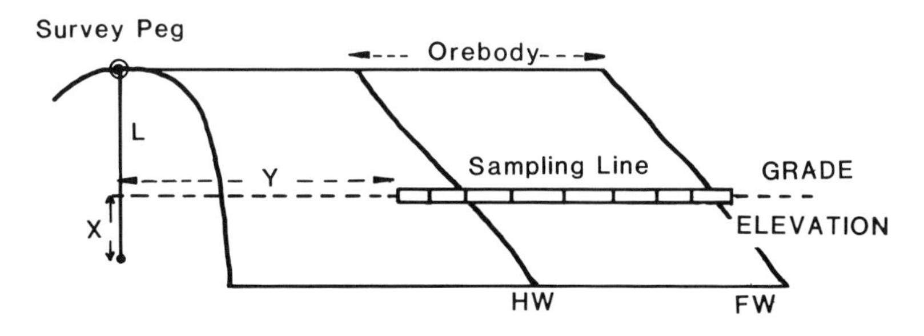

Fig. 2.1 Cross-cut showing grade elevation and sample locations across an orebody intersection.

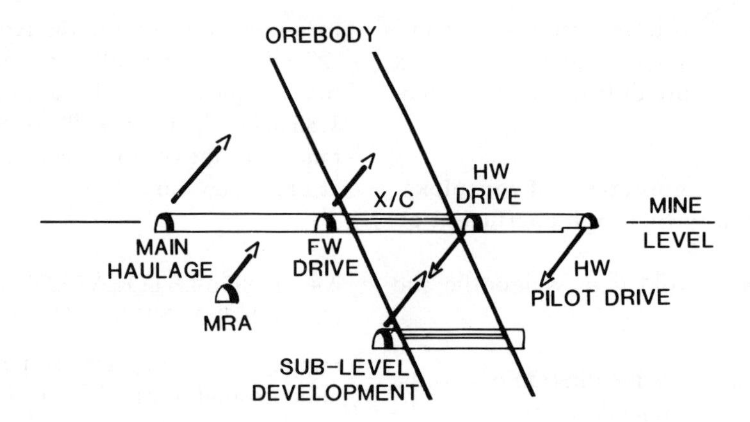

Fig. 2.2 Locations from which samples can be obtained underground by drilling or cross-cut sampling.

Roof or back

In cross-cuts, sampling can follow the centre-line of the back across the mineralized zone but this is much more difficult because of the height of the sampling (Figure 2.4(a)). All values would then be projected down-dip on to grade elevation. In drives which are following the lode (e.g. thin veins), back sampling may be undertaken either because the face or blast muckpiles were not sampled at the time of drive advance, or because resampling at a later date is deemed necessary perhaps to recheck earlier sampling. In this case, the sampling is at regular intervals (e.g. 10 m) across the back at right angles to the lode (Figure 2.4(b)). Generally both ore-zone and waste are sampled.

Faces

Figure 2.5(a) shows vertical sampling of an advancing face to determine the exact location of the orebody footwall while Figure 2.5(b) shows horizontal sampling of the face in a lode drive. Alternatively, face advance muckpiles can be grab sampled (section 2.7).

Raises

One or both walls of a raise can be sampled at right angles to the lode at regular intervals or

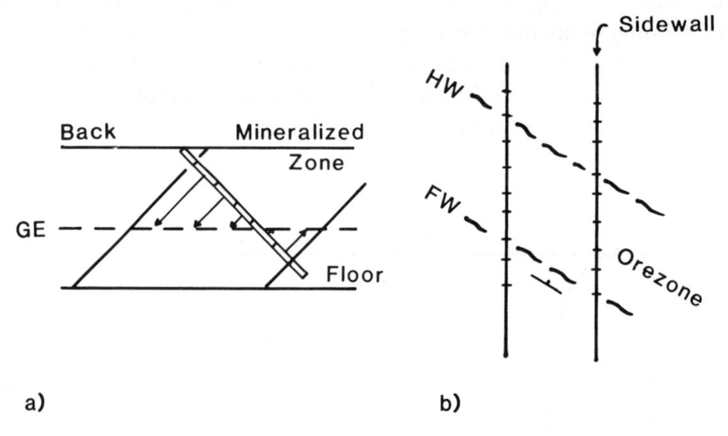

Fig. 2.3 Sidewall sampling. (a) Cross-cut section showing projection of samples on to grade elevation (GE). (b) Plan of cross-cut showing assay contacts defined by sampling.

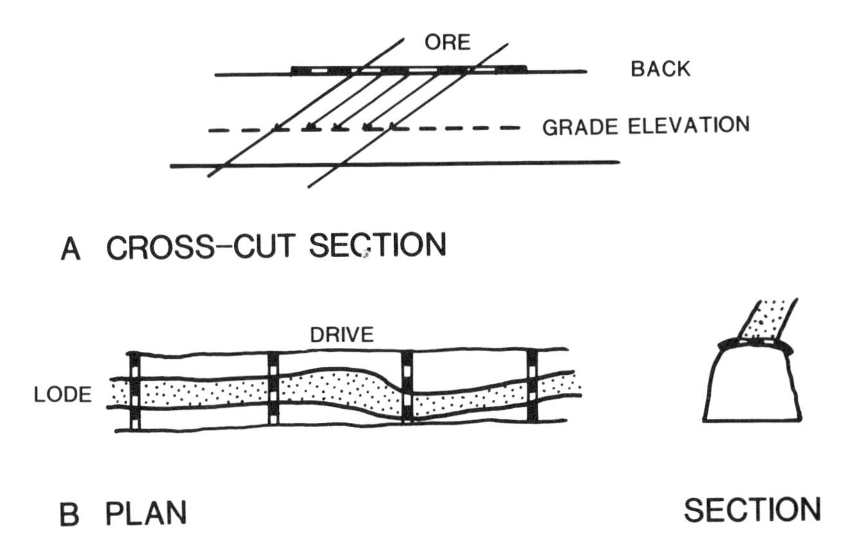

Fig. 2.4 Back sampling. (a) in a cross-cut. (b) along a drive.

horizontally at each platform level (section 2.6.3).

2.4.2 Sampling locations and orebody morphology and inclination

The most suitable type of sampling and the combination of methods used, depend to some extent on the type of deposit being evaluated as is demonstrated by the following examples.

Thin vein – on-lode development
Figure 2.6 shows the typical sampling layout for a thin, steeply inclined, vein. It includes back, raise, sub-level drift and development face sampling. Additional sampling may include blasthole cuttings and muckpile sampling.

Thick vein – on-lode development
Where the drive dimensions are less than the thickness of the vein or mineralized horizon, the sampling strategy may include a combination of methods. The drive is developed so that the hangingwall is just exposed (e.g. top left corner in Figure 2.7). Face sampling and combined back and sidewall sampling are thus possible to gain a partial intersection of the ore-zone. This must thus be augmented by cross-cutting at regular

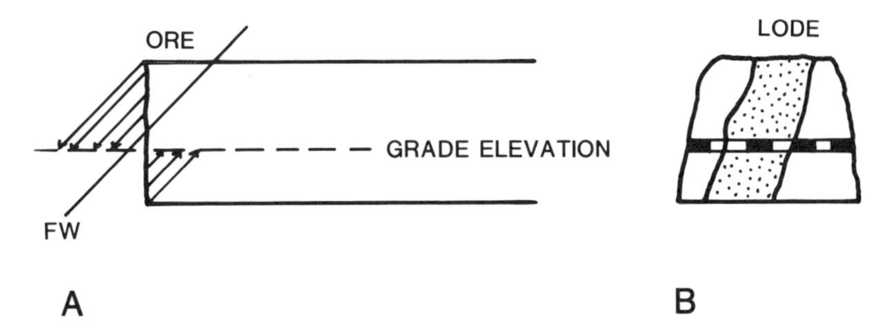

Fig. 2.5 Face sampling. (a) Longitudinal section. (b) Cross-section showing face.

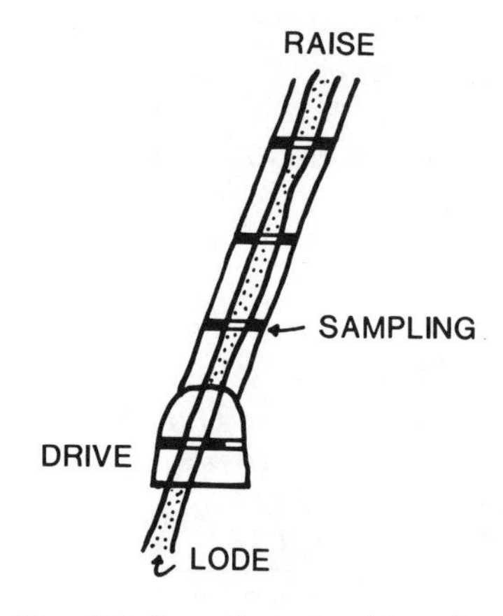

Fig. 2.6 Sampling of a thin vein.

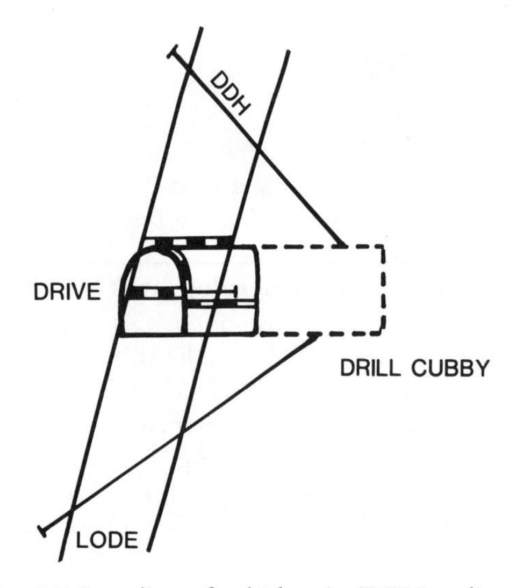

Fig. 2.7 Sampling of a thick vein (DDH = diamond drill hole).

intervals to expose the footwall so that the sidewall or back sampling can be undertaken to complete the intersection. Some of these cross-cuts can then be extended to provide drill-hole cubbies to gain intersections above and below the drive level and so prove continuity of ore. Alternatively, short drill-holes using a lightweight diamond drill (e.g. Bazooka) could be drilled to the footwall to complete the intersection.

Where the orebody is flat lying, as in Figure 2.8, both hangingwall and footwall cross-cuts could be developed at regular intervals along strike so that drilling can give greater dip-length penetration of the ore.

Off-lode development

Figure 2.9 shows the situation where the haulage or footwall drive is separated from the orebody by a pillar. In this case, drilling from a sidewall cubby can be used to determine grade and thickness together with sidewall sampling in cross-cuts driven on block-mids or boundaries.

Thick tabular or podiform deposits

Here sampling is dominantly by diamond drilling; either long-hole from haulages or main return airways (MRAs) or short-hole from footwall gathering drives (Figure 2.10). Occasional cross-cuts also exist from which sidewall sampling can be undertaken or muckpile/tram-car grab sampling. In some mines, e.g. Mufulira in Zambia, these cross-cuts are extended as drainage cross-cuts and then linked along strike by hangingwall pilot drives. These give the opportunity for deep intersections of the orebody as well as drainage holes for hangingwall aquifers. Here drilling is closely tied in with mine development as illustrated in Figure 2.11. Drilling follows the various advancing faces to ensure that grade and orebody thickness information is readily available to the stope-planning engineers. In practice, it has been found desirable for drilling to be kept no further than one mining block (100 m at Mufulira) behind the face. A phased sequence of development advance and drilling is thus maintained in which hangingwall pilot drives are advanced from drainage cross-cuts ahead of the haulage and MRA so that information is available to redirect the haulage should it be necessary. Drilling from the hangingwall pilot drives is generally on block-boundaries. The MRA is developed approximately half a block ahead of, but 24 m below, the haulage, and drilling on both is on boundaries and mids.

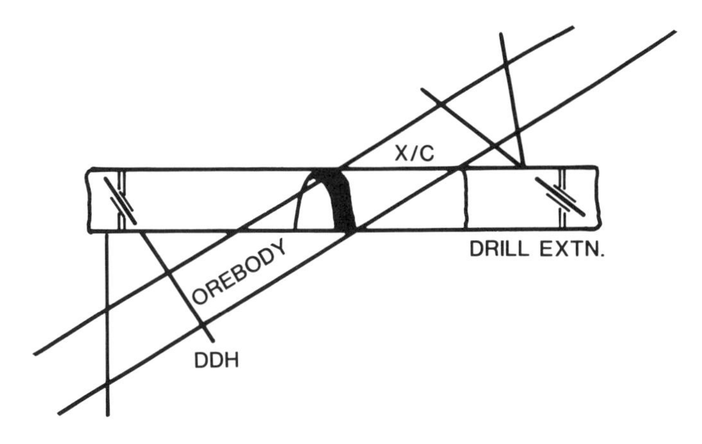

Fig. 2.8 Drilling pattern for a flat-lying deposit with on-lode development.

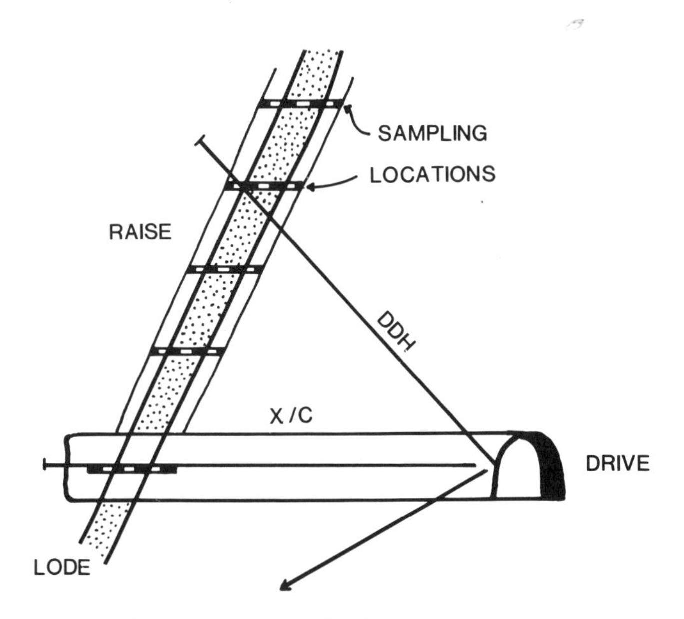

Fig. 2.9 Sampling from remote development, cross-cuts and raises.

Thick irregular podiform orebodies

Figure 3.20 shows a classic example of this type of orebody at the now closed Craigmont Mine in British Columbia. This mine was exploiting a chalcopyrite-magnetite skarn deposit at the margins of the Guichon Creek Batholith. Here orebody sampling was largely via fan drilling on closely spaced (50 ft) sections from exploration drives centred on ore pods, or lying between pods. Additional sampling information was gained from access cross-cuts and the drives themselves. In those operations using vertical crater retreat mining methods, the sampling of the blast-hole cuttings also gives valuable grade information.

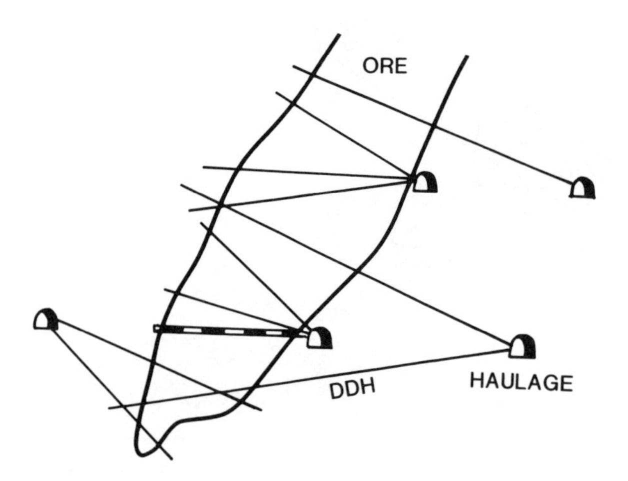

Fig. 2.10 Drilling from footwall and hangingwall for thick tabular or podiform bodies.

Flat-lying tabular deposits
These are usually mined by room and pillar methods (e.g. Troy, Montana and Navan, Eire) and hence sampling can be undertaken at the advancing stope-faces or around the walls of pillars. Where the deposits are particularly thick, sampling of blast-hole cuttings in benches or in breasts of 'cut and fill' stope lifts (Figure 2.12(b)) may also be undertaken. The former is the case in the 'top slicing' method where the sampling situation resembles that in an open-pit operation (Figure 2.12(a)) and mining progresses downwards from the top of the stope via benches.

Porphyry-copper-molybdenum deposits
These are large volume deposits with erratically dispersed mineralization and it is not unusual, where topographical conditions are suitable, to drive an exploration adit into the centre of the deposit prior to the final feasibility studies for an open-pit operation. This method allows bulk sampling of each blast advance together with sidewall chip sampling and fan drilling around the adit at regular intervals. This was undertaken at Newmont's Ingerbelle operation in British Columbia.

2.5 CHANNEL SAMPLING

This method involves the cutting of a narrow channel across the exposure of the ore, either horizontally, vertically or perpendicular to the dip of the ore. As far as possible, this channel is kept at a uniform width (e.g. 3–10 cm) and depth (e.g. 5 cm) and the exercise is thus very time consuming and hence expensive. Variations in rock hardness and ease of fragmentation make

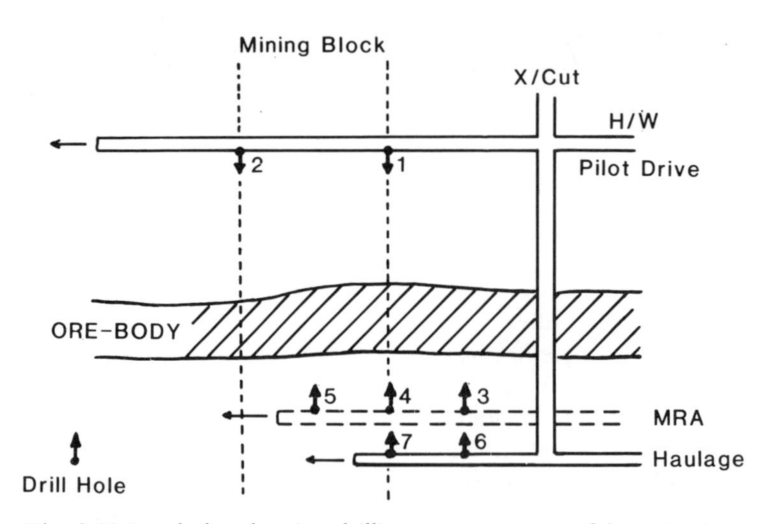

Fig. 2.11 Level plan showing drilling sequences at Mufulira, Zambia.

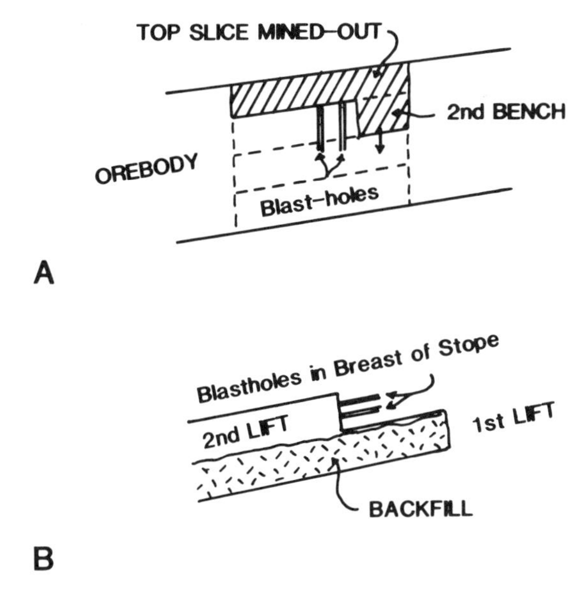

Fig. 2.12 Sampling of flat-lying deposits. (a) Top slicing and blast-hole sampling. (b) Breast sampling via blast-holes.

sampling difficult and, as a result, the use of this method has declined dramatically over the last two decades. It does, however, produce a superior result to the chip sampling described in section 2.6.

The wall to be sampled should be cleaned with a high-pressure hose, and a yard-brush if necessary, and then a plastic sheet or tarpaulin placed at the base to collect each sample as it falls. The channel is cut using a hammer and bolster chisel, a pneumatic hammer with chisel bit (e.g. Kango) or a masonry saw (tungsten carbide or diamond). Both the latter methods are particularly unpleasant for the sampler due to the dust and noise generated. In the case of horizontal channels, sampling proceeds from left to right either from the start of ore-grade material (based on visual estimates) or in waste if dilution is to be allowed for. Where the channel is vertical, sampling should begin at the base, working upwards so that contamination of the surface of each successive sample does not occur. The length of each sample can be based on a constant length set for the mine, or on geological, mineralogical or grade changes. This matter will be dealt with in more detail in section 2.9 during discussion of diamond drilling.

Information that should be recorded for each sample line includes:

(1) Cross-cut, driveage or raise sampled (including level, and mining block) plus sidewall (right or left);
(2) Distance and offset from nearest survey point;
(3) Elevation of grade elevation;
(4) Total width of ore-zone (horizontal, true, etc.);
(5) Number of samples collected plus range of sample numbers;
(6) Sampler and date.

Information that should be recorded for each sample includes:

(1) Sample number;
(2) Length;
(3) Ore mineralogy – relative proportions of different ore minerals/sulphides/oxides;
(4) Visual estimate of grade of metal;
(5) Apparent dip or true dip of mineralization in the case of a bedded deposit or vein;
(6) Host rock type (or lithological code).

Once the analyses have been received from the assay laboratory, the results will be plotted on, or projected down the dip on to, a grade elevation plan showing drives, cross-cuts and survey points. An assay section is also useful. In both cases, each sample length is accompanied by its grade (including acid-soluble metal if relevant) and a colour scheme is used to highlight zones of waste, low grade or high grade mineralization.

In the Cornish tin mines, e.g. Wheal Jane, it was standard practice to collect channel samples at 8–10 m intervals at the face on every other bench up the dip of the stope. Approximately 2 kg of material were collected to represent a length of channel not exceeding 50 cm. However, in most mines around the world, channel sampling has given way to chip sampling as described below.

2.6 CHIP SAMPLING

In this technique, rock chips are taken over a continuous band across the exposure approximately 15 cm wide using a sharp-pointed hammer or an air-pick. This band is usually horizontal and samples are collected over set lengths into a cloth bag which is usually 15 cm by 35 cm and which is equipped with a tie to seal it. The requirement is to collect small chips of equal size (or in some cases coarser lumps) at uniform intervals over the sampling band; a difficult undertaking at the best of times, especially when sampling across a hard, brittle quartz vein. The possibilities for unintentional or intentional bias due to variable chip sizes, and the over-sampling of higher grade patches or zones, are high. An alternative approach is to chip sample the entire face between blasts on a fixed grid to reduce the chance of sampling bias. At Sigma Mine, Val d'Or, rock chips are taken at intervals of 0.25–0.5 m along horizontal lines marked on the face. Each line is spaced at 0.75 m from its neighbour, and provides between 3.5 kg and 5.0 kg of material which is sent for assay. Chip sampling is used to define assay cut-offs in developmental headings and is particularly useful for grade control in producing mines. The latter subject is considered in greater detail in Chapter 7 but case histories of its application are given in the following sections.

2.6.1 Case history – Anglo American reef sampling, South Africa

Sampling of the Witwatersrand banket reefs (polymictic conglomerates) in South Africa, involves not only the chip sampling of all reef development at 2 m intervals (to provide block ore-reserve estimates) and of stope faces at 5 m intervals (to check the accuracy of previous block estimates), but also the recording of important sedimentological information. This includes lithofacies type and thickness; colour; grain-size; maturity (ratio of fragments of quartz plus chert to total rock fragments); sorting (particle size range about the mean); clay content; clast assemblage and percentage of each type using a 1 cm counting grid; clast size (the arithmetic mean of the lengths of the long axes of the 10 largest quartz clasts within 1 m^2 of complete reef exposure); pebble packing density (number of 1 cm grid intersections falling on a clast divided by 2); sedimentary structures present; abundance, mode of occurrence and size range of pyrite; presence of carbon specks or veneers and finally, palaeo-current direction if measurable. The geologist also draws reef profiles, approximately 10 cm apart, on which the true thickness of each lithological unit in the reef is recorded.

After clearing the exposure the chip samples are taken in a 10 cm wide band at right angles to the reef contacts over lengths which vary between 7 cm and 30 cm depending on the lithology, mineralization and pebble population. However, the first and last samples at the assay cut-offs are extended 2 cm into the footwall and hangingwall and are kept to lengths of 7–13 cm. This practice has been found to be necessary because high grades tend to occur at the sharp contacts between the reef and surrounding country rock. Approximately 0.5 kg is collected for each sample using a hammer and chisel and its position relative to the base of the reef is recorded. The sampling record sheet includes: length of each sampled section, sample number, geological information, gold grade in g/t, uranium in kg/t and the two metal accumulations (cm g/t, cm kg/t) together with a running thickness total.

2.6.2 Case history – Dome Mines Ltd, Ontario, Canada

Dome Mine is a major producer of gold from a variety of sources including ankerite and quartz-tourmaline veins, podiform en-echelon quartz veins in massive lavas, stockworks in conglomerates, porphyries and associated highly altered rocks, pyritized country rock and silicified greenstone. After each round is blasted on a development face, it is quartered (Figure 2.13) and four samples are taken over a band width of 3–4 in. Each sampling band is thus approximately 5 ft long and cuts the central section of each quarter yielding approximately 3 lb of material. The location of the face with respect to the nearest survey station is recorded so that the assay information can be plotted on the level plan. A description of each sample is made which includes the total length of quartz vein cut in each sample, an estimate of grade (good, fair, poor) and the presence, or otherwise, of visible gold (VG). Great care is taken to ensure that the sample taken is representative of the face quarter and that no sampling bias is introduced. At the same time it is realized that the preparation of the

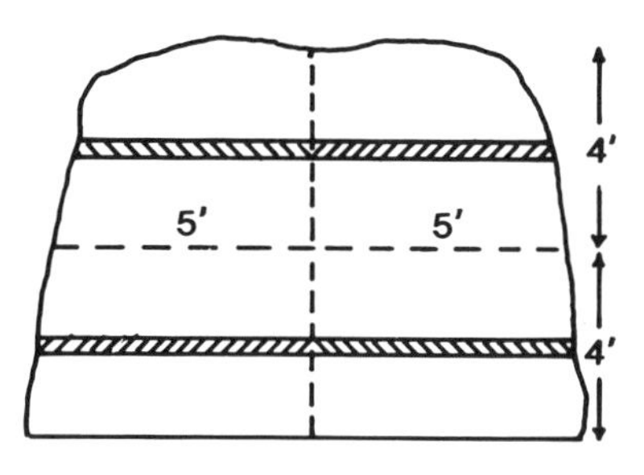

Fig. 2.13 Face quartering and sampling at Dome Mine, Ontario.

sample for analysis is also important for here considerable errors of contamination could be introduced.

Each sample is crushed in a cone-crusher to −0.125 in diameter chips and then riffled three times to produce a 0.5 lb sample which is then dried in an oven. Two separate pulverizers are then used to pulp the samples; one takes all the even-numbered assay samples and the other the odd-numbered samples. Each machine is cleaned with compressed air between each sample. The reason for employing this method is that if contamination is occurring in one machine during pulping, then this will appear in alternate samples and will be quickly apparent. The problem is generally due to the presence of free VG which tends to smear on to the plates of the pulverizer and can cause heavy contamination. It is for this reason that the presence of VG is recorded during sampling so that such samples can be removed from the set and examined under hand lens. This allows a grade estimate to be assigned as below:

abundant VG	10 oz/t
moderate VG	5 oz/t
a few specks of VG	2.5 oz/t

This rating method does not affect the ore-reserve calculation as these values are above the level of cutting applied at the present time (section 3.3.2). The final stage of sample preparation involves the homogenization of the −200# sample with a mechanical tumbler for approximately 15 minutes. The pulp is then spread evenly on a mat and random samples taken to produce 5 g for AAS analysis.

2.6.3 Case history – raise sampling at Galena Mine, Idaho

Raises to the cut and fill stopes at Asarco's Galena Mine, near Wallace, Idaho, are of the three-compartment variety with a central man-way and outer ore-chutes. These raises are driven at 200 ft strike intervals up the dip (75–80°) of the tetrahedrite-siderite silver-bearing veins. As the raise progresses, timber sets are installed at 6 ft intervals and at every second set (even numbers), both left and right sidewalls are chip sampled over the wall width working from footwall to hangingwall and including waste. At the same time, a set plan is drawn as in Figure 2.14. This shows the location of the main vein relative to the raise and also any subsidiary veins that may be present. Dips of bedding, joints and veins are also recorded along with other items of particular geological or mineralogical interest. From these set plans a raise cross-section can be compiled. When the

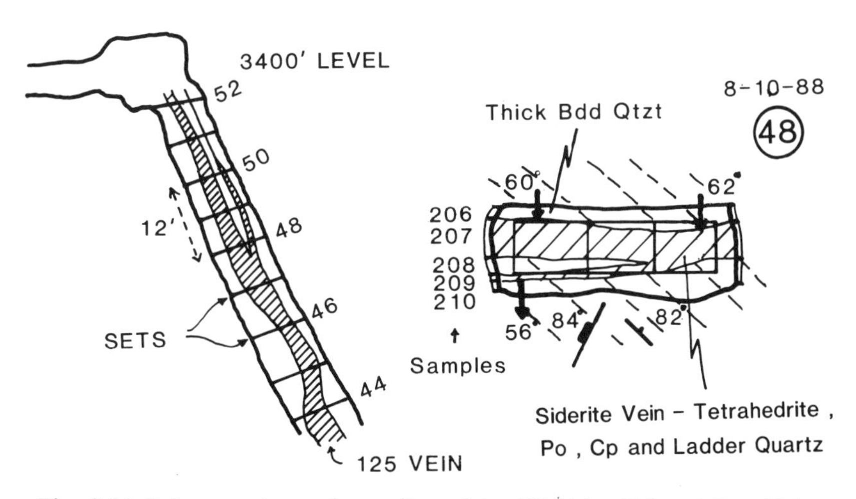

Fig. 2.14 Raise mapping and sampling of the 125 Vein, Galena Mine, Idaho.

reserves of the adjacent stope-blocks are calculated the left sidewall samples are used to evaluate the block to the left and the right sidewall samples the block to the right.

2.7 GRAB SAMPLING

This involves collecting a large sample from the muckpile at a face or at a drawpoint, or from the tram cars or conveyor belts transporting the ore from these points. The accuracy of such sampling is frequently in doubt and sampling bias is known to be large. Bias can be due to the natural tendency of the sampler to be drawn to richer fragments or to the fact that fines tend to be enriched in metal, especially in gold mines (e.g. Dome). The main problem is that the material in muckpiles, or the material loaded into tram cars, is rarely sufficiently mixed to be representative of the block of ground from which it was drawn. Also, material collected will be from the surface of the pile and rarely from its interior. In order to obtain a reliable sample, a large quantity of broken rock would have to be collected on a regular day-to-day basis and transported to surface for assay. This is generally impracticable in a mining situation. Sampling would have to be throughout the pile which is impossible and hence tram car sampling is preferred.

The amount that should ideally be taken is dependent on the size of the largest fragments in the pile and on the nature of the contained mineralization. The latter includes whether the ore minerals are homogeneous and evenly distributed and how fine grained they are. Generally, the coarser the rock fragments and the coarser and more erratic the ore minerals, the larger the sample that is required. Studies have shown that for 'average ore' anywhere between 100 and 1000 kg should be taken which is obviously impossible for routine sampling.

One approach to muckpile sampling is that employed at the gold mines in Val d'Or, Quebec, where the 'string and knot' method is used. The broken ground from each blast at the face is transported to surface and spread over a concrete pad. Three of four strings, with knots at 0.5 m intervals, are then placed over the pile at 3 m intervals. At each knot a sample is taken and its weight recorded, along with the position of the knot. Each sample is assayed and the result weighted by the relevant weight to obtain the overall grade. In this way, it is hoped that a more representative grade is obtained. A similar method is also used for tram car sampling. A knotted string, as above, is placed diagonally across each 10 tonne truck and samples taken at each knot position. These are combined and then composited on the basis of pairs of trucks. The weight and grade of each composite sample is then used to obtain a weighted grade for the current round. Comparisons of the two methods indicate that the pad method is more reliable but it is considerably more time consuming, involving as it does more handling of the ores. The mean and standard deviation values of a large number of weighted pad grades are lower than those found from the truck sampling.

At Dome Gold Mine, Ontario, one handful of ore is taken from each of six 3-ton tram cars and combined to make a composite sample. After crushing, a 3–4 lb sample is taken from the original sample and sent for assay. The result is thus meant to be representative of 20 tons of broken ground. At Wheal Jane Tin Mine in Cornwall much larger samples are collected, ranging between 150 lb and 200 lb, which represent the shift production from a face advancing on lode. The length and grade of each shift advance would then be plotted on a plan drawn in the plane of the dip of the vein so that values from raises could also be included (Figure 2.15).

At Kerr Addison Gold Mine, muck sampling practice varies with ore type. 'Flow Ore', which consists of carbonatized lavas and interlayered tuffs impregnated with pyrite and gold, is sampled by taking one lemon-sized piece from each of up to 15 cars (5 tons) and then combining them into one composite which is taken to be representative of 75 tons or less. In the case of 'Green Carbonate Ore', however, which is fuchsite bearing and derived from ultramafic volca-

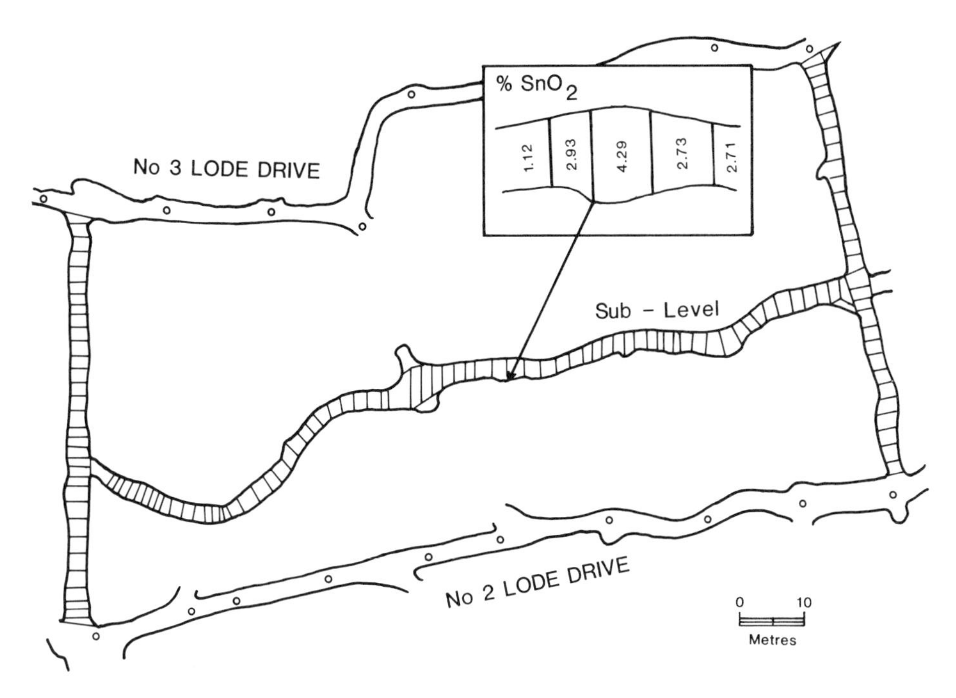

Fig. 2.15 Raise and sub-level muckpile sampling for shift advance, Wheal Jane tin mine, Cornwall. Assay plan plotted in plane of lode.

nics and which is invaded by auriferous quartz veins, the erratic gold content is taken into account by taking three lemon-sized pieces of rock from each of no more than five cars at a time. Fifteen pieces are thus taken to represent 25 tons of broken ground.

2.8 PERCUSSION/BLAST-HOLE SAMPLING

Where rotary-percussive techniques (see Reedman, 1979, Chapter 7, pp 322–7) are to be used at an early stage of the exploration of a deposit, especially where ground conditions are poor, it is essential that the weight of cuttings recovered, per unit length of drilling, is monitored carefully in case spalling/caving of the sidewalls of the hole is occurring. This will result in a change in sample volume per unit length and thus bias the weighted overall grade for the hole. Alternatively, it will cause contamination if the collapse is at a higher level in the hole. The weight of sample recovered over a drilling length is determined and compared with the theoretical volume of the hole over this length. Tonnage factors, expressed in terms of tonnes per cubic metre, would also have to be measured for the various horizons intersected, including the weathered zone, hydrothermally altered rock, 'fresh' rock and mineralized rock. This is best done by test coring in a few holes over the prospect area. A comparison of the theoretical weight and the recovered weight can thus be made and if discrepancies become evident, then caliper logging of holes may be considered necessary to obtain an accurate hole profile and hence volume. Consideration should be given at this stage as to whether open-hole techniques are giving sufficiently reliable information and whether perhaps an alternative, although often more expensive, technique should be used such as reverse-circulation drilling (Reedman, 1979, p. 340).

The sampling of cuttings from air-flush percussion drilling of mineral deposits involves the use of cyclones, 'auto-samplers', trowels, shovels or sample cutters. The latter include wedge, box, profile and pipe cutters. This subject is, however, covered in detail in section 7.2.1 under the heading of grade-control in open-pit operations and thus will not be considered further here.

2.9 DIAMOND DRILL SAMPLING

Although the mining geologist tends to play only a supervisory role in chip, channel and grab sampling in a mine, a direct involvement in the logging and assaying of drill cores will be necessary. A considerable proportion of the mining geologist's time will be spent in the core sheds keeping up to date with core brought up from underground during the previous shift.

Diamond drilling, and in some cases rotary percussive drilling, is the main sampling technique available to the geologist at an early stage in the exploration of a prospect. This may also be the case in an operating mine, but more commonly it is used to complement or augment chip and muckpile sampling. Diamond drilling may be used for many different purposes underground, some of which may not be directly relevant to the subject of this book; these include:

(1) Aquifer drainage;
(2) Shaft pilot holes;
(3) Cover holes for advancing haulages, etc., where water is deemed a hazard;
(4) Geological holes drilled to locate contacts, faults, shear zones, etc;
(5) Rock mechanics/geotechnical surveys.

The last of these has an indirect relevance to orebody evaluation in that ground conditions, particularly on the orebody hangingwall, will govern the amount of unintentional dilution. If allowance is to be made for such dilution in the calculation of stope reserves, then the grade of weak ground beyond the orebody limits should be determined so that a realistic estimate can be made of its impact on stope grades.

Diamond drilling, designed to intersect mineralized zones underground, may be for one of the following purposes:

(1) Exploration – to extend the known limits of mineralization.

(2) Determination of the attitude or structure of the mineralized zone and its thickness and grade at various cut-off-grades.

(3) To up-grade mineralization to the status of ore-reserve or to increase the confidence in reserve estimates, so that 'possible ore' can be upgraded to 'probable', and 'probable' to 'proven'.

(4) Grade control – infill short-hole drilling to ensure continuity of ore-grade material, above a minimum stoping thickness, ahead of stoping.

(5) Bulk sampling for metallurgical pilot plant studies – this usually involves larger diameter cores.

2.9.1 Layout of drill-holes underground

A review of drilling practice at mines generally reveals that in most cases 'rule of thumb' criteria are applied. Typical drill-hole spacings tend to be in the range 30–60 m, with closer spaced drilling, especially in the case of gold mines, at 15–30 m intervals. At a spacing of 30 m, the sample area in the plane of the orebody is approximately 1/542 000 of the zone of influence assigned to a 46 mm diameter drill-hole. This sample can only be representative of the mineralization if the mineralization is very uniformly disseminated and predictable. If financial constraints do not allow closer sampling, then it must be accepted that the confidence, or precision, of the estimate will be low and that we are taking a gamble in accepting the calculated grades. A rule of thumb that has been applied in the past is that the ratio of drill-hole spacing to orebody thickness should lie between 10 : 1 and 30 : 1, i.e. a 2 m thick orebody should have a spacing between 20 m and 60 m. This seems to be based on a 'gut feeling' that the thinner orebodies, e.g. veins, tend to have more erratic grades and need closer sampling than those which are much thicker. This is a very dangerous assumption to make as it totally ignores the nature of the contained mineralization. Few mines have made in-depth studies to determine whether the size of core being drilled is the optimum or whether an increase from say AX to NX warrants the additional cost. Such a decision can only be made by close study of the nature of the mineralization by the application of statistical techniques, as in section 2.15.4, or geostatistical techniques, as described in section 4.15.

Ignoring, for the time being, the problem of optimum drill spacing, we can review some of the criteria that should be applied when planning a drilling programme underground. These are summarized below:

(1) Holes should be collared and completed on the same structure section for ease of plotting and geological interpretation, i.e. avoid having to project off-line holes on to section.

(2) Rig movements should be planned to minimize down-time and the movement from level to level.

(3) All services (water, compressed air, ventilation and power) should be made available well in advance and, if drilling operations are likely to interfere with mining activities, e.g. tramming, a drill-hole cubby should be blasted in the sidewall.

(4) Primary drilling should aim to intersect the orebody footwall as close as possible to a mine level. This will ensure that the exact location of this contact is known for advance planning of the haulage and footwall drives on this level.

(5) Secondary infill drilling should then divide the dip-length between levels into equal portions (e.g. three).

(6) As many holes as possible should be drilled at each collar location to avoid unnecessary rig movement. Although the ideal intersection angle to the orebody is 90°, this will mean a wide range of intersection angles.

(7) Low intersection angles (< 30°) should be avoided in long holes. This will reduce the chance of deflection of the hole into line with the orebody dip and will also reduce the dip-length between the intersection of the orebody hanging-

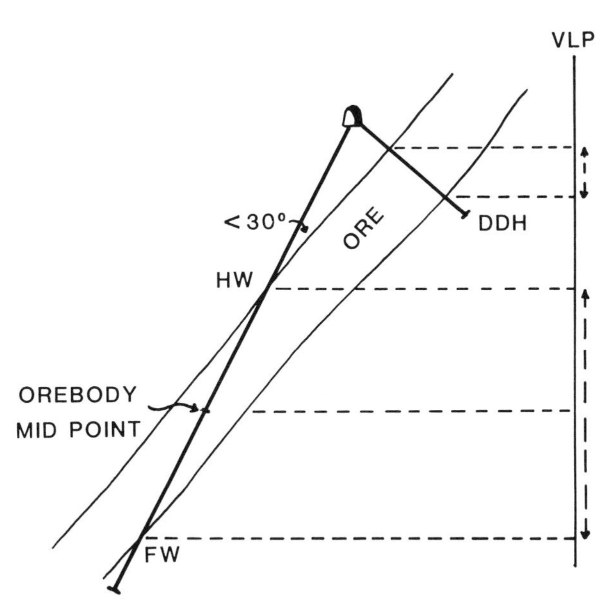

Fig. 2.16 Drill section showing the increased dip and VLP lengths incurred by drilling holes with low intersection angles.

wall and the intersection of the footwall. Figure 2.16 shows that, in the case of a low angle, the weighted grade assigned to the mid-point of the intersection could be produced by combining mineralization over a wide dip-length range and would thus not be representative of the mineralization at this point (calculated over the true thickness section through this point).

(8) When drilling from the hangingwall, holes should be allowed to continue 5 m into the footwall to ensure that no additional potentially ore-grade zones exist and to provide geotechnical information on future sill or safety pillars.

(9) When drilling from the footwall, holes should be allowed to penetrate at least 2–3 m into the hangingwall to determine the probability of spalling of the stope back and also the grade of this material if it were to take place.

2.9.2 Layout of surface drill holes

When an inclined tabular deposit is to be drilled from the surface, the following guidelines should be applied to the siting of the drill-holes:

(1) Two holes should be drilled on the down-dip side of the outcrop/suboutcrop of the mineralization as detected by shallow follow-up sampling on a geochemical and/or geophysical anomaly. These are located so that they intersect the mineralization at as high an angle as possible (ideally at right angles) at a depth which is below the rock-head and also just below the zone of oxidation and weathering (Figure 2.17). Usually these holes are drilled down-dip of the 'eye' of the anomaly on alternate drill lines as shown in Figure 2.18, thus allowing later intermediate drilling if it is warranted. Typically, drill lines would be 50–100 m apart. If the inclination of the mineralization is unknown, then a drill-hole will be needed on each side of the anomaly so that hopefully one will make an intersection and reveal the dip direction.

(2) The mineralization should be intersected at at least NQ size and hence the hole should be collared at a larger size to allow for casing through bad ground.

(3) In deciding the optimum angle of inclination of the drill-hole determine the intersection angle of bedding, schistosity and cleavage to the drill-hole to ensure that acute angles do not exist which may result in a 'rasher of bacon' effect. In this, the core splits up into thin slices which glide over one another in the barrel and eventually block it before the full capacity of the barrel has been drilled.

(4) Drill at right angles to the strike keeping the hole, as far as possible, on the drill line so that accurate geological sections can be drawn.

(5) The drilling schedule should follow a scheme such as that outlined in Figure 2.18. If holes 1 and 2 are encouraging, then hole 3 should be drilled on the intermediate line to test for down-dip continuity of mineralization. If this is

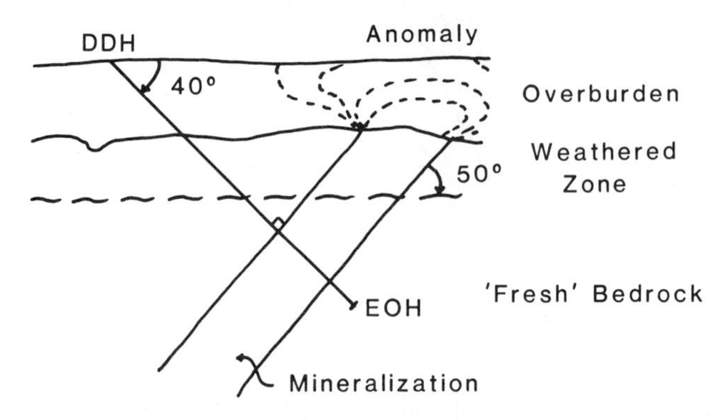

Fig. 2.17 Section showing layout of first exploratory drill-holes.

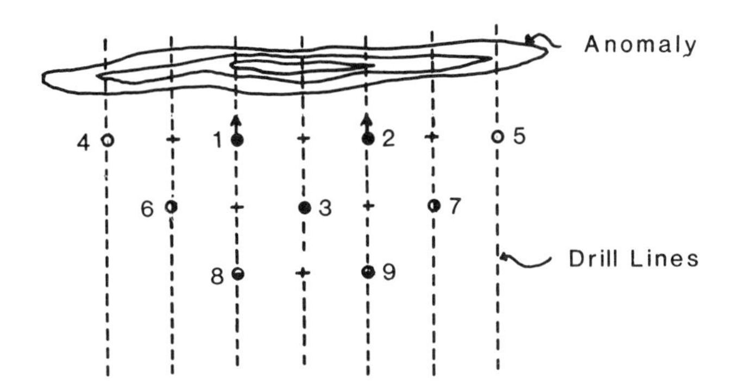

Fig. 2.18 Plan showing surface geochemical anomaly and layout of exploratory drilling.

equally encouraging, holes 4 and 5 are then drilled to extend the strike continuity. This in turn would be followed up with holes 6 and 7, and then 8 and 9, so that a pyramid of holes is built up on an offset grid.

(6) Once the stratigraphy and the location and dip of the ore-zone is well established, infill drilling ('+' locations on Figure 2.18) can be undertaken. Consideration should be given at this stage to the drilling of the ore-zone using rotary percussive techniques and then reverting to coring to make the required intersection. Also consideration should be given to the possibility of drilling larger diameter holes to obtain samples for metallurgical pilot plant studies.

The drilling strategy outlined above uses an offset grid for the primary drilling campaign as opposed to a square or rectangular grid. The logic behind this is illustrated in Figure 2.19. In both layouts, the drill-hole spacing is such that the least-known point is the same distance 'a' from the nearest drill holes. In the case of the square grid, the drill hole spacing and area of the rectangular zone of influence centred on each drill hole are $a\sqrt{2}$ and $2a^2$ respectively. The equivalent values for the offset grid are $a\sqrt{3}$ and $1.5\sqrt{3}a^2$. The offset grid is thus more efficient in that the drill-holes are further apart. Expressed as a percentage, they are 22.5% further apart in the offset pattern and 29.9% fewer holes are needed

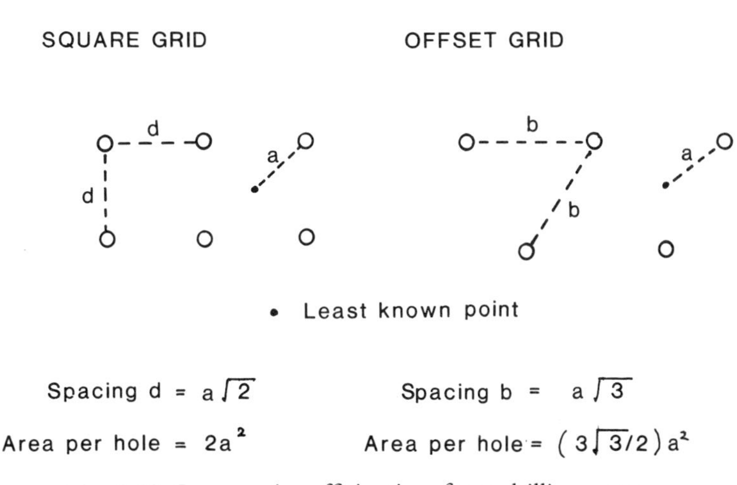

Fig. 2.19 Comparative efficiencies of two drilling patterns.

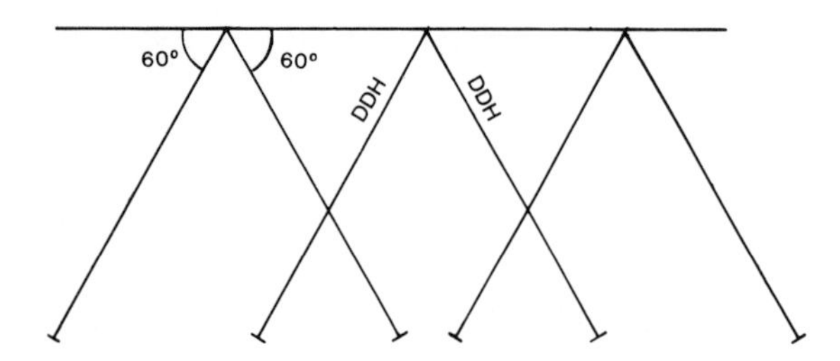

Fig. 2.20 Drilling pattern for deposits for which the orientation of mineralized fractures is unknown or subvertical.

to cover the same area with the same degree of confidence.

When a drilling programme is being devised for a non-tabular deposit, the strategy is somewhat different for the first holes will be drilled at the centre of the anomaly or over the suboutcrop of the mineralization. Later holes will be added to expand the grid outwards in all directions. There are three main options which could be used:

(1) Square, rectangular or offset grids of vertical holes which are suited to those deposits where there is no preferred orientation of mineralized fissures/veins.

(2) Square, rectangular or offset grids of inclined but parallel holes drilled to perpendicularly intersect mineralized fissures with a known orientation and dip.

(3) Square, rectangular or offset grids with two holes at each point inclined in opposing directions (Figure 2.20). This is suitable for those situations where the orientation of stringer vein systems is unknown or where two opposing systems are present. This method ensures that bias due to the orientation of drill holes will be avoided. It is also applicable to subvertical vein systems which would not (or rarely) be intersected using option 1.

2.9.3 Sampling of drill cores

Although expensive, diamond drilling has many advantages over other sampling techniques in that:

(1) A continuous sample is obtained through the mineralized zone.

(2) Constant volume per unit length is maintained; this is very difficult to achieve in both chip and channel sampling.

(3) The core can be logged and assayed on surface under conditions which allow more care to be exercised in both respects.

(4) Good geological, mineralogical and geotechnical information can be obtained, as well as assay information.

(5) Problems of contamination, which plague sidewall and muckpile sampling, are minimal for the core has good clean surfaces. Where contamination does exist the core can be easily cleaned using water, dilute HCl or industrial solvents.

(6) Drilling allows samples to be taken in areas remote from physical access, i.e. between mine levels and raises or within ore-blocks.

Once the core has been brought up from underground, it should be washed and then examined to ensure that all the sections of core fit together and that none have been misplaced or accidentally inverted in the box. As well as using geological way-up indicators, the geologist should also look out for ground ends that fit together, one of which may be concave, the other convex. Core-spring marks and grooves also indicate the bottom end of a core run. It is also important to ensure that the first piece of core from a run fits with the last piece of the previous run, both physically and geologically. Once the core is in the correct order, the core recovery is measured throughout the mineralized interval (Figure 2.21) and, where losses have occurred, an attempt is made to assign these to specific depth ranges in the core boxes. It is essential that the above examination is made prior to marking up the core for assay. In the long run it saves time and embarrassment. There is nothing worse than getting half-way through sampling a zone and then finding that the core has been mixed up by the drillers or during transport to the surface.

Fig. 2.21 Measuring core recoveries at Ogofau Gold Mine, Dyfed, Wales.

A line is then drawn longitudinally down the length of core with a wax crayon so that the whole width of the mineralized zone, and at least 1–2 m each of the apparently barren or low grade hangingwall and footwall material, is covered. In order to minimize the possibility of biased sampling, this line should be run through the low points of bedding traces (in the case of sedimentary hosted deposits), or down the low points of the vein contacts, when all adjacent core pieces are correctly orientated relative to one another (Figure 2.22). Where these planes are not present, other reference planes could be used such as cleavage, schistosity, or flow lamination in igneous or volcanic rocks. Where the core is totally homogeneous and massive then one can only pick an arbitary starting point and continue down the core from this point ensuring that, as far as possible, all the pieces are correctly positioned relative to each other. Barbs should be marked on this line pointing in the down-hole direction and on one side of the core only (Figure 2.22). This prevents confusion as to which way up each piece is when it is returned to the box after cutting.

At this stage the core can be split longitudinally using a capstan or hydraulic core splitter or a diamond saw. The splitter has a chisel edge which impacts against the core and then (hopefully!) splits it evenly along its length (Figure 2.23). In practice, the split is very jagged and influenced by structures in the core. This introduces bias because the volume of sample per unit length of core is variable. The diamond saw (Figure 2.24), however, allows an accurately centred longitudinal cut to be made using a 'V'-shaped sample guide. It also gives a flat surface on which the mineralization can be examined with a hand lens, and on which intersection angles of bedding or vein contacts can be measured with ease (Figure 2.25). The maximum intersection angle is displayed on the surface because the cut is through both the high and low points of the plane being measured (Figure 2.26). Figures 2.27 and 2.28 show the measurement of intersection angles using a protractor and a clinorule respectively.

An alternative technique for sampling core is groove sampling which involves passing the core along a guide so that a revolving diamond-studded wheel can cut a concave groove into the side of the core over the whole length of the core. The cuttings are collected in a tray beneath the machine and then bagged.

The next stage is the subdivision of the split core into sample intervals. There are many criteria which could be taken into account and a decision has to be made as to what information is the most important and what can be lost without too great an impact. To some extent, the method that will be used to compute the ore reserves will also play a role in the final decision. Ideally,

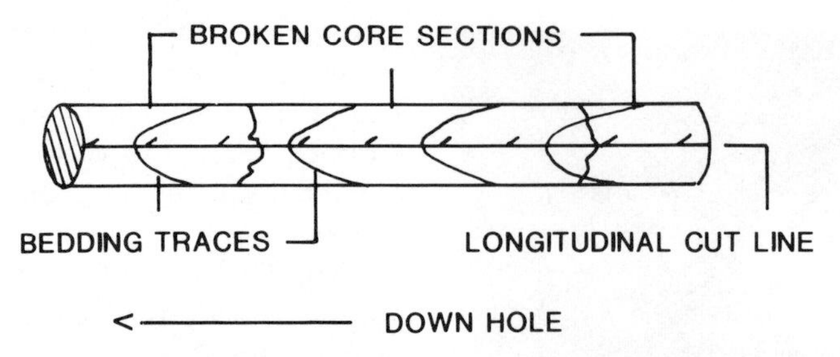

Fig. 2.22 Longitudinal line on core through bedding low points.

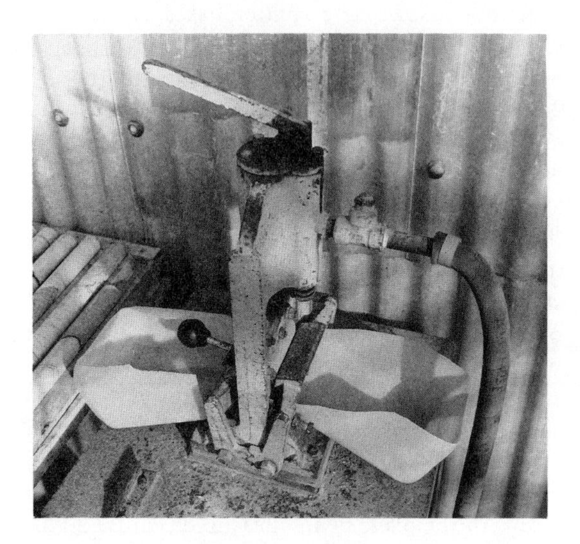

Fig. 2.23 Hydraulic core splitter, Navan Mine, Eire, (Courtesy of Tara Mines Ltd.)

samples should not exceed 1.5 m in length or drop below 0.2 m. Sampling should coincide with the visible hangingwall and footwall contacts but should be extended into the wall rocks by taking two samples of uniform length on either side of the potential ore zone. In the case of South African 'banket' reefs, where the upper and lower limits of the reef are clearly marked and where these same contacts could host high grade concentrations of gold, the sampling is commenced 2–3 cm into the foot wall and extended through the reef into 2–3 cm of hangingwall. Gencor Ltd subdivide this interval into individual samples that are greater than 15 cm but less than 40 cm long and which are based on sedimentological changes in the reef. Other factors that should influence sampling are as follows:

(1) Sudden changes in grade of valuable metal – within each sample interval the grade should be uniform.

(2) Changes in mineralogy of ore minerals, e.g. chalcopyrite and pyrite to chalcopyrite and bornite.

(3) Changes from recoverable to refractory minerals, e.g. changes from sulphides to oxides, silicates and carbonates which are not amenable to beneficiation in the existing plant.

(4) Changes in host rock lithology, e.g. quartz vein to altered wallrocks, shale to quartzite, basalt to serpentinite, 'granite' to hornfels, etc.

(5) Changes in style of mineralization, e.g. from massive to disseminated or stringer ore.

(6) Width of the mineralization – thin veins tend to be sampled over shorter lengths than thick deposits.

(7) The degree of smoothing required of erratic grades to reduce nugget effect – the longer/larger the sample volume the less effect localized high grade patches will have on the overall sample grade.

(8) Changes in rock strength produced by shearing, leaching, hydrothermal activity, weathering, etc.

(9) Whether sample values will be composited at a later stage as might be the case for an open-pit operation – here the sampling length should relate to bench heights to be employed.

Fig. 2.24 Cutting core with a diamond saw.

Fig. 2.25 Cut core from a Western Australian gold operation. Intersection angles are clearly evident on the cut surfaces.

Statistical and geostatistical ore-reserve techniques ideally require that the sample lengths be equal and thus we have a conflict of interests between the geologist who wishes to gain a better understanding of the controls of mineralization, and the geologist who wishes to evaluate the deposit using these techniques. Generally, a compromise can be reached in which variability in sample length is kept to a minimum or within set limits, e.g. ±50%. Thus in standard 1 m sampling it would be permissible to take samples varying in length from 0.5 to 1.5 m but not beyond these limits. This is important for the production of mine semi-variograms (Chapter 4). Such a system also removes the natural bias of the geologist to take long samples in low grade material and short samples in high grade material. Carras (1984) also suggests that sample volume/length should be adjusted until the coefficient of variation (C) of the assay population is less than, or equal to, 1. The optimum length would be estimated by using previous, or specially collected, samples that have been taken at close intervals. Adjacent samples are systematically composited into larger and larger lengths and the coefficient of variation for each new population determined. However, account should be taken of the smallest (or selective) mining unit (SMU) to be used so that the length selected also relates to the minimum stoping width underground, or to the bench height in an

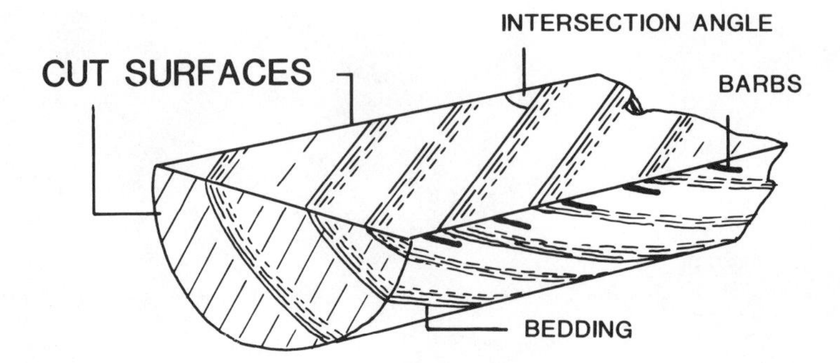

Fig. 2.26 Intersection angles on cut surface of core.

open-pit. Taking a larger sample smoothes out the more localized erratic fluctuations in grade reducing the variance of the assays and hence the coefficient of variation. This results in an improvement in the precision of the grade estimate.

An alternative sampling strategy may be applied to mineralization which is stratigraphically extensive and which is basically stratiform in nature. This involves the recognition of stratigraphic markers within, and outside, the orezones which allow the intervening strata (mineralized or not) to be subdivided into an equal number of intervals whose length will be variable from drill-hole to drill-hole due to changes in orebody dip, inclination of the drill-hole and true thickness fluctuations related to sedimentological conditions. This technique ensures that individual samples can be compared to stratigraphically equivalent samples in adjacent drill-holes. It allows the use of an ore evaluation technique called stratigraphic slicing, which has been used at Navan in Eire and White Pine in Michigan, USA, and which is particularly suited to the geostatistical evaluation of thick zones of mineralization (Chapter 4).

Once the core has been subdivided as indicated above, the top half of the core for each sample

Fig. 2.27 Measuring intersection angles at Ogofau Gold Mine, Dyfed, Wales.

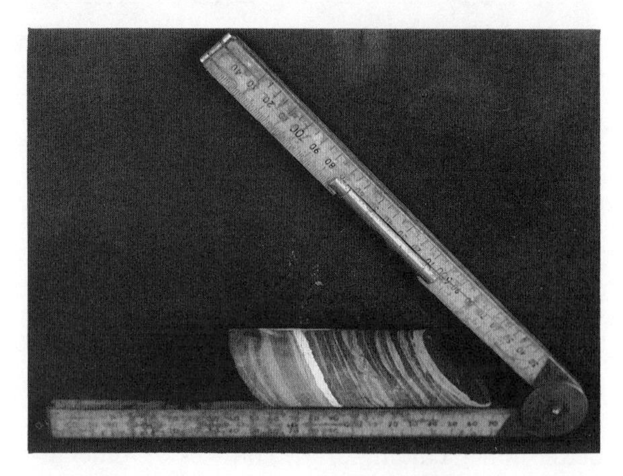

Fig. 2.28 Use of a clinorule for the determination of intersection angles.

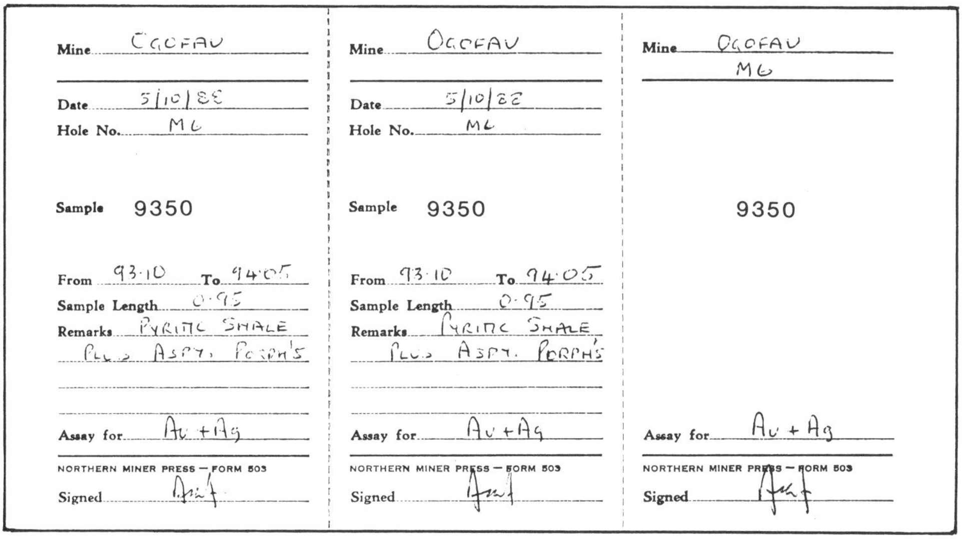

Ticket 1	Ticket 2	Ticket 3
Mine OGOFAU	Mine OGOFAU	Mine OGOFAU
		M6
Date 5/10/88	Date 5/10/88	
Hole No. M6	Hole No. M6	
Sample 9350	Sample 9350	9350
From 93·10 To 94·05	From 93·10 To 94·05	
Sample Length 0·95	Sample Length 0·95	
Remarks PYRITIC SHALE PLUS ASPY. PORPH'S	Remarks PYRITIC SHALE PLUS ASPY. PORPH'S	
Assay for Au + Ag	Assay for Au + Ag	Assay for Au + Ag
NORTHERN MINER PRESS — FORM 503	NORTHERN MINER PRESS — FORM 503	NORTHERN MINER PRESS — FORM 503
Signed	Signed	Signed

Fig. 2.29 Assay tickets, Ogofau Gold Mine, Dyfed, Wales.

Fig. 2.30 A Jones riffle sample splitter.

interval (i.e. the side without the barbs) is bagged ready for assay. Two undetached assay tickets (Figure 2.29) are placed in the bag and an aluminium tag is stamped with the assay number and placed in the core box at the end of the sample interval. The use of aluminium tags is preferable to the use of paper tickets, or wooden blocks marked with felt-tip pen, as both tend to become illegible with time or disintegrate. In the sample preparation laboratory, the core sections are crushed to −0.25 in and the sample is then split using a Jones' riffle box (Figure 2.30). One portion (with one of the two assay tickets) is sent for pulping and further splitting ready for analysis, while the other with the second ticket is bagged and stored as a 'coarse reject' in case repeat analysis is required at a later stage.

If a diamond saw has been used a clean-cut face of the lower section of core will now be exposed and a detailed mineralization log can be made using a logging sheet such as that in Figure 2.31. This gives the assay ticket number for each sample, its depth range, its represented and recovered lengths, the average intersection angle, if

MINERALIZATION LOG SHEET

NAME OF SAMPLERS ____________ DATE STARTED ____________ CO-ORDS OF COLLAR ____________ INCLINATION ____________

LOGGED BY ____________ DATE COMPLETED ____________ ELEVATION OF COLLAR ____________ METRES RECOVERED ____________

Sample No.	Depth		Sample Length		Assays			Intersection Angle	Mineralisation	Minerals Present								Rock Strength	Horizon
	From	To	Represented	Recovered	TOTAL CU.	A.SOL.CU.	METALL.CU.			Chalcocite	Bornite	Chalcopyrite	Pyrite	Malachite	Cuprite	Metallic Cu.	Chrysocolla		

Depth	Pressure	Yield	Depth	Pressure	Yield
Drilling Size					

M129

Fig. 2.31 Mineralization logging sheet.

Table 2.1 Mineralization codes

% Cu	Description	Category	Symbol
>3	Rich	High grade	+++
2–3	Good	Mine grade (above minimum mining grade)	++
1–2	Moderate	Above cut-off grade*	+
0.5–1	Minor	Subeconomic	–
0.3–0.5	Sparse	Subeconomic	– –
<0.3	Meagre	Waste	– – –
	Trace	Waste	Tr
	No visible mineralization	Waste	NVM

*For definition see section 3.2.1.

applicable, and details of the minerals present. Visual estimates are made of the amount of sulphide or metal in each assay interval so that a check can be made against the values from the assay laboratory. Other significant items of geological information could also be recorded at this stage such as the percentage of quartz vein material, ground conditions, degree of alteration or oxidation, etc. The amount of metal present in each interval can also be recorded using a scheme of symbols consisting of '+'s and '−'s. The symbols used are directly related to various grade categories used in the mine. Table 2.1 is an example which is relevant to a typical Zambian Copperbelt deposit. The same method can be used to reflect the amount of metal in a sample interval related to a particular sulphide/ore mineral, e.g. bornite.

2.9.4 Diamond drill sampling of gold ores

It is generally recognized that diamond drill sampling of deposits which are low grade, heterogeneous and discordant, such as gold vein deposits, presents a major problem. This is particularly acute for shear zone gold deposits where hard quartz veins and veinlets are interspersed with either hydrothermally altered, or sheared, country rocks. Variable recovery of the core is thus likely, which to some extent can be overcome by the use of reverse-circulation (RC) techniques. It is not just the softer or sheared rocks which may suffer from poor core recovery, for hard brittle quartz may disintegrate during drilling giving serious errors or bias in grade estimation. Thus grade estimates for such a deposit, produced by diamond drilling, should be treated with caution for, apart from the core recovery problems discussed above, the sample volumes may be totally inadequate to assess deposits whose metal content is a few parts per million. Larger diameter open-hole drilling is a possibility but this suffers from depth uncertainties and the mixing of material from adjacent sampling intervals due to the variable time lag for material of different grain size, or density, to reach the surface.

At an early stage in the investigation of a gold deposit, however, we have little option but to use diamond or RC drilling. We must thus optimize the way we drill by ensuring that the orientation of the drill-holes relative to the vein or dominant vein system is as close as possible to right angles, that the largest possible sample volume is obtained and that the spacing of drill-holes is dense enough to provide reliable information. In the latter case, it may be necessary to wedge certain holes to obtain multiple intersections or to drill a series of closed-spaced holes to assess the local variability in grade. Geostatistical methods (Chapter 4) are available to help in this respect. Above all, we must ensure that the crushing, pulping, splitting and assaying of the samples is undertaken with great care to reduce the possibilities of contamination and bias.

Diamond drilling is best used for the determination of the attitude and structure of the gold-bearing zone, its thickness and the continuity of the host structure, or horizon, with depth and also laterally. Grade information from drilling can be augmented and confirmed (or otherwise) by bulk sampling (section 2.16).

2.10 PROSPECT SAMPLING

Many different techniques are available for the subsurface sampling of a prospect under active exploration. Those described below are designed to allow both bulk sampling and direct access to the mineralization and thus do not include those sampling techniques which are designed to detect the associated geochemical anomaly.

Fig. 2.32 An exploration pit in the Zambian Copperbelt.

2.10.1 Pitting

The sinking of 1 m diameter pits through the overburden into weathered bedrock (Figure 2.32) has been standard practice in Central Africa where exposure is very poor due to the depth of weathering. Circular pits, 5–10 m apart, are sunk to depths of 10–15 m along lines crossing the strike of geochemical anomalies to allow the geologist to cut sampling channels in the pit wall and to identify the bedrock type, structure and mineralization, if present. The information gained allows the production of geological sections on which geochemical values are plotted. These sections then allow the more accurate siting of drill-holes for deeper investigations of the mineralized zone.

Pitting is a slow, labour-intensive exercise and the depth of penetration may be limited by B-zone laterite, a high water-table, the presence of gas (CO_2, H_2S), collapse due to loose friable rubble zones in the soil profile and hard bedrock. It can only be used where a plentiful supply of cheap labour is available and where the geologist is willing to overcome the psychological barrier of being lowered on the end of a rope into an unshored pit which is continually crumbling and collapsing.

2.10.2 Trenching

Trenches can be cut by hand, by back-hoe or by bucket excavator (JCB or Hymac) to expose

Fig. 2.33 An exposure of an auriferous quartz vein in a sampling trench at Loksi Hatti, Surinam.

mineralized bedrock where the overburden thickness is not great (i.e. < 5 m). Most trenches are less than 3 m deep because of their narrow width (< 1 m) and their tendency to collapse. Alternatively, a continuous trencher can be used where the soil cover is less than 1.5 m deep and where the terrain is not too steep. One such machine is described in more detail in section 2.11, while Figure 2.36 shows it in operation. The trench is cut across the strike of the anomaly to expose the full width of the mineralization. Channel sampling is then possible in the sides of the trench or in the weathered bedrock floor and bulk sampling of the spoil can be undertaken. Figure 2.33 shows a gold vein exposed in a trench at Loksi Hatti in Suriname.

2.10.3 Overburden stripping

A bulldozer can be used to strip off the overburden over the entire prospect area exposing the bedrock surface. This is then cleaned with high-pressure hoses to allow detailed mapping and examination. Traverse lines are then marked with spray paint across the strike of the mineralized zone at regular intervals (e.g. 5–10 m) and then 1.5 m long samples are taken either by hammer and chisel or by using a masonry saw. In the latter case, a 'V' notch is cut with the saw or two parallel vertical cuts are made to depths of 5 cm before removing the intervening material with a bolster chisel, a hand-held percussion drill or air-pick. Figure 2.34 shows such a sampling line near Noranda in Quebec with the start of each sample length marked by a cross-cut with the saw.

Bedrock sampling has been used by Riofinex North Ltd in their Lack prospect, Northern Ireland, where auriferous quartz veins occur in Dalradian mica-schists. Once the overburden has been stripped, parallel sampling lines are marked with string at 1–2 m intervals. One metre long samples are then cut with a masonry saw (Figure 2.35) by making three parallel cuts, each 5 cm apart over a 10 cm wide zone, to a depth of 10 cm. The material between these cuts is then chiselled out and bagged for assay. This method of sampling has been chosen because it is

Fig. 2.34 A bedrock channel sample cut in a gold prospect near Noranda, Canada, using a masonry saw.

relatively fast and because it gives good regular coverage of the bedrock surface and uniform sample volumes.

2.10.4 Winzing/trial adits

Bulk samples may be obtained by sinking a small exploration shaft on the suboutcrop of the mineralization to depths of 20–30 m so that fresh bedrock is exposed. In some instances, short levels may be driven from it to allow additional sampling which may include chip, channel and blast muckpile sampling. This approach is particularly important for gold exploration where the

Fig. 2.35 Bedrock sampling using a masonry saw at the Lack gold prospect, Co. Tyrone, N. Ireland. (Courtesy of Riofinex North Ltd.)

results of diamond drilling can be misleading. Where the topography is favourable, short exploration adits could also be driven into the hillside so that both bulk sampling of broken rock, and fan drilling around the adit, can be accomplished.

2.11 CONTINUOUS SAMPLING FOR OPEN-PIT OPERATIONS

Where open-pit operations are exploiting near-surface deposits of alluvial or eluvial material, or where the depth of weathering or the degree of hydrothermal alteration is great, the ore is soft and may be easily ripped or trenched by mechanical means. This particularly obviates the need for blast-hole drilling and sampling. Operations of this type are common in Western Australia where a gold rush, which commenced in 1981, has led to the working of at least 100 shallow open-pits. More detail on the evaluation and grade control practices in these pits is to be found in section 7.2.2 – Case History VI.

Dozer rip-lines (e.g. D6/7 bulldozer ripper) can be cut across the strike of the ore-zone at regular intervals (5–10 m). These allow samples (4–5 kg) to be obtained at regular intervals (1–2 m) from a depth of approximately 0.3 m but the results are not always reliable because of the blockiness of the broken fragments and the bias induced by selective sampling of either the harder or softer fragments. The sample is by no means thoroughly homogenized. The other problem is that the penetration depth is only 10% of the average flitch thickness (a flitch is a subdivision of a bench and represents the thickness which can be mined in one pass by the current mining method, e.g. there may be five 3 m flitches in a 15 m bench). The samples may thus be unrepresentative of the underlying 90% of the flitch. Considerable difficulties are thus faced in reconciling calculated grades with those recorded in the mill, a problem compounded by the high nugget effect of the gold mineralization itself.

The *Ditch Witch Trencher* has to some extent revolutionized sampling in soft rock conditions and overcome some of the problems highlighted for the dozer rip-line method. This machine was first produced in the USA in 1950 where it was designed for laying cables and pipes. It has been further developed by Mole Engineering Pty. Ltd for use in Australian gold mines where it has gained increased acceptance since 1985 (Archer, 1987; Bird, 1988). There are two main types of trencher (plus variations in tractor unit power). The first is a boom and chain trencher which operates like a chain saw (Figure 2.36) and which is better suited to softer conditions. The chain cuts a slot 15 cm wide and up to 1.5 m deep (Figures 2.37, 7.24, 7.25). In doing so, the spoil tribute the spoil laterally from the trench and create two longitudinal spoil heaps or windrows (Figures 2.37, 7.25, 7.25). In doing so, the spoil from one portion of the face in the trench is displaced approximately 0.5–1.0 m forward from the face position in the direction of advance. As a result, all trenches must be cut in the same direction and allowance made for this

Fig. 2.36 Ditch Witch R100 model using a trenching chain to cut a trench and an auger to distribute the spoil into windrows. (Courtesy of The Charles Machine Works, Inc.)

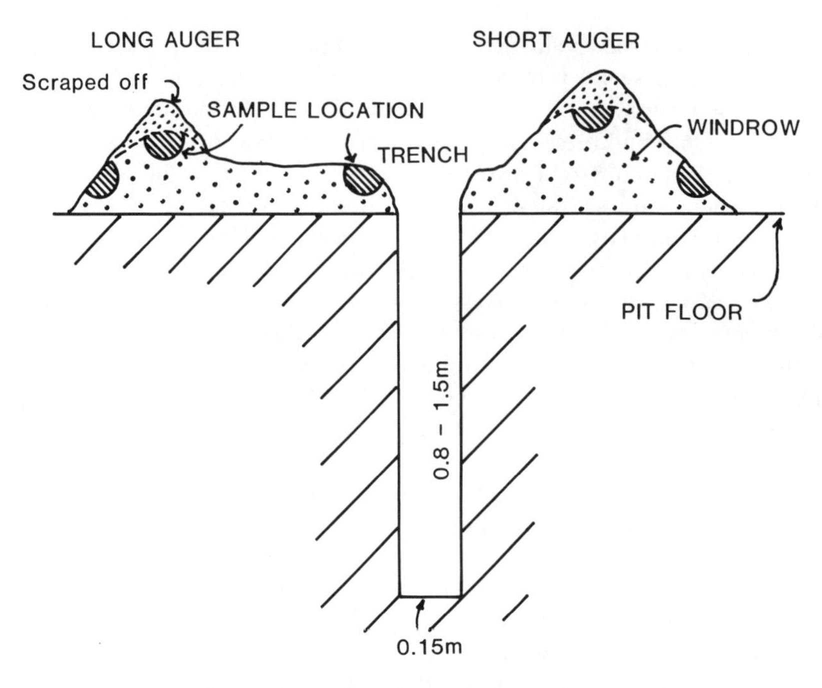

Fig. 2.37 Ditch Witch trench and sample locations in windrows.

Fig. 2.38 Saw trencher for Ditch Witch sampling. Note chute discharge for the spoil. (Courtesy of The Charles Machine Works, Inc.)

displacement in plotting the assay results on the flitch grade control plan. The second type of trencher is a saw-like device (Figure 2.38) which is more suitable for harder conditions and which cuts an 11 cm slot 0.8 m deep and which has a chute discharge for the spoil resulting in zero displacement along the line of the trench.

As a sampling technique, the Ditch Witch has many advantages. The material produced is finely ground (< 2 cm) and jaw crushing is not required. It is also thoroughly homogenized and thus the location of the sample in the pile at any given point is not too critical. Because the sampling is continuous and the material homogenized over 1 m sampling lengths (0.225 m^3), the assay grades are thought to be more representative of the flitch at the sample location. As sample mixing is limited along the line of the trench, geological contacts can be easily defined and plotted on the grade control plan. The presence of thin stringers and veinlets of mineralization, easily missed by drilling, will be reflected in the sample grade values. Productivity ranges between 70 and 200 m/hour and hence the ability of the assay laboratory to cope with the workload is critical, for ideally a 24 hour turn-around is required. The main limitation of the trencher is its inability to cope with hard quartz veins and thus the cutter has to be partially withdrawn resulting in a variability in the sample volume produced.

Prior to the use of the Ditch Witch, the surface of the flitch is cleaned and the traverse lines marked out from a baseline using lime, crushed limestone, paint or coloured tape. Line spacings are usually 5 m for very erratically mineralized ores, but up to 15 m for more uniform ores. The traverses are then cut across the strike of the orezone in one direction only. According to Schwann (1987), this should be hangingwall to footwall to minimize smearing of the grades but some mines work in the opposite direction to improve penetration at the expense of smearing. The surface of the windrow (spoil heap) is scraped clean to remove fly rock contamination and samples are then taken to represent 1–1.5 m lengths of trench. These usually weigh 1.5–2.0 kg. Figure 2.37 shows the possible locations of the samples. The sampling itself can be accomplished by hand grab sampling, or using a scoop or trowel, or guttering or a longitudinally split 3 in PVC pipe. The grades of the samples can be plotted on lines and coloured on the basis of grade categories, thus allowing delineation of ore types in the bench and their overall tonnages and grades. It should be noted that the sampling depth is still not through the entire flitch thickness, but at 25–50% it is a marked improvement on dozer rip-lines.

2.12 SAMPLING OF UNCONSOLIDATED SURFICIAL DEPOSITS

2.12.1 Sand and gravel

The satisfactory sampling of unconsolidated surficial deposits is a problem which has always been with the aggregates industry. Developments in recent years have allowed better samples to be obtained but the difficulty in accurately identifying these subsurface materials still remains a problem in certain circumstances.

This section outlines the various methods currently employed in the industry, with an assessment of the limitations and advantages of each method.

Sampling methods

Test pitting

The simplest method of sampling unconsolidated surficial deposits is to excavate a test pit using a hydraulic excavator. This has a fundamental advantage over any borehole in that the sequence can clearly be seen *in situ* and large cobbles can be identified. However, it has two main disadvantages. Most importantly, it is strictly limited in depth to deposits above the water level, and secondly, it can cause substantial disturbance to an area.

This method is commonly used in the industry for very shallow (generally less than 3 m depth) dry deposits, or for obtaining large bulk samples during a final, detailed investigation stage. Depths of up to 7 m or 8 m are, however, possible.

'Shell and auger' (cable percussion) drilling

This technique has been, and to a lesser extent still is, widely used for obtaining samples of sand and gravel, etc. The reason for this has little to do with quality or speed of drilling but relates to its low cost. The technique uses simple, relatively inexpensive equipment which can be towed behind a Land Rover and is, therefore, commonly operated by 'one-man' outfits.

It is, however, generally accepted within the aggregates industry that this drilling method produces poor quality samples and that productivity is low. Samples commonly show a heavy loss (10–15%) of the finer fraction when drilling below water, and drilling in gravels with the addition of water is slow and problematical.

Continuous flight augers

Like the shell and auger method, the continuous flight augering method has been widely used throughout the industry, and it also produces results of variable quality. It is, however, able to drill sand and gravel deposits at a significantly faster rate.

One variant of this technique uses a cutting head attached to spirally grooved rods which are rotated to allow penetration of the ground. The cuttings reaching the surface are then sampled. This represents a very inaccurate sampling method and is of little use for precise commercial purposes.

A better approach, and one which is commonly used today, is to drill the auger into the ground for a fixed distance and then withdraw the rods without further rotation. The sample can then be taken from the 'flights'. Practice and experience with this technique allows a judgement to be made as to the optimum sampling method and the amount of travel of the sample up the 'flights'.

Using this approach, the augering method can generally produce reasonably representative samples below the water table, and at least some indication of the materials below it. The method is also relatively fast and cost effective, and is, therefore, well suited for preliminary exploratory drilling of a site and for the identification of thickness variations throughout a deposit. Depths of up to 25 m can be attained in favourable situations.

Reverse circulation

This technique is a relatively recent advance for the sampling of unconsolidated, superficial deposits. It has evolved because of a requirement to produce better quality samples, particularly below water level, than can be obtained from the more traditional augering techniques. Although still not a widely used technique, it is proving highly successful and is, therefore, becoming popular within the aggregates industry.

The equipment consists of a twin-walled string of drill rods with a cutting bit at the base. Air is used as the flushing medium and this is delivered to the bit via the annulus between the two walls. The cuttings are transported by the air up the centre of the rods and then, via a flexible pipe, to a sample collection point.

Unlike 'shell and auger' or augering, this method recovers all the sample, including any water, with the exception of any cobbles larger than the internal diameter of the rods. This, in theory, produces a better sample, but only if the solid materials are carefully separated from the water before bagging. This is neither simple nor fast, and is, unfortunately, not always carried out with due care, thus defeating the main advantage of the method.

An experienced crew, using the correct equipment for the material to be sampled, can drill superficial deposits at a relatively fast rate. The continuous sampling removes the need to remove the drill string from the borehole as with augering. However, the added requirement for a compressor and additional equipment significantly increases the cost of the operation. The technique is, therefore, best suited for detailed 'second phase' drilling, where greater precision, in terms of sample quality, is required.

Conclusion

Although the techniques outlined above are not the only ones available, they are the main methods used in the Industry at the present time for the sampling of unconsolidated deposits. The choice of method depends on many factors including, cost, speed required, site access and quality of sample required, but the ideal approach for a detailed study is probably a combination of test pitting, augering and reverse circulation. For example, the investigation of a sand and gravel deposit up to 10 m thick, and partly below water, may involve the following stages:

(1) Rapid, preliminary investigation on a widely spaced grid (e.g. 150 m) using continuous flight augers to identify and delineate the deposit.

(2) Second phase, 'follow-up' drilling with augers on a close-spaced grid, (e.g. 75 m) to accurately delineate the deposit and assess the reserve.

(3) Reverse circulation drilling on a regular, widely spaced grid in order to increase confidence in sample quality, particularly below water.

(4) Limited test pitting at specific points in order to obtain information on the abundance, or otherwise, of 'oversize' cobbles or boulders. This technique is obviously limited in depth, but can be used to remove large bulk samples for trial processing through a nearby plant.

A case history of the evaluation of a sand and gravel deposit is given in Chapter 8 (section 8.7).

2.12.2 'Heavy mineral' deposits

Under the heading of 'heavy mineral' deposits are included continental placers containing gold, platinum, tin, tungsten and gemstones (diamonds) whose volumes represent a very small fraction of the total sediment volume and whose particle size range may be large, together with heavy mineral sands related to present-day shorelines and also to palaeo-shorelines. This latter group usually contains variable amounts of rutile, zircon, ilmenite, magnetite, monazite, RE minerals and apatite.

The sampling of alluvial or littoral 'heavy mineral' deposits has many problems in common with those encountered during the exploration of sand and gravel deposits. These largely centre around the difficulty of obtaining representative samples without loss, bias or contamination. Difficulties arise where the sampling is below the water-table; where the material is dry, unconsolidated and free running; where boulders and cobbles are present in the deposit whose dimensions exceed the diameter of the borehole; and where a large proportion of fines exist which may be lost during recovery of material from a slurry. Over and above the practical difficulties of sample collection, additional errors may be incurred due to the collection of an inadequate sample volume; use of an unsuitable sampling pattern, in which sampling spacing may be unnecessarily close in one direction but too wide in another; where valuable mineral is lost during sample handling, reduction and preparation; where sample sizes vary per unit depth of the borehole or pit; where inadequate geological control of the sampling results in excessive dilution of potentially payable horizons/lenses by barren overburden or underlying sediment or weathered bedrock. Problems are also faced due to the natural variability in valuable mineral or metal content of these deposits from the point of view of the spatial distribution or the wide range in grain size. Valuable minerals may be localized in narrow paystreaks and channels, in potholes, depressions or crevices in the bedrock surface and may thus be missed by drilling and sampling.

Sampling methods

The variety of methods that are available for the exploration and evaluation of 'heavy mineral' sands/gravels is discussed in detail in MacDonald (1983) to which the reader is referred. What follows is thus a summary of these methods:

(1) **Hand pitting**. Timber- or steel-lined pits (caissons) are ideal for the collection of bulk samples from dry unconsolidated or wet unstable ground to depths of up to 15 m.

(2) **Mechanical pitting**. Hydraulic excavators with jointed booms and hydraulically operated clamshell grabs, 1 m in diameter, are able to dig pits to depths of up to 16 m.

(3) **Trenching**. Hydraulic back-actors can be used to cut trenches across a deposit to depths of 9–10 m over widths of up to 1 m. Vertical channels can be cut into the walls to obtain samples at regular intervals, or an entire wall of the trench can be slyped to produce a more representative composite, or all the material excavated from a set length of trench can be processed/dressed where grades are low (e.g. diamonds).

(4) **Hand-operated Banka drilling**. In this method, 10–15 cm diameter casing, fitted with a casing shoe, is driven into the ground using a manually operated capstan or a pile-driver, although penetration is often achieved under the weight of the casing alone. A steel cutting tool is then used to chop the sediment trapped inside the string of casing, which is then removed by a flap-valve bailer, adding water if necessary.

(5) **Hand augering**. Here the hand auger is manually screwed into the ground and then removed at regular intervals to retrieve the sample. Depths achieved rarely exceed 10 m.

(6) **Hand sludging**. This method is used for waterlogged sediments. Casing is inserted into an auger hole and the sample is retrieved by the use of a sludge pump. Depths of 5–10 m are typical.

(7) **Percussion or churn drilling**. In this method, 20 cm casing is driven into the ground to depths of 30–50 m by drop hammers. The core inside is then recovered using special bits and bailers.

(8) **Machine auger drilling**. Solid flights bring the sample up to surface and thus contamination is inevitable. Also, the depth range from which each sample was derived is uncertain. Hollow auger rods, however, allow a core to be cut and retrieved.

(9) **Sandrill sampling**. This is a power-driven auger specifically designed for aeolian material. The Sandrill uses casing fitted with cutters and also counter-rotating flights to feed the sample into a core barrel. Depths of up to 100 m are possible.

(10) **Vibro drilling**. These machines use a vibrator which delivers impulses to screw or shell augers. Chisel bits can also be used if hard ground or boulders are encountered. Depths of up to 20 m are typical.

(11) **Hammer drilling**. Hammer drills are basically pile drivers which use reverse circulation air flushing methods to retrieve the sample. Cyclones allow recovery of the sample into a container.

(12) **Caldwell drilling**. This method uses a drive kelly and bucket, the latter possessing cutting teeth at its base. Downward pressure is maintained by the weight of the kelly and the bucket and holes of 45 cm to 2.5 m diameter can be drilled.

(13) **Conrad drilling**. The Conrad drill is a mechanical version of the hand operated Banka drill. The casing is rotated under its own weight into the ground so that samples can be recovered using clamshells or buckets. Plate 2.40 shows a Yost clam excavator being used to recover material from inside 0.9 m casing sunk into a placer gold deposit in Bolivia.

Sampling patterns and density

Three basic strategies have been employed in the past for the evaluation of heavy mineral deposits:

(1) Random drilling at approximately 100–200 m centres.

(2) Line drilling transverse to the elongation of the deposit at line spacings of 50–250 m and at 15–50 m centres.

(3) Grid drilling on regular 400 × 400 m grids reducing systematically down to 50 × 50 m grids

at an advanced stage of exploration and evaluation.

In the Offin River placer gold deposit of Central Ghana, Dunkwa Goldfields Ltd initially used line drilling at 1220 m (4000 ft) spacings to outline the limits of the placer and of the potential payzone. Boreholes were at 15 m (50 ft) centres except where the alluvial flat was wide, when only alternate holes were drilled initially. Later, the remaining holes were drilled, but within the potential payzone limits only. Having proved the existence of economically viable concentrations of gold, a major exploration drilling programme was then undertaken at 240 m (800 ft) line spacings and with holes at 15 m centres on lines. These lines were located between those recce lines which appeared promising. Once dredge mining had begun, a dredge control drilling programme was undertaken on 80 m lines and at 15 m centres. This allowed a more precise definition of the boundaries and grades of payzones immediately ahead of the area being dredged. These data were thus not available for the economic feasibility studies that were made prior to exploitation of the deposit.

A geostatistical study of this deposit by Annels and Boakye (1990) led the authors to the conclusion that the 15 m spacing of boreholes along lines was unnecessarily small but it did provide an excellent database for the production of reliable semi-variograms. Ranges of 145 m and 285 m were obtained for thickness and log gold accumulation respectively, although a small scale structure of 30–50 m was detected at low lags on the semi-variograms. In the down-stream direction, the equivalent ranges were 400 m and 480 m respectively, giving anisotropy coefficients of 1.4 and 3.3 respectively. Figure 2.39 shows that an increase in the drill-hole spacing along lines to 50 m, and of the line spacing to 120 m, would not only retain a precision of less than 20%, but would produce a marked drop in drilling costs per unit area. At the same time, sufficient data points would be retained to produce reliable semi-variograms.

Studies, such as the one reported above, suggest that the optimum drilling pattern is a rectangular grid, whose dimensions are such that the ratio of the sides approximates to the geometric anisotropism of the deposit as determined from geostatistical studies (section 4.7.3). In alluvial gold deposits, this usually varies between 1.5 and 3.5. The initial recce sampling should involve close-spaced drilling (e.g. 25 m) along both transverse and longitudinal drill-lines to facilitate the production of semi-variograms in these directions from which the dimensions of the exploration sampling grid can be determined. The use of a rectangular grid will then allow the construction of a matrix of ore-blocks with drill-holes at each of the four corners, rather than the usual matrix containing single centred drill-holes. This will considerably reduce the extension variances and thus improve the precision of the estimates of individual blocks.

Down-hole sampling

Bulk samples covering the entire depth of the hole or pit are not really recommended for they do not allow changes in the physical or geological characteristics of each horizon in the deposit to determine the depth ranges of individual samples. Serial samples should thus be taken from the collar of the hole to the bedrock interface. Typical payzone sample lengths used in the past have ranged from 300 mm to 500 mm, but in thick aeolian deposits, sample lengths may be considerably longer, e.g. 3 m, as at Richards Bay in Natal, South Africa. Careful measurement of the sample volume is essential to allow comparison with the theoretical volume of the drill-hole. Drill-holes are typically 200–250 mm in diameter, although clamshell mechanized pitting machines allow considerably larger diameter shafts to be sunk (Figure 2.40).

Typically, 'heavy mineral' deposits display considerable variability in terms of degree of compaction, grain-size distribution (grading)

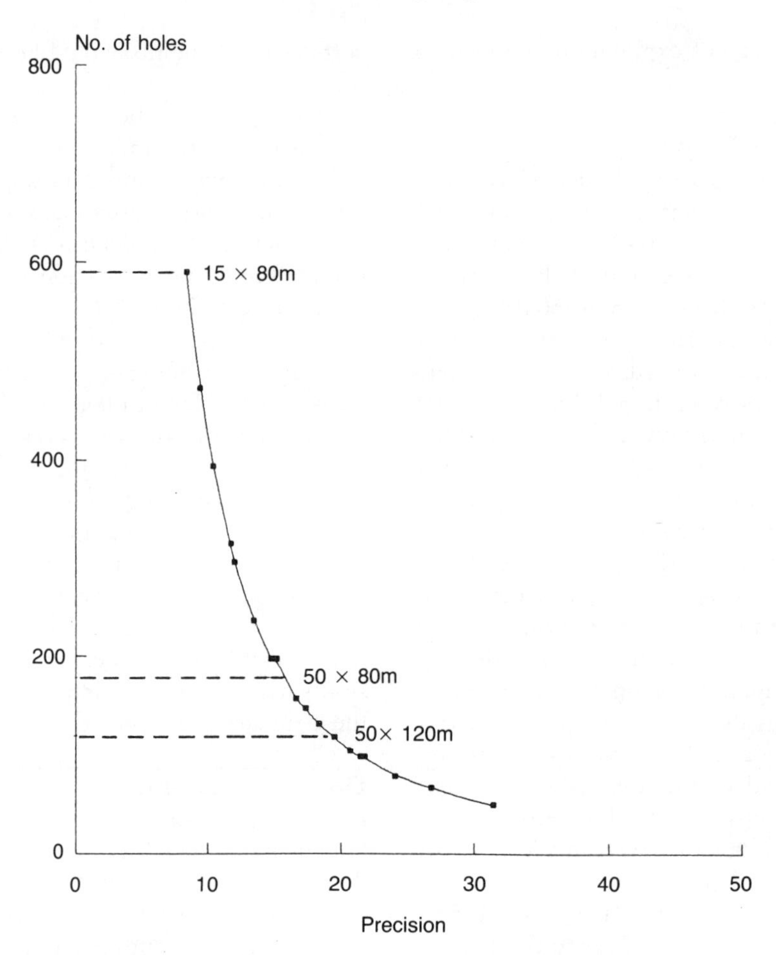

Fig. 2.39 The relationship between drilling grid size, number of holes drilled and the precision of reserve estimates for the Offin River placer, Ghana.

and water content, therefore the use of tonnage factors, to convert volumes to tonnages, is not advised. Generally, therefore, the variables used for reserve estimation purposes are expressed in terms of volumes, e.g. g/m^3 or kg/m^3, or areas, e.g. g/m^2 or kg/m^2.

Mineral dressing

Because of the large amount of sample produced, it is dressed on site not only to reduce the sample volume, but to determine:

(1) The particle size distribution of valuable and non-valuable components;

(2) The washability characteristics;

(3) The slimes content;

(4) The anticipated mineral recoveries.

The method used varies from deposit to deposit, but could involve:

(1) Puddling and screening;

(2) Hand panning;

(3) Sluicing;

(4) Gravity separation by jigging, by using shaking tables, or superpanners;

(5) Heavy liquid separation;

(6) Low and high intensity magnetic separation.

Fig. 2.40 A Yost clam excavator at work on the Tipuani placer gold deposit, Bolivia. (Courtesy of Watts, Griffis and McOuat Ltd, Toronto.)

Mineral concentrates produced can be weighed in the field so that they can be expressed as percentages of the total sample weight. These can then be examined under the binocular microscope and then point or grain counted. In the case of gold, amalgamation can also take place in the field followed by later extraction in the laboratory. Further details of mineral dressing can be obtained from MacDonald (1983, Chapter 4). In some cases, such as for cassiterite-bearing sands, it may be necessary to classify the sand on the basis of grain size and to quote a %Sn value for each size range. Fine clay fractions may have a low metal recovery during processing. At Richards Bay Minerals, RSA, the samples are fractionated into sand fractions (63 μm–1.44 mm) and into clay/mud fractions (< 63 μm). The latter is then reported as a % slimes (typically 1.2–8.6%). Also determined for each sample are % valuable heavy mineral, $\%TiO_2$ (magnetic), $\%TiO_2$ (non-magnetic) and $\%ZrO_2$.

2.13 THE APPLICATION OF COPPER-SENSITIVE PAINTS

The use of copper-sensitive paints to determine the location of assay hangingwalls and footwalls in sidewall exposures of copper sulphide mineralization, or in cores, was first developed by MTD (Mangula) Ltd. These paints consist of ammonium molybdate (40 g), sodium pyrophosphate (20 g) and titanium dioxide (170 g) mixed in 200 ml of HCl (50%). Such paints should be made up immediately prior to their use for they cease to be effective after 10 days. The paint is applied by brush on a previously cleaned surface in stripes 5 cm wide across the suspected cut-out of the mineralization. After a period of 30–60 seconds, specks of blue colouration appear reflecting the presence of chalcopyrite, bornite or chalcocite. The density of this speckling can then be used to locate the limits of economic sulphides. Eventually, any pyrite also present reacts with the paint (usually after 4 minutes) as it is sulphide sulphur sensitive. The paint liberates H_2S from the sulphides which then causes reduction of molybdate ions (Mo^{4+}) to molybdenum blue (Mo^{2+} or Mo^{3+}).

This technique has been used effectively by ZCCM Ltd at Konkola Mine in Zambia, where the following benefits have been highlighted:

(1) Reduction of unnecessary cross-cut development beyond the assay hangingwall.
(2) Delays whilst sample analysis is undertaken in the mine assay laboratories are avoided.
(3) Better detection of unpay zones in an active stope.
(4) The amount of subgrade sampling and analysis is reduced.
(5) Very fine grained disseminated mineral-

ization does not go undetected (e.g. chalcocite in a dark host rock).

(6) The sampler is able to vary his sample length on the basis of sudden changes of grade.

(7) If used on core during drilling, it can result in a reduction in unnecessary drilling in barren ground.

The main disadvantages are that, unless the paint is examined soon after application any pyrite present will give misleading results. Also, if 'oxide' copper coexists with the sulphide, its presence will not be detected, resulting in an underestimation of grade. However, such 'oxide' copper may not be recoverable in the existing processing circuit and thus its non-detection may not be too serious.

2.14 GRADE ANALYSIS BY FLUORESCENCE AND SPECTROMETRIC TECHNIQUES

The cost of diamond drilling, channel or chip sampling and analysis of the samples produced, not to mention the time involved, have led to the development of techniques which provide an immediate *in situ* analysis of mineralized zones and the definition of assay cut-offs. Two main methods have been used, i.e. gamma spectrometry and X-ray fluorescence spectrometry. These are described below.

2.14.1 Gamma spectrometers

Where the mineralization under investigation contains a radioactive element, such as U_3O_8 or K (uraninite or potash, for example), gamma spectrometers can be used to evaluate *in situ* grades of faces or boreholes underground. The instruments used are directly calibrated in % U_3O_8 (or lb/ton) or in % K. Faces or borehole lengths can be scanned initially to locate the most interesting areas prior to spot measurements of grade.

ICI's GammaTrol Potash Probe (PRI 85/86) is an example of a borehole gamma probe. It was initially developed for use at the Boulby Potash Mine, Cleveland, UK (section 8.6). This consists of a 44 cm long steel probe (Figure 2.41) containing a GM tube to measure the K^{40} gamma radiation emitted by the potash. The instrument is calibrated using a series of ore samples of known grade over the range 30–80% K. The probe can be inserted into boreholes via a series of fibreglass drain rods. A cable links the probe with the

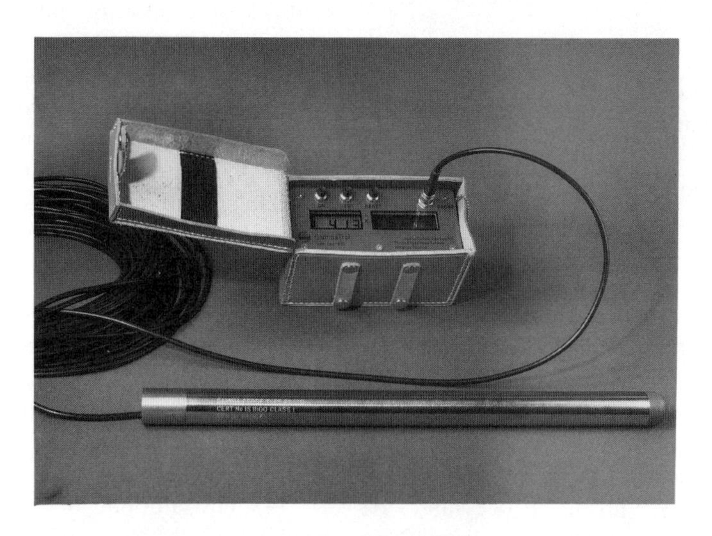

Fig. 2.41 The GammaTrol Potash Probe (PR1 85/86). (Courtesy of ICI Tracerco Ltd.)

control/scaler unit which possess a liquid crystal display and an analogue bar graph display ratemeter. The latter is able to show count rate levels, with a 10 second time constant, as the probe is inserted into the borehole with the scaler unit switched on. Spot readings can be taken at regular depth intervals, as required, over the range 10–80% K with a precision of ± 7% at the 50% level. Background readings are taken prior to insertion in the hole to allow correction of the *in situ* assay readings.

2.14.2 Portable XRF analysers

These instruments contain a radioactive source which 'excites' elements in the rock mass or sample (soil, rock, chip, sediment, liquid, pulps, concentrates or tailings). The characteristic X-ray fluorescence energy emitted by the elements can be measured by a solid state detector and converted to a metal value expressed in parts per million, grams or ounces per ton or percentage by weight.

Scitec Corporation's 'MAP' (Mineral Analysis Probe) Portable XRF Analyser is unique in its ability to provide highly quantitative analyses of those elements with an atomic number of 28 and above. Metals, such as copper, zinc, lead, silver, the rare earths, molybdenum, nickel, arsenic antimony, uranium, tungsten, gold and platinum, can be detected and measured with two types of sensors; the 'scanner' (Figure 2.42), which is a hand-held pistol-shaped device, or the borehole 'probe' (Figure 2.43), which is 1 m long and 30 mm in diameter. A winch or push-rods can be used with the probe to allow its insertion into horizontal, vertical or angled holes.

The scanner can be applied in the following situations:

(1) In exploration programmes, for the analysis of soil and sediments, outcrops, drill cuttings or cores, and bulk and grab samples from trenches, exploration adits or winzes.

(2) In mining operations, for the analysis of grab samples from underground muckpiles, mine dumps and stockpiles (Figure 2.44), of chip samples from rotary of percussion drilling, of mineralized pillars, backs and stope-faces (Figure 2.45), and of bench-faces in open pits (Figure 2.46).

(3) In the processing plant, for the analysis of belt feed stock, concentrates and tailings.

The scanner is highly versatile in that it can be used to make spot analyses (sample weight > 15 g) or it can be applied in a scanning (paint brush) mode which generates a cumulative aver-

Fig. 2.42 The MAP-2 face scanner and control console. (Courtesy of Scitec Corporation.)

Fig. 2.43 MAP-1 borehole probe being inserted into an exploration drill hole. (Courtesy of Scitec Corporation.)

age grade. The assays are shown on a digital display and are also stored in memory with an ID or reference number, depth (for boreholes) or heading number and other coded data as required. All data can easily be down-loaded to a printer or computer for further processing. Analyses of samples can be made rapidly, in 30–120 seconds in most cases, without having to prepare the sample in any way. Other advantages are its portability (it weighs less than 6 lb) and its utilization of a 'low level' radioactive source, which makes it safe.

The control console of the current version (MAP-3) can also be used with a cryogenic scanner (Figure 2.47), thus allowing the analysis of gold, an important addition to the range of elements covered by the instrument.

Preussag's Slimhole Analyser (Figure 2.48) is a vehicle-mounted computerized unit specifically designed for use in holes of 40 mm diameter (e.g. blast holes). It consists of a 73 cm long, 32 mm diameter, probe which is hydraulically positioned in front of the hole and then inserted. This is linked via a 60 m long steel spiral armoured coaxial cable to the computer data processing unit. The standard radiation source in the probe allows the detection and analysis of Pb, Zn, Cu and Fe in the hole, although other sources are

Fig. 2.44 Dump sampling with the MAP-2 face scanner. (Courtesy of Scitec Corporation.)

available for other metals. Analyses are made as the probe is withdrawn from the hole at speeds varying from 0.1 m/minute to 5 m/minute, the detection limit being lower at the lower speeds. For example, at the 0.1 m/min. speed the limits for both Pb and Zn are 0.5%. The concentrations of metals are plotted on site and average metal values calculated over preselected depth intervals. The results can be produced as a numerical readout or as an analogue printout, at a scale of 1 : 10, against borehole depth (Figure 2.49).

The use of borehole XRF probes thus reduces the need for expensive diamond drilling for grade control purposes and allows the use of rotary percussive holes instead. Face scanners will also obviate the need for channel/chip sampling. Of particular importance in both cases is the rapid acquisition of results and the immediate definition of assay limits. The latter allows flexi-

Fig. 2.45 MAP-2 face scanner in use for grade control purposes underground. (Courtesy of Scitec Corporation.)

Fig. 2.46 Bench face sampling in a Nevada gold-silver open pit. (Courtesy Scitec Corporation.)

bility in the application of mining cut-off grades and thus the rapid selection of new mining cuts.

2.15 SAMPLING THEORY

In this section we will examine the sources of error that may occur during the sampling of an orebody. This will be followed by a discussion of the methods that can be used to determine the optimum sample size and sample density. These methods will be considered on the basis of the sampling of *in situ* mineralization and then the sampling of broken ores, i.e. those that have been blasted and are ready for tramming/ trucking.

2.15.1 Sampling errors

Errors can be introduced at many stages during the sampling of an ore-deposit and also during crushing, pulping and splitting of the sample in preparation for analysis. In the first instance, the sample taken may be too small to be truly representative of the large block of ground to which its value will be assigned. This error is difficult to evaluate for it is dependent on the homogeneity of the mineralization and on the continuity or predictability of grades. This error cannot be evaluated by classical statistics and thus it is necessary to apply geostatistics. The production of semi-variograms for a deposit, as described in Chapter 4, allows an insight to be gained into this problem for the nugget variance and the geostatistical range can both be estimated. The former quantifies the random component controlling the spatial variability of grades and, through the nugget effect 'ε', allows a comparison with the regionalized variance which reflects the existence of predictable trends in grade distribution. The geostatistical range indicates the distance to which a sample can be extrapolated (zone of influence) and hence indicates whether the sampling density is adequate.

In the case of diamond drill sampling, we have seen (section 2.9.3) that the core is split longitudinally into two halves. Two possible errors are introduced at this stage. The first is because the split may not be accurate (especially when core

Fig. 2.47 A MAP-3 control console with cryogenic scanner for in-situ gold analyses. (Courtesy of Scitec Corporation.)

Fig. 2.48 The Preussag Slimhole Analyser at Bad Grund lead-zinc mine. (Courtesy of World Mining Equipment Ltd.)

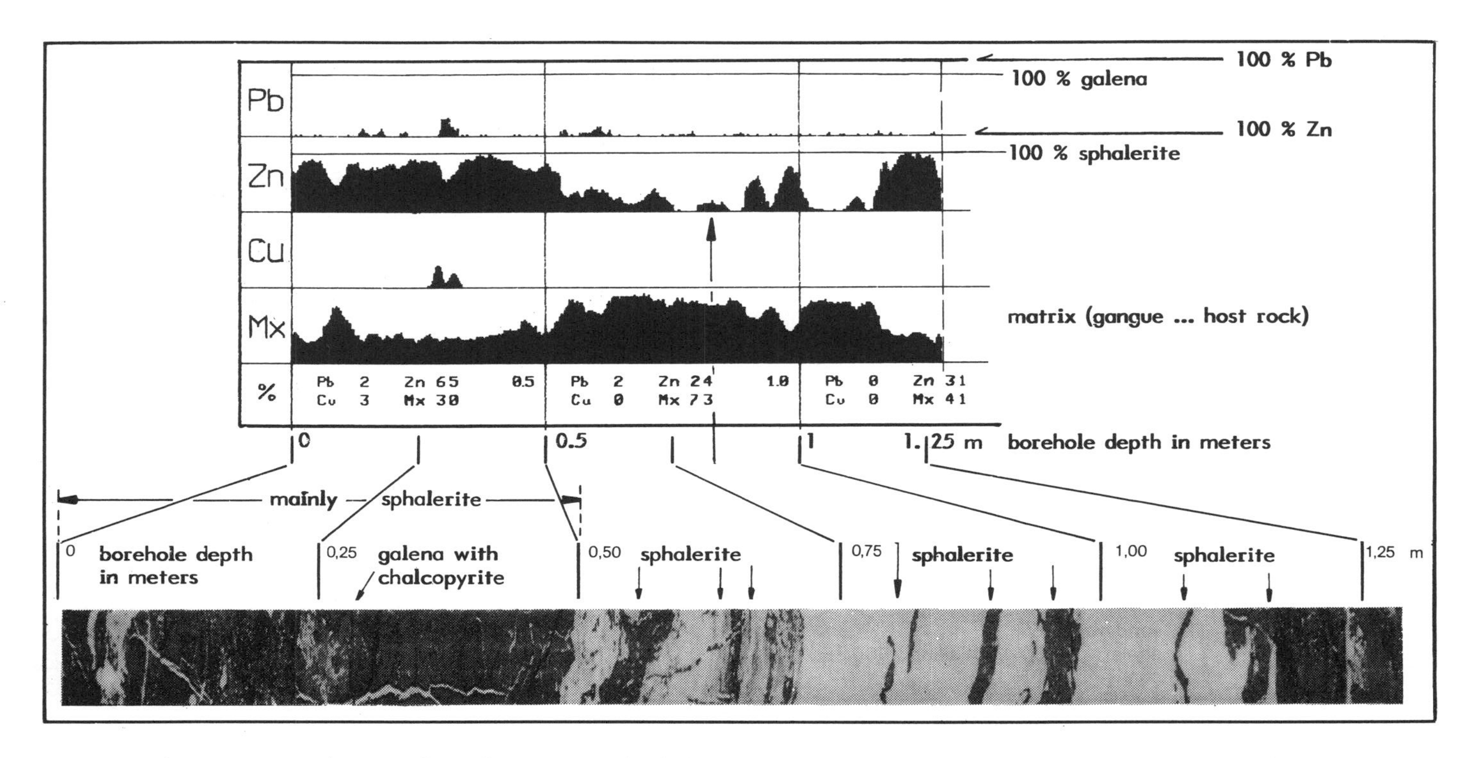

Fig. 2.49 Typical output from the Preussag Slimhole Analyser compared with a core section. (Reproduced from World Mining Equipment, Dec. 1987, p.70.)

splitters are used) so that constant volume per unit length is not achieved. The second is because the two halves of the core may contain different concentrations of mineralization, i.e. a high nugget effect exists. In the case of channel sampling, the depth and/or width of the channel may be variable so that once again constant volume per unit length is not achieved.

Each sample taken from the drill core, or from a channel, is crushed separately through a jaw-crusher, usually to −0.25 in, and then thoroughly mixed. It is then passed through a sample splitter (Figure 2.30) to obtain a smaller sample; perhaps 10–50% of the original. In other cases, we may 'cone and quarter' the sample so that 25% of the cone of crushed material is passed on to the next stage in the process. During this sample reduction process, another error may be introduced which may be due to contamination or inadequate mixing/homogenization of the sample before splitting. Doubt now exists as to how representative our new sample is of the original.

The above sample may now be pulped through a Tema or gyratory mill of some kind, so that another small sample can be cut from it and sent for analysis. In the laboratory, a small (e.g. 5 g) sample is then removed from this in turn. Each time we take a small sample from a larger one we run the risk of introducing another error. We express this error in terms of variance or relative variance (variance ÷ (accepted value)^2). The variance at each stage in the reduction process is determined and all these values are then combined to determine the overall error.

2.15.2 Sampling reduction error variance (Gy's formula)

Gy's formula gives an expression for the relative variance at each stage in the sample reduction process. Using this formula we can either (a) calculate the variance/error for a given sample size split from the original, or (b) calculate what subsample size should be used to obtain a specified variance at a 95% confidence level. Gy's formula is as follows:

$$s^2 = d^3 \cdot f \cdot g \cdot l \cdot m \cdot (1/M_1 - 1/M_2)$$

where:

M_2 is the weight of the initial sample (grams).

M_1 is the weight of the subsample (grams).

d is the dimension (cm) of the largest particle in the initial sample (the mesh size which passes 90–95% of the material).

f is the particle shape factor. It has a maximum value of 1.0 for particles which are perfect cubes. It has been found in practice that the shape factor deviates very little from an average of 0.5 irrespective of ore type and size fraction (f is dimensionless).

g is the particle size range factor which ranges from 0.25 (i.e. a large range of particle sizes) to 1.0 (all particles of identical size). In practice, g varies from 0.17 to 0.40 with an average of 0.25 (g is dimensionless).

l is the particle liberation factor which takes values from zero (completely homogeneous material) to 1.0 (completely heterogeneous material). In practice, values of L are estimated by the square root of the ratio of the practical liberation size L to the largest particle size d:

$$L = (L/d)^{1/2}$$

when $L/d > 1$, l is taken as 1.0 (l is dimensionless).

L the practical liberation size is defined as the maximum particle diameter with ensures the complete liberation of the valuable component (L is expressed in cm).

m is the mineralogical composition factor defined as:

$$m = (1 - a)/a[(1 - a)r + at]$$

where: r and t are the mean densities of the valuable mineral and gangue minerals respectively, while a is the average mineral content (not metal content) expressed as a decimal part of 1.0 (m is expressed in g/cm^3).

s^2 is the relative variance of the error when a

sample of M_1 grams is taken from a sample of M_2 grams characterized by the parameters d, f, g, l, m.

The use of Gy's formula can be best illustrated by examination of a worked example. Let us assume that the initial sample was a 0.5 m length of BX core (diameter 3.5 cm) which contained 5% Zn as sphalerite (2/3 Zinc).

$$\text{Volume of core} = \pi r^2 h = \pi(1.75)^2 \times 50 = 481 \text{ cm}^3$$

If we assume sphalerite has an SG = 4.0 and the host rock SG = 3.2 and that we have 7.5% sphalerite and 92.5% gangue, then the SG of the core =

$$100(7.5/4 + 92.5/3.2) = 3.25 \text{ g/cm}^3$$

and the weight of the core = 1563 g

The sample sent for crushing etc. will thus weigh approximately 800 g. Given that 90% of the discharge from the crusher used (e.g. rolls crusher) is −10 mesh (i.e. < 1.65 mm) and that the particle liberation size to release our valuable mineral is 0.2 mm, we can calculate the error involved when: (a) 200 g is split from the 800 g after crushing and sent direct for assay and (b) when this 200 g is first pulverized to −200 mesh (< 74 μm) and 5 g taken for analysis. The mineralogical composition factor (m) is:

$$(1 - 0.075)/0.075\ [(1 - 0.075) \times 4.0 + 0.075 \times 3.2] = 48.6 \text{ g/cm}^3$$

while the particle liberation factor (l) is:

$$(L/d)^{1/2} = (0.02/0.165)^{1/2} = 0.35$$

We may assume that $f = 0.5$ and $g = 0.25$.

Case (a)

$$\begin{aligned} S_1^2 &= (0.165)^3 \times 0.5 \times 0.25 \times 0.35 \times 48.6\ (1/200 - 1/800) \\ &= 9.55138 \times 10^{-3} \times (5 \times 10^{-3} - 1.25 \times 10^{-3}) \\ &= 3.58 \times 10^{-5} \end{aligned}$$

Hence $S_1 = 6 \times 10^{-3}$

As the absolute error = relative standard deviation times the accepted estimate then, in this case, the absolute error is $6 \times 10^{-3} \times 7.5$ % sphalerite, i.e. 0.045% or expressed in terms of Zn it is $2/3 \times 0.045 = 0.03$% Zn and our value for the sample is 5 ± 0.06% Zn at the 95% confidence level.

Case (b)

In the case of the second sample split, d = 0.0074 cm and $l = (0.2/0.0074)^{1/2}$, i.e. > 1, therefore we let l = 1. The relative variance for this second split is thus:

$$\begin{aligned} S_2^2 &= (0.0074)^3 \times 0.5 \times 0.25 \times 1 \times 48.6 \times (1/5 - 1/200) \\ &= 4.8 \times 10^{-7} \end{aligned}$$

Hence $S_2 = 6.93 \times 10^{-4}$

The absolute error is thus $6.93 \times 10^{-4} \times 7.5$ % sphalerite (5.2×10^{-3}) or 3.5×10^{-3} % Zn.

The combined error for the two splits is then $S_1^2 + S_2^2 = 3.58 \times 10^{-5} + 4.8 \times 10^{-7} = 3.628 \times 10^{-5}$. Hence $S = 6.02 \times 10^{-3}$ % ZnS.

The absolute combined error = $6.02 \times 10^{-3} \times 7.5 = 4.52 \times 10^{-2}$ ZnS, or = 3.01×10^{-2} % Zn.

Hence the value at the 95% confidence level is 5.0 ± 0.0602% Zn. Thus the improvement in analytical precision obtained by using a finer sample has been achieved with only a minor increase in sample reduction error.

Had we decided that an accuracy of ± 0.1% Zn was adequate then it is possible to calculate the new weight of our sample cut in case (a) as follows:

$$2\sigma = 0.1\%, \text{ therefore } \sigma = 0.05\% \text{ Zn or } 0.075\% \text{ ZnS}$$

Relative standard deviation = 0.075/7.5 = 0.01.

Hence relative variance = 1×10^{-4}.

Transposing Gy's formula:

$$\begin{aligned} 1/M_1 - 1/800 &= 1 \times 10^{-4}/[(0.165)^3 \times 0.5 \times 0.25 \times 0.35 \times 48.6] \\ &= 0.0105 \\ 1/M_1 &= 0.0105 - 0.00125 \\ &= 0.00925 \\ \text{Hence } M_1 &= 108 \text{ g} \end{aligned}$$

2.15.3 Sampling broken ores

Attempts to put the sampling of broken ores on a sound scientific footing eventually led to the publication in 1967 of Volume 1 of Pierre Gy's treatise 'L'échantillonage des minerais en vrac'. The main aspects of the theory were later summarized in a paper entitled 'Theory and practice of sampling broken ores' (1968). The presentation of this theory was long overdue as grab sampling practice at many mines was, and unfortunately still is, a rather haphazard process leading to serious miscalculations and systematic bias in grade estimations.

The derivation of Gy's formula is covered in the above references to which the interested reader is referred. A simplifed version of the formula presented in the previous section can be expressed as follows:

$$S^2 = C\,d^3/M$$

where S^2 is the relative estimation variance, or the error incurred in using a sample of weight M to estimate a bulk sample whose largest ore fragments have dimensions d cm. C represents a combination of factors which include shape factor, a mineralogical composition factor, a liberation factor and a grain-size factor as defined in section 2.15.2. The error is increased if the quantity of the economic phase in the ore fragments decreases relative to gangue, and also if there is increased liberation of the economic mineral from the gangue; the worst situation being where the fragments consist of either gangue or of the ore mineral. The error also decreases as the range in grain size within the fragments increases so a uniform grain size for the ore minerals will give the highest error.

The simplified version of Gy's formula also indicates that the relative error is proportional to the volume of the largest fragments and inversely proportional to the weight of the sample. In other words, to keep the error at a constant value a larger weight of sample is required from a muckpile containing larger than normal fragments. Ideally, the muckpile should be homogeneous from the point of view of the shape, density and size of the contained fragments otherwise the total sampling error is significantly enhanced by what Gy refers to as the grouping-segregation error. This is produced where certain fragments are segregated in the pile on the basis of the parameters (i.e. factors) listed above.

Summarizing, therefore, for a well size-homogenized muckpile, the sampling error will be least if the bulk ore is high grade, the broken ore well fragmented and the mineralization fine grained. Deviations from this require an increase in sample weight.

2.15.4 Sampling *in situ* ores

Perhaps the most worrying question that the geologist has to answer is whether a deposit is being under- or over-drilled. Often there may be a 'gut feeling' as to what is the best sampling interval, based on knowledge of the nature of the deposit and on empirical studies of predicted and realized grades in blocks of ground. Various statistical, and now geostatistical, methods, have been used in an attempt to resolve this problem, some of which are reviewed below.

Correlograms

This method involves the calculation of correlation coefficients (ρ) between the grades of pairs of samples which are initially spaced at a distance equivalent to the minimum sampling intervals along, for example, a level or a raise. Ideally for this study, close-spaced drilling or sampling should have been undertaken to produce a data set of at least 100 pairs at this spacing. Pairs of samples are then selected which are separated by twice the minimum sampling interval and once again the correlation coefficient is calculated. This exercise is repeated so that the sample separation gradually increases. The results are then plotted as a graph of ρ against sample separation (h) – a correlogram. In Figure 2.50, hypothetical results of sampling a raise and a drive are compared. Here it is seen that the best fit curve for the raise data reaches a value of zero before that

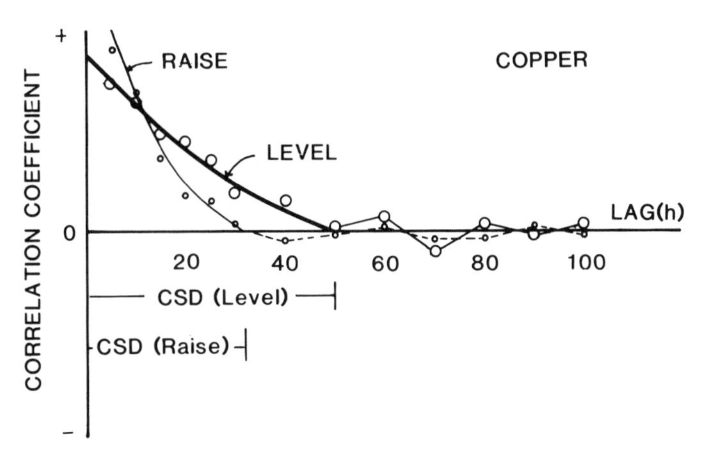

Fig. 2.50 A correlogram for copper grades in levels versus raises with the relevant critical sampling distances (CSDs).

for the drive. After this point the correlation coefficient fluctuates randomly about a mean close to zero. The correlograms thus define a critical sampling distance (CSD) for one element (e.g. Cu) beyond which the correlation coefficient cannot be distinguished from zero. If several elements are present in the ore, then CSDs can be determined for each element for both the drive and raise sampling. A safe sampling distance is then taken as the nearest multiple of 5 m which is less than the smallest CSD for all the elements.

Basically the correlogram shows the point at which adjacent assay values are totally unrelated or, put in another way, the point at which covariance has ceased to exist. For example, if the following results had been obtained:

	Drive	**Raise**
Cu	50 m	32 m
Zn	43 m	27 m
Pb	31 m	24 m
Cd	53 m	38 m

then a sampling distance of 30 m would be chosen for the drives and 20 m for the raises. Unfortunately, this method does not always work as the values of 'ρ' may be totally erratic and it is thus not possible to construct a correlogram.

Student's *t*-distribution

Barnes (1980, pp. 31–32) demonstrates the use of Student's *t* distribution tables to determine the precision of our estimate of ore grade based on a set sampling density and to determine what the sampling density should be, given a set level of precision. The relevant formula is:

$$r/2 = t{\cdot}S/\sqrt{n}$$

where r is the total width of the confidence interval for a set level of confidence, S is the standard deviation determined from previous sampling, and n is the number of samples taken. Barnes gives as an example the situation where 45 holes have been drilled into a vein and for which the mean grade was 9.4 oz Ag/ton, with a standard deviation of 4.9 at the 90% confidence level. Applying the equation we have:

$$r/2 = [(1.681)(4.90)]/\sqrt{45} = 1.23$$

1.681 is obtained from the table (Table 2.2) for $F = 0.95$ (one-sided distribution) and n between 40 and 60.

Hence our estimate is 9.40 ± 1.23 oz Ag/ton, with a precision of 13% (1.23 × 100/9.40). If a 10% precision had been required at the 90% confidence level then, by assuming that the

Table 2.2 Student's t values

n	F: 0.75	0.90	0.95	0.975	0.99	0.995	0.9995
1	1.000	3.078	6.314	12.706	31.821	63.657	636.619
2	0.816	1.886	2.920	4.303	6.965	9.925	31.598
3	0.765	1.638	2.353	3.182	4.541	5.841	12.941
4	0.741	1.533	2.132	2.776	3.747	4.604	8.610
5	0.727	1.476	2.015	2.571	3.365	4.032	6.859
6	0.718	1.440	1.943	2.447	3.143	3.707	5.959
7	0.711	1.415	1.895	2.365	2.998	3.499	5.405
8	0.706	1.397	1.860	2.306	2.896	3.355	5.041
9	0.703	1.383	1.833	2.262	2.821	3.250	4.781
10	0.700	1.372	1.812	2.228	2.764	3.169	4.587
11	0.697	1.363	1.796	2.201	2.718	3.106	4.437
12	0.695	1.356	1.782	2.179	2.681	3.055	4.318
13	0.694	1.350	1.771	2.160	2.650	3.012	4.221
14	0.692	1.345	1.761	2.145	2.624	2.977	4.140
15	0.691	1.351	1.753	2.131	2.602	2.947	4.073
16	0.690	1.337	1.746	2.120	2.583	2.921	4.015
17	0.689	1.333	1.740	2.110	2.567	2.898	3.965
18	0.688	1.330	1.734	2.101	2.552	2.878	3.922
19	0.688	1.328	1.729	2.093	2.539	2.861	3.883
20	0.687	1.325	1.725	2.086	2.528	2.845	3.850
21	0.686	1.323	1.721	2.080	2.518	2.831	3.819
22	0.686	1.321	1.717	2.074	2.508	2.819	3.792
23	0.685	1.319	1.714	2.069	2.500	2.807	3.767
24	0.685	1.318	1.711	2.064	2.492	2.797	3.745
25	0.684	1.316	1.708	0.060	2.485	2.787	3.725
26	0.684	1.315	1.706	2.056	2.479	2.779	3.707
27	0.684	1.314	1.703	2.052	2.473	2.771	3.690
28	0.683	1.313	1.701	2.048	2.467	2.763	3.674
29	0.683	1.311	1.699	2.045	2.462	2.756	3.659
30	0.683	1.310	1.697	2.042	2.457	2.750	3.646
40	0.681	1.303	1.684	2.021	2.423	2.704	3.551
60	0.679	1.296	1.671	2.000	2.390	2.660	3.460
120	0.677	1.289	1.658	1.980	2.358	2.617	3.373
∞	0.674	1.282	1.645	1.960	2.326	2.576	3.291

previous estimate of the mean (9.4) was accurate, $r/2 = 0.94$. Rewriting the equation we have:

$$n = [t \cdot S/(r/2)]^2 = \{[(1.671 \times 4.90)]/0.94\}^2 = 75.9$$

The t value was taken from Table 2.2 assuming that n would be of the order of 60. On the basis of this method, therefore, we would need an extra 31 holes to improve the confidence by only 3%!

Method of successive differences

De Wijs (1972) devised this method to determine the optimum sample spacing, a method which has much in common with geostatistical techniques. He examined the differences between adjacent samples using:

$$S_D{}^2 = [\textstyle\sum_{i=1}^{n-1} (x_{i+1} - x_i)^2]/2(n - 1)$$

He recognized that the serial interdependence of values was important, i.e. whether adjacent values were clearly part of a trend or whether they were totally erratic and independent variables. He measured the degree of interdependence with the coefficient of correlation (f), where $f = (S - S_D)/S$ and S is the classical standard deviation of the data.

To illustrate the use of this formula, he gave an example in which 41 samples were taken at 2 m intervals from a Bolivian tin deposit. The mean of the %Sn values was 1.9% with a standard deviation (S) of 0.87. $S_D = 0.5$ and hence $f = +0.43$. The data set was then split into two by taking alternate samples. The separation of samples now being compared was 4 m. The mean S_D of the two sets was calculated as 0.54 while the standard deviation remained at 0.87. The new value of f was thus +0.379. This process was continued so that the distance between sample pairs increased. At a spacing of 12 m, it was found that $S_D \approx S$, i.e. 0.87, and hence at this point autocorrelation ceased ($f \approx 0$). Samples should be close enough to produce autocorrelation, hence a spacing above 12 m is too wide.

Geostatistics

Geostatistical theory demonstrates that the size of sample and sample spacing can be determined from knowledge of the nugget variance and of the geostatistical range which are obtained from semi-variogram modelling of the mineral deposit. This subject is covered in section 4.17 and will thus not be considered further here. The effect of changing sample size will be discussed in terms of the volume–variance relationship which involves an understanding of extension variances and the application of Krige's formula.

2.16 BULK SAMPLING OF GOLD ORES

The bulk sampling of gold ore in the preproduction phase of a mining operation is extremely important for, not only does it provide a large volume for metallurgical test work (to determine % recovery), but it also allows a comparison to be made between the estimated value of blocks of ground and those obtained by bulk sampling. In other words, it allows the efficiency of the ore-reserves technique to be tested.

Such an exercise was undertaken in 1981 at the Augmitto Pit, near Noranda in Canada. Material from the initial box cut in the open-pit was crushed to −0.625 in using a quarry crusher. Batches of the order of 400 tons of −11 in feed were crushed in this fashion. The crushed material was then fed to a Snyder sample splitter which removed a 40 ton sample from the 400 tons. This was then reduced to 4 tons using a 10% Vezin sample splitter before passing it through a rolls crusher to produce a −0.25 in feed to a second Vezin sample splitter. This produced an 800 lb (0.4 ton) sample. This was then reduced to 200 lb by two passes through a Jones type riffle (50% reduction each time). A rolls crusher in the laboratory then reduced this sample to −10# before passing it through a riffle once again to collect a 100 lb sample. This was split into two 50 lb samples each of which was split into two 25 lb samples. The weight of each was carefully recorded.

One 25 lb sample was assayed and the other three retained to be used for repeat analysis or metallurgical tests at a later date. The sample sent for assay was pulverized to −48# and two 0.5 lb samples removed, both of which were assayed for gold and one for SiO_2. If one of the assays exceeded 0.05 oz/t gold, then 4 × 6 lb samples were taken from the original by passing through a riffle twice. Each 6 lb sample was pulverized to −200# and then screened to produce +100# and −100# fractions which were weighed. The entire +100# sample was then assayed for gold while the −100# sample was assayed in triplicate and the average of the three assays calculated.

The final assay value for each 6 lb sample was later calculated from:

$$[(Au_{+100} \times Wt_{+100}) + (Au_{-100} \times Wt_{-100})] / (Wt_{+100} + Wt_{-100})$$

The average of all four 6 lb samples was thus assigned to the 400 ton bulk sample originally obtained from the pit. As can be seen from the above, the procedure is complex and great care is needed to ensure that sample reduction produces a final small sample which is representative of the original bulk sample. Repeating the sample reduction process on samples split, but retained, at an earlier stage will allow a check on the accuracy of the whole procedure.

REFERENCES

Annels, A. E. and Boakye, E. B. (1990) Evaluation of the Offin River gold placer, central region of Ghana. *Trans. A IMM*, **99**, A15–A25.

Archer, W. L. (1987) Development of the Ditch Witch trenching machine as a grade control sampling tool. *Aus. IMM Kalgoorlie Branch, Equipment in the Minerals Industry: Exploration, Mining and Processing Conference, Kalgoorlie WA, October 1987.* pp. 21–3.

Barnes, M. P. (1980) *Computer-Assisted Mineral Appraisal and Feasibility*, Society of Mining Engineers, American Institute of Mining, Metallurgical and Petroleum Engineers, New York, 167 pp.

Bird, R. G. (1988) Rapid reconnaissance trench sampling for the exploration industry. Aus. IMM Sydney Branch, *Minerals and Exploration at the Crossroads*, Sydney NSW, July 1988, pp. 161–3.

Carras, S. N. (1984) Comparative ore reserve methodologies for gold mine evaluation. *Aus. IMM Perth and Kalgoorlie Branches Regional Conference on 'Gold-Mining, Metallurgy and Geology', October 1984.* Aus. IMM, pp. 59–70.

Carras, S. N. (1987) Ore reserve estimation methods and their relationship to recoverable reserves, in *Management and Economics of Mining Operations* (Symposium Proceedings, Western Australian School of Mines, Kalgoorlie, October 1986), pp. 57–67.

De Wijs, H. J. (1972) Method of successive differences applied to mine sampling. *Trans. A IMM*, **81**, 78–81.

Gy, P. (1968) Theory and practise of sampling broken ores, Ore Reserve Estimation and Grade Control. *Canad. IMM*, **9**, (spec. vol.) 5–10.

MacDonald, E. H. (1983) *Alluvial Mining: The Geology, Technology and Economics of Placers*, Chapman and Hall, London, 508 pp.

Reedman, J. H. (1979) *Techniques in Mineral Exploration*, Applied Science, London, 533 pp.

Schwann, B. B. (1987) The application of Ditch Witch sampling in oxidized open-cut gold mines, in *Equipment in the Minerals Industry* (Exploration, Mining and Processing Conference, AusIMM, Kalgoorlie Branch, Kalgoorlie, WA), pp. 25–31.

BIBLIOGRAPHY

Anglo-American Corporation of South Africa Ltd (1985) *Reference Manual for Mine Geologists*, Vol. 1 (Gold and Uranium Division), Anglo-American, Corporation of South Africa Ltd, Johannesburg RSA.

Clifton, H. E. *et al.* (1969) Sample Size and Meaningful Gold Analysis, Prof. Paper 625C, US Geological Survey, 17 pp.

Demming, W. E. (1950) *Some Theory of Sampling*, Wiley and Sons, New York, 602 pp.

Gy, P. (1967) L'échantillonnage des minerais en vrac; Tome I: Théorie Génerale, *Revue de l'Industrie Minérale*, Saint-Etienne, 186 pp.

Bibliography

Jones, M. J. (ed.) (1974) *Geological, Mining and Metallurgical Sampling*, IMM, London, 268 pp.

Parks, R. D. (1949) *Sampling, Examination and Valuation of Mineral Property*, 3rd edn, Addison-Wesley, Cambridge, MA, pp. 30–43.

Rowland, R.St J. and Sichel, H. S. (1960) Statistical quality of control of routine underground sampling. *J. South Afr. IMM*, **60**, 251–284.

World Mining Equipment (1987) Preussag Slimhole Analyser, **11**(12), p. 70.

3

Ore-reserves by 'Classical Methods'

3.1 INTRODUCTION

This chapter presents definitions of the various factors and criteria used in the production of mine metal, or mineral, inventories and ore-reserves. Descriptions are also provided of the methods available for the production of an overall estimate of grade, tonnage and metal/mineral content of a deposit. These 'classical methods' have stood the test of time but are now being gradually superseded by geostatistical techniques which are described in the following chapter. They are, however, still applicable in many situations and may well produce an end-result superior to that possible by a geostatistical method called kriging. A critical assessment of the use of kriging should always be undertaken before dismissing the classical methods. Too often, attempts to apply kriging are based on the use of mathematical parameters which have not been adequately tested or proven, perhaps due to time or information constraints. Geostatistical methods will only work satisfactorily if sufficient sampling is available to allow the production of a mathematical model which is adequate to describe the nature of the mineralization in the deposit under evaluation. Where this is not the case, we are well advised to use one of the classical methods outlined in this chapter. Prior to a detailed examination of these methods, a review of the classifications of reserves and resources is necessary.

3.2 CLASSIFICATION OF RESERVES AND RESOURCES

Although several attempts have been made over the last decade to classify reserves, no one method has yet gained international acceptance. Indeed, the whole matter is an area of considerable debate and disagreement. The problem lies in the attempt to devise a system, and phraseology, which is acceptable to all parties concerned in the evaluation, exploitation and financing of mining projects. The matter is further complicated by attempts to create classifications which are applicable over a range of natural resources which include oil and gas, industrial minerals, aggregates, metal deposits and coal. Very often it is difficult to arrive at a method which is suitable for metal deposits alone, never mind the whole spectrum listed above.

In recent years, there has been an increasing reluctance, especially in North America, to use the term 'ore-reserve'. The feeling is growing that these two words should be split and defined separately with the term 'ore' being used to describe the material in the ground which is currently being mined by a specific method and which is deemed economically viable at present costs and prices. However, although so-called 'ore' is defined on the basis of an economic 'breakeven-grade', some material at less than this grade could still be mined at a profit if local conditions in the mine are favourable. It is thus very dangerous to attempt to rigidly define 'ore' for this is a very subjective matter and based on a value judgement by the mine at the time of mining. Lane (1988) presents an economist's viewpoint in suggesting that the economic definition of ore should be that definition which maximizes the net present values (NPV; Chapter 6) of a mining operation. Net annual cash flows (revenues – costs) are combined into a NPV by discounting future flows back at an appropriate cost of capital and totalling them. Those interested in a more detailed analysis of this method are referred to this publication.

It is also proposed that the term 'reserve' should be used to define the amount of material available for extraction on the basis of economic studies and mine planning. However, whatever definition is eventually accepted, it must be possible to include in a reserve category those reserves estimated during the exploration phase of mineral deposit evaluation. It may be worthwhile considering the use of the term reserve base in which material is included which meets physical and chemical criteria and is technically mineable but which may not necessarily be economically mineable at present. Reserves would then represent part of the reserve base which is economically mineable under specified conditions. Further subdivision would be then made on the basis of the density of sampling or drilling and on the confidence in the reserve estimate (both grade and tonnage).

Some of the assumptions that have to be made in re-classifying a resource as a reserve on the basis of a mine plan, are summarized below. There are no doubt many others that could be added to the list, but these are the most important. The assumptions are:

(1) That there is a market for the product and that a contract would be forthcoming.

(2) That the location of the deposit is such that the transport costs would not be prohibitive and that environmental constraints would not produce an insurmountable financial burden.

(3) That the physical, metallurgical and chemical make-up of the deposit is such that metallurgical recoveries will be acceptable and that the product meets user specifications.

(4) That sufficient drilling and sampling are available to allow the geologist to calculate the confidence of his estimates, or at least make a realistic estimate on the basis of past experience.

(5) That the demand and price of the metal or mineral will be maintained and that economic, political, fiscal and environmental factors will not change dramatically over the anticipated life of the mine – mineral incorporated in the reserve could be relegated to the resource category overnight due to adverse changes in one or more of these factors.

Definitions of resources and reserves suitable for metalliferous deposits have been provided by many different bodies, including, (a) the US Bureau of Mines (USBM) and the US Geological Survey (USGS) (1980) with the most recent statement on the matter in USGS Circular 831, (b) the Association of Professional Engineers of the Province of Ontario (APEO) (1976) and, most recently, (c) by the Australasian Institution of Mining and Metallurgy (AIMM) and the Australian Mining Industry Council (AMIC) (1989). These are briefly reviewed in the following sections to allow comparison, and a more detailed presentation is provided in the appendix to this chapter. Definitions applied in the Zambian mining industry are also described together with classifications which are specifically applicable to coal deposits.

3.2.1 USBM/USGS Classification

This classification is largely based on that proposed by McKelvey (1972) and which is summarized in Table 3.1 ('The McKelvey Box'). This box shows how resources and reserves can be categorized depending on the increasing certainty of existence and on increasing feasibility of economic recovery. The USBM/USGS classification (Table 3.2) has come the closest to being accepted internationally but this dominance is now being challenged by the AIMM/AMIC classification.

Resources (for definitions see Appendix 3.1) in the USBM/USGS scheme include minerals that have yet to be discovered but are considered likely to be present on geological grounds, plus minerals that have been located and sampled but which, under present economic/political/technical conditions, are not commercially viable at the present time. Reserves, on the other hand, are those that have been identified by sampling

Table 3.1 'The McKelvey Box' – a classification of resources and reserves (McKelvey, 1972)

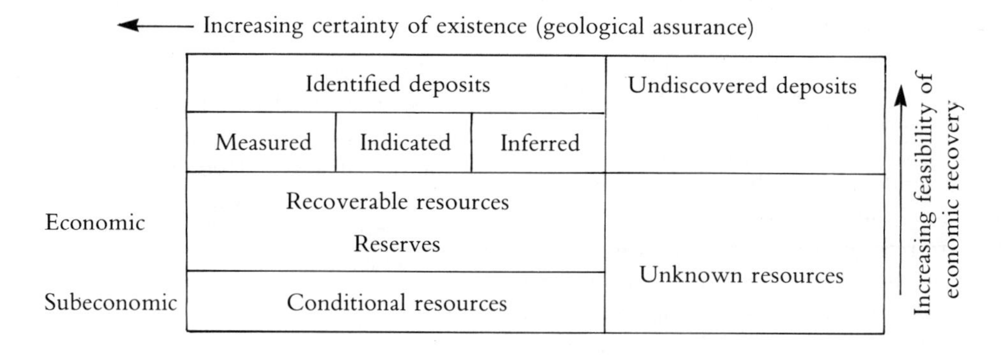

← Increasing certainty of existence (geological assurance)

	Identified deposits			Undiscovered deposits
	Measured	Indicated	Inferred	
Economic	Recoverable resources Reserves			Unknown resources
Subeconomic	Conditional resources			

Increasing feasibility of economic recovery ↑

Table 3.2 USBM/USGS Classification of identified mineral resources and reserves

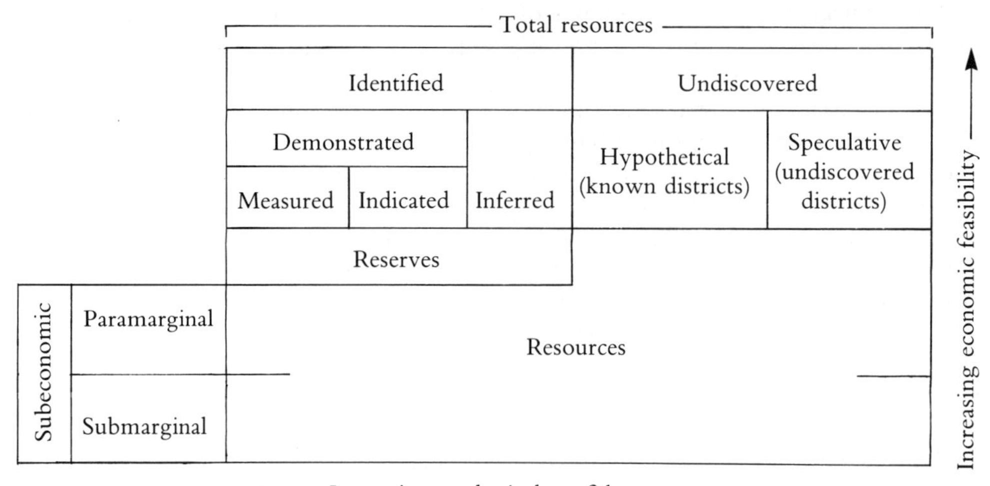

Total resources

		Identified			Undiscovered	
		Demonstrated		Inferred	Hypothetical (known districts)	Speculative (undiscovered districts)
		Measured	Indicated			
		Reserves				
Subeconomic	Paramarginal	Resources				
	Submarginal					

← Increasing geological confidence

Increasing economic feasibility →

and which are potentially economic/mineable. They are subdivided into three main categories: Measured, Indicated and Inferred, on the basis of degree of sampling and confidence in their estimated tonnages and grades.

3.2.2 APEO Classification

The Association of Professional Engineers of the Province of Ontario (APEO) (1976) have proposed a simplified set of definitions for reserves. This recognizes Geological Reserves, which do not take into account mining dilution or method, *In Situ* Reserves, which are those that could be mined but which have not yet been proven to be viable by a specific mining method or under specific economic conditions, and finally, Mineable Reserves, which are those that have been proved to be mineable in relation to a mining plan. More detailed definitions are provided in Appendix 3.2.

3.2.3 Australasian IMM/AMIC Classification

Howe and McCarthy (1987) have critically reviewed the USBM/USGS classification and have compared it with that proposed by the Australasian IMM and the Australian Mining Industry Council (1987). Although similar, the Australian guidelines would retain a mineral deposit in the resource category until a feasibility study had been completed and all factors affecting production from that deposit have been dealt with. Most importantly, however, they propose the dropping of the 'Inferred Ore Reserve' category which they consider archaic. They prefer either 'mineralization' or 'inferred resources'. The recommendations of the 1989 AIMM/AMIC report on identified mineral resource and ore-reserves are summarized in Table 3.3 which shows that the USGS terminology has only been retained for mineral resources and that, as suggested by Howe and McCarthy, ore-reserves are subdivided into only the Proved and Probable categories. Detailed definitions are presented in Appendix 3.3 which are taken verbatim from this report.

3.2.4 Coal resources and reserves

Classifications suitable for coal are somewhat different to those described above for metalliferous deposits. The Australian Code for Reporting Identified Coal Resources and Reserves (1986) is taken as an example and is reproduced in

Table 3.3 AIMM/AMIC Classification of identified mineral resources and reserves

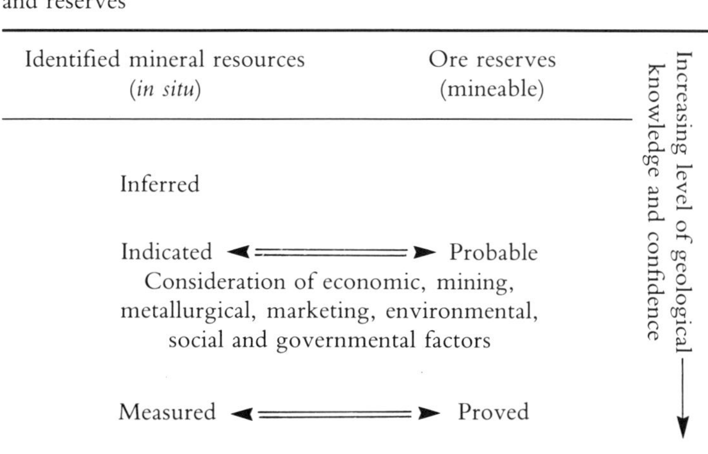

Appendix 3.4. In the various categories, the required spacing of drill-holes or exposures (surface or underground) is specified. Resources are subdivided into measured, indicated and inferred as in the AIMM/AMIC classification but, Reserves are defined on the basis of Mineable *In Situ* Reserves, Recoverable Reserves and Marketable Reserves.

3.2.5 Requirements of banks, financial institutions and stock exchanges

In preparing a document or report specifically aimed at these bodies, it is essential that the mining company involved is aware of their requirements. Full information must be provided and all reserve classifications must be precisely defined with a clear statement of all assumptions made. In the absence of an internationally accepted reserve classification, nationally agreed definitions should be used. *In Situ* and Mineable Reserves must be quoted separately and must be clearly distinguished. The report should be prepared by a 'competent person', who is generally considered to be a corporate member of a professional institution, and should contain the following information.

(1) The location of the deposit relative to transport systems, in-house or custom smelters, markets, labour supply, etc.

(2) Whether planning permission has been obtained and mineral rights secured.

(3) The nature of the deposit.

(4) The dimensions, attitude and depth of the deposit. This includes a structural interpretation as uncertainty here could have considerable impact on the accuracy of the quoted reserves.

(5) The Mineable Reserves based on Measured plus Indicated Reserves. Some banks prefer this terminology to Proved and Probable as Proved implies a high degree of precision whereas, in fact, there is still a high degree of uncertainty. Reserves can only be proven after mining is complete.

(6) The method used to estimate the reserves.

(7) The mining plan – method of working.

(8) The production rate and life expectancy based on 5. Ideally, the latter should be one and a half to two times the proposed loan life.

(9) The confidence limits of the reserve estimations. Ideally, the financial institutions would like the various reserve categories defined on the basis of specified confidence levels but, at the moment, only the APEO definitions involve a statement of accuracy for proven ore (within 20%). Confidence limits are important in that they allow banks, etc., to test whether the debt can be serviced from the annual cash flow of an operation.

(10) The drilling and sampling technique used.

(11) The number of holes drilled or samples taken, their distribution, and the accuracy of their location. It is necessary to be able to demonstrate that the Mineable Reserves are substantiated by physical sampling and analysis.

(12) The quality of the assay data.

(13) The average core recovery (and range) over the ore-zone.

(14) Weighted grade estimates, perhaps broken down on the basis of depth or sub-area in the orebody. An overall global grade is not adequate unless it persists throughout the orebody. Short-term financial viability is affected if the grades are lower than average at shallow depth and increase with depth.

(15) The cut-off grade applied and how this was derived.

(16) Whether the tonnage factor was measured or estimated.

(17) Whether mining dilution has been applied and if so, the basis for the estimate.

(18) The unit operating costs.

(19) The unit sales price and the basis for future projections.

(20) The results of bench and pilot plant processing tests, including the methods used, expected recoveries and methods of analysis. Apparently high grade ore may be relegated to waste

on the basis of the poor recovery of the valuable mineral.

(21) In the case of an operating mine, the reserve statements should be based on the audit principle and include a statement of the change between the end of one year and the next. In financial institutions there is a strong feeling that reserves should be audited in the same way as the fixed assets of a company. At present, not all companies quote their reserves in annual reports to shareholders.

3.2.6 Ore-reserve classification in the Zambian copperbelt

ZCCM Ltd use a method of ore-reserve classification which is more specific to the practical mining situation in the Zambian Copperbelt. Here, ore-reserves are defined as '. . . those estimated in-situ tonnages of an indentified mineral resource which are committed or planned to be mined and processed through existing or approved planned facilities'. Ore, in the case of underground mines, is furthermore limited in vertical extent to the lowest existing or approved planned extraction level of the mine. Thus the tonnage of ore in the various categories will not include material at greater depth than this lowest production level. Ore-reserves are declared to assay cut-off grades of 1% in the case of copper, and 0.1% in the case of cobalt, with local exceptions where well-defined geological contacts are used. ZCCM Ltd recognize three main categories for ore underground:

Undeveloped ore-reserves

These reserves are calculated for those areas of an orebody which have been intersected by surface drill-holes and/or long underground exploration holes and in which little or no mining development exists. As stated earlier, however, they must be accessible from existing or approved planned development (shafts or haulages or declines).

Partly developed ore-reserves

These reserves are those tonnages within defined stoping blocks where extraction haulages have been mined and sufficient evaluation information is available to allow stope preparation development to proceed. Generally, they cover areas where haulages have been developed on two levels from which drill-holes have intersected the orebody at both chamber (final back of proposed stope) and haulage elevations. Additional, but less accurate, information is provided by earlier exploration holes from surface or underground.

Fully developed ore-reserves

These reserves are from defined stoping blocks where at least 95% of the development and sampling has been completed and the reserve is available for extraction. Generally the tonnage takes into account that ore removed during development (e.g. on sub-levels) and is thus a net tonnage.

At Mufulira, three categories were further defined on the basis of intensity of sampling as follows:

Class I (fully developed)

These reserves are calculated in areas of the mine where mining development, e.g. footwall and hangingwall drives, haulages, cross-cuts and sub-levels, is complete and sampling is intense. Short-hole drilling has been completed from the immediate footwall of the orebody on all boundaries, mids and panels (100 m, 50 m and 10 m intervals respectively). Fan drilling has thus allowed the infilling of the gaps between earlier long-hole drilling from the main haulages (Figure 3.1). This is augmented by chip sampling across all exposures of the orebody in cross-cuts (e.g. drainage, hangingwall exploration or stope development).

Class II (partly developed)

These are calculated where the orebody, above the lowest level of mine development, has been sampled by diamond drilling from haulages, main return airways, hangingwall exploration

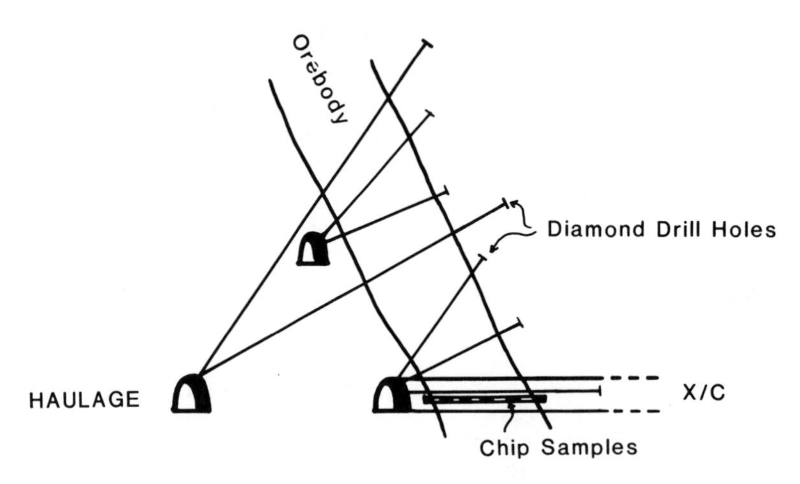

Fig. 3.1 Class I sampling, Mufulira Mine, Zambia.

pilot drives on block boundaries and mids (100 m and 50 m intervals). Some chip sampling may also exist in cross-cuts. No sampling has been undertaken from the immediate footwall of the orebody (Figure 3.2).

Class III (undeveloped)

These have been sampled by long drill-holes from hangingwall pilot drives and haulages. The intersections have been made below the lowest current level of mine development but are in areas which will be eventually accessible from existing shaft systems. Intersections are widely spaced down-dip and drilling has only been undertaken on mine boundaries (100 m intervals). These reserves are sufficiently accurate for advance planning of mine development.

Class IV (inferred reserves)

These have been located by deep underground or surface exploration drill-holes and lie outside the categories outlined above. Intersections are widely spaced, e.g. 250 m, and the continuity of ore is still to be proven. These reserves thus have very low confidence levels.

Open-Pit Reserves

In the case of open-pits, the Fully Developed Reserves are those tonnages that are fully exposed and available for extraction. Partly Developed Reserves are those bench tonnages that can be made available for extraction by removal of ore and waste. Finally Undeveloped Reserves are those drill indicated tonnages not included in the partly developed category which are within the approved final pit design.

3.2.7 Additional reserve classifications

These include:

Mining block

The reserves of a block of ore defined by mine levels on the one hand, and by block boundaries or raises, on the other. A proportion of these reserves may be left behind as pillars.

Stope

The reserves of a stope as defined by stope-planning engineers. Hopefully all these reserves will be extracted. They may be defined on the basis of *In Situ* Reserves or Diluted Reserves.

Orebody

The combined Proved plus Probable Reserve of each orebody in a mine, several of which may be under development at any one time.

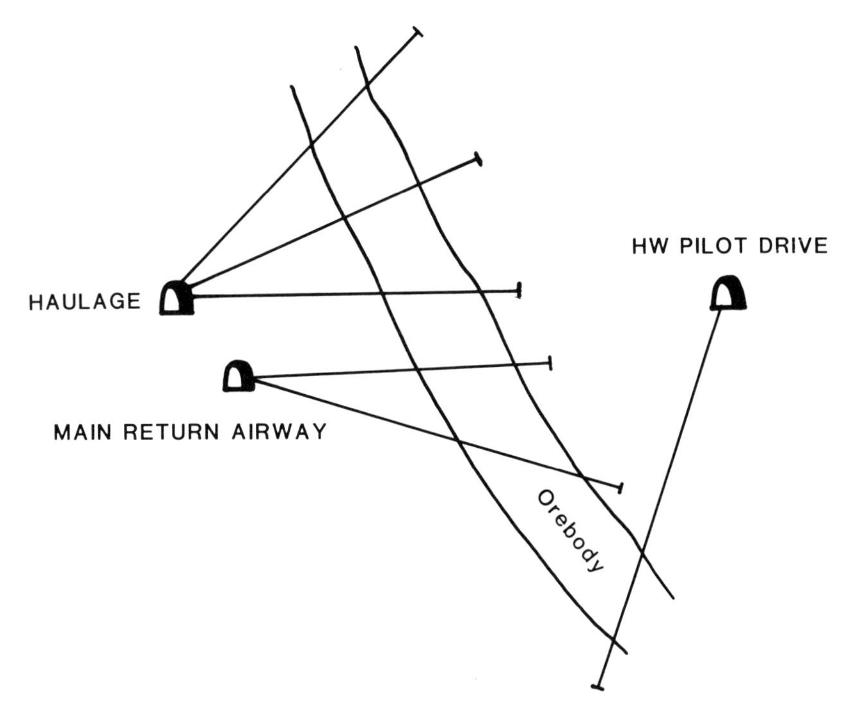

Fig. 3.2 Class II sampling, Mufulira Mine, Zambia.

Bench
The reserves of ore-grade material in an open-pit bench together with tonnages and grades of material falling into other classifications, e.g. low grade, 'oxide', leach, waste, etc.

Stockpile
These include high grade and low grade stockpiles awaiting blending prior to milling and also stockpiles of marginal or 'oxide' material awaiting improvement in metal prices, introduction of new technology or the construction of new concentrator circuits capable of processing them at a profit.

3.3 DETERMINATION OF POTENTIALLY ECONOMIC INTERSECTIONS

Each intersection of a mineral deposit, be it by chip sampling, channelling or drilling (diamond core, rotary, percussion) must be examined to determine whether it meets specific criteria for economic extraction. Of particular importance here is whether the weighted grade of all individual assay samples in the intersection exceeds a set level for a thickness equal to, or greater than, the minimum mineable thickness. Before explaining how this can be achieved, it is necessary to define some of the concepts involved in more detail.

3.3.1 Definitions

Thickness components (for tabular deposits)
The intersected thickness (IT) of a mineral deposit is of little value in this form as it is dependent on the dip of the body and on the attitude of the borehole. However, if the intersection angle between the body and the axis of the drill-hole is known, or can be measured in core, then IT can

be converted to true thickness (TT), i.e. the thickness perpendicular to the upper and lower contacts of the mineralized horizon.

True thickness (TT) = intersected thickness (IT) × sine intersection angle (θ)

This can be calculated separately for each assay sample in the intersection and then accumulated, or an average intersection angle calculated and applied to the total intersected thickness. It should be noted that TT is only applied to situations where the assay information is to be plotted, and the reserves calculated, on unrolled (palinspastic) plans.

Where the intersection angle cannot be measured, the true thickness of an intersection can be determined from the inclination of the drill-hole at the mid-point of the intersection (α) and the dip of the orebody (β) as measured from sections. Hence:

$$TT = IT \cdot \sin(\alpha + \beta)$$

However, the assumption made so far is that the plane containing the drill-hole is perpendicular to the strike of the orebody. Where this is not the case, then allowance must be made for this discrepancy. Wellmer (1989) derives the necessary formula for the thickness reduction factor which he calls R_m.

$$R_m = \sin(\alpha + \delta) \cdot \cos\beta / \cos\delta$$

where α and β are as defined above and δ is the apparent dip of the orebody in the vertical plane containing the drill-hole. Hence:

$$TT = IT \times R_m$$

The vertical thickness (VT) component of thickness can be calculated from:

$$VT = TT/\cos\beta$$

where β = the dip of the mineral deposit.

This component is only used when the data are to be plotted on plan projections of the deposit, or on mine level plans, or bench plans in an open-pit operation. Plan projections are particularly suitable where the deposit has a flat dip. The reason for the use of vertical thickness in this case can be ascertained from Figure 3.3. The area of the side of the parallelepiped (outlined by stippling) in the plane of the deposit is identical to that of the side ($ABCD$) of a rectangular prism of height VT. Thus the volume of the orebody below the rectangle $WXYZ$ on the plan is thus equal to VT × area $WXYZ$.

The horizontal thickness (HT) component of thickness can be calculated from:

$$HT = TT/\sin\beta$$

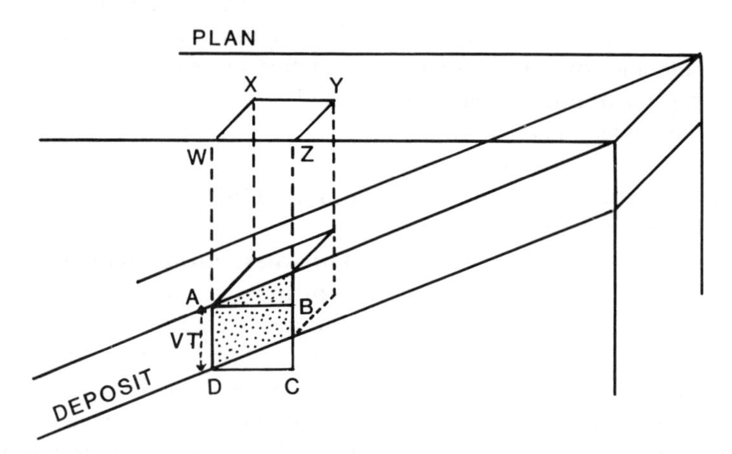

Fig. 3.3 Use of vertical thickness for plan projections of mineral deposits.

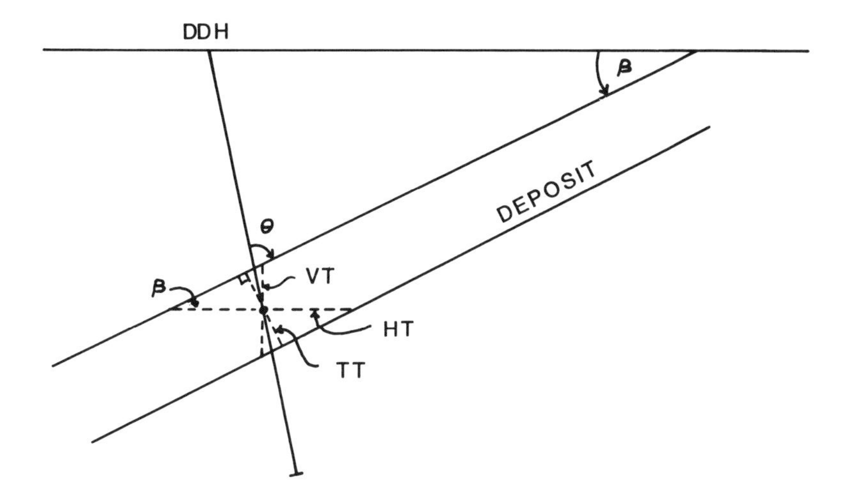

Fig. 3.4 Thickness components used in reserve evaluation.

This is used when the data are to be plotted on vertical longitudinal projections (VLPs) which are used for underground operations exploiting steeply dipping tabular deposits. The logic behind the use of HT is similar to that described for VT. The relationship of all these components is illustrated in Figure 3.4.

Grade

This is a measure of the quantity of mineral/metal per unit volume or weight and can be expressed in the form of kg/m^3, g/tonne or as a weight percentage (metal, ash, volatiles, etc.). In the case of diamonds, this would be quoted as carats per 100 tonnes where 1 carat = 0.2 g.

Bench composite grade

Where a deposit, such as a porphyry copper-molybdenum deposit, is to be evaluated using horizontal slices equivalent to benches in an open-pit, all sample grades in each borehole falling within the slice limits are composited together to produce a composite grade for that intersection of the bench. This usually involves allocation of portions of samples cut by the slice limits to adjacent slices. Samples, or portions of samples, are weighted by length to gain the overall bench intersection grade as explained later in section 3.3.2.

Metal accumulation

This is a measure of the quantity of metal in a particular section of an orebody and is the service variable used to determine the reserves of a deposit whose thickness is variable. Where the thickness is constant, as in a bench in an open-pit, bench composite grades on their own can be used to produce an overall bench grade. However, when separate benches are to be combined to produce a global reserve, bench metal accumulations (tonnes % or m^3%) must be calculated.

In the case of individual drill-holes, metal accumulation is the sum of the products of sample length and grade, i.e:

$$MA = \sum_{i=1}^{n} G_i \times L_i$$

However, the lengths in this equation are expressed as their horizontal, vertical or true thickness values so that we produce a HT m%, a VT m% or a TT m% value depending on the method to be used to compile the information. The units in this case thus may be in the form of m%, as above, or cm·g/tonne. Metal accumulations can also be expressed as the amount of metal beneath each square metre of the deposit in plan or vertical longitudinal projection, e.g. kg/m^2.

Minimum stoping width (MSW)

This represents the thinnest ore-zone that can be mined by the mining method employed. If the orebody has a width less than this then some barren or low grade material has to be mined with the mineralized interval causing dilution of the overall grade. MSW can be expressed in terms of true, vertical or horizontal thickness. In the case of the Nchanga Lower Orebody in Zambia, a 6 m vertical MSW is required for the block caving method employed, whereas in Canadian gold mines true thickness stoping widths may be as low as 1.2 m. Typical stoping widths in South African banket reef gold mines range between 1.2 and 1.4 m, although 1.0 m is possible.

Cut-off grade (COG) and minimum mining grade (MMG)

Confusion exists in the utilization of the term cut-off grade for in some mines this is distinguished from a minimum mining grade. One definition of COG is the lowest grade material that can be included in a potentially economic intersection without dropping the overall grade below a specified level, referred to as the minimum mining grade. This can be applied rigidly, or in a more flexible way, allowing the incorporation of internal low grade (< COG) intervals so that outlying higher grade zones can be taken into the ore-zone. However, once these internal waste zones exceed a specified thickness, the two higher grade zones must be treated as separate potential stopes, or one must be rejected. Minimum mining grade is thus defined as the overall grade of an intersection or ore-block which can be mined and processed at a profit. It is thus the minimum average/weighted grade allowable over one economic interval when material down to the stipulated cut-off grade is included. It should be noted, however, that this differentiation between COG and MMG is only really applicable where clear-cut hangingwall and footwall assay contacts exist. In more diffuse, nontabular deposits only a COG is defined (Chapter 5).

Camisani-Calzolari, De Klerk and Van der Merwe (1985) define cut-off grade, or pay limit, as being that grade at which potential revenue balances all costs. This is thus equivalent to MMG but no allowance is built in for a specific profit margin. Also, the grade of the mineralized zone being compared with this COG should contain an allowance for both intentional and unintentional dilution (see later). The costs which must be met by the revenue include those related to:

(1) sampling;
(2) mining;
(3) transport;
(4) processing;
(5) underground access development in waste.

To these must be added administrative overheads which include interest charges, amortization of capital, etc.

There are many factors which influence the COG, some of which are summarized under the categories geological, economic and mining method.

Geological factors

(1) mineralogy – relative proportions of treatable to refractory phases;

(2) grain size and the 'free to lock' ratio of the valuable phase in gangue – amenability to processing;

(3) grade of deposit and presence of deleterious (penalty) elements;

(4) grade/mineralogical homogeneity of the deposit;

(5) shape, size and thickness of the deposit;

(6) structural complexity;

(7) ground conditions – dilution;

(8) water problems.

Economic factors

(1) accessibility to markets/smelter;
(2) labour – availability, cost, skill, man shift production rates;
(3) current metal price versus mining/processing cost;
(4) political and fiscal factors;
(5) anticipated recovery of *in situ* ore and of contained metal in the processing plant;
(6) cost of waste disposal, reclamation;
(7) capital costs and interest rates.

Mining methods

Open-pit operations will obviously be able to work to much lower cut-offs than those possible in underground mines. In the case of copper, open-pits may work down to 0.2–0.3% compared to 1–2% in underground mines. At Mufulira Mine, Zambia, drill-hole intersections of the lower (C) mineralized horizon are defined as being of ore grade on the basis of a metal accumulation of greater than 6 m% and a thickness of 3 m when calculated on the basis of a 1% COG. This thus implies a minimum mining grade of 2% Cu. Large differences also exist between superficial gold deposits (alluvial and eluvial) and those in banket reefs or shear zones worked by underground methods. In the former case, cut-off grades of 0.7 g/t or even lower have been applied as in the case in the Royal Hill lateritic gold deposits of Suriname (Smith, 1987). COGs for underground gold mines may vary from 1.5 g/t to 5 g/t dependent on local mining conditions, depths, etc. At Kerr Addison Gold Mine, Ontario, minimum mining grades can be varied depending both on the current gold price and on the mining method used. A series of graphs is produced of minimum grade against gold price, each representing one of the seven main methods used. For example, at a gold price of $470 Canadian, blast-hole stopes require a grade of 0.051 oz/ton while overhand square set stoping methods require a grade of 0.115 oz/ton. Should the price drop to $400 Canadian then these grades rise to 0.06 oz/ton and 0.135 oz/ton respectively (figures based on 1986 costs). At Falconbridge, Sudbury, different COGs are applied to shrinkage stopes (1% Ni), square-set (1.4% Ni) and underhand 'cut and fill' stopes (1.9–2.0% Ni) reflecting the relative costs of the technique and production rates.

More information on the determination of cut-off grades for open-pit operations is presented in Chapter 5, where a distinction between break-even and operational COGs will be made.

Intentional dilution

This is low grade or barren material added to a mineralized zone to make the thickness up to the MSW. It is expressed as a decimal, or a percentage, of the width of the mineralization.

Unintentional dilution

This is the product of over-blasting or collapse of weak hangingwall material into the stope. Generally it is of the order of 10% but values exceeding 40% have been recorded. Again it can be expressed as a decimal, or percentage, of the ore width.

Allowance is often made for the anticipated amount of dilution in calculating stope reserves. A dilution factor can be applied to individual intersections or to an average thickness and weighted grade calculated for all drill-holes within the limits of the stope. The method used to estimate the magnitude of this factor is usually based on past observation and experience. This may be a fixed percentage (e.g. 12%) or the type and thickness of ore may be taken into account. At Stillwater Pt-Pd Mine in Montana, the dilution is estimated by use of a power curve equation (see also section 8.3.4) based on studies of the actual dilution associated with different ore thicknesses. The dilution factor (DF) equation is:

$$DF = 1 + 25.788\ (\text{ore thickness})^{-2.802}$$

Hence for a 4 ft stope, this factor would be 1.53 whilst for an 8 ft stope it would be 1.08. Hence the thicker the ore, the lower the percentage dilution.

At Hemlo Gold Mine, Ontario, the dilution is also related to ore thickness but with fixed values being applied to specific thickness ranges (< 3 m = 30%, 3–5 m = 20%, > 5 m = 10%; section 8.4.4). However, at Kerr Addison Gold Mine, also in Ontario, fixed dilutions are applied on the basis of ore type. Ore in pyritized rhyolitic and dacitic lavas ('Flow Ore') is diluted by 10% relative to the geological *in situ* reserves although, where this ore is graphitic, a 10% tonnage increase is accompanied by a 20% dilution in grade to allow for gold losses due to the presence of graphite. Ore in fuchsite-bearing carbonated basic volcanics ('Green Carbonate Ore') is diluted by 35% reflecting the more erratic distribution of the gold and the weaker ground conditions.

Extraction ratio

This is the proportion of the *in situ* reserve which is actually extracted during stoping, after allowance for pillars (safety, sill, rib, crown, stope, etc.). At Panasqueiras tin-tungsten mine in Portugal, the mining of sub-horizontal veins by room and pillar methods allows an extraction ratio of 0.8–0.85 to be achieved for a MSW of 2.2 m (locally 1.75 m). In some cases, the overall percentage extraction of coal seams may drop to as low as 35% because of the geometric inflexibility of long-wall mining methods and ground sterilized by fault pillars and by geological disturbances.

Metallurgical recovery

During the milling and concentration (processing), a proportion of the valuable mineral or metal in the mill feed (diluted ore) is inevitably lost and is disposed of with the tailings, i.e. the fine silt of clayey material representing the gangue components of the ore. Such losses are partly due to the fine-grained nature of some of the ore minerals and their intimate intergrowth with the gangue minerals. Excessive grinding would be necessary to release these grains thus increasing costs. Losses are also incurred by inefficient plant, by the lack of operational control in the concentrator and by the oxidation of some metallic particles. Typical recoveries are listed in Chapter 5 (section 5.3.7).

The formula for the determination of metallurgical recovery (MR) can be derived from first principles by working through an example. Let us assume that the following values apply:

Mill feed grade = 5% Cu
Concentrate grade = 25% Cu
Tailings grade = 0.5% Cu

One tonne of feed thus contains 0.05 t of copper metal. If the tonnage of concentrate produced from this feed is T then the amount of contained metal is $T \times 25/100$ and the amount of metal lost to the tailings is $(1 - T) \times 0.5/100$. Thus we have an expression for the total amount of metal which must equal 0.05 tonnes. Hence:

$$T \times \text{concentrate grade}/100 + (1 - T) \times \text{tailings grade}/100 = \text{mill feed grade}/100$$

or

$$T = (\text{mill feed grade} - \text{tailings grade}) / (\text{concentrate grade} - \text{tailings grade}) \quad (1)$$

thus in the present example,

$$T = (5 - 0.5)/(25 - 0.5)$$

and

$$T = 0.184\ \text{t}$$

Metallurgical recovery, however, equals tonnes of metal recovered divided by tonnes of metal in the original feed. Hence, in this case, we have:

$$MR = (T \times \text{concentrate grade}) / \text{feed grade} \quad (2)$$

or,

$$MR = (0.184 \times 25/100)/0.05 = 0.92 \text{ or } 92\%$$

Using formulae (1) and (2) we can produce a general formula:

$$MR = \frac{(\text{feed grade} - \text{tailings grade})}{(\text{concentrate grade} - \text{tailings grade})} \times \frac{\text{concentrate grade}}{\text{feed grade}}$$

Many of the factors described in this section are applied in the calculation of the reserves of a gold block presented in Appendix 3.5.

3.3.2 Weighted overall grade of an intersection or groups of intersections

Prior to the calculation of the overall grade of a series of samples through a mineral deposit, it is necessary to confront the problems of what to do where the intersection contains refractory and non-refractory ore minerals and also, whether anomalously high grade samples should be accepted or reduced in some way.

Refractory versus non-refractory minerals
If the concentrator is designed to process only sulphide minerals (by froth flotation) and has no capacity for dealing with the 'oxide' or refractory minerals, then the grades of individual samples should be corrected to allow for this. Under the term 'oxide', we generally include a range of secondary supergene minerals which range from carbonates, phosphates, silicates to oxides. All of these can be represented by an analysis of acid-soluble metal. A total metal analysis is then undertaken and the difference between the two is considered to be the recoverable metal, in this case the sulphide metal. In the case of copper:

$$\text{Sulphide copper} = \text{total copper} - \text{acid-soluble copper}$$

The weighted grade of the intersection is thus calculated on the basis of the sulphide metal values.

Abnormal assays
Very high metal values may appear in a sequence of assays which, if not due to contamination, reflect very localized random phenomena such as gash veins, concretions/accretions, or coarsely crystalline aggregates of the valuable mineral. Repeat analysis will discriminate between analytical errors and a real phenomenon. These high assay values contribute to the nugget effect of an orebody which will be discussed in Chapter 4. We have to decide whether to accept them, even though they are very localized and will probably heavily weight or bias the final results, or whether to reduce them in some way.

There are several different ways of resolving this problem. We could cut the grade to the average of the adjacent samples, or to the mine average grade, or to an arbitrary percentile value based on a cumulative frequency or log probability plot of mine assays. Alternatively, the mean plus 3 standard deviation value of the mine assay population could be calculated and applied as the level of cut. Some mines, especially gold mines, cut to an arbitrary level (e.g. 1 oz/ton or 30 g/t) but monitor the relationship between annual mined grade, based on concentrator recovery, and the grade of the mined ore as calculated from the sampling information. In the case of gold ores, sampling grades usually exceed mined grades and thus the level of cut is decreased to bring the ratio to as close to unity as possible. This is thus an empirical method based on historical grade reconciliation. Another approach is by visual inspection of the grade distribution of large populations of assay data from the section of the mine under evaluation. This will quickly allow discrimination of 'outliers' (i.e. abnormally high values) from the main population. The level of cut can then be applied at this point. Such a subjective method is perhaps to be recommended for there is little basis for the application of mathematical techniques. Finally, each assay could be weighted by its frequency of occurrence in the mine assay data file as a whole, as described later in this section ('Weighting by zone of influence (face sampling)') so that abnormally low or high values are given very low weighting factors. Values close to the mean would then be given the greatest weighting. Finally, all assays could be weighted by zone of influence and length, except for the abnormal assay, where the zone of influence is considered to be the width of the sample. This method will be described later in this section ('Weighting by zone of influence (abnormal assays)').

Basically, it must be determined whether high values are typical of the orebody as a whole and should thus be accepted unmodified. In most cases their bias effect is minimized by the fact that such high grade patches are sampled over a shorter than normal length and thus one of the weighting factors is lower than usual. If no cut is to be applied to assays, there are two ways of overcoming the positive bias that may result. First, by taking larger samples so that an 'averaging out' effect takes place, or secondly, by taking samples of equal length and then log-transforming all the assays before assigning the geometric mean value (see section 3.5) as the intersection grade.

Weighting techniques

Sampling and assaying of drill core through a mineralized intersection will be based on the appearance and disappearance of visible quantities of the valuable mineral (section 2.9.3). Additional samples may also be taken beyond these limits, in apparently barren ground, in case unsuspected values are present and in case this material is needed as intentional dilution. Within the visibly mineralized zone, the first and last samples above COG will be located and the first attempt at the definition of a potentially mineable (ore) zone made (Figure 1.5(b)). These two samples, and the intervening samples, must now be used to determine the overall weighted grade and thickness of this zone to see if it meets the necessary criteria for a potentially economic intersection. This weighted grade must exceed the MMG (section 3.3.1, 'Cut-off grade and minimum mining grade') specific to the mine in question, and the true thickness ($IT \times \sin\theta$) must be equal to, or greater than, the TT MSW (section 3.3.1, 'Minimum stoping width'). If the grade is less than the MMG then attempts must be made to group smaller numbers of assay samples in such a way as to produce an overall grade which exceeds the MMG while maintaining a thickness in excess of the MSW. On the other hand, if the initial grade is above the MMG but the thickness is less than the MSW, then low grade or barren material must be added from the immediate hangingwall or footwall (whichever has the highest grade) to make up for the deficiency. Recalculation of the overall grade will show whether it still exceeds the MMG. If not, then the intersection is not viable in its own right. It may, however, be carried by adjacent intersections if these are of sufficiently high grade and thickness.

Each sample grade in an intersection of a deposit can be weighted in a variety of ways. The first is simply by length-weighting, in which the sum of the products of intersected length and grade are divided by the sum of the intersected thickness. This method can be expressed mathematically as follows:

$$G = \Sigma_{i=1}^{n} (G_i \times L_i) / \Sigma_{i=1}^{n} (L_i) \qquad (1)$$

Where G indicates weighted grade, n is the number of samples combined and G_i and L_i are the grades and lengths of each sample.

If the specific gravity of individual intervals changes significantly in response to changes in rock type, mineralogy or contained metal, then equation (1) should be modified as follows:

$$G = \Sigma_{i=1}^{n} (G_i \times L_i \times SG_i) / \Sigma_{i=1}^{n} (L_i \times SG_i) \qquad (2)$$

Although this would appear to require the measurement of specific gravity on each sample, this can, in fact, be estimated from knowledge of host rock specific gravity and the information provided by the mineralogical and assay logs. The calculation of specific gravity by this method is discussed in detail in section 3.4.3. It should be noted that specific gravity weighting is only applicable where the values to be weighted are reported on a weight % basis. Those that are

Table 3.4 Assay and mineralogical information from a hypothetical drill-hole

			Volume (%)			
Sample	Length	Cu grade	Cc	Bn	Cp	Calculated *SG*
A	1.30	1.58	0	20	80	2.571
B	1.55	2.39	5	30	65	2.587
C	1.50	1.79	0	40	60	2.572
D	1.40	8.65	20	20	60	2.737
E	1.60	15.19	50	30	20	3.853
F	1.50	3.51	10	90	0	2.599

expressed on a volume % basis, such as porosity, must not be weighted in this way.

The impact of SG weighting can be illustrated by using the program SGORE, listed in Appendix 3.6, which calculates the SG, for each of the samples from a drill-hole intersection listed in Table 3.4.

Note: Host rock is shale with $SG = 2.53$

Weighted grade using equation (1):

$$G = 50.12/8.85 = 5.66\%$$

Weighted grade using equation (2):

$$G = 137.938/23.505 = 5.87\%$$

It is evident that, in this case, if we neglect to use specific gravities, an underestimation of the weighted grade results. We are not dealing with abnormal assays here and there is thus no logical basis for cutting the high grade intervals using one of the methods described earlier.

If individual samples have poor core recoveries then there is an uncertainty as to the true grade. Although most mines measure recovery, very little use is made of this information. In order to produce a more conservative estimate of weighted grade, it may thus be worth considering the application of the following modification of equation (2):

$$G = \Sigma_{i=1}^{n} (G_i \times \text{recovered length}_i \times SG_i)$$

$$/ \ \Sigma_{i=1}^{n} (\text{represented length}_i \times SG_i) \qquad (3)$$

One criticism of the use of equation (3) is that it assumes that the non-recovered core has zero grade which is unlikely but at least there is less chance of overestimation of grade. This is important as grade is the most sensitive parameter when considering the economic viability of a deposit.

Baker and Binns (1987) used a similar approach at the Golden Plateau orebody at Cracow, Queensland where gold grades were reduced, on the basis of core recovery, to a nominal 95% recovery. For example, a 1 m interval, with 50% recovery and an assay value of 1 g Au/t, would be assigned a grade of 0.55 g/t. Application of equation (3) in this case, but ignoring SG, would have given a grade as follows:

$$G = (1_{g/t} \times 0.5\,\text{m})/(1_m \times 0.95) = 0.53\,\text{g/t}$$

Assay weighting – worked example

Some of the concepts described earlier can be illustrated in the following worked example. Table 3.5 represents the assay information for a drill-hole intersection of a copper deposit. In this table, the assays have been subdivided, on the basis of a strict 1% COG, into intervals A, B and C. For a potentially economic intersection the following requirements must be met:

(1) MMG = 2% Cu;
(2) MSW (TT) = 2.5 m.

Applying a strict 1% COG for interval A:

Table 3.5 Sampling information from a drill-hole

Interval	Waste	A	B	C	Waste
% Copper	0.1 0.3	2.1 3.6 1.3	0.7 0.5	1.8 2.3 3.6 4.1 3.2 1.1	0.5
Intersected thickness	0.5 0.5	0.5 0.5 0.5	0.5 0.3	0.5 0.5 0.2 0.2 0.3 0.3	0.5
Average intersection angle		←	— 68° —	→	

Metal accumulation = $\Sigma(G \times L) = 3.5$ m %
Intersected thickness (IT) = $\Sigma L = 1.5$ m
Therefore $G_A = 2.33\%$ Cu $TT = 1.5 \times \sin 68° = 1.39$ m

$G_A < MMG$ but $TT < MSW$

Applying a strict 1% COG for *interval C*:

Metal accumulation = $\Sigma(G \times L) = 4.88$ m %
$IT = \Sigma L = 2.0$ m
Therefore $G_C = 2.44\%$ Cu and $TT = 2.0 \times 0.9272 = 1.85$ m

$G_C > MMG$ but $TT < MSW$

Hence, neither interval on its own constitutes a potential ore-zone and a more flexible approach is warranted.

Flexible 1% COG for intervals A + B + C:

Total metal accumulation = 3.5 + 4.88 + 0.5 = 8.88 m %
Intersected thickness = 1.5 + 2.0 + 0.8 = 4.3 m

Therefore G = 2.07% Cu and $TT = 1.39 + 1.85 + 0.74 = 3.98$ m

This now meets the requirements for an economic intersection. Note also that the metal accumulation indicates a marked increase in the amount of metal available for recovery.

Note: At Nchanga in Zambia, a slightly different approach is used for the calculation of potentially economic intersections in the Lower Banded Shale (Lower Orebody). This results from the fact that higher grades tend to exist close to the base of the Ore-shale and in the underlying footwall arenites. In this case, a COG (1.5%) is only applied to the footwall contact and samples are progressively added, working upwards (stratigraphically) until an assay hangingwall is reached which must lie at least 6 m vertically above the footwall, the requirement for block caving at this mine. If the overall grade is too heavily diluted relative to MMG by the time this is attained, then the intersection is classed as subeconomic.

Frequency weighting

This method was originally developed by Watermeyer (1919) for the evaluation of the reserves of Witwatersrand gold ores. It requires the production of a frequency histogram or curve from a large assay data base which is assumed to be representative of the deposit from which the intersection has been made. For each assay value (G_i) obtained during the sampling, the corresponding frequency of occurrence (F_i) is read off and used to weight the assay as follows:

$$G = \Sigma_{i=1}^{n} (G_i \times L_i \times F_i) / \Sigma_{i=1}^{n} (L_i \times F_i) \quad (4)$$

Very high assay values, which only occur infrequently, are thus assigned a very low frequency weighting factor and thus their tendency to bias the overall grade is reduced. For this reason this

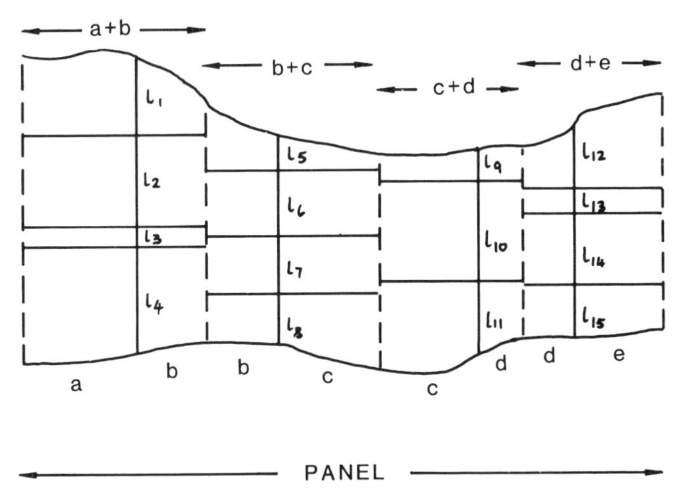

Fig. 3.5 Face sampling and zones of influence.

technique is applicable to the situation discussed in section 3.3.2 'Abnormal assays' above.

Weighting by zone of influence (face sampling)

Figure 3.5 shows a face which has been sampled by vertical channels at irregular intervals and by samples of variable length.

The weighted grade assigned to the panel is thus calculated by multiplying the grade of each sample by its area of influence, which is based on the sum of half the distances to the adjacent channels (the ZOI) times its sample length, i.e.

$$\begin{aligned} G_p &= \Sigma(L_i \times ZOI_i \times G_i)/\Sigma(L_i \times ZOI_i) \\ &= [(L_1 \cdot G_1 + L_2 \cdot G_2 + L_3 \cdot G_3 + L_4 \cdot G_4)(a + b) \\ &\quad + (L_5 \cdot G_5 + \ldots + L_8 \cdot G_8)(b + C) \\ &\quad + (L_9 \cdot G_9 + \ldots + L_{11} \cdot G_{11})(c + d) \\ &\quad + (L_{12} \cdot G_{12} + \ldots + L_{15} \cdot G_{15})(d + e)] \\ &\quad /[(L_1 + L_2 \ldots + L_4)(a + b) \\ &\quad + (L_5 + \ldots + L_8)(b + c) \\ &\quad + (L_9 + \ldots + L_{11})(c + d) \\ &\quad + (L_{12} + \ldots + L_{15})(d + e)] \end{aligned}$$

Weighting by zone of influence (block grades)

If a block of ore has been sampled around its margins (in levels and raises) by irregularly spaced channel samples, then each channel grade (G_i), determined by length weighting methods, must be weighted by its length (L_1) and its zone of influence (ZOI_i) and SG_i, if required. The zone of influence in this case is the sum of half the distances to the two adjacent channels. We can thus express this mathematically as follows:

$$G_{BLOCK} = \Sigma_{i=1}^{n} (G_i \times L_i \times ZOI_i \times SG_i) / \Sigma_{i=1}^{n} (L_i \times ZOI_i \times SG_i)$$

Although not strictly relevant here, the thickness estimate applied to this block can be produced in a similar way by weighting by zone of influence. We thus have:

$$Th_{BLOCK} = \Sigma_{i=1}^{n} (L_i \times ZOI_i)/\Sigma_{i=1}^{n} (ZOI_i)$$

where the denominator is the total length of the block sides.

Weighting by zone of influence (abnormal assays)

Figure 3.6 shows three adjacent channels, 10 cm wide, which have been sampled over varying lengths. The individual assay grades and lengths

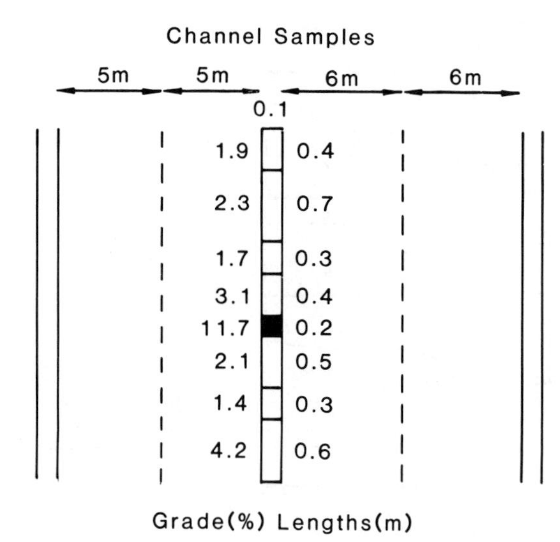

Fig. 3.6 A channel containing an abnormally high grade (11.7%). Not to scale.

for the central channel, which contains an anomalously high grade (11.7%), are listed on the diagram. The spacing between the channels is also variable and hence the zone of influence (ZOI) of the channel is arbitrarily assumed, as above, to be the sum of half the distances to the adjacent channels (in this case 5 + 6 + 0.1 = 11.1 m). We can thus weight all the sample grades by ZOI and length, with the exception of the abnormal assay, which is weighted by the channel width and length only. We are thus assuming that the high grade is very localized. The formula for grade thus becomes:

$$G = [\ \Sigma_{i=1}^{n-1}\ (G_i \times L_i \times ZOI) \qquad (5)$$

$$+ \ (G_{ABN} \times L_{ABN} \times W)]/\ \Sigma_{i=1}^{n}\ (L_i \times ZOI)$$

where W = the channel width and G_{ABN} and L_{ABN} are the grade and length respectively of the abnormal assay sample.

The use of the full ZOI for all samples in the denominator effectively dilutes the high grade value over the rest of the ZOI which is assumed to have zero grade. We can now compare the effect of accepting an uncut abnormal grade with the weighted grade calculated using equation (5).

Using equation (1):

$$G = 10.45/3.4 = 3.07\%$$

Using equation (5):

$$G = [(8.11 \times 11.1) + (2.34 \times 0.1)]/37.74$$
$$= 90.255/37.74 = 2.39\%$$

As can be seen, the difference is considerable and a serious overestimation of grade could result if such high grades are a common phenomenon. It is always best to err on the conservative side.

Evaluation of polymetallic intersections

Where two or more metals occur in an intersection, and all contribute to the viability of the mineralization, it is necessary to combine them in some way. In the case of lead and zinc, this is straightforward as the two grades are usually added together. However, at the Tynagh Mine in Ireland (now closed) a different approach was used in that dollar values were assigned to the various metal grades. Immediately prior to the mine closure in 1980, a total value of 12$ per ton was deemed to be the critical level and the drill-hole or stope) value was calculated using the following:

each 1% Pb	≡	3$
each 1 oz Ag/t	≡	1.5$
each 1% Zn	≡	2.5$
each 1% Cu	≡	−0.5$

Copper was assigned a minus value as, in this case, it is a penalty element as it is hosted by chalcopyrite containing inclusions of tennantite. Smelter penalties were applicable for any As, Sb and Hg retained in the concentrate and hence a high copper grade could imply subeconomic mineralization. An intersection grading 2.1% Pb, 2.3% Zn, 0.9% Cu and 1.0 oz Ag/ton could thus be valued at:

$$(2.1 \times 3) + (2.3 \times 2.5) - (0.9 \times 0.5)$$
$$+ (1 \times 1.5) = 13.1\$$$

This would thus be economic.

The critical $ level was, in fact, variable depending on the metallurgical classification of the ore (lead, lead + zinc, zinc, etc.) and on whether it contained a high MgO level (reflecting dolomitization) which affected the extraction of zinc from the concentrate and necessitated an initial acid leach.

Intersections in a polymetallic deposit can be evaluated on the basis of a value per tonne calculated from the net smelter return (NSR). NSR represents the total value of metals recovered from each tonne of ore minus the cost of smelting where applicable. Mining and milling costs are not considered at this stage as these apply irrespective of the grades of the various metals. For example, if we consider a polymetallic volcanogenic sulphide deposit containing Cu, Zn, Ag and Au which is to be mined by open-pit methods (bench dilution = 0%), we need to know the NSR for each 1% Cu, 1% Zn and 1 g Ag or Au in each tonne of the ore. The information needed for the NSR calculations is listed below:

Current metal prices (August 1989)

Cu .... 2880 $/t
Zn .... 1807 $/t
Ag .... 6.0 $/troy oz
Au .... 363.6 $/troy oz

Estimated smelting costs

Copper concentrate containing 20% Cu .... 185$
Zinc concentrate containing 52% Zn .... 350$

Concentrator recoveries

Copper 60.0%
Zinc 80.0%
Silver 14.0%
Gold 55.0%

Note: The above smelting costs are estimates only for actual values are difficult to obtain, being subject to contractual confidentiality. Costs are variable depending on the nature and amount of concentrate supplied to the smelter. The recoveries are typical values for such a deposit.

Calculations

One tonne of ore containing 1% Cu will yield 0.6/100 t of Cu metal worth 0.006 × 2880$, i.e. 17.28$. This will be contained in 0.6/20 t of concentrate costing (0.6/20) × 185$ to smelt. Hence the NSR for the copper = 17.28 − 5.55 = 11.73$. Similar calculations can be made for Zn, Ag and Au (in the latter two metals no smelter costs are incurred as recovery takes place in the concentrator). The results are tabulated below, together with the combined NSR:

Copper 1%	Zinc 1%	Silver 1 g/t	Gold 1 g/t	Combined NSR
$11.73	$9.08	$0.03	$6.43	$27.27

Thus, an intersection grading 0.8% Cu, 3.1% Zn, 30 g/t Ag and 2 g/t Au would have an NSR value of:

$$0.8 \times 11.73 + 3.1 \times 9.08 + 30 \times 0.03 + 2 \times 6.43 = \$51.29$$

This would then be compared with a cut-off grade expressed in terms of a fixed NSR value or a range of different values. The above exercise is best performed through a spreadsheet, such as Lotus 1-2-3, so that rapid recalculation can be achieved if any of the input variables are changed. Each sample and eventually each ore zone intersection can then be assigned a code representing the NSR range in which its value falls. Table 3.6 shows an example of such a calculation for an ore-zone intersected in a drill-hole. Lafleur (1986) presents an example of the application of this method at the Mobrun deposit in Quebec.

An alternative approach involves the determination of metal equivalent values in which the grade of one metal is expressed in terms of another, after allowance has been made for the difference in metal prices and recovery factors. In some porphyry copper-molybdenum deposits a

Table 3.6 Portion of a Lotus worksheet showing the calculation of net smelter return (NSR) for a polymetallic intersection

User defined variables: Date 13 Sept. 1989

Prices

Cu	1810.8 £/tonne	Zn	1807 $/tonne
Ag	6 $/troy oz.	Au	363.6 $/troy oz.

Current exchange rate (dollars to sterling = 1.59)

Concentrator recoveries (%):

Cu	60	Zn	80
Ag	14	Au	55

Smelting costs ($/tonne concentrate):

Cu	185	conc. grade (%)	20
Zn	350	conc. grade (%)	52

NSR cut-off value 30$ av. int. angle = 67

Drill-hole database

	From	To	Int. thick	Cu	Zn	Ag	Au	NSR
	73.20	75.10	1.90	0.30	1.10	2.10	0.70	18.05
O/Z	75.10	76.80	1.70	0.85	3.10	30.00	2.00	51.76
O/Z	76.80	79.00	2.20	0.65	2.43	18.20	4.20	57.16
O/Z	79.00	80.10	1.10	1.31	0.99	24.30	6.11	64.28
O/Z	80.00	81.20	1.10	0.66	2.77	44.57	6.34	74.83
O/Z	81.20	82.00	0.80	0.17	0.82	7.80	1.00	16.07
O/Z	82.00	84.00	2.00	0.13	6.91	39.77	4.46	93.96
	84.00	85.10	1.10	0.07	1.30	4.24	0.05	13.05

Weighted NSR of ore zone = 63.77
True thickness of ore zone = 8.19

copper equivalent grade is calculated as follows:

$$Cu_{eq} = (\% \text{ total Cu} - \% \text{ acid soluble Cu}) + F \times \% \text{ Mo}$$

where F is a variable factor dependent on current prices and processing costs and is calculated as below:

$$F = \{(NSR_{Mo}) \times \text{molybdenum recovery}\} / \{(NSR_{Cu}) \times \text{concentrator copper recovery} \times \text{smelter recovery for copper}$$

where NSR is the net smelter return, i.e. selling

price minus the smelting cost for copper (e.g. 30 c/lb) or the roasting cost for molybdenum (e.g. 50 c/lb). Thus, if the current copper and molybdenum prices are 134 c/lb and 677 c/lb (based on $3.25/lb MnO as a 56% concentrate) respectively, the molybdenum recovery is 70%, the concentrator/mill and smelter recoveries for copper are 87% and 96% respectively then:

$$F = (677 - 50) \times 0.70/(134 - 30) \times 0.87 \times 0.96 = 5.05$$

Thus an intersection grading 0.25% Cu and 0.08% Mo would have a Cu_{eq} value of 0.65%. The copper value on its own would be very marginal but the molybdenum has upgraded the intersection to an acceptable level. An example of the calculation of gold equivalent grades at the Montana Tunnels operation, Montana, is given in section 7.2.2, Case history II. In this case, the deposit contains Zn, Pb and Ag as well as gold. Diehl and David (1982) quote an example of a massive Cu–Ni deposit in the Churchill Province of N. Quebec in which copper equivalent values are calculated by multiplying the nickel grade by 3.2316 and adding the result to the copper grade.

In the Sudbury mining area in Canada, Falconbridge have used nickel units (NU) to determine whether an intersection meets the criteria set for an economic intersection. The set break-point changes with metal price and mining method but could be of the order of 120 NUs. Each 0.01% Ni is made equivalent to 1 NU. If the copper grade is > 0.9% then each 0.02% Cu is made equivalent to 1 NU. Thus, if the weighted grades of an intersection were 1.1% Cu and 1.2% Ni, then this represents 55 + 120 = 175 NUs. This would thus be considered payable.

Evaluation of a drill-hole intersection for a potential open-pit

In a deposit for which open-pit feasibility studies are being undertaken, an approximate assessment of the economic viability of each drill-hole intersection can be made using the following method, which assumes vertical drilling. In the case of the borehole illustrated in Figure 3.7, the ore value can be represented by:

$$T \times SG_{ORE} \times G \times MR \times P$$

Where MR = metallurgical recovery, P = metal price, T = thickness and G = weighted grade of the mineralization at a specific COG. This must at least balance the cost of mining (TC) which is the combined cost of overburden stripping, of mining waste and of mining ore, i.e.

$$TC = (S \times SG_{OB} \times SC) + (D \times SG_W \times WC) + (T \times SG_{ORE} \times OC)$$

Where SG_{OB}, SG_W and SG_{ORE} are the specific gravities/tonnage factors for overburden, waste and ore and S, D and T are the respective thicknesses. The cost of sampling, mining and processing each tonne of ore = OC, the cost of overburden striping = SC per tonne and the cost of sampling and mining each tonne of waste = WC.

Thus at the breakeven point:

$$T \times SG_{ORE} \times G \times MR \times P = TC$$

$$\text{or } T \times G \text{ (i.e. metal accumulation)} = TC/(SG_{ORE} \times MR \times P)$$

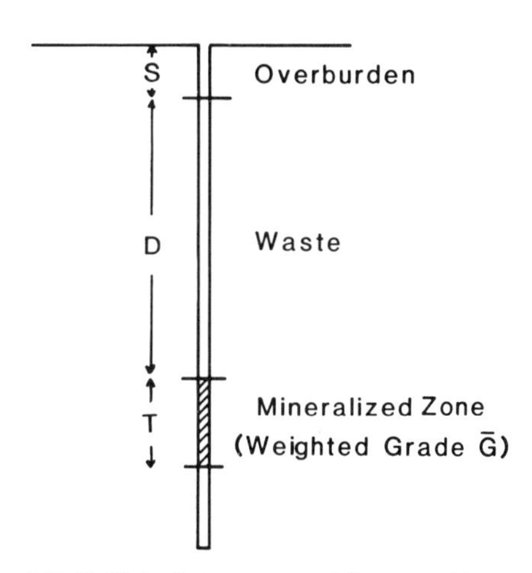

Fig. 3.7 Drill-hole in a potential open-pit operation.

Table 3.7 Assay summary sheet

Drillhole DH635D1 Checked by .... Date ....

From	To	Rep.	Rec.	Av. int. angle	COG-CU	TT	SulCu	AsCu	SulCo	AsCo	TTm%Cu	TTm%Co
371.5	380.7	9.2	8.95	67	0.5	8.47	1.95	0.38	0.18	0.05	16.52	1.61
371.5*	378.6	7.1	6.9	67	1.0	6.35	2.38	0.35	0.17	0.05	15.11	1.08
373.0	377.5	4.5	4.4	66	1.5	4.02	3.17	0.17	0.17	0.05	12.74	0.68
403.7	408.1	4.4	3.75	73	1.0	4.21	3.62	Tr	0.05	Tr	15.24	0.21

*Mid-point coordinates: X = 32 175.86, Y = 16 371.42, Z = 823.17.

The actual metal accumulation value, for the drill-hole under consideration, must be greater than this calculated value to allow for profit margins and the pit slope (batter), for vertical prisms of rock have been assumed. Also, no allowance is made for the fact that the effect of poor holes can be offset by adjacent rich holes. However, the method does allow an initial delineation of orebody fringes from which a more detailed study can be made, as described in Chapter 5.

Assay summary sheets

It is standard practice to calculate several different intervals in a borehole, perhaps because there are distinct ore zones or because different COGs have been applied. The results are thus tabulated in an assay summary sheet so that, at a later date, a selection can be made on the basis of current operating costs/methods. Table 3.7 shows a typical layout of such a sheet. In this case, allowance is made for the presence of 'acid-soluble' copper and cobalt.

3.4 MINE/DEPOSIT RESERVES

3.4.1 Data preparation and representation

The first step in the process usually involves the decision as to how to present the weighted assay data in a form suitable for the computation of the overall or global reserve. To some extent this depends on the method by which the ore-deposit is to be mined. In the case of open-pit methods, it is obviously best if the reserves are calculated on a bench by bench basis. Where the orebody is highly irregular, or folded in cross-section, it is often better to undertake the calculation on mine structure sections drawn at regular intervals along the strike of the deposit. Each will show the traces of any drill-holes drilled on, or close to, the section line. Alternatively, level and sub-level plans could be used on which all assay data are plotted and projected, if necessary. Where the deposit is tabular in form, the choice is between a vertical longitudinal projection (VLP) and a surface plan, depending on whether the deposit is steeply dipping (VLP) or relatively flat (plan). A final possibility is the use of an unrolled plan (palinspastic map) which is particularly useful where an orebody is heavily folded and faulted. This allows it to be re-assembled in its original shape prior to deformation. It allows the global reserves to be calculated more accurately, although it is less useful to the mining engineer who is more interested in the grade and tonnages of individual blocks of ground defined by mine levels and mining block boundaries. The unrolled plan can, of course, show the intersection lines (curves) of all mine levels on the median plane of the deposit as well as mining block boundaries and the location of faults and fold axes whose effects were removed in the production of the plan. Individual mining blocks will thus appear as distorted shapes which mining engineers find difficult to accept and visualize.

Once the decision is made as to how to present the information, this naturally leads to the selection of the necessary thickness component to be used as explained in section 3.3.1. The only case where intersected lengths of individual samples or overall intersections are used is when the reserves are to be calculated using sectional areas. The thickness component, the equivalent metal accumulation and grade are then plotted at the point representing the drill-hole intersection of the mid-point of the deposit or the mid-point of the bench in the case of an open-pit. Where the object is to produce a global reserve alone, without any constraints applied by the mining method(s) to be employed, a wide range of computational techniques are available which will be explained later in section 3.5. However, where the reserves must be related to mineable blocks of ground, it is necessary to produce local reserves of regularly shaped blocks whose limits are related to mine levels and block boundaries. An exact number of ore-evaluation blocks must fit into a mining block. Where mining is to be based on pit benches, or lifts in a 'cut and fill' stope, then individual blocks will be evaluated in each of a series of stacked horizontal slices whose thicknesses relate to bench heights (e.g. 10–15 m) or stope lifts (e.g. 4 m). It becomes necessary, therefore, to superimpose a matrix of ore-blocks over the VLP or plan prior to proceeding with the reserve calculation. The smaller the blocks the better the fit that will be eventually gained to the economic fringes of the deposit, however, this will result in longer computational times as each block has to be evaluated separately. Also, if the drill-hole spacing is wide then the reliability of the grade estimates of small blocks becomes suspect. An initial evaluation should really be undertaken using blocks whose X and Y coordinate dimensions are roughly equivalent to the average X and Y direction spacing of borehole intersection points. At the same time, allowance must be made for the size of mining blocks to be used in the mine, as explained earlier.

3.4.2 Production of data files

The calculation of the reserves of a mineral deposit, whether by manual or computer methods, requires the production of data-files in which all available information is brought together in a form suitable for the proposed method. Basically, there are two main types of file to be considered, (a) the survey file and (b) the assay file. The former may contain all, or most of, the following elements: drill-hole no., hole-type, X-Y-Z coordinates of collar and end-of-hole depth. The 'hole-type' is usually a code which indicates the type of drilling, e.g. RC – reverse circulation, or size, e.g. BQ, in the case of diamond drilling. A second subsidiary file will contain 'down-the-hole' directional survey information, e.g. depth of survey, inclination and bearing. The assay file may contain such details as drill-hole no., assay sample no., depth range (from, to), intersection angle, grade information and finally a geological or lithological indicator code. Derived from this file, and from the survey file, a third 'Input' file may be produced which includes the following: drill-hole no., assay sample no., X-Y-Z coordinates of sample mid-point, vertical or horizontal thickness and grade.

In order to speed up the compilation, vetting and amendment of these files, spreadsheet programs such as Lotus 1-2-3 and SuperCalc 3 are used these days. From the worksheet files produced (Figure 3.8) selected information can be extracted and saved separately as an ASCII file suitable for direct input into the ore-reserve computer program. The added benefit of this approach is that the spreadsheet can calculate the statistical parameters of the data population using built-in commands (@ VAR etc.) and also print out the histogram. Similar commands also exist for producing log-transformed data for specified ranges of columns and rows in the spreadsheet, a useful facility where the histogram shows evidence of a strong positive skew (section 3.5).

A3: ^DDH NO. [MENU]
Worksheet Range Copy Move File [Print] Graph Data System Quit
Output a range to the printer or a print file

	A	B	C	D	E	F	G	H
1	KALULUSHI EAST DRILLHOLE DATA FILE							
2								
3	DDH NO.	X-COORD	Y-COORD	V.DEPTH	T.T	TOT.Cu	AS.Cu	TOT.Co
4								
5	KLE 1	724.00	245.50	95.00	7.60	5.63	0.15	0.02
6	KLE 2	736.00	413.00	122.00	15.00	4.05	0.01	0.06
7	KLE 3	621.00	127.50	142.00	0.00	0.00	0.00	0.00
8	KLE 4	622.50	309.00	188.00	5.90	1.41	0.01	0.02
9	KLE 5	736.50	587.50	120.00	0.00	0.00	0.00	0.00
10	KLE 6	621.00	479.00	177.00	0.30	3.51	0.06	0.01
11	KLE 7	723.50	68.00	39.00	5.00	1.39	0.78	0.04
12	KLE 8	492.00	358.00	322.00	8.60	5.43	0.01	0.01
13	KLE 9	503.50	550.00	290.00	3.30	1.52	0.01	0.04
14	KLE 10	519.00	198.00	262.00	0.00	0.00	0.00	0.00
15	KLE 11	703.00	344.00	142.00	2.80	4.18	0.09	0.03
16	KLE 12	774.50	307.00	81.00	9.40	2.27	0.02	0.07
17	KLE 13	792.50	364.00	62.00	1.10	3.71	0.02	0.03
18	KLE 13D1	771.00	391.00	83.10	7.50	1.36	0.01	0.21
19	KLE 14	388.50	440.00	415.00	4.30	4.82	0.01	0.02
20	KLE 15	789.00	469.50	77.20	13.00	4.53	1.81	0.08

Fig. 3.8 Lotus 1-2-3 worksheet file showing assay information.

Some of the more sophisticated commercial packages for the evaluation of mineral deposits, such as SURPAC, DATAMINE and LYNX (the updated version of GEOMIN), have a built-in file creation and editing facility.

3.4.3 Components of a reserve calculation

Tonnage and tonnage factors

The calculation of tonnage requires the determination of the area of a deposit, its weighted thickness and the tonnage factor to be applied. The area can be produced in a variety of ways which include:

(1) digital planimeter (Figure 3.9);
(2) digitizer (Figure 3.10);
(3) counting squares;
(4) summing areas of ore-blocks;
(5) applying Simpson's Rule.

All of these require definition of the outer economic fringe of the deposit which can be done on the basis of half the distance between ore and non-ore intersections, or by contouring techniques based on linear interpolation between drill-holes. In the latter case, the relevant contour is that for metal accumulation (e.g. 6 m% in the case of Zambian mines). Methods 1–4 are self-explanatory. Figure 3.11 shows the method by which Simpson's Rule is applied. A baseline is drawn along the long axis of the deposit and a series of equally spaced perpendiculars erected on it so that an odd number of lines exist giving the

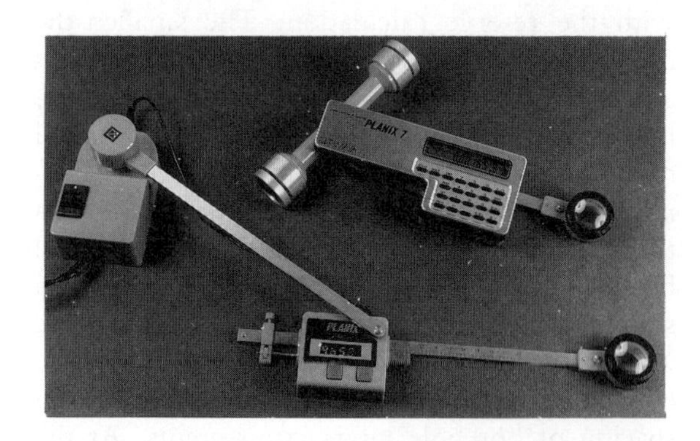

Fig. 3.9 Planix digital planimeters.

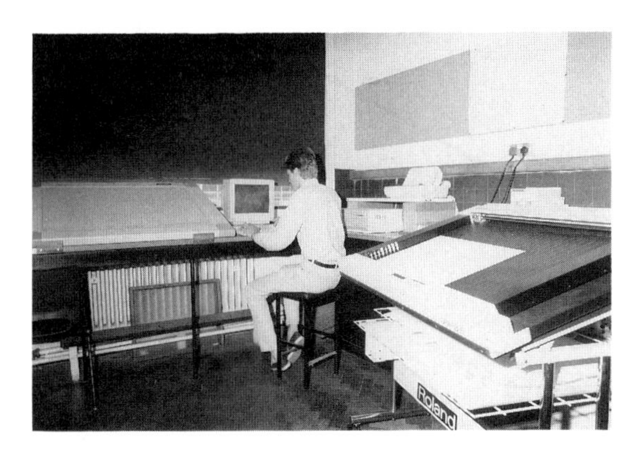

Fig. 3.10 Calcomp digitizing table and Roland A1 plotter.

width of the deposit at each point on the baseline (l_1 to l_{11}). The formula below is then applied to produce an estimate of the area:

$$\text{Area} = d/3[l_1 + l_n + 2(l_3 + l_5 \ldots l_{n-2}) + 4(l_2 + l_4 \ldots l_{n-1})]$$

In the case of Figure 3.11, $n = 11$.

Weighted thicknesses are normally produced by straight arithmetic averages of all intersections in the deposit, although sometimes these may be weighted by areas (zones of influence). Other methods will be explained under section 3.5.

Most mining operations tend to apply a constant tonnage factor (t/m^3) which has been derived from the average of a large number of specific gravity determinations, often undertaken

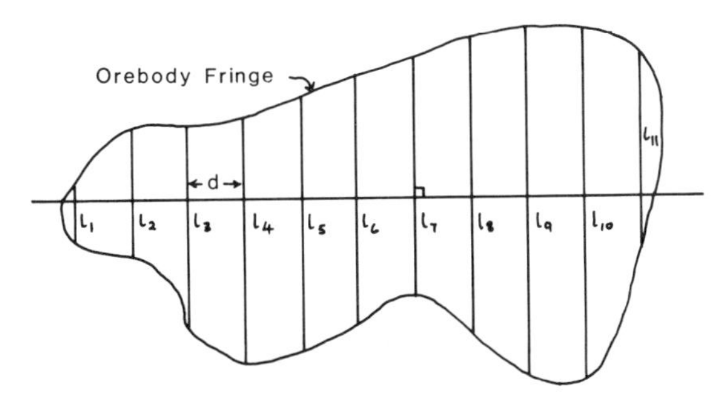

Fig. 3.11 Area of an orebody using Simpson's Rule.

at an early stage in the mine's operational life. The method employed involves the use of drill core and the displacement of mercury or sand. However, this method may lead to considerable errors in the determination of tonnages of ore and contained metal, especially if metal grades are highly variable, if the host rock lithology changes, if the degree of alteration or depth of weathering is variable, and if the mineralogy of the valuable components changes. There are three possible ways to overcome this problem, the first of which assumes no significant change in mineralogy.

1. Linear regression

This method involves the determination of *SG* from a large number of variably mineralized samples whose grade is accurately known and which contain only one valuable metal (e.g. copper). This data set contains information for one single host lithology. Similar sets can be created for other lithologies encountered in the mine. Scattergrams are then plotted of *SG* versus metal grade and a 'best-fit' line determined.

Just such an exercise has been undertaken at Nchanga Division of ZCCM Ltd, in the Zambian Copperbelt, where SGs for chalcocite ± malachite mineralization in the Lower Orebody were determined. These were grouped on the basis of host rock type and the equations ($y = mx + c$) of the best-fit lines calculated. These are listed below and illustrated in Figure 3.12.

(1) Arkose (FW)
$SG = 0.0276 \times \%T\text{·}Cu + 2.643$

(2) Transition Zone
$SG = 0.0229 \times \%T\text{·}Cu + 2.403$

(3) Lower Banded Shale
$SG = 0.0218 \times \%T\text{·}Cu + 2.528$

The first figure, on the right hand side of each equation, represents the slope of the line while the last is the Y-axis intercept and represents the specific gravity of barren rock. In each drill-hole intersection of an orebody, therefore, it is necessary to record a lithological indicator code so that

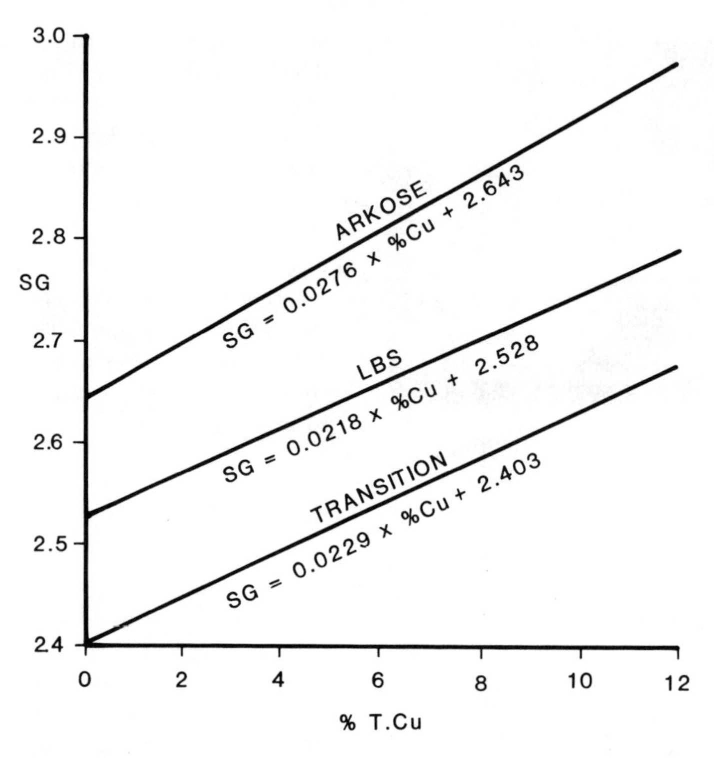

Fig. 3.12 Specific gravities of mineralized lithologies at Nchanga, Zambia.

the correct equation can be applied to each sample grade to convert it to a specific gravity. From this information, the overall weighted *SG* can be determined for the orebody intersection as well as a more accurate weighted grade (section 3.3.2 'Weighting techniques'). Each intersection in a block of ground can thus be weighted by its relevant *SG* to determine the overall tonnage and grade. It has been found that, by not using variable *SG* values at Nchanga, the grade of ore is seriously undervalued compared with that obtained by *SG* weighting. The discrepancy worsens for higher grade blocks, as might be expected.

In Western Australian gold operations which are working ore at shallow depths and within the zone of weathering, tonnage factors have been determined by undertaking a regression analysis of specific gravity against depth. Tonnage factors can then be applied on the basis of depth below the original topographic surface.

2. *Mineralogical weighting*

This method can be used for those orebodies which have more than one contained metal and where the mineralogy of each intersection, or sample, is known. It also requires that the specific gravities of barren host rocks have been determined. The formula used to compute the SG (or tonnage factor) of the sample or intersection is as follows:

$$SG = 100/[\Sigma_{i=1}^{n} W_i⅓SG_i + (100- \Sigma_{i=1}^{n} W_i)/SG_{HR}]$$

Where:

n = number of ore minerals present
SG_i = specific gravity of each ore mineral
SG_{HR} = specific gravity of the host rock
W_i = weight % of each mineral present determined from the volume % estimate by the geologist or from assay data.

Table 3.8 Specific gravities and tonnage factors

Chalcopyrite	4.2	Shale	1.6–2.9
Bornite	4.9	Sandstone	2.0–3.2
Chalcocite	5.6	Dolomite	2.7–2.8
Malachite	3.9	Andesite	2.4–2.8
Sphalerite	4.1	Basalt	2.7–3.2
Galena	7.5	Gabbro	2.9–3.1
Pyrite	5.0	Granite	2.6–2.7
Pyrrhotite	4.6	Rhyolite	2.2–2.7
Magnetite	5.2	Biotite Schist	2.6–3.0
Carrollite	4.9	Ultrabasic	3.2–3.4
Pentlandite	4.8		
Arsenopyrite	6.0	Gneiss	2.7–2.8

Typical tonnage factors for bulk materials:

Limestone	2.7–2.8
Kaolinite	2.67
Bauxite	2.81–2.86
Haematitic Iron Ores	2.5–4.1
Sand (Dry)	1.7–2.0
Gravel (Dry)	1.6–2.0
Coal	1.0–1.8
Clay	2.2–2.6

Typical SGs for ore minerals and host rocks are to be found in Table 3.8, but a more comprehensive listing can be found in Peters (1978).

A listing of a computer program SGORE, to determine the weighted specific gravity for a sulphide-bearing rock, is presented in Appendix 3.6. This has been written in Locomotive Basic II for IBM compatible PCs. If iron sulphides are present in the ore, then this program requires that sulphide-sulphur analyses are also undertaken. The program is designed to contend with an ore which contains one or more of the metals Cu, Co, Pb, Zn, Fe with a mineralogy represented by chalcopyrite, bornite, chalcocite, malachite, carrollite, galena, sphalerite, pyrite and pyrrhotite. However, it could be modified to deal with any specific situation.

3. Standard graphs – nomograms

These allow the direct reading of estimated tonnage factors based on the assay values for each sample or intersection. Figure 3.13 represents an example of one such graph from Sullivan Pb-Zn Mine in British Columbia, Canada. The lead and iron assays are combined and the value located on the right-hand scale. Similarly the zinc grade is located on the left-hand scale. A ruler is then placed between these two points and, where it crosses the SG/TF scale, the values are read off. If an intersection grades 8% Zn, 5% Pb and 30% Fe then the tonnage factors are 3.74 t/m^3 or 8.55 ft^3/ton (Figure 3.13).

Grade

The overall grade of a deposit can be determined by statistical methods (mean, median, geometric mean and Sichel's t estimators – section 3.5) or by weighting methods applied to blocks to which a grade, metal accumulation or thickness have been assigned by methods described in sections 3.9–3.11. Grade on its own can only be used where the support is constant; in other words, where the thickness, and area to which the grade is assigned, are constant. This is only the case where a bench or horizontal slice is evaluated by equal-sized ore-blocks and where bench composite grades are used. Where the thickness is variable, the weighted grade of the orebody is determined from the sum of the metal accumulations divided by the sum of the thicknesses or:

$$G_{OB} = \Sigma m\% / \Sigma m$$

Where the block sizes and thicknesses vary, then the weighted grade is the sum of the products of block area × metal accumulation divided by the sum of the block volumes, i.e.

$$G_{OB} = \Sigma_{i=1}^{n} (m_i^2 \cdot m\%_i) / \Sigma_{i=1}^{n} (m_i^3)$$
$$\equiv \text{total volume } \% / \text{total volume}$$

Where the tonnage factor varies between blocks then the above becomes:

$$G_{OB} = \Sigma_{i=1}^{n} (m_i \cdot m\%_i \cdot SG_i) / \Sigma_{i=1}^{n} (m_i^3 \cdot SG_i)$$
$$\equiv \text{overall tonnage } \% / \text{overall tonnage}$$

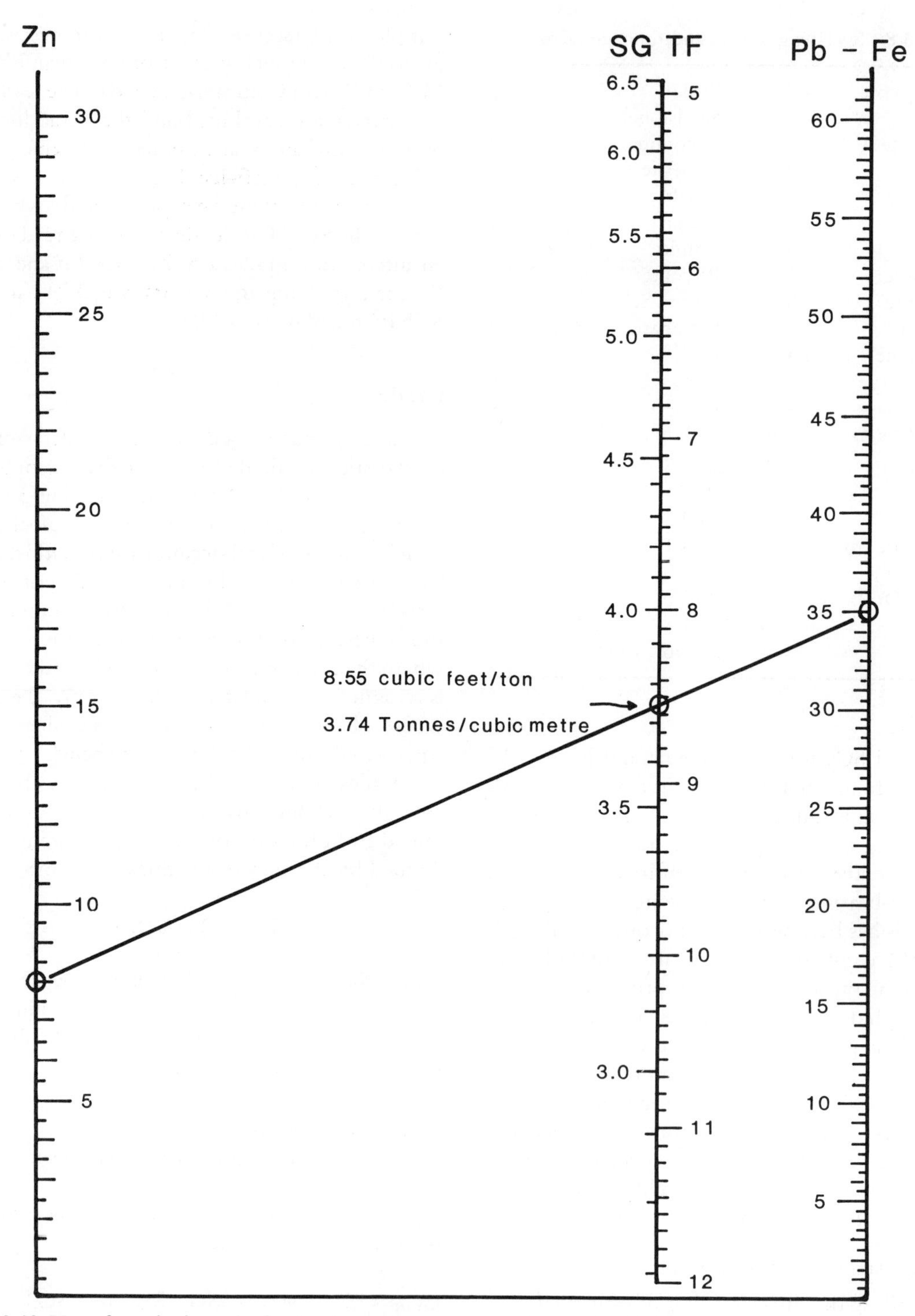

Fig. 3.13 Use of standard graphs for the determination of specific gravity (*SG*) or tonnage factor (*TF*) – Sullivan Pb-Zn mine, British Columbia.

3.5 STATISTICAL ESTIMATORS OF GRADE

Statistical estimators of the grade of a deposit are not always applicable as they require that the samples are randomly, but uniformly, distributed throughout the area being evaluated and that the values are far enough apart to be independent variables. Ideally, the sample volume or length should also be constant. They also require that the distribution of grades be gaussian or 'normal'.

Statistical methods are usually used on sample grade values, especially where a reserve evaluation is to be accomplished by cross-sectional area methods (i.e. on drill sections). Only rarely are they used on weighted intersection grades. An example of the latter, however, is the statistical evaluation of bench grade using bench composite grades.

It is not the intention here to present an exhaustive discussion of statistical methods and thus some prior knowledge is assumed in what follows. The first stage in the process is the production of histograms or frequency curves so that an overall impression of the nature of the assay distribution can be gained. The approach to normality of this population can also be assessed by producing a cumulative frequency diagram. Once the arithmetic mean and associated variance or standard deviation are calculated, then the shape of the assay distribution can also be described in terms of skewness and kurtosis.

$$\text{Skewness} = (\Sigma_{i=1}^{n} (x_i - m)^3)/\delta^3$$

where:

δ = standard deviation
m = mean
x_i = individual assay values

This measures the departure from symmetry for a population. A positive value indicates a positive skew (i.e. an excess of high values compared to a normal population) while a symmetrical distribution should have a value approaching zero.

$$\text{Kurtosis} = (\Sigma_{i=1}^{n} (x_i - m)^4)/v^2$$

where:

v = variance of values
m = mean

The degree of kurtosis represents the amount of peakedness of the distribution relative to a normal gaussian population. For a normal population, this value should be close to 3.0.

Filtering of the data may be necessary at this stage to remove 'outliers' from the main population. If the data now conform to a normal population, the arithmetic mean or the 50 percentile (median) values can be used as grade estimators.

The coefficient of variation C (standard deviation ÷ mean) if often used to describe the variability of assays in a deposit. Koch and Link (1970) suggest that, for a data population to be considered as normal, the coefficient of variation should be less than 0.5. Larger values indicate either log-normality or an erratically distributed data set. However, this value is very much a function of sample length, for the variance of assays decreases as sample length increases, due to an averaging out, or smoothing, of localized high grade patches by increased sample volume. Carras (1984) advises that data should be composited until the C value is close to 1.0. He points out that, for most Australian gold deposits, C is > 2 for a standard 1 m sampling interval. Once compositing has achieved the necessary reduction in C, then the average grade can be calculated. The presence of outliers may, however, make this exercise impossible.

Where a population is positively skewed, it is generally advisable to undertake a log-transformation of the data and then replot the histogram to see if the population is normalized by this process. If it is, then we can describe the population as being a two-parameter log-normal population (the parameters being log mean and log variance). We can test the approach to normality by using the Chi-square test or by plotting a log-probability diagram. From the latter (Figure 3.14) we can thus read off the 50

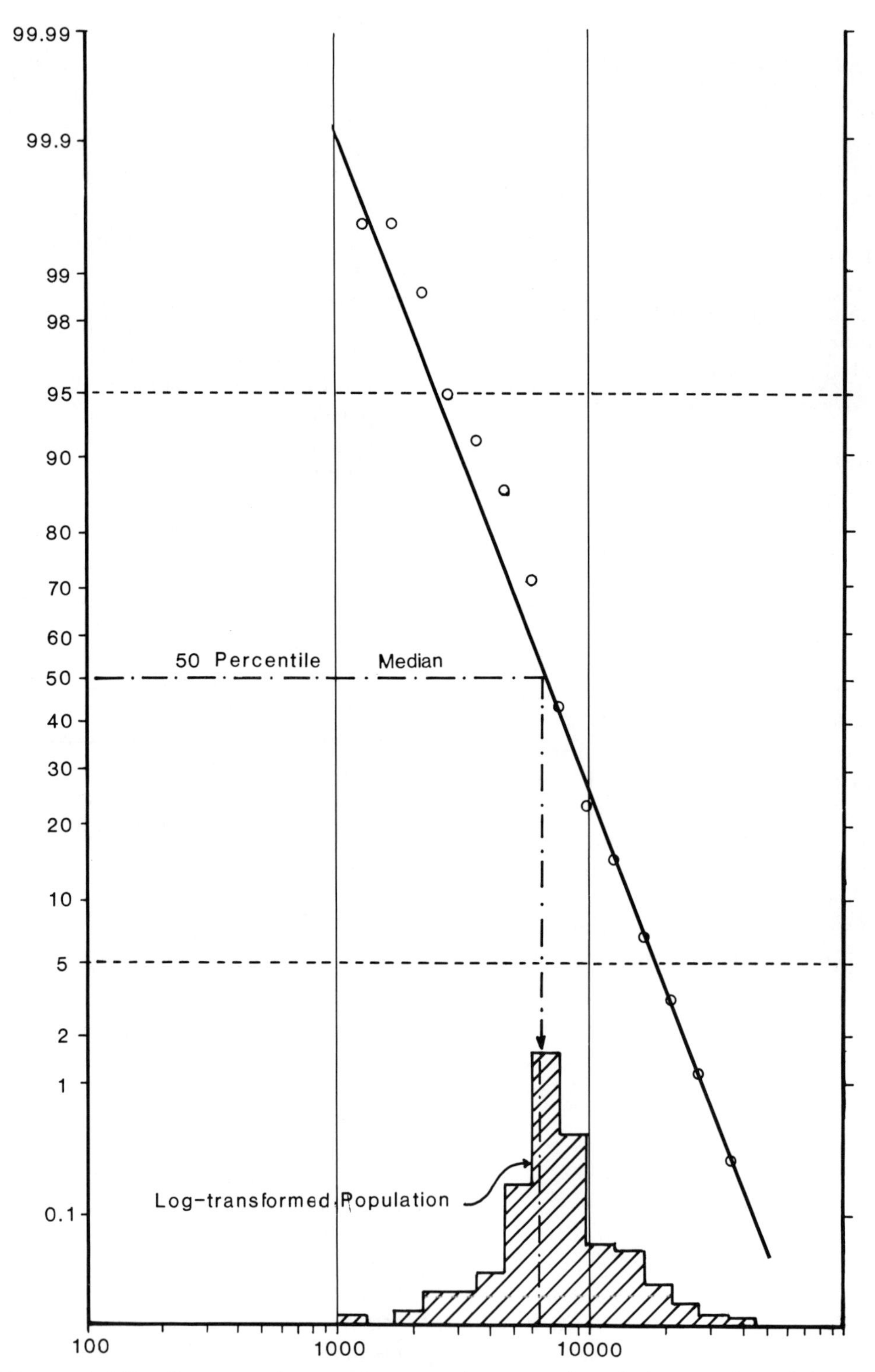

Fig. 3.14 Log-probability diagram used to test the approach to normality of a log-transformed data population (see histogram).

percentile value (median or geometric mean). More information on the log-probability diagram can be obtained from Lepeltier (1969). The true mean of a log-normal population (μ) can be calculated from:

$$\mu = e^{(\alpha + var/2)}$$

where α = mean of the logarithms of raw data ($\alpha = \ln \phi$ from the log-probability plot in Figure 3.15(a) and var = variance of the logarithms of raw data. The latter can be calculated by reading from the graph the 84 and 16 percentile values which give the range mean plus one- and mean minus one-standard deviation. Hence standard deviation = $0.5[\ln(x_{16}) - \ln(x_{84})]$.

The true mean of the log-normal population μ is a better estimator of the deposit than the median or the arithmetic mean, especially if the coefficient of variation is > 1.2.

If a gentle curve can be drawn through the log-

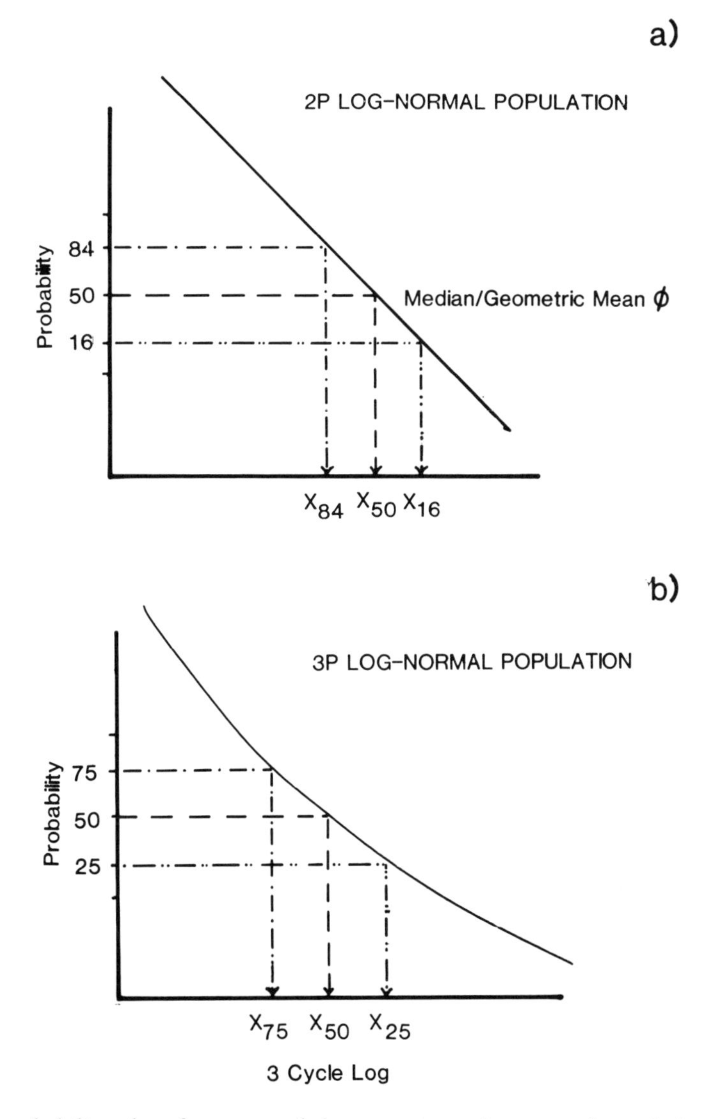

Fig. 3.15 Log-probability plots for two- and three-parameter log-normal populations (not to scale).

probability plot, rather than a straight line, then a three-parameter log-normal population is suspected (Figure 3.15(b)). The third parameter is called the additive constant α which is calculated as follows:

$$\alpha = (x_{50} - x_{75} \cdot x_{25})/(x_{25} + x_{75} - 2 \cdot x_{50})$$

where x_{25}, x_{50} and x_{75} are the 25, 50 and 75 percentile values respectively.

This value α is then added to the raw data values and the results log-transformed giving a new population ($\ln(x_i + \alpha)$) which, when plotted on the log-probability graph, gives a straight line. The procedure explained for the two-parameter population is then repeated for this population with α being subtracted from the final result. Both processes are described as normalization techniques for they create normal populations for which statistics can be calculated without risk of strong bias. Another method that has been applied to achieve normalization is by taking the square root of the positively skewed raw data and applying statistical techniques to this new data population. The mean value obtained is then squared at the end of the procedure.

Where an assay population is small, for example at the early feasibility stage of deposit evaluation, and where the raw data population has a high coefficient of variation and is log-normal, Sichel's t estimator can be used to estimate its mean. This involves grouping the assay data on the basis of natural log class intervals, as used in the Chi square test, and then calculating the variance of the natural logs from:

$$V = [\Sigma_{i=1} n_i(\ln x_i)^2/(N - 1)$$

$$- (\Sigma_{i=1} n_i \cdot \ln x_i)^2/N(N - 1)]$$

where:

n_i = number of samples in each class
$\ln x_i$ = natural log of lower class limit
N = number of samples in total population

Alternatively the variance can be calculated using:

$$V = \Sigma y_i^2/n - (y')^2$$

where $y_i = \ln x_i$, x_i are the raw data values, n is the number of values and y' is the mean of the log-transformed data.

Or, by using:

$$V = \{\Sigma(y_i^2) - (\Sigma y_i)^2/n\}/(n - 1)$$

The above can be calculated from a Lotus worksheet or determined directly, using Lotus's @VAR command.

Using n and V, we can then look up values of $f(V)_n$ from tables (Table 3.9). Interpolation may be necessary between tabulated values of V and n. Sichel's t estimator can then be calculated from:

$$t = m \times f(V)_n$$

where $m = e^{y'}$, the geometric mean of the log-transformed data.

The use of this estimator can be demonstrated using the 3170 ft bench assay data from the Ingerbelle Pit, British Columbia (Figure 3.30, section 3.10.2). A total of 181 copper values are present in this bench which, in the raw form (x_i), display a positively skewed distribution (Figure 3.16(a)). Log-transformation, using Lotus 1-2-3, and replotting of the histogram (Figure 3.16(b)), shows that a better approximation can be gained to normality. The following statistical parameters were calculated:

Mean of the log-transformed data (y')	=	−1.67048
Geometric mean (m)	=	0.188155
Variance of the log-transformed data V	=	1.33277

Using Sichel's tables (Table 3.9), and interpolating between $V = 1.3$ and $V = 1.4$, and $n = 100$ and $n = 1000$, we find $f(V)_n = 1.9404$. Hence:

$$t = m \times f(V)_n$$
$$t = 0.188155 \times 1.9404 = 0.3651\%$$

The 95% confidence limits can now be

Table 3.9 Sichel's $f(V)n$ function

	n																
V	2	3	4	5	6	7	8	9	10	12	14	16	18	20	50	100	1000
0.00	1.000	1.000	1.000	1.000	1.000	1.000	1.000	1.000	1.000	1.000	1.000	1.000	1.000	1.000	1.000	1.000	1.000
0.02	1.010	1.010	1.010	1.010	1.010	1.010	1.010	1.010	1.010	1.010	1.010	1.010	1.010	1.010	1.010	1.010	1.010
0.04	1.020	1.020	1.020	1.020	1.020	1.020	1.020	1.020	1.020	1.020	1.020	1.020	1.020	1.020	1.020	1.020	1.020
0.06	1.030	1.030	1.030	1.030	1.030	1.030	1.030	1.030	1.030	1.030	1.030	1.030	1.030	1.030	1.030	1.030	1.030
0.08	1.040	1.040	1.040	1.040	1.040	1.041	1.041	1.041	1.041	1.041	1.041	1.041	1.041	1.041	1.041	1.041	1.041
0.10	1.050	1.051	1.051	1.051	1.051	1.051	1.051	1.051	1.051	1.051	1.051	1.051	1.051	1.051	1.051	1.051	1.051
0.12	1.061	1.061	1.061	1.061	1.061	1.061	1.061	1.061	1.061	1.062	1.062	1.062	1.062	1.062	1.062	1.062	1.062
0.14	1.071	1.071	1.071	1.072	1.072	1.072	1.072	1.072	1.072	1.072	1.072	1.072	1.072	1.072	1.072	1.072	1.072
0.16	1.081	1.082	1.082	1.082	1.082	1.082	1.082	1.083	1.083	1.083	1.083	1.083	1.083	1.083	1.083	1.083	1.083
0.18	1.091	1.092	1.092	1.093	1.093	1.093	1.093	1.093	1.093	1.094	1.094	1.094	1.094	1.094	1.094	1.094	1.094
0.20	1.102	1.102	1.103	1.103	1.104	1.104	1.104	1.104	1.104	1.104	1.104	1.104	1.104	1.105	1.105	1.105	1.105
0.3	1.154	1.156	1.157	1.158	1.158	1.159	1.159	1.159	1.160	1.160	1.160	1.160	1.160	1.161	1.161	1.162	1.162
0.4	1.207	1.210	1.212	1.214	1.215	1.216	1.216	1.217	1.217	1.218	1.218	1.219	1.219	1.219	1.220	1.221	1.221
0.5	1.260	1.266	1.269	1.272	1.273	1.275	1.276	1.276	1.277	1.278	1.279	1.279	1.280	1.280	1.282	1.283	1.284
0.6	1.315	1.323	1.328	1.332	1.334	1.336	1.337	1.338	1.339	1.341	1.342	1.343	1.344	1.344	1.348	1.349	1.350
0.7	1.371	1.382	1.389	1.393	1.397	1.399	1.401	1.403	1.404	1.406	1.408	1.409	1.410	1.411	1.416	1.417	1.419
0.8	1.427	1.442	1.451	1.457	1.462	1.465	1.468	1.470	1.472	1.475	1.477	1.478	1.480	1.481	1.487	1.490	1.492
0.9	1.485	1.503	1.515	1.523	1.529	1.533	1.537	1.540	1.542	1.546	1.549	1.551	1.552	1.554	1.562	1.565	1.568
1.0	1.543	1.566	1.580	1.591	1.598	1.604	1.608	1.612	1.615	1.620	1.623	1.626	1.628	1.630	1.641	1.645	1.649
1.1	1.602	1.630	1.648	1.661	1.670	1.677	1.682	1.687	1.691	1.697	1.701	1.705	1.708	1.710	1.723	1.728	1.733
1.2	1.662	1.696	1.718	1.733	1.744	1.752	1.759	1.765	1.770	1.777	1.782	1.787	1.790	1.793	1.810	1.816	1.822
1.3	1.724	1.764	1.789	1.807	1.820	1.831	1.839	1.846	1.851	1.860	1.867	1.872	1.876	1.880	1.900	1.908	1.916
1.4	1.786	1.832	1.862	1.884	1.900	1.912	1.922	1.930	1.936	1.947	1.955	1.961	1.966	1.971	1.995	2.004	2.014
1.5	1.848	1.903	1.938	1.963	1.981	1.996	2.007	2.017	2.025	2.037	2.047	2.054	2.060	2.065	2.095	2.106	2.117
1.6	1.912	1.975	2.015	2.044	2.066	2.082	2.096	2.107	2.116	2.131	2.142	2.151	2.158	2.164	2.199	2.212	2.226
1.7	1.977	2.049	2.095	2.128	2.153	2.172	2.188	2.201	2.212	2.229	2.242	2.252	2.260	2.267	2.308	2.323	2.340
1.8	2.043	2.124	2.177	2.214	2.243	2.265	2.283	2.298	2.310	2.330	2.345	2.357	2.367	2.375	2.422	2.440	2.460
1.9	2.110	2.201	2.260	2.303	2.336	2.361	2.382	2.399	2.413	2.436	2.453	2.467	2.478	2.487	2.542	2.563	2.586
2.0	2.178	2.280	2.347	2.395	2.431	2.460	2.484	2.503	2.519	2.545	2.565	2.581	2.594	2.604	2.688	2.692	2.718
2.1	2.247	2.360	2.435	2.489	2.530	2.563	2.589	2.611	2.630	2.659	2.682	2.700	2.714	2.726	2.800	2.827	2.858
2.2	2.317	2.442	2.526	2.586	2.632	2.669	2.698	2.723	2.744	2.778	2.803	2.824	2.840	2.854	2.937	2.969	3.004
2.3	2.388	2.526	2.618	2.686	2.737	2.778	2.811	2.839	2.863	2.900	2.929	2.952	2.971	2.987	3.082	3.118	3.158
2.4	2.460	2.612	2.714	2.788	2.846	2.891	2.928	2.959	2.986	3.028	3.060	3.086	3.108	3.125	3.233	3.274	3.320
2.5	2.533	2.699	2.812	2.894	2.957	3.008	3.049	3.084	3.113	3.160	3.197	3.226	3.250	3.270	3.391	3.438	3.490
2.6	2.607	2.789	2.912	3.003	3.073	3.128	3.174	3.213	3.245	3.298	3.339	3.371	3.398	3.420	3.557	3.610	3.669
2.7	2.682	2.880	3.015	3.114	3.191	3.253	3.304	3.346	3.382	3.441	3.486	3.522	3.552	3.577	3.730	3.791	3.857
2.8	2.759	2.973	3.120	3.229	3.314	3.382	3.437	3.484	3.524	3.589	3.639	3.680	3.713	3.740	3.912	3.980	4.055
2.9	2.836	3.068	3.228	3.347	3.440	3.514	3.576	3.627	3.671	3.743	3.799	3.843	3.880	3.911	4.102	4.178	4.263
3.0	2.914	3.166	3.339	3.469	3.570	3.651	3.718	3.775	3.824	3.902	3.964	4.013	4.054	4.088	4.301	4.387	4.482

Sichel (1966).

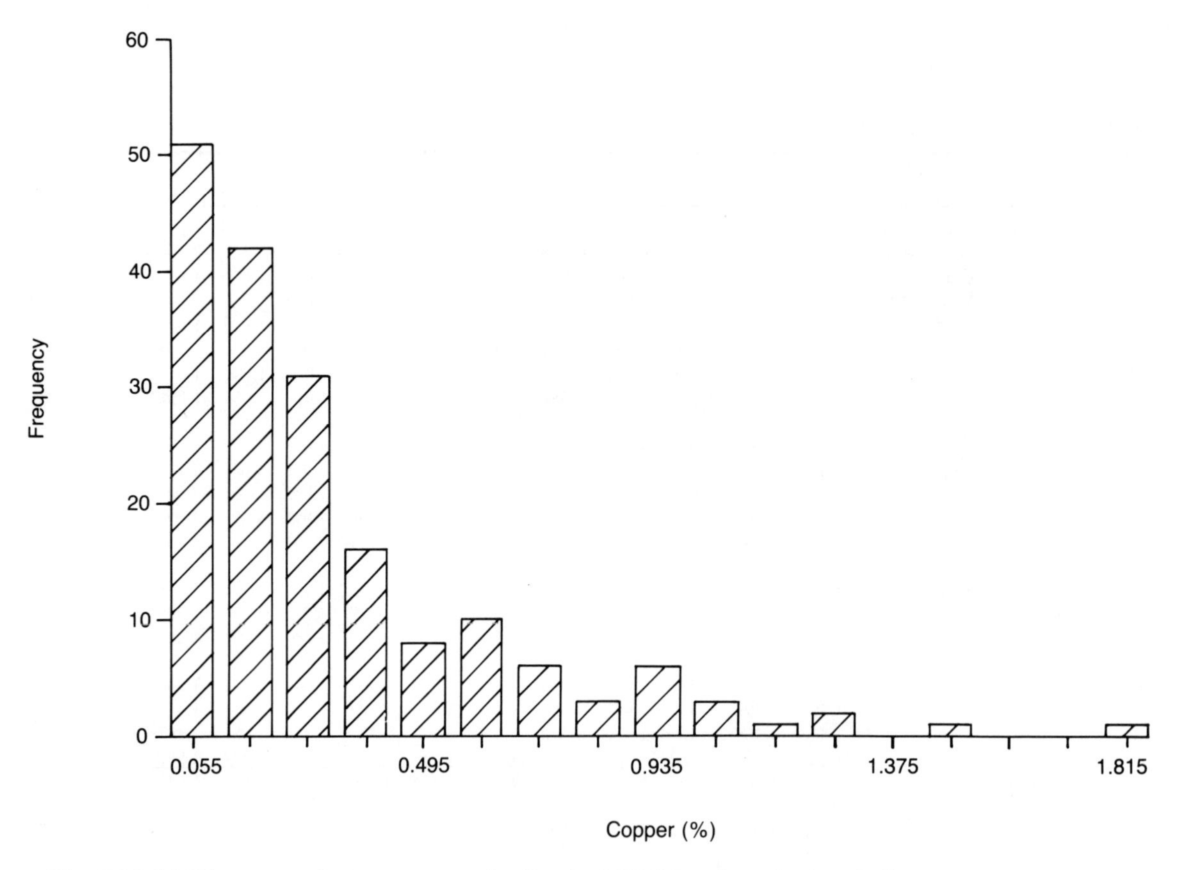

Fig. 3.16 (a) Histogram of raw copper grades for the 3170 ft bench at the Ingerbelle Pit, British Columbia.

determined using tables provided by Rendu (1981) and presented here as Tables 3.10 and 3.11. These give the values $\phi_{0.95}(V; n)$ and $\phi_{0.05}(V; n)$ which, when multiplied by t, give the upper and lower confidence limits respectively. In this case, we have:

$$\text{Upper limit} = 0.3651 \times 1.2994 = 0.474$$
$$\text{Lower limit} = 0.3651 \times 0.8124 = 0.2966$$

Thus our estimate of the statistical grade of this bench is 0.36% with a 95% probability that this estimate lies between 0.30% and 0.47%.

It should be realized, however, that if the log-transformed assay population deviates from normality then Sichel's t estimator will also be biased.

The best estimator of a deposit is the one which gives the lowest variance when the variance of the data about the estimator is calculated.

A cumulative frequency curve could be used to provide a rough estimate of the relative proportions of ore and waste in an open-pit bench where grades are highly erratic but the sampling pattern is believed to give a reliable coverage of the bench concerned. The class intervals are set so that one coincides with the COG for the pit. From the cumulative frequency curve of the grades, the percentage of values above this cut-off can then be determined. This percentage is then applied to the bench tonnage to determine the tonnage > COG in the bench. The grade of the 'ore' can then be calculated using a quasi-statistical method as follows:

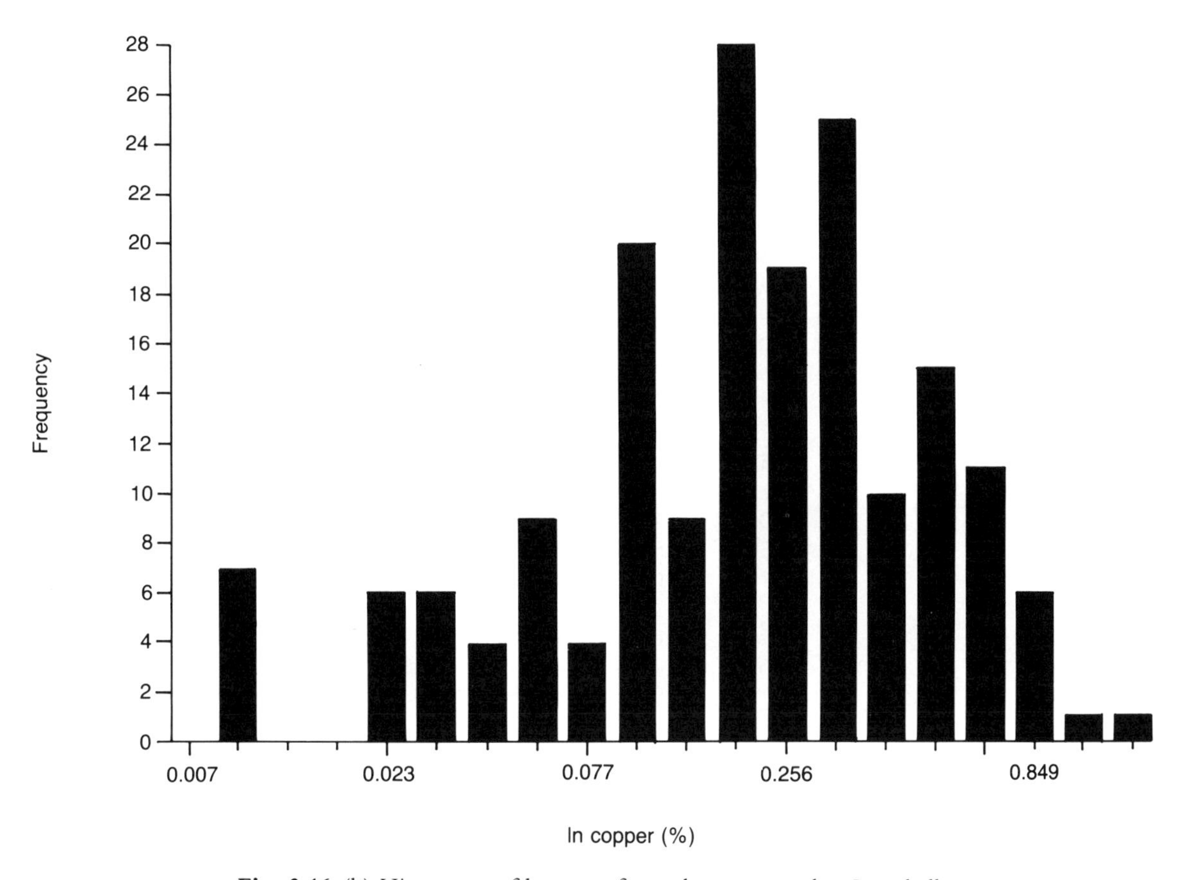

Fig. 3.16 (b) Histogram of log-transformed copper grades, Ingerbelle.

$$G_{ORE} = \Sigma G_i \times n_i / \Sigma_{ni}$$

where G_i represents the mid-point grade of each class interval **above COG** and n_i is the number of values in each class. Similarly, the tonnage and grade of the 'waste' in the bench can be determined. This method assumes that the values in each class are uniformly distributed about the mid-point of each class. It gives the global reserve of the bench only, for it is not possible to state where ore and waste areas lie. These are the reserves that could exist if selective mining were possible.

The above method is, however, flawed in that it assumes that the selection of ore and waste will be made on the basis of core-sized volumes. In fact, this should be based on SMU (selective mining unit) volumes. These could be simulated by subdividing the bench into blocks of this size and averaging the contained values. Cumulative frequency curves would then be produced from these block grades.

There are many problems that may be faced in applying statistical estimators which include:

(1) Variable sample sizes in the data set. The assay distribution is a function of orebody geometry, sample length/volume and the level at which high assays have been cut. Different populations exist for different sample sizes and confusion could exist if these vary significantly. Here the requirement for a homogeneous data set is not met.

(2) Variable density of sampling.

(3) Areas in which samples are so close that

Table 3.10 Upper confidence limit (factor $\phi_{0.95}(V,n)$)

	n						
V	5	10	15	20	50	100	1000
0.00	1.000	1.000	1.000	1.000	1.000	1.000	1.000
0.02	1.241	1.117	1.084	1.067	1.038	1.026	1.007
0.04	1.362	1.171	1.122	1.099	1.055	1.037	1.011
0.06	1.466	1.216	1.154	1.124	1.069	1.046	1.013
0.08	1.561	1.256	1.181	1.146	1.080	1.053	1.015
0.10	1.652	1.293	1.207	1.166	1.091	1.060	1.017
0.12	1.740	1.327	1.230	1.184	1.100	1.066	1.019
0.14	1.827	1.361	1.253	1.202	1.109	1.072	1.020
0.16	1.914	1.393	1.274	1.219	1.118	1.078	1.022
0.18	1.999	1.425	1.295	1.236	1.126	1.084	1.023
0.20	2.087	1.455	1.316	1.252	1.135	1.089	1.025
0.30	2.532	1.606	1.415	1.328	1.172	1.113	1.031
0.40	3.019	1.756	1.509	1.399	1.207	1.135	1.037
0.50	3.563	1.910	1.603	1.470	1.240	1.156	1.042
0.60	4.176	2.070	1.682	1.541	1.273	1.175	1.047
0.70	4.870	2.237	1.798	1.614	1.306	1.196	1.052
0.80	5.663	2.415	1.901	1.688	1.338	1.215	1.057
0.90	6.570	2.604	2.006	1.763	1.371	1.235	1.062
1.00	7.605	2.805	2.117	1.842	1.404	1.254	1.067
1.10	8.795	3.019	2.233	1.924	1.437	1.274	1.071
1.20	10.155	3.250	2.355	2.008	1.471	1.294	1.076
1.30	11.718	3.497	2.483	2.096	1.506	1.314	1.080
1.40	13.513	3.761	2.617	2.187	1.540	1.334	1.085
1.50	15.569	4.045	2.758	2.282	1.576	1.354	1.089
1.60	17.928	4.351	2.907	2.380	1.613	1.374	1.094
1.70	20.639	4.680	3.064	2.484	1.650	1.395	1.098
1.80	23.749	5.034	3.229	2.592	1.688	1.416	1.103
1.90	27.318	5.414	3.403	2.704	1.728	1.438	1.107
2.00	31.398	5.825	3.588	2.822	1.767	1.459	1.112
2.10	36.079	6.268	3.783	2.945	1.808	1.481	1.116
2.20	41.444	6.745	3.989	3.074	1.850	1.504	1.121
2.30	47.586	7.260	4.208	3.209	1.893	1.526	1.125
2.40	54.611	7.815	4.438	3.351	1.937	1.549	1.130
2.50	62.661	8.415	4.683	3.498	1.982	1.572	1.134
2.60	71.861	9.061	4.941	3.670	2.029	1.596	1.139
2.70	82.366	9.759	5.214	3.816	2.076	1.620	1.144
2.80	94.377	10.512	5.504	3.986	2.125	1.645	1.148
2.90	108.115	11.326	5.811	4.164	2.175	1.670	1.153
3.00	123.750	12.206	6.137	4.351	2.226	1.695	1.158

From Rendu (1981), based on Sichel (1966) and Wainstein (1975).

Table 3.11 Lower confidence limit (factor $\phi_{0.05}(V,n)$)

	n						
V	5	10	15	20	50	100	1000
0.00	1.000	1.000	1.000	1.0000	1.0000	1.0000	1.0000
0.02	0.8978	0.9333	0.9458	0.9540	0.9697	0.9782	0.9927
0.04	0.8589	0.9071	0.9246	0.9344	0.9573	0.9692	0.9895
0.06	0.8302	0.8874	0.9079	0.9200	0.9478	0.9622	0.9872
0.08	0.8070	0.8708	0.8943	0.9077	0.9398	0.9564	0.9852
0.10	0.7870	0.8563	0.8821	0.8972	0.9328	0.9512	0.9833
0.12	0.7693	0.8439	0.8716	0.8878	0.9264	0.9564	0.9817
0.14	0.7535	0.8323	0.8617	0.8790	0.9204	0.9420	0.9801
0.16	0.7389	0.8216	0.8527	0.8709	0.9149	0.9380	0.9787
0.18	0.7255	0.8116	0.8442	0.8632	0.9097	0.9341	0.9773
0.20	0.7129	0.8023	0.8360	0.8558	0.9048	0.9304	0.9760
0.30	0.6605	0.7618	0.8008	0.8243	0.8828	0.9139	0.9701
0.40	0.6187	0.7284	0.7717	0.7981	0.8639	0.8996	0.9648
0.50	0.5838	0.6995	0.7462	0.7744	0.8470	0.8867	0.9600
0.60	0.5538	0.6739	0.7270	0.7534	0.8313	0.8741	0.9554
0.70	0.5277	0.6508	0.7020	0.7338	0.8168	0.8632	0.9511
0.80	0.5044	0.6297	0.6825	0.7156	0.8030	0.8525	0.9470
0.90	0.4836	0.6103	0.6646	0.6987	0.7899	0.8421	0.9429
1.00	0.4650	0.5923	0.6476	0.6826	0.7774	0.8322	0.9389
1.10	0.4481	0.5756	0.6317	0.6674	0.7654	0.8226	0.9351
1.20	0.4328	0.5599	0.6165	0.6530	0.7538	0.8133	0.9313
1.30	0.4189	0.5452	0.6023	0.6393	0.7426	0.8042	0.9276
1.40	0.4062	0.5315	0.5888	0.6262	0.7318	0.7954	0.9240
1.50	0.3946	0.5186	0.5760	0.6137	0.7214	0.7868	0.9203
1.60	0.3840	0.5065	0.5637	0.6018	0.7112	0.7784	0.9168
1.70	0.3743	0.4950	0.5521	0.5904	0.7014	0.7702	0.9133
1.80	0.3655	0.4842	0.5410	0.5794	0.6918	0.7622	0.9098
1.90	0.3574	0.4740	0.5305	0.5688	0.6825	0.7544	0.9064
2.00	0.3501	0.4644	0.5203	0.5587	0.6734	0.7466	0.9030
2.10	0.3433	0.4552	0.5106	0.5489	0.6646	0.7391	0.8996
2.20	0.3372	0.4466	0.5014	0.5395	0.6560	0.7317	0.8962
2.30	0.3316	0.4385	0.4925	0.5304	0.6476	0.7245	0.8929
2.40	0.3266	0.4308	0.4840	0.5217	0.6394	0.7173	0.8896
2.50	0.3220	0.4234	0.4759	0.5133	0.6314	0.7104	0.8864
2.60	0.3179	0.4166	0.4681	0.5044	0.6236	0.7035	0.8831
2.70	0.3142	0.4100	0.4606	0.4974	0.6160	0.6967	0.8799
2.80	0.3110	0.4039	0.4535	0.4899	0.6085	0.6901	0.8767
2.90	0.3081	0.3981	0.4467	0.4826	0.6012	0.6836	0.8736
3.00	0.3055	0.3926	0.4401	0.4756	0.5941	0.6772	0.8704

From Rendu (1981), based on Sichel (1966) and Wainstein (1975).

strong covariance exists between samples for which allowance should be made.

(4) Assay populations may be totally erratic and no normalization process can be applied.

(5) Assay populations may be composite (e.g. bimodal).

(6) Populations may be truncated due to the application of a COG to define the outer limits of the orebody. In this case a deficit of low values may exist.

(7) Populations may be truncated due to the presence of a chemical barrier. In this case, an apparent negative skew exists because no values above a certain threshold are possible. Here the ore consists of 100% of the ore mineral.

(8) The population may possess an exponential distribution, although this may be the product of the superimposition of a population with a strong positive skew, reflecting sulphide or precious metal mineralization, and a strong, but low grade, population related to the existence of internal areas of waste in the ore envelope.

3.6 ORE-RESERVES BY PANEL/SECTION METHODS (UNDERGROUND OPERATIONS)

The methods to be outlined in this section are those that are particularly applicable to non-tabular orebodies or those with a somewhat irregular outline which have been evaluated by drilling on mine sections or panel boundaries and are to be exploited by underground methods. Tabular deposits could be evaluated in this way but other methods are preferable.

3.6.1 Volume calculation

The areas of each section or panel boundary can be obtained by manual planimetry, or computer-aided digitization, or by the use of Simpson's rule as described in section 3.4.3. The volume associated with each section can then be determined by extending its area for half a section spacing on either side, i.e.

$$\text{total orebody volume} = \Sigma_{i=1}^{n} (A_i \times D)$$

where D is the section spacing and A_i is the area of each section.

Alternatively, the average area of adjacent sections is used and this is multiplied by the separation of two sections. The global volume is then:

$$\Sigma_{i=1}^{n-1} ((A_i + A_{i+1})/2 \times D)$$

Figure 3.17 demonstrates this method but also reveals than an end correction (C) is necessary for the volumes at the extremities of the orebody (in this case before A_1 and after A_5).

$$C = (A_1 \times d_1)/2 + (A_5 \times d_2)/2$$

A third method is the application of the prismoidal formula in which three sections are used to determine the volume of the block of ground between the outer sections. In this case, the central section is given a dominant weighting (Figure 3.18). The relevant formula is thus:

$$\text{Volume} = (A_1 + 4A_2 + A_3)h/6$$

where h is the distance between A_1 and A_3. This process would be repeated for A_3, A_4 and A_5 then A_5, A_6 and A_7, etc. until the whole orebody is covered. Obviously an end correction would also have to be applied as above.

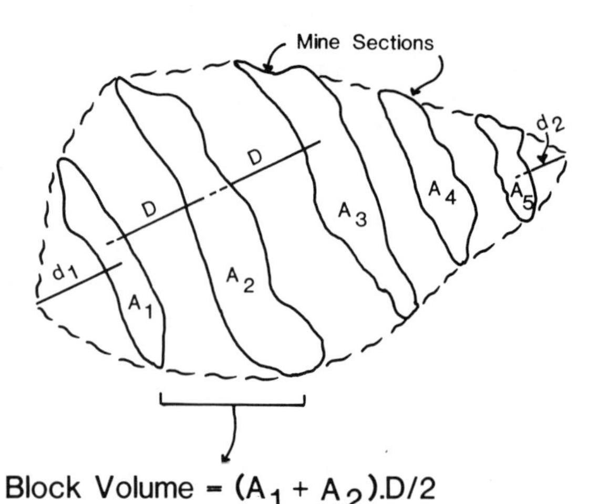

Fig. 3.17 Determination of orebody volume by sections.

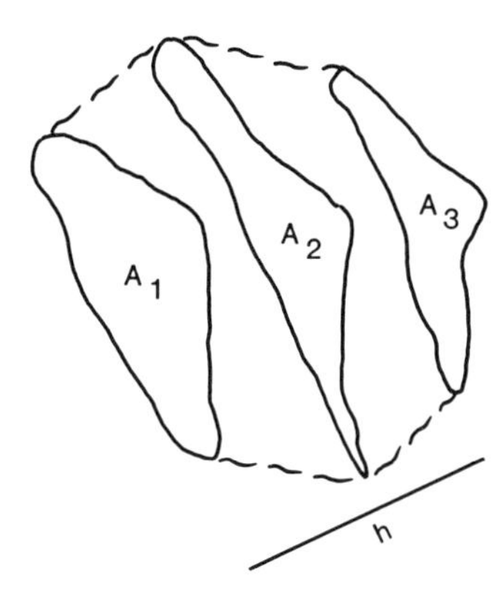

Fig. 3.18 Application of the prismoidal formula for orebody volume.

3.6.2 Grade calculations

A range of methods is available to determine grade which can be illustrated with reference to Figure 3.19. This represents a cross-section of an orebody which has been intersected by five drill-holes (DDH1–DDH5). The panel width is assumed to be 20 m. Thickness and assay data for each sample are recorded on the section. The thickness would be the intersected thickness in the case of vertical holes, or the vertical component in the case of inclined holes, plotted at the mid-point position for the intersection. For each hole, an overall vertical thickness (m), a grade (%) and a metal accumulation (m%) are tabulated.

1. Statistical methods

Here the arithmetic mean could be taken of all the assay data ignoring the variable sample support and area of influence. Similarly medians, geometric means, log means and Sichel's t estimators could be used. The arithmetic mean in this case (18 samples) is 2.34%. This grade would therefore be assigned to the volume (or tonnage) associated with this section (20 × 169.78 = 3396 m^3). To determine the overall grade of the orebody, each such grade (G_i) would then be weighted by volume (Vol_i) associated with each section, i.e.

$$G_{OB} = \Sigma_{i=1}^{n} (Vol_i \times G_i) / \Sigma_{i=1}^{n} Vol_i$$

2. Metal accumulation method

Here each sample grade in the section would be weighted by its length (and SG if required) in the usual way:

$$G_{Sect} = \Sigma_{i=1}^{n} (G_i \times L_i(\times SG_i)) / \Sigma_{i=1}^{n} L_i(\times SG_i)$$

The weighted grade in the example is thus 63.98/26.9 = 2.38% which is only slightly different from the arithmetic mean because only minor variations in sample length exist.

The global grade is then calculated as in 1 above.

3. Polygon method

The overall grade of each intersection is assigned to a polygonal block of ore defined on the basis of half the distances to the adjacent boreholes (e.g. A_1–A_5 in Figure 3.19). Each of these grades is then weighted by the area of the host polygon, i.e.

$$\Sigma_{i=1}^{n} (A_i \times G_i) / \Sigma_{i=1}^{n} A_i$$

Here the denominator should equal the area of the whole cross-section by planimetry. In the example quoted, the calculation is as follows:

$$\begin{aligned} G &= [(16.56 \times 1.89) + (37.65 \times 3.22) \\ &\quad + (46.68 \times 2.22) + (50.07 \times 2.17) \\ &\quad + (18.82 \times 1.45)]/(16.56 + 37.65 \\ &\quad + 46.68 + 50.07 + 18.82) \\ &= 392.113/169.78 \\ &= \mathbf{2.31\%} \end{aligned}$$

Total volume = 169.78 × 20 = 3396 m^3

4. Area under the curve method (using drill-hole values)

The lower portion of Figure 3.19 contains graphs of drill-hole grades, thicknesses (vertical) and metal accumulation plotted against horizontal distances between drill-holes. Such a plot allows the geologist to interpret trends between holes

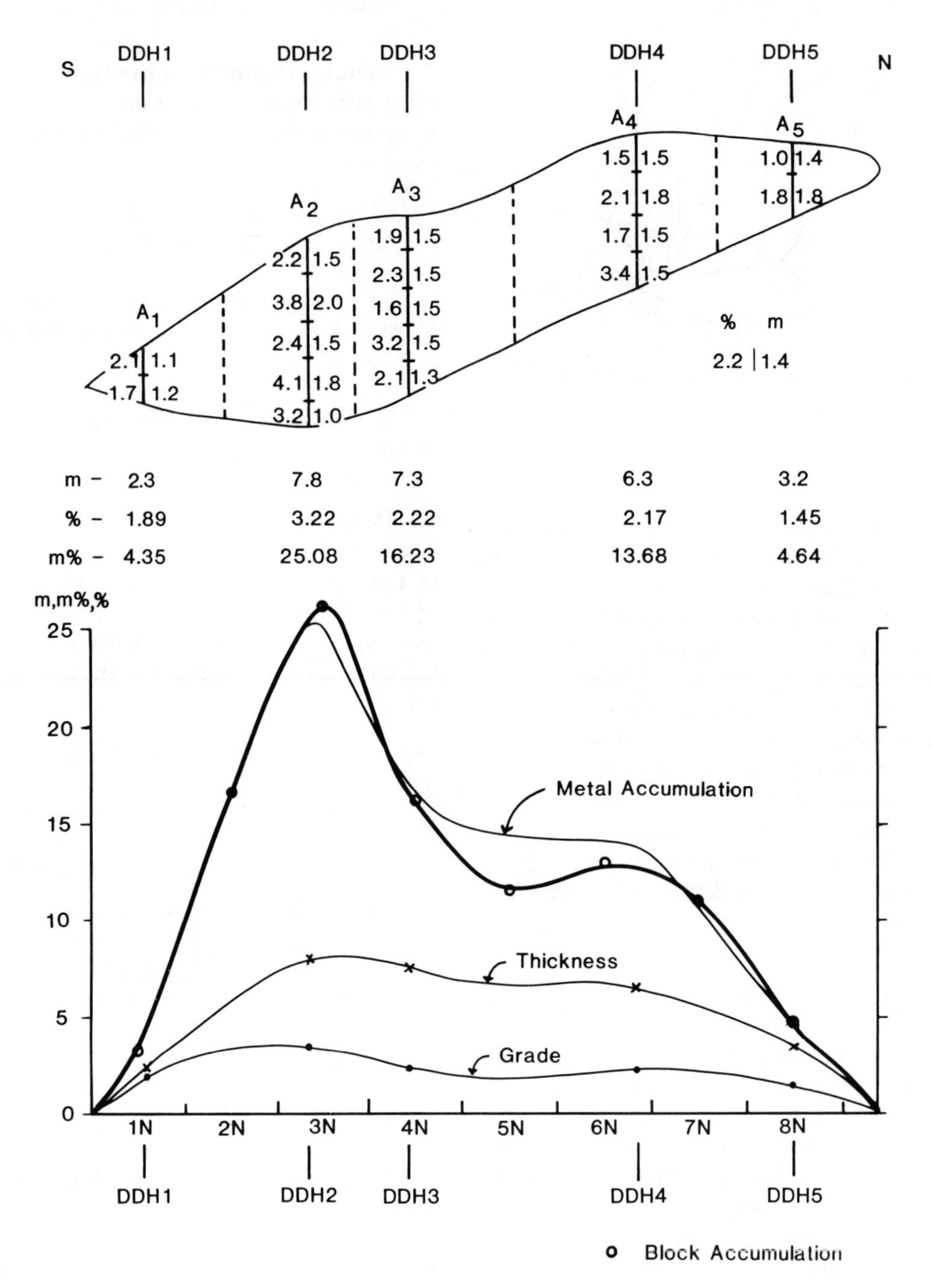

Fig. 3.19 Ore reserves by section/panel.

on a non-linear basis. To calculate the grade of the section, the area beneath the metal accumulation curve is divided by the area under the thickness curve. In this case:

G_{Sect} = 844 planimeter units/360.5 planimeter units
= 2.34%

5. Area under the curve method (using block values)

This is a variation of (*d*) in which the section is divided into equal length blocks which could be related to selective mining units (SMUs) for the mine concerned. At the mid-point of each block (e.g. 1*N* to 8*N*) the values of thickness and grade are read off and multiplied to give a new metal accumulation curve (Figure 3.19). This should be similar to that produced directly from the drill-hole data but variations could exist because the geologist has extrapolated non-linear trends for both thickness and grade. Note, for example, the differences for blocks 5*N* and 6*N*. This further highlights the problem facing the geologist for all these methods, viz. how should grades be extrapolated between holes. Differences are often the product of the subjective nature of this process.

One advantage of the current technique is that the area under the metal accumulation and thickness curves can be determined over the width of each block so a block grade can be produced. The overall grade produced by this method is now 817 units/360.5 units = **2.27%**, i.e. slightly less than that calculated in 4 above.

As can be seen from Table 3.12, only minor differences exist between the different methods although situations may exist where larger discrepancies may become evident. Allowing the geologist to extrapolate grades and thicknesses (methods 4 and 5), using his intuition and geological knowledge, results in the calculation of a lower grade suggesting that some of the other methods suffer a positive bias. It is almost impossible to prove, however, which gives the most accurate result although one would suspect that Method 5 is to be favoured, being the most logical.

Table 3.12 Summary of results for grade estimates

Methods	%
Statistical mean	2.34
Length weighted mean	2.38
Polygon method	2.31
Area under curve (using drill-hole values)	2.34
Area under curve (using block values)	2.27

6. Areas of influence based on fan drilling

Where an orebody has been sampled by fan drilling, as at Craigmont Mine in Figure 3.20, the weighted grade of each intersection can be assigned to an area produced by bisecting the angles between adjacent drill-holes and by joining up adjacent hangingwall and footwall assay contacts. Figure 3.21 shows one such area which is shaded and assigned to drill-hole DDH2. This has been subdivided into four triangles to allow ease of calculation of the total area of influence, A^0 $(A_1 + A_2 + A_3 + A_4)$. The area of each triangle is thus:

$$\sqrt{S(S - a)(S - b)(S - c)}$$

where $S = (a + b + c)/2$ and a, b and c are the lengths of the sides of the triangle.

The overall grade of the section drilled would then be:

$$G_{Sect} = \Sigma_{i=1}^{n} (A^0{}_i \times G_i)(/ \Sigma_{i=1}^{n} A^0{}_i)$$

3.6.3 Stope reserves by section methods – case history

The Sullivan stratabound lead-zinc deposit in British Columbia provides an ideal example of the application of sections to the calculation of stope reserves and also to a typical pragmatic or empirical approach to grade weighting. The orebody has a flat dip and hence the Cascade mining method has to be adopted to generate gravity

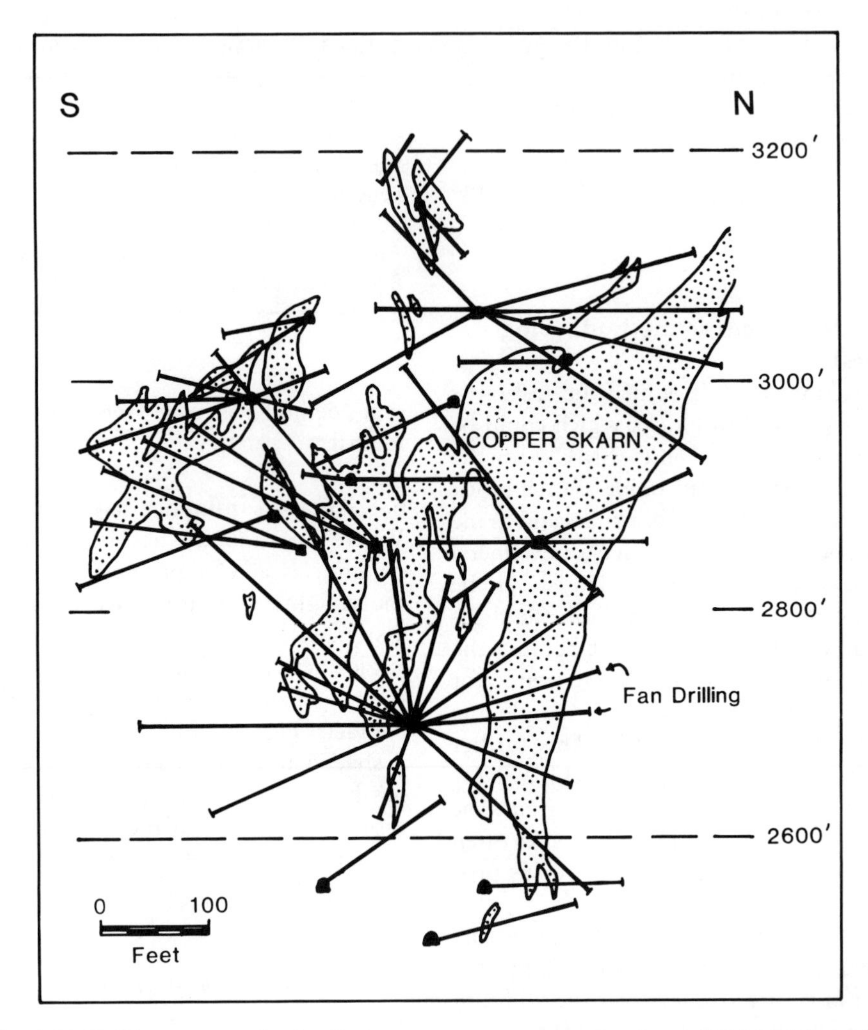

Fig. 3.20 Drilling patterns in the Craigmont copper-magnetite skarn, British Columbia.

feed through the drawpoints (Figure 3.22). This diagram also shows the fan drilling required to exploit each stope and create the underlying drawpoints. Figure 3.23 shows a typical stope outline in plan which, in this case, is transected by four geological sections at 15 m intervals. This particular stope also has a total of 25 lines of blast-hole drilling for which sections have been drawn. The spacing of these sections (the burden) is slightly variable but averages 2.5 m. For each section in turn, the areas of drilled ore, undrilled ore (to be left) and sub-grade footwall in drawpoints, are planimetered and then converted to volumes by multiplying each by the sum of half the distances to the adjacent sections.

The next stage in the calculation is the listing of all drill-hole assay data from the geological sections lying in the stope and at its margins, and also from the two immediately adjacent section lines. To this list is added any chip sample information in the stope (e.g. from raises) and also any immediately outside. These assays include Pb, Zn, Ag, Fe and Sn. Each assay for each sample is then multiplied by a weighting factor

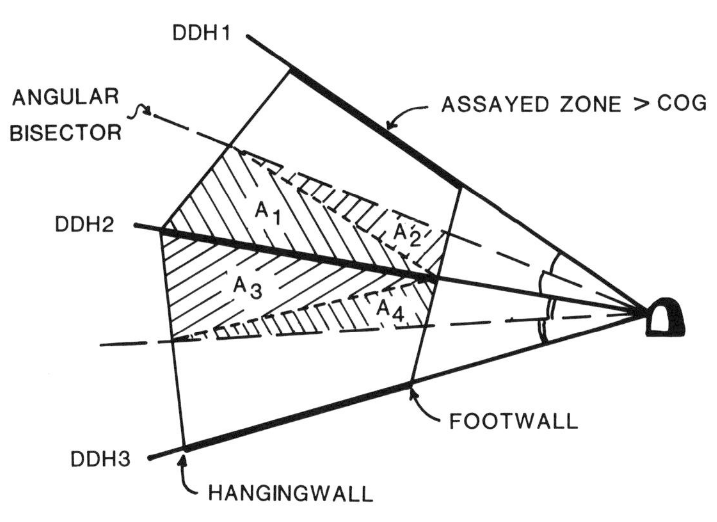

Fig. 3.21 Areas of influence based on fan drilling.

depending on its location/type as follows:

Sample	**Weighting factor**
Drill-hole in block	5
Drill-hole at margin	3
Drill-hole outside block	1
Chip sample in block	2
Chip sample outside block	1

A new listing is thus produced. All the weighted values for each element in turn are summed and divided by the sum of the weighting factors to determine the weighted grade for each element. The same procedure is applied to the sub-grade footwall drawpoints. On the basis of the Zn and Pb + Fe values, tonnage factors are then obtained from charts (Figure 3.13, section

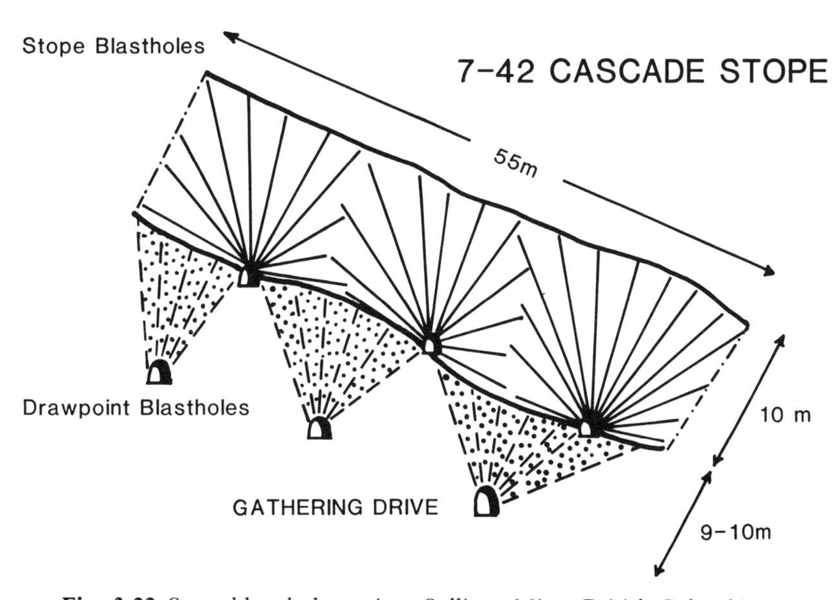

Fig. 3.22 Stops blast-hole section, Sullivan Mine, British Columbia.

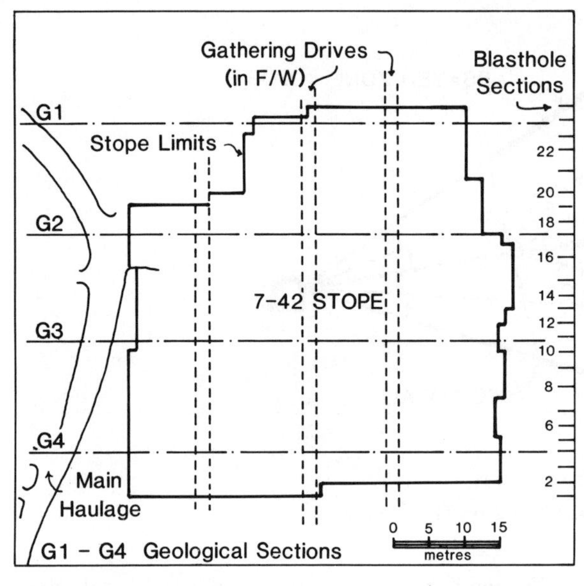

Fig. 3.23 Typical stope layout plan, Sullivan Mine, showing the location of geological and blast-hole section lines.

3.4.3a). Typical values would be as follows:

	Tonnage	Pb	Zn	Fe	Ag(g/t)
Drilled ore	100 000	4.2	10.9	20.1	15.4
Undrilled ore	3 000	4.2	10.9	20.1	15.4
Footwall Mineralization	36 000	0.3	0.6	5.6	3.4

Tonnage factor ore = 3.45 t/m^3
Tonnage factor FW = 2.78 t/m^3

From this information, the intentionally diluted grade and tonnage of ore to be mined would be:

136 000 t @ 3.2% Pb and 8.2% Zn

The combined Pb + Zn grade of 11.4% is still well above the critical level for mine planning purposes, although this of course does not take into account the effects of unintentional hanging-wall dilution, itself weakly mineralized.

3.7 ORE-RESERVES BY TRIANGULATION

This method requires that all intersections of a tabular orebody are plotted on a suitable plan (e.g. a VLP, a horizontal plan or a palinspastic map). If a VLP is chosen, then the mid-point of each intersection is projected horizontally onto the VLP and the horizontal thickness, grade and horizontal metal accumulation plotted at each point. For plans, the projection is vertical and the vertical thickness, grade and vertical metal accumulation are plotted as in Figure 3.24. Adjacent boreholes are then connected by tie-lines to produce a series of triangles, ideally with angles as close to 60° as possible but certainly avoiding acute-angled triangles. Calculating ore-reserves by this method then involves the determination of the area of each triangle, its weighted thickness and its grade. The method produces a conservative estimate of the reserves, for it does not allow extrapolation of mineralization beyond the area of drilling.

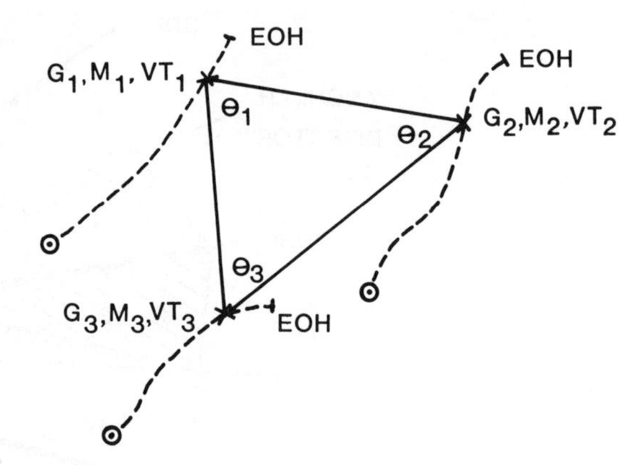

Fig. 3.24 Plan showing drill-hole collars and traces, orebody intersection mid-points and ore-reserve triangle.

Where the support of the grades is constant, as in the bench of an open-pit, then there are two main methods of estimating the grade:

(1) Arithmetic mean $\Sigma_{i=1}^{3} G_i$

(2) Included angle weighting $\Sigma_{i=1}^{3} (G_i \times \theta_i) / \Sigma\theta_i$ where $\Sigma\theta_i = 180$

If we use the data from the two triangles illustrated in Figure 3.25 and ignore for the time being the thicknesses, the results of using these two methods are as follows:

	Case A	Case B
Mean	1.53	1.53
Included angle	1.55	1.36

The discrepancy between the two values increases as the corner angles deviate from 60°.

Where the thickness at each intersection is variable, then there are another three methods which can be used to determine grade.

(1) Thickness weighting

$$\Sigma_{i=1}^{3} (G_i \times Th_i)/\ \Sigma_{i=1}^{3} Th_i$$

(2) Thickness and included angle

$$\Sigma_{i=1}^{3} (G_i \times Th_i \times \theta_i)/\ \Sigma_{i=1}^{3} (Th_i \times \theta_i)$$

(3) The percentage method

$$\tfrac{1}{4}\ (\Sigma G_i \times Th_i/\Sigma Th_i + \Sigma G_i)$$

Again using the two cases illustrated in Figure 3.25, we have:

	Case A	Case B
Thickness weighting	1.56	1.56
Thickness and included angle weighting	1.57	1.40
Percentage method	1.54	1.54

The agreement is good in case A, but considerably worse in case B, indicating that bias may occur if acute angle triangles are mixed with triangles that are almost equiangular. The grade of triangle B has dropped considerably by using method 2, for the drill-hole at the obtuse angle has a lower grade than the others and is also closer to the centre of the triangle. It should be noted that methods 1 and 3 give the same result in each case, even though the triangles are markedly different. This lends further support to the suggestion that method 2 is the best.

To compute the global reserve, each triangle grade is weighted by its area or volume and the combined total is divided by the sum of the areas or volumes to produce a global trade. The tonnage is computed from the sum of the volumes times the tonnage factor/bulk density. Note that each triangle could have its own tonnage factor (TF_i) based on the rock type and grade, or type of contained mineralization, in the three corner holes. Hence:

$$G_{OB} = \Sigma(Area_i \times G_i \times TF_i)/(\Sigma Area_i \times TF_i)$$

(for bench composite grades)

or

$$G_{OB} = \Sigma(Vol_i \times G_i \times TF_i)/\Sigma(Vol_i \times TF_i)$$

(for variable thickness ore)

An example of the use of triangulation in the computation of coal reserves and extraction volumes is presented in the case history in section 8.5 in Chapter 8.

Fence diagrams have been used to determine the tonnage of inclined tabular deposits. The intersection positions of the mid-point (or FW) of the orebody are plotted on plan and the elevations of these, above a suitable datum, are plotted at the corners as in Figure 3.26. Triangles are drawn with sides of length x, y and z. Perpendiculars are erected on a baseline ($\equiv$ to the datum) whose scale lengths are equivalent to the elevation in each hole and whose separations are the distances x, y and z. Joining up the tops of these perpendiculars then gives the sides of a triangle x', y', z' lying in the plane of the orebody. The area can thus be computed from the formula for triangle area, using the length of the sides of the triangle, and the volume from the true thickness weighted by one of the methods described earlier.

3.8 ORE-RESERVES BY POLYGONS

Where drill-holes are randomly distributed, i.e. not on a regular grid, the grade and thickness of each hole can be assigned to an irregular polygon. These are assumed to remain constant throughout the area of the polygon, which can be defined on the basis of two methods: perpendicular bisectors and angular bisectors.

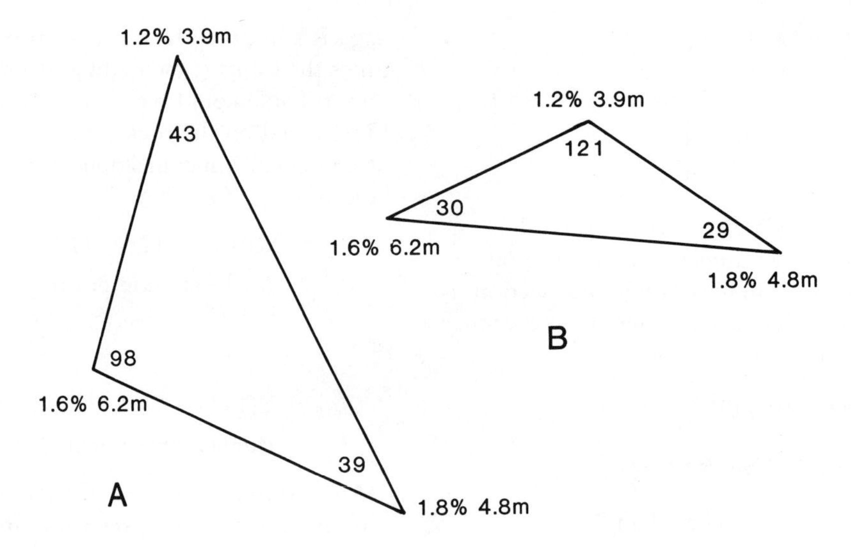

Fig. 3.25 Ore-reserve triangles to illustrate the impact of acute angled triangles on weighted grades.

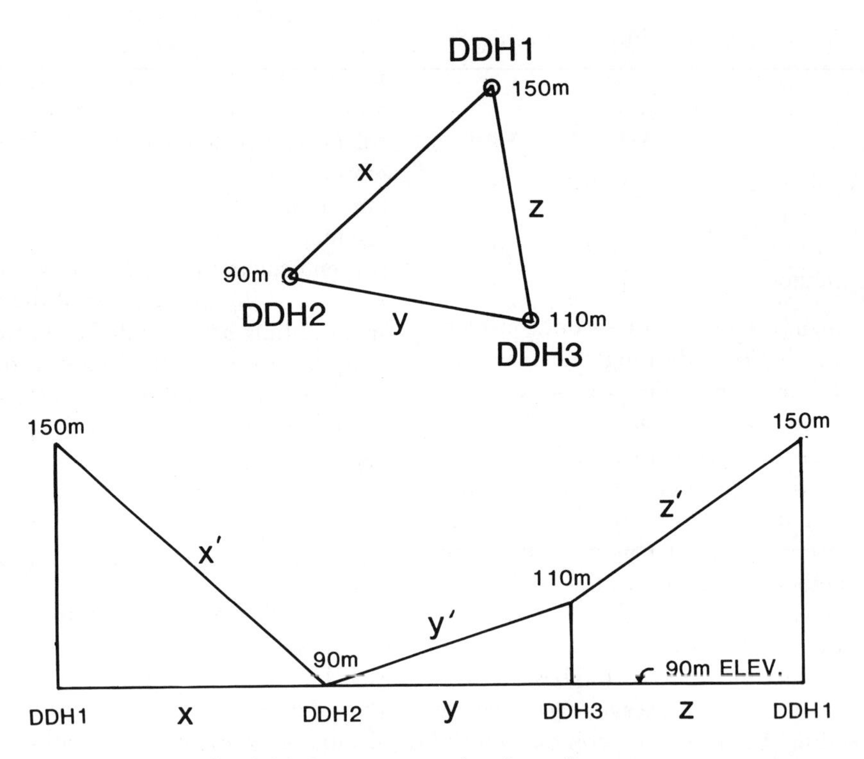

Fig. 3.26 Ore reserves using fence diagrams.

Perpendicular bisectors
Here polygonal mosaics are established by constructing perpendicular bisectors to tie-lines linking adjacent holes to the hole under consideration. The method is illustrated in Figure 3.27(a) where the grade, *G*, and thickness, *Th*, are extrapolated to the polygon ABCDE.

Angular bisectors
Here each polygon is established by linking drillholes with tie-lines and then constructing angular bisectors between these lines to define a central polygon, as illustrated in Figure 3.27(b).

Camisani-Calzolari (1983) suggests an alternative to single grade weighting by polygon. His method involves allocating 50% of the weight to the central drill-hole and the remaining 50% to the surrounding samples, in equal proportions. For example, in Figure 3.27(a), the polygon ABCDE would be assigned a grade as follows:

$$G_{ABCDE} = G_1 \times 0.5 + G_2 \times 0.1 + G_3 \times 0.1 + G_4 \times 0.1 + G_5 \times 0.1 + G_6 \times 0.1$$

where G_1 is the grade of the central hole and G_2 to G_6 are the grades of the peripheral holes.

These weighting coefficients are entirely arbitrary and no allowance is made for thickness. It is, however, an attempt to overcome one of the main criticisms of the method that polygons (sometimes very large in areas of sparse drilling)

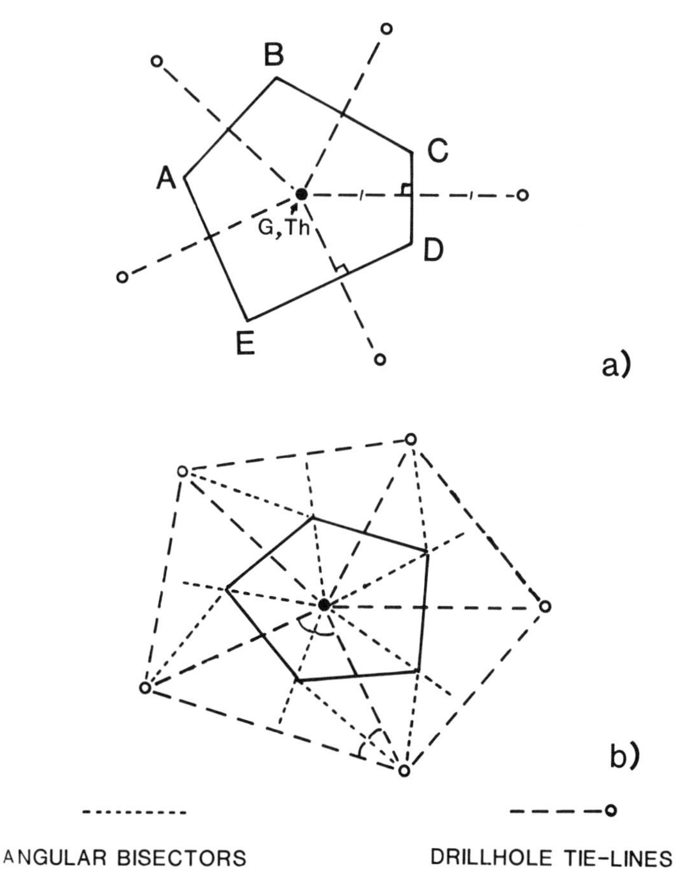

Fig. 3.27 Polygon construction by (a) perpendicular bisectors of tie-lines and (b) by angular bisectors.

are evaluated by only one drill-hole, totally ignoring adjacent holes.

As with triangulation, the global grade is calculated by volume or area weighting of each polygon grade.

The case history in section 8.4 presents an example of the application of the polygon technique for both the classification and calculation of ore-reserves.

3.9 ORE-RESERVES BY BLOCK MATRICES

Where the data are on lines, or on rectangular or regular offset grids, regular blocks can be fitted to the drill-holes as in Figure 3.28. The method is basically similar to that used in the polygon method and is particularly suited to the exploratory phase of drilling of a prospect where rapid updating of the reserve can be undertaken as each new hole is drilled and where precision of the estimates is not as crucial as at a later feasibility or mining stage. In Figure 3.28, methods (a), (b) and (d) allow extrapolation of mineralization beyond drilling but only use one hole to evaluate each block. On the other hand, methods (c) and (e) give conservative reserves (tonnage) but use four holes to evaluate both grade and tonnage and are thus somewhat more reliable. Generally, the thickness applied in the latter two cases is the arithmetic mean while the grade is thickness weighted (plus *SG* if required) between the four holes.

3.10 CONTOUR METHODS

These are particularly suited to manual calculations of ore-reserves when spatial trends can be recognized in the data, thus facilitating contouring by hand. Four main methods exist which are described in the following sections.

3.10.1 The grid superimposition method

Here drill-hole intersection points are plotted on plan, VLP or bench plan, along with the relevant component of thickness, metal accumulation or grade (in the case of the bench plan). Contour plans are then produced of two of the three variables – in the discussion that follows we will assume that the deposit varies in thickness and is plotted on to VLP. On the horizontal thickness metal accumulation VLP, the limits of the deposit are defined using the relevant criteria and then a matrix of ore-blocks is superimposed whose dimensions allow them to fit exactly within mining blocks. Their size may reflect the drill-hole spacing vertically and horizontally on the VLP, or the spacing of sub-levels in the mine, or the height of each lift in a 'cut and fill' stope. This is repeated for the horizontal thickness VLP using the same block size and orebody limits.

For all blocks within the ore limits, values are assigned to the mid-point of each block by interpolation between contours; first for metal accumulation, then for thickness. Where blocks overlap the boundary, an estimate of the proportion of ore in the block is made (Figure 3.29) together with an estimate of metal accumulation and thickness at the centre of gravity of this section of ore. Alternatively, the exercise can be simplified by allowing a certain amount of ore loss and gain at the margins, by defining the fringe on the basis of stepped block sides. In this case, no estimate of ore-fraction, or centre of gravity values, are needed.

Assuming that the ore-fraction method is used, the global reserve can now be calculated, as below, for all those blocks lying within the defined limits.

$$\text{GRADE} = (\Sigma MA_{CB} + \Sigma(MA \times O.F)_{ICB}) / (\Sigma HT_{CB} + \Sigma(HT \times O.F)_{ICB})$$

$$\text{VOLUME} = (\Sigma HT_{CB} + \Sigma(HT \times O.F)_{ICB}) \text{ x block area}$$

where:

CB = complete blocks
ICB = incomplete blocks
MA = horizontal metal accumulation
HT = horizontal thickness
O.F = ore-fraction in incomplete blocks

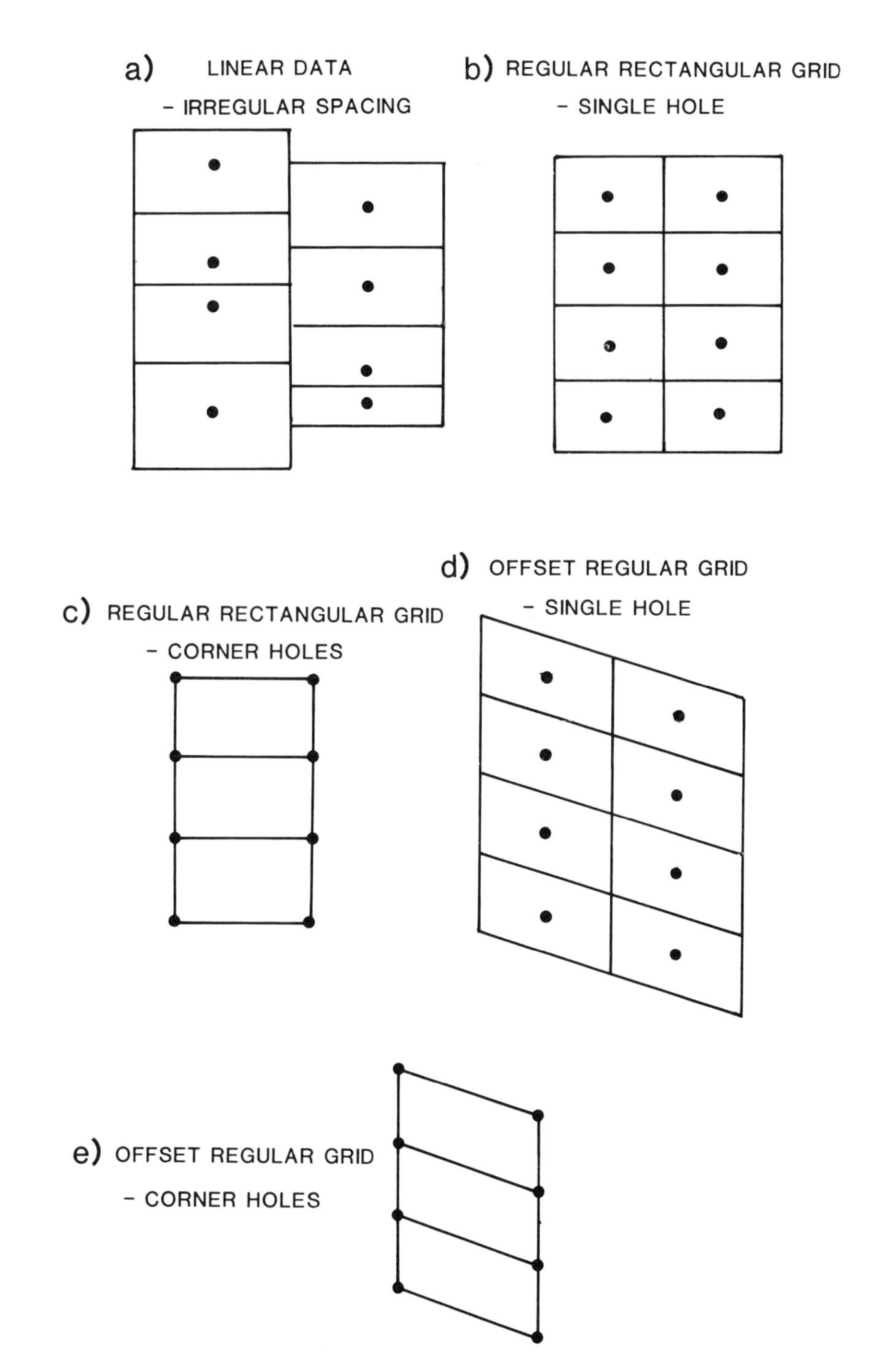

Fig. 3.28 Block matrices for ore-reserve calculations.

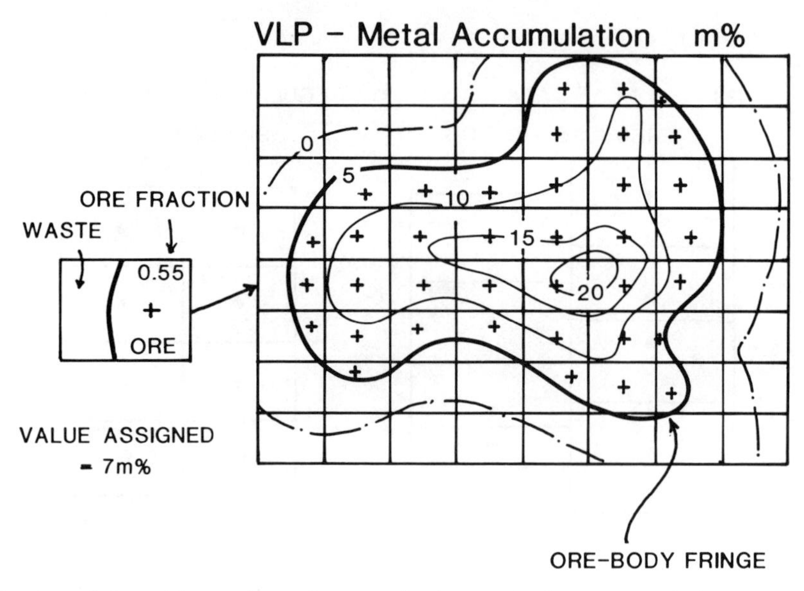

Fig. 3.29 Application of the grid superimposition method to orebody reserves. '+' represents the centre of a block or the centre of gravity of ore in a block.

In the case of an open-pit bench, where the thickness is constant, the global grade can be calculated from:

$$\text{GRADE} = (\Sigma G_{CB} + \Sigma(G \times O.F)_{ICB}) / (\text{no. of complete blocks} + \Sigma(O.F)_{ICB})$$

$$\text{VOLUME} = (\text{no. of complete blocks} + \Sigma(O.F)_{ICB}) \times \text{block area}$$

3.10.2 Moving window method

This is a smoothing technique, particularly suited to the calculation of reserves of an open-pit bench which has been intersected by a series of irregularly spaced drill-holes, or blast-holes, which have revealed a highly erratic fluctuation in bench composite grades. This situation was found to exist in the 3170 ft bench in the Ingerbelle Open Pit, near Penticton in British Columbia. A cursory examination of the grades revealed that contouring of the data was impossible due to the patchy distribution of chalcopyrite in this copper porphyry deposit. As a result, the usual grid superimposition and grade interpolation method could not be applied.

The moving window method involves the fitting of a grid of ore-blocks to the outline of the deposit in the bench under evaluation. A search window is then drawn whose dimensions are twice those of each ore-block. This can be done via a computer program or manually. In the latter case, the window, the ore-block and the block centre are drawn on a piece of tracing paper or draughting film, as in the top left-hand corner of Figure 3.30, so that it can be superimposed on the gridded bench plan. Ideally, at least 15 drill-holes should fall in the search area, so the dimensions can be modified to achieve this number if required. The window is then positioned so that its centre falls over the first block to be evaluated and the arithmetic mean of all the raw data values falling in the window, or their log-transformed equivalents, is then calculated and the result plotted at the block centre. The window is then moved laterally to the next block and the above calculation repeated. Each line of blocks is then covered until the whole bench has been completed. If the window dimensions are

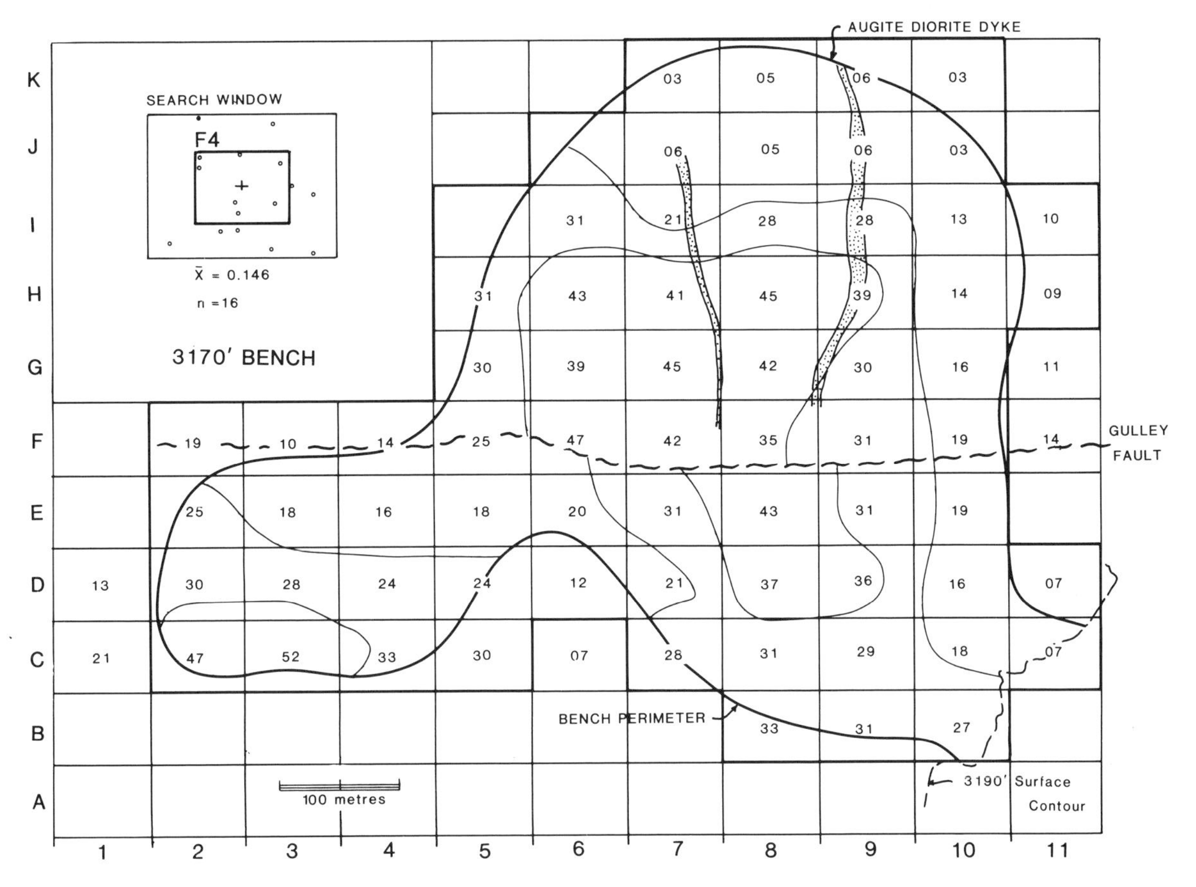

Fig. 3.30 The 3170 ft bench in the Ingerbelle Pit, British Columbia, demonstrating the application of the moving average smoothing technique to a deposit with erratic grades. Contour intervals at the cut-off grade and at the bench average grade for ore (0.34% Cu).

twice those of the ore-block, then a 50% overlap is achieved for each successive window position. The larger the window relative to the block, the greater degree of smoothing of the data achieved.

Figure 3.30 shows the result of such a procedure at Ingerbelle where ore-blocks 81.3 × 59.6 m have been used. Details of block F4 are shown as an inset. It is now evident that the grades can be contoured indicating the existence of three separate areas in which the grade exceeds the cut-off grade of 0.22% copper. From such a plan the bench reserves can be calculated as described earlier in section 3.10.1. Smaller blocks than those used in the text figure would normally be used to allow more accurate delineation of ore fringes. The results of the bench reserve calculation are listed below, together with those produced by the quasi-statistical technique explained in section 3.5.

Method	Bench tonnage	Bench grade	Ore tonnage*	Ore grade*
Moving average	8930 108	0.26	5167 890	0.34
Statistical	8630 924†	0.32	4315 462‡	0.53

*Based on COG of 0.22% Cu.
†Determined by electronic planimetry.
‡Based on the % of assays > 0.22% Cu (i.e. 50%).

The bench tonnages produced by ore-blocks (and ore-fractions) and by planimetry, are similar but the tonnages of ore in the bench show a big discrepancy as the statistical method takes no account of the location of the samples in the bench. The occurrence of low grade values with high grade values in each position of the search window has resulted in the production of a lower overall grade, but a higher tonnage. The statistical method has assumed that all the samples above COG are grouped together producing a higher grade and a lower tonnage. It is thus evident that the moving window method produces a more realistic view of the distribution of mill feed ore in the bench but the grades are only those that would be achieved by blending material from a large area of the bench.

A variation of the above moving window method can be achieved by calculating the co-ordinates of the data centroid and plotting the mean of the data at this point and not at the block centre.

$$x = \Sigma_{i=1}^{n} (x_i \cdot z_i) / \Sigma_{i=1}^{n} z_i$$

$$y = \Sigma_{i=1}^{n} (y_i \cdot z_i) / \Sigma_{i=1}^{n} z_i$$

where:

z_i = data value at each point
x_i = X coordinate of each point
y_i = Y coordinate of each point

The position of the centroid is thus influenced by the magnitude of the assay value as well as the distribution of the holes. The smoothed data values can now be contoured (one hopes!) and the grid superimposition and grade interpolation method applied to evaluate each ore-block centre.

3.10.3 Graticule method

Where no correlation exists between thickness and grade, the graticule method could be employed. Contour maps of these variables are superimposed and the area of each graticule (e.g. *A* in Figure 3.31) determined by manual planimetry, or by use of a digitizer and suitable software. The thickness (Th_i) and grade (G_i) assigned to each graticule within the orebody limits, is then the mean of the bounding contours, e.g. 6.75 m and 2.25% respectively, for graticule *A*. The total volume of the deposit is then:

$$\Sigma_{i=1}^{n} A_i \cdot Th_i$$

where A_i is the area of each graticule. The overall grade is then:

$$\Sigma_{i=1}^{n} (A_i \cdot Th_i \cdot G_i) / \Sigma_{i=1}^{n} (A_i \cdot Th_i)$$

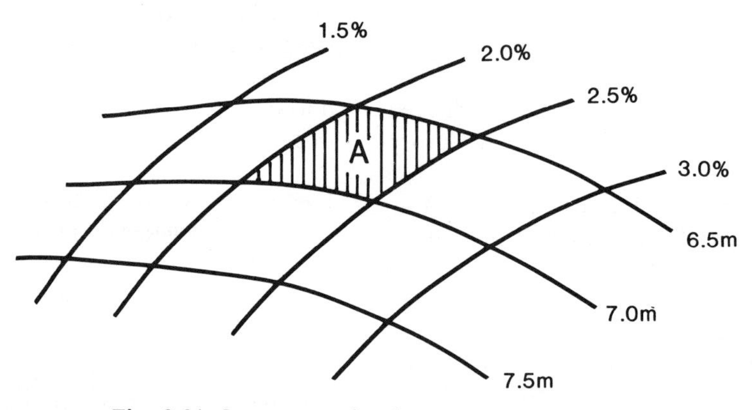

Fig. 3.31 Ore reserves by the graticule method.

3.10.4 Metal accumulation and thickness contours

This requires the contouring of the relevant thickness component of the deposit and then the determination of the area (A_i) between each adjacent set of contours, as in the shaded portion of Figure 3.32. This area is then multiplied by the average (Th_i) of the bounding contours. The total volume is then:

$$\Sigma_{i=1}^{n} (Th_i \cdot A_i)$$

to which a small additional volume is added to allow for that portion of the deposit exceeding the highest contour (e.g. Th_4 in Figure 3.32). In this case, this would be $(Th_4 + Th_5)/2 \times$ area within Th_4. The same procedure is repeated for metal accumulation. The total area weighted metal accumulation is then divided by the total volume to obtain the global grade of the deposit.

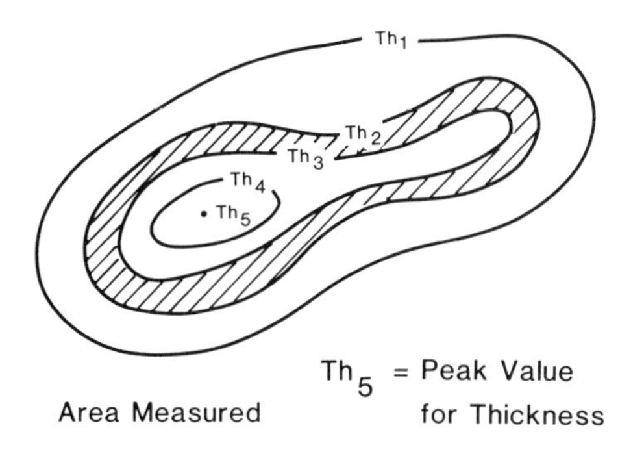

Fig. 3.32 Thickness contours for an ore deposit.

3.10.5 Reserves of stockpiles or coal tips

Horizontal substrate

The initial step, having surveyed the surface of the stockpile or tip, is to contour the elevation above a datum – in this case the tip floor. The contour intervals may be set on an arbitrary basis, or on a bench height, if the material is to be moved by horizontal slicing. The bench height would be determined by the machinery to be used. The tip volume can now be calculated by planimetry of each contour. The volume of each horizontal slice is determined by averaging the areas of the bounding surfaces and mutliplying by the contour interval/slice thickness (t). The sum of individual slice volumes gives the total volume, although once again a correction will be necessary for the tip summit (Figure 3.33) above the uppermost contour, as explained in section 3.10.4.

Alternatively, the contours are drawn at the mid-slice height and the area of each determined. The sum of these areas, times the slice height (t), gives the total volume, which must again be corrected for the summit volume. Here the contour is placed half-way between the peak height, and the top of the underlying slice, and its area multiplied by the elevation difference between the two.

Calculations on the basis of a conical tip show that the averaging of contour areas method slightly overexaggerates the volume while the

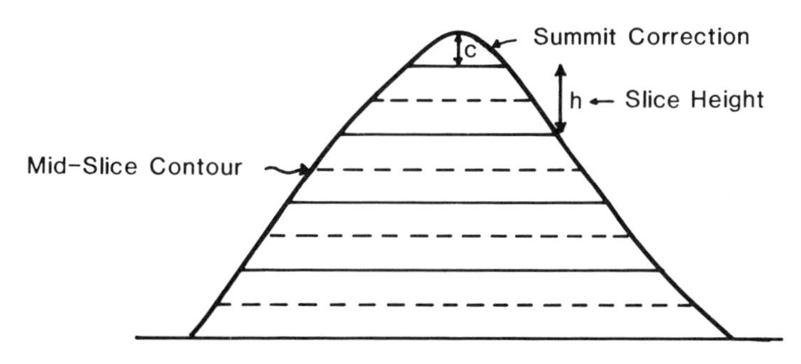

Fig. 3.33 Stockpile or coal-tip with horizontal substrate.

slice-mid contour method undervalues the volume, but in neither case is the discrepancy severe.

Inclined substrate

On the basis of drilling results (e.g. shell and auger), or old topographic maps, the palaeo-surface on which the tip rests is contoured at elevations equivalent to the slice-mid contours selected for the tip itself. A contour plan of the present tip surface is superimposed so that a third map can be produced on which, for those slices abutting against the old hillside, each mid-slice contour is linked with the equivalent topographic contour, as in Figure 3.34.

An alternative method to determine the total tip reserve requires the superimposition of a rectangular grid on the two plans (palaeotopography and tip surface). At the centre of each grid panel, the elevations of the footwall and tip surface are read off by interpolation between contours and subtracted. This value is the vertical thickness of the tip beneath this point or, put in another way, the height of a rectangular prism of tip material. The sum of the products of vertical thickness and grid area, for all grids within the limits of the tip, gives the tip volume. The application of a bulk density then allows the calculation of the total tonnage. Obviously, the smaller the grid dimensions the better fit there will be to the tip limits.

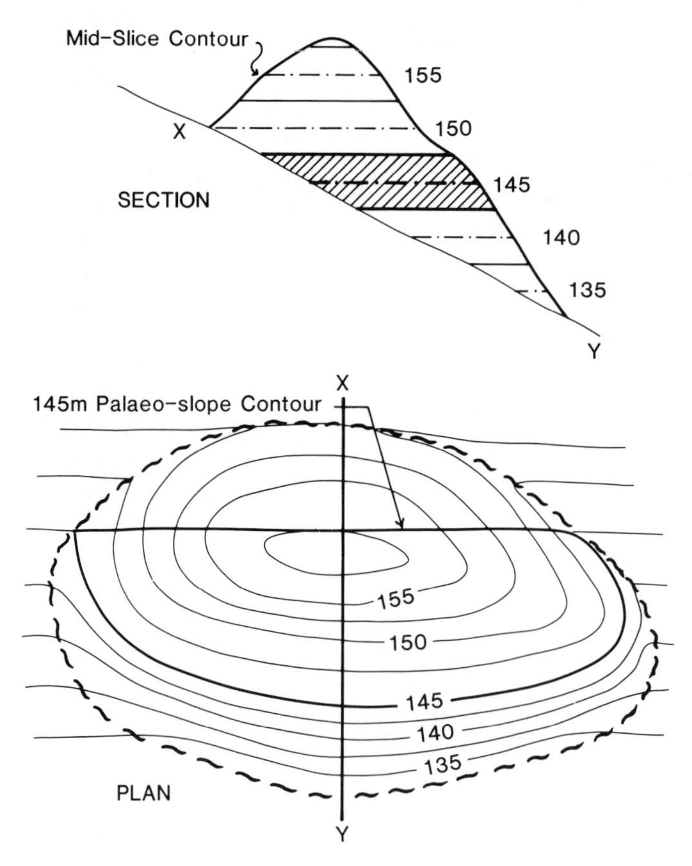

Fig. 3.34 Stockpile or coal-tip with inclined substrate. Diagram shows the mid-slice area on the 145 m level delimited ready for planimetry.

Exactly the same method, as described above, can be applied to an oil reservoir in a structural/stratigraphic trap. In this case, the oil–water interface (often inclined due to hydraulic gradients) is equivalent to the inclined topographic surface while the base of the cap-rock is equivalent to the tip surface.

3.11 INVERSE DISTANCE WEIGHTING METHODS (IDW)

This technique applies a weighting factor, which is based on a linear or exponential distance function, to each sample surrounding the central point of an ore-block. The weighting factor is the inverse of the distance between each sample and the block centre, raised to the power 'n', where 'n' usually varies between 1 and 3. Only samples falling within a specified search area, or volume, are weighted in this way. Because the method is laborious and repetitive, it is usually computerized.

3.11.1 2D search areas

2D search areas, such as circles or ellipses, are used when the assay data from a tabular inclined orebody have been projected on to VLP or plan, or composite grades have been calculated for individual benches in an open-pit. In these cases, the data are irregularly distributed. A grid of ore-blocks is superimposed on the plan or VLP so that the value of the central point of each block can be determined (Figure 3.35). This value will be extrapolated throughout the entire block. The computer program initially locates a search area, whose dimensions and shape are specified by the user, around the first block to be evaluated. Where the deposit is considered to be isotropic, in that grade or thickness variations are constant in all directions or the drilling grid is square, a circle is used. The radius is chosen to capture a minimum number of samples within its perimeter at every location in the orebody. Where circular search areas are not deemed to be suitable, on the basis of the drilling pattern or knowledge of the nature or continuity of the mineralization as determined from geostatistical studies, then an elliptical search area can be selected instead. This will allow for anisotropism in a mineral deposit or the use of a rectangular drilling grid.

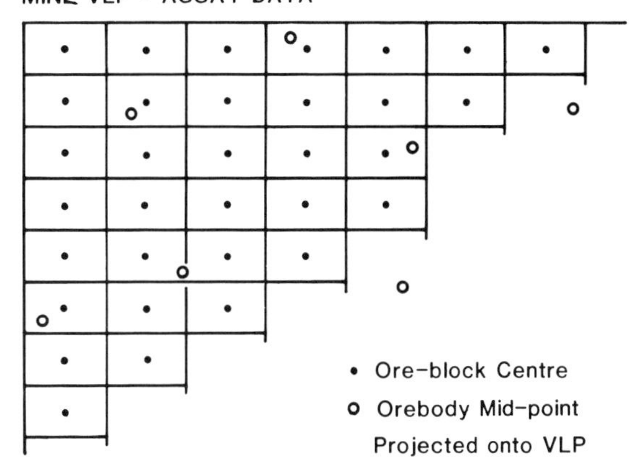

Fig. 3.35 Portion of an assay VLP showing drill-hole intersection points and part of the ore-block matrix.

Circular search area

In this case, the computer checks the distance of each sample in the data set from the current block centre (d_1) and rejects those that are outside the specified radius. Those that are accepted have their grades (for an open-pit bench), or metal accumulations and thicknesses, weighted by the inverse of this distance as below:

$$Z_B = \Sigma_{i=1}^{n} (Z_i/d_i^{n}) / \Sigma_{i=1}^{n} (1/d_i^{n})$$

where Z_B is the estimate of block grade, metal accumulation or thickness based on the values of each of these (Z_i) at each sample location in the search area. The exponent 'n' can be chosen by the user but the higher it is, the greater the weight given to close samples at the expense of the more distant ones. Inverse square distance (ISD, n = 2) methods seem to be the most favoured amongst those operators who use this technique. Examples are given in Chapter 8, case histories of White Pine and Boulby Potash Mine in sections 8.2 and 8.6 respectively. Barnes (1980) shows that an exponent of 5 closely approximates to the nearest point rule, i.e. all other values have been effectively screened out by being given a very low weighting factor.

The method can be illustrated by reference to Figure 3.36. In this instance, the distance between samples and the block centre (d_i) is expressed in terms of the differences in X-coordinates (XD_i) and the differences in the Y-coordinates (YD_i), as is required by the computer program. The general formula now becomes:

$$Z_B = \{\Sigma_{i=1}^{n} Z_i/(XD_i^2 + YD_i^2)^{n/2}\}/ \{\Sigma_{i=1}^{n} 1/(XD_1^2 + YD_i^2)^{n/2}\}$$

Three-drill holes fall in the search area and their grades (%) are given in the diagram,

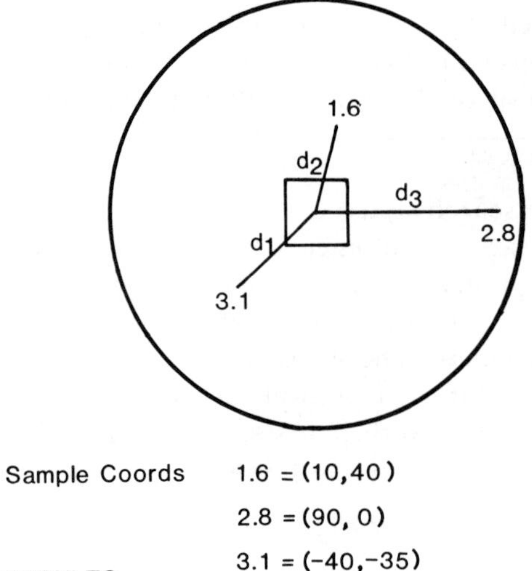

Sample Coords 1.6 = (10,40)
2.8 = (90, 0)
3.1 = (-40,-35)

RESULTS

% Weighting to each Grade				Weighted Block Grade
n	1.6 %	3.1%	2.8 %	
1	44.8	34.7	20.5	2.37
2	55.2	33.2	11.6	2.34
3	64.0	29.9	6.1	2.12

Fig. 3.36 Inverse distance weighting – circular search area.

together with the XD_i and YD_i values. As can be seen in the results tabulation, the weighting given to the nearest sample (1.6%) increases with 'n' while that given to the others decreases. This results in a drop in the estimated block grade as the nearest sample has the lowest grade of the three. Note that in this example, the grade was only weighted by inverse distance and not thickness as well, indicating that the grades have constant support (bench composite grades). If the thickness is variable, then the above equation should be modified as follows:

$$Z_B = \{\Sigma_{i=1}^{n} Z_i \times Th_i/(XD_i^2 + YD_i^2)^{n/2}\}/ \{\Sigma_{i=1}^{n} Th_i/(XD_1^2 + YD_i^2)^{n/2}\}$$

Elliptical search area

The selection of holes or samples which lie within an elliptical search area is achieved by an initial transformation of the coordinates of the samples (relative to the block centre). This is undertaken by calculating the differences in the X-coordinates (XD_i) of the sample and the block centre and then multiplying the difference in the Y-coordinates (YD_i) by K, the ratio of the axes of the chosen ellipse (a/b). This ratio can be determined on the basis of the ratio of the average drill-hole spacings in the two grid directions or on the basis of the anisotropy coefficient as determined from directional semi-variograms, a subject which will be dealt with in more detail in Chapter 4 (section 4.7.3). If we consider the situation where the axes of the ellipse are parallel to the local coordinate grid, then the square root of the sum of the squares of the differences in coordinates (after transformation) of each sample (d in Figure 3.37 A) is then compared to 'a', the length of the axis of the ellipse in the X direction. If the result is less, then the sample lies within the search area. All such samples are then weighted by inverse distance methods as below:

$$Z_B = \{\Sigma\ Z_i/(XD_i^2 + (K \cdot YD_i)^2)^{n/2}\} / \{\Sigma\ 1/(XD_i^2 + (K \cdot YD_i)^2)^{n/2}\}$$

The transformation has effectively transformed the search area to a circle of radius 'a'. In

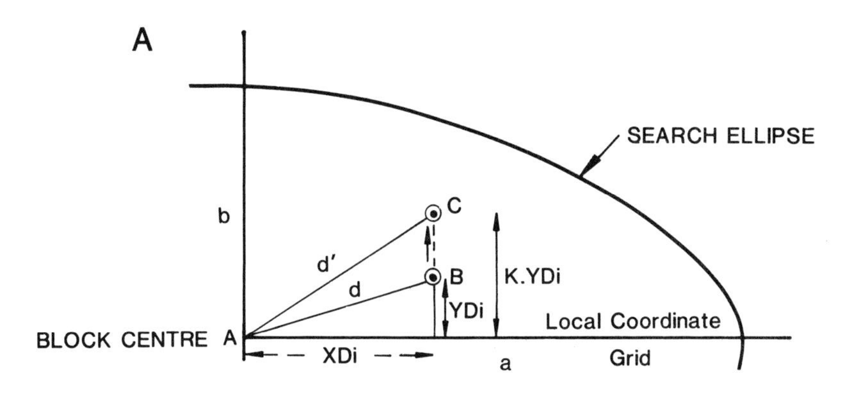

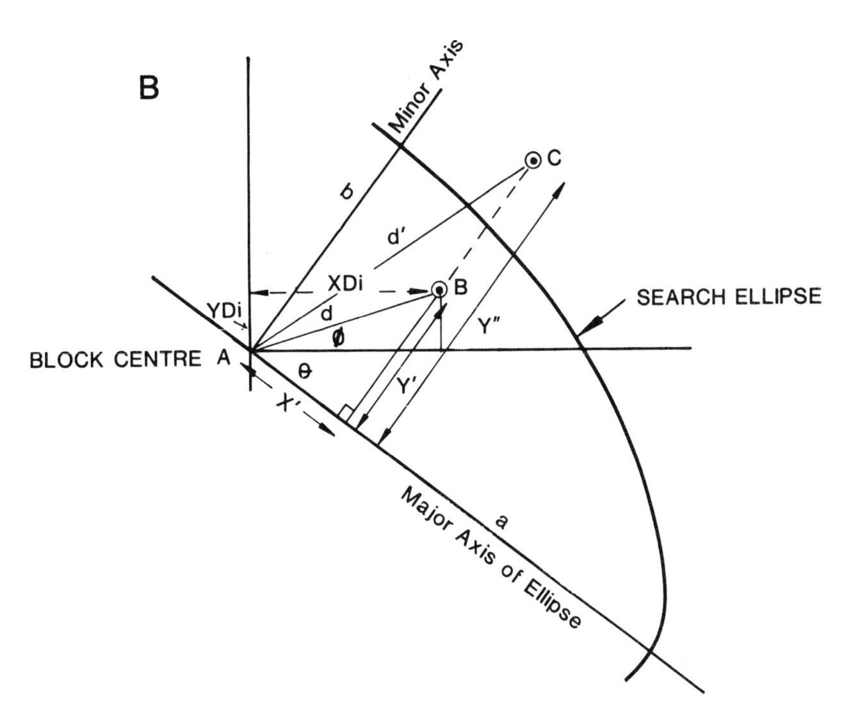

Fig. 3.37 (a) Search ellipse axes parallel to grid axes; (b) axes inclined. C = location of sample B after coordinate transformation.

doing so, the application of inverse distance weighting will thus give a greater weighting to those samples which lie closer to the long axis of the ellipse.

When the axis of the ellipse (i.e. the axis of geometric or geostatistical anisotropy) is inclined to the local coordinate grid, then the distance d' used for inverse distance weighting of a sample at point B (Figure 3.37(b) will be:

$d' = (X'^2 + Y''^2)^{1/2}$ assuming d' is less than 'a'

where $Y'' = Y' \times a/b$ and $x' = d \times \cos(\phi + \theta)$, using $Y' = d \times \sin(\phi + \theta)$ and $d = (XD_i + YD_i)^{1/2}$.

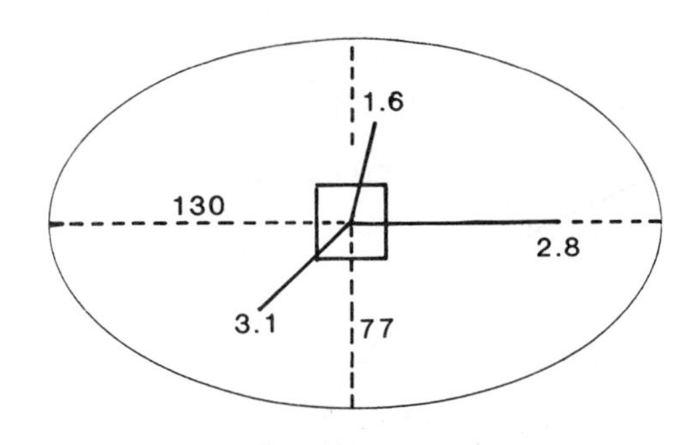

λ = 130/77 = 1.69

RESULTS

% Weighting to each Grade				Weighted Block Grade
n	1.6%	3.1%	2.8%	
1	36.8	35.2	28.0	2.46
2	40.1	36.7	23.2	2.43
3	43.2	37.9	18.9	2.40

Fig. 3.38 Inverse distance weighting – elliptical search area.

The effects of applying elliptical search areas can be seen in Figure 3.38 which uses the same data as was used for the circular search (Figure 3.36). As can be seen, there is little difference in the block grades as '*n*' is varied but the grade is significantly higher than that obtained using a circular search area. More weight has been given to the more remote value of 2.8% as it lies on the axis of anisotropy. Again thickness has been ignored as a weighting factor.

Sector search methods

This variant of inverse weighting divides the area around the block centre to be evaluated into eight sectors and then proceeds to search for the nearest specified number of samples in each sector in turn. Usually, an eight-point sector search is used but this can be varied by the user who also has the ability to set a fixed limit to the distance it may search in each sector. Each sample is then weighted, as before, to obtain the block value. A maximum of 64 samples would be used in an eight-point sector search, although some sectors may reach the set distance limit before eight points are located and thus a smaller data set would be used.

This method reduces the bias incurred when denser sampling exists to one side of the block under evaluation. Problems still exist, however, for blocks at the orebody fringes where some sectors will be totally empty.

Alternative weighting procedure

If a circular search area of radius S is established around an ore-block, and the distance of a sample from the centre is D, then an alternative weighting factor of this sample value is as follows:

$$W = (S^2 - D^2)/D^2$$

Thus, if the sample lies on the perimeter, the weighting factor is set to zero. In a computer program, therefore, once the calculated value of W becomes negative, the sample is rejected and not used to evaluate the current block.

Cross-validation for inverse distance weighting

To test how well inverse distance methods are able to estimate points in an orebody, the process of cross-validation can be used. This is effectively the equivalent of point-kriging (Chapter 4) in which the value at each sample point/drill-hole is removed in turn and the value of this point estimated from the surrounding samples. The same search area and '*n*' value are used as was the case in the block evaluation. The calculated values are then compared to the actual drill-hole values by calculating the correlation coefficient (p) and plotting a regression of one against the other, to which a best-fit line can be fitted. The latter should have a slope of 45° and p should be strongly positive. Figure 3.39 shows an example of such an exercise in the Offin River placer gold deposit, Ghana. The effect of using different

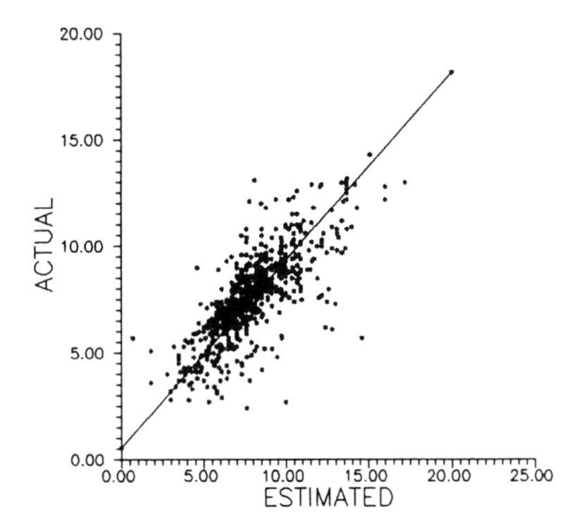

Fig. 3.39 Cross-validation by IDW weighting of thickness data from a portion of the Offin River placer gold deposit, Ghana. (Search radius 100 m, weighting factor $1/d^2$, correlation coefficient +0.792; from Boakye (1989).)

search radii, 'n' values and search areas can also be assessed.

3.11.2 3D search volumes

The use of 3D search volumes is most applicable to large equi-dimensional bodies such as porphyry copper-molybdenum deposits to be mined by open-pit methods. The method obviates the need to calculate bench composite grades, for each individual assay sample can play a role in determining the block grade by inverse distance weighting in three dimensions. The search volume selected can be varied by the geologist depending on the local geological conditions. The choice rests between a rectangular prism, a parallelopiped, a sphere, an ellipsoid or a discoid. The end product of the exercise is a 3D block model for the mineral deposit from which a mineral inventory or ore-reserve can be calculated, or an open-pit design produced (Chapter 5).

Table 3.13 shows an example of a small data set (S_1–S_5) which has been treated by spherical and ellipsoidal distance weighting methods. In the case of the former method, all samples were found to lie in the search sphere because their distances from the centre are less than its radius (100 m). As might be expected, sample S_2 has a very small impact on the final result. In the case of the ellipsoidal search volume, the sample coordinates were transformed by multiplying the YD_i values (section 3.11.1) by the anisotropy coefficient in the X–Y plane (a/b = 100/60 = 1.67) and by multiplying the ZD_i values (difference of elevation of each sample and the block centre) by the anisotropy coefficient in the X–Z plane (a/c = 100/40 = 2.50). From these new coordinates, the inclined distance to the block centre can be calculated. Table 3.13 shows that the distance calculated for S_5 is greater than 100 m, the length of the long axis of the ellipse, and this sample was thus excluded. The remaining samples thus give a higher block grade of 2.87%, as S_5 had a low grade of 1.30%. Notice also that the weighting given to S_2 has shown a marked increase as, though it is a long way from the block centre compared to the others, it lies exactly on the major axis of the ellipsoid (or on the axis of anisotropism).

At Brenda Open-pit, at Peachland in British Columbia, the copper-molybdenum mineralization is largely concentrated in a series of subvertical fractures, striking 60°T, cutting the host biotite-hornblende-quartz diorite. For ore-reserve purposes, each bench was subdivided into 50 ft (15 m) cubes. Those blocks which were intersected by a drill-hole were assigned a grade equal to the weighted grade of that part of the hole within the block so long as its length exceeded 8 ft (2.4 m). If this was not the case, or no intersection existed, then a parallelopiped search volume was erected around the block. The parallelopiped (Figure 3.40) was orientated by computer so that its long axis was parallel to the main fracture direction (e.g. 60°) and the short axis was then orientated down-dip (in this case vertical, so producing a rectangular prismatic search volume). The dimensions usually used are shown on the diagram and are based on drill-hole spacing and local geology. All assays in the

Table 3.13 Comparison of spherical and ellipsoidal weighting techniques

Sample data	X	Y	Z	Grade (%)
S_1	10.00	40.00	0.00	1.60
S_2	90.00	0.00	0.00	2.80
S_3	−40.00	−35.00	0.00	3.10
S_4	−35.00	20.00	22.00	4.10
S_5	50.00	−40.00	30.00	1.30

	Spherical IDW*			Ellipsoidal IDW†		
	d	$1/d^2$	WF	d	$1/d^2$	WF
S_1	41.23	0.00059	0.34	67.41	0.00022	0.30
S_2	90.00	0.00012	0.07	90.00	0.00012	0.17
S_3	53.15	0.00035	0.20	70.73	0.00020	0.27
S_4	45.92	0.00047	0.27	73.22	0.00019	0.26
S_5	70.71	0.00020	0.11	112.11	—	—
Σ		0.00174	1.00		0.00073	1.00
Weighted grade		2.64			2.87	

*Radius = 100 m.
†a = 100 m, anisotropy X–Y = 1.67.
b = 60 m, anisotropy X–Z = 2.50.
c = 40 m.

search volume were then weighted by the inverse square method.

Prior to the introduction of geostatistical ore-reserve techniques at Lornex (now merged with Valley Copper) in the Highland Valley of British Columbia, the bench reserves were calculated using computerized ellipsoidal search methods combined with inverse cube weighting. Benches 40 ft (12.2 m) thick were subdivided into 100 ft by 100 ft (30.5 × 30.5 m) squares and bench composite grades determined from four 10 ft (3.05 m) samples in each drill-hole. The assay grade was expressed in copper equivalents based on (1.1 × Cu% + 3 × Mo%) to allow for the presence of molybdenite in the ore and for a 10% discrepancy between drill-hole grades and mill-head grades that had become apparent over a period of time. As mineralization is strongly fracture controlled, each bench was subdivided into structural domains each containing a homogeneous fracture set (Figure 3.41). Each block, in each domain, then became the centre of an initially discoid search volume whose equal axes lay in the plane of the dominant fracture and the shorter axis perpendicular to it. Typical dimensions were 500 ft (152 m) in the plane of the fracture and 300 ft (91.4 m) at right angles. All bench composite grades within the search area were then weighted by the inverse cube of the distance method. If only one bench intersection grade was found in the discoid then a larger (e.g. 350 ft or 107 m radius) spherical search volume was applied.

In both examples cited, a geostatistical study

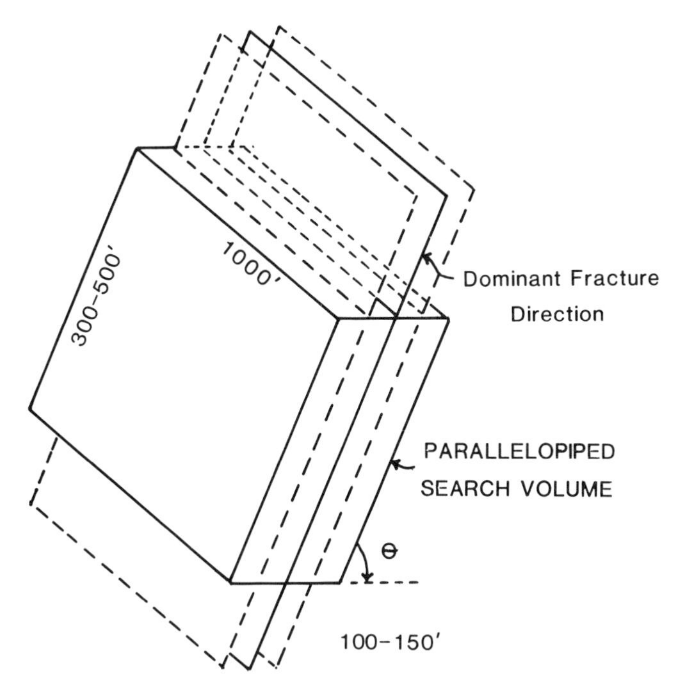

Fig. 3.40 Typical 3D search volume as applied at Brenda Open-pit, British Columbia. The parallelopiped becomes a rectangular prism if $\theta = 90°$.

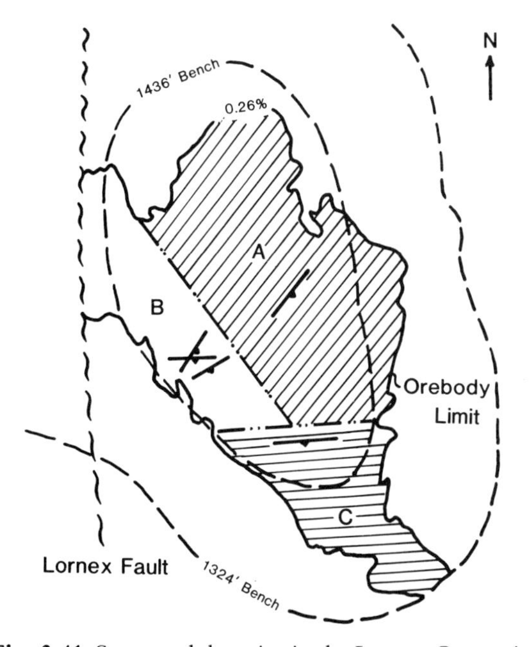

Fig. 3.41 Structural domains in the Lornex Open-pit, British Columbia. Diagram shows the dominant fractures in each domain.

would allow the dimensions and orientation of the 3D search volume to be determined by a more rigorous analysis of the nature of the mineralization and thus avoid the use of largely arbitrary values.

3.11.3 General comments

Inverse weighting is a smoothing technique, and as such is unsuited to deposits that have sharply defined boundaries and a very sudden fall-off in grade. In these situations, the method tends to produce larger tonnages at lower grade than actually exist, which may thus seriously affect the results of any economic feasibility study. This smearing-out effect is particularly marked if 2D search areas are used to evaluate an inclined tabular orebody on the basis of open-pit benches. It is evident, therefore, that inverse distance weighting works best for mineralization which displays a gradual decline in grade across its economic fringes. It is thus ideal for porphyry deposits, some alluvial or eluvial deposits and for limestones. Barnes (1980) also stresses that the computer program should be designed to recognize that the samples it is using to produce a block grade are taken from the same mineralogical population as the block itself. In other words, changes in host-rock lithology, or metallurgical type, should be recorded with assay data so that the program can accept, or reject, values depending on the geological location of the block being evaluated.

3.12 OREBODY MODELLING USING IDW METHODS

A variant of IDW techniques can be used to determine the shape, and hence volume, of an inclined tabular orebody, or portion thereof. It is particularly suited to the estimation of stope or mining block reserves.

Initially the computer program calculates the 3D coordinates of a series of regularly spaced points down each drill-hole together with the assay hangingwall, footwall and orebody mid-point. This is accomplished from the collar coordinates, collar inclination and bearing, bearing and depth of all internal surveys and finally the end-of-hole depth (EOH). The program then calculates the horizontal projection distances of the hangingwall and footwall intersection points from the mine VLP, together with the VLP X-coordinates of these points. The elevation remains unchanged.

From this irregularly spaced information, IDW block matrices are produced, using a circular search area, of the hangingwall horizontal projection distances, and then the footwall horizontal projection distances (i.e. D_1 and D_2 respectively, in Figure 3.42). Once these two VLP plans are produced, they can be carefully vetted by the geologist and modified to improve the structural interpretation. This procedure is aided by the production of contour maps from the grid data. Many contouring packages (e.g. Calcomp GPCP) use IDW weighting to produce a regular data matrix from irregular point information, from which contour plots are then produced. The next stage in the operation is to undertake a 'grid to grid' subtraction which yields a new data matrix which corresponds to the horizontal thickness of the orebody as in the lower portion of Figure 3.42. This can be used to determine the volume, and hence tonnage, of specific areas on the VLP. Metal accumulation values from each drill-hole are projected in the usual way on to VLPs from the orebody mid-points and gridded by IDW methods, so that each block on the horizontal thickness VLP now has a metal accumulation assigned to it. From this information, the weighted grade of the specific area can be determined as in section 3.10.1.

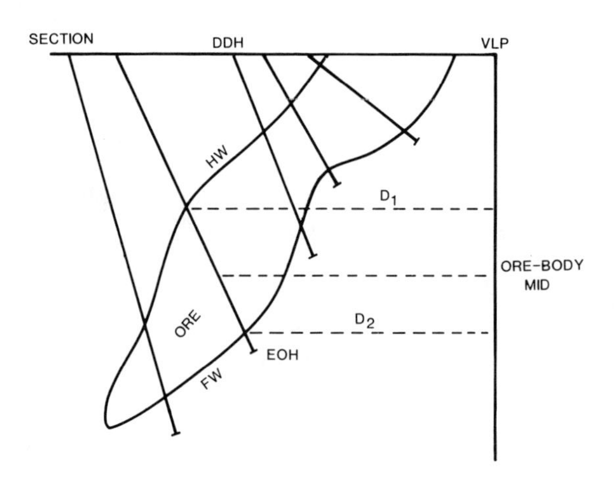

GRID TO GRID SUBTRACTION

VLP – HW

115	110	107	111
121	117	125	130
129	127		

VLP – FW

73	71	65	57
77	82	71	62
84	87		

VLP – HORIZ. THICKNESS

42	39	42	54
44	35	54	68
45	40		

Fig. 3.42 Orebody modelling using IDW block matrices.

Bibliography

REFERENCES

Association of Professional Engineers of the Province of Ontario (1976) *Performance Standards for Professional Engineers Advising on and Reporting on Oil, Gas and Mineral Properties*, APEO, pp. 1–15.

Australasian and Australian Mining Industry Council (Joint Committee) (1987) *Reporting of Mineral Resources and Ore Reserves* (Revised March 1987).

Australasian IMM and Australian Mining Industry Council (1989) Australasian Code for Reporting Identified Mineral Resources and Ore Reserves, *(Report of the Joint Committee, February 1989)*, 8 pp.

Baker, C. K. and Binns, M. J. (1987) Resource Estimation from a Diverse Data Source – Golden Plateau Ore Body, Cracow, Queensland, in *Resources and Reserve Symposium, Sydney Branch, Aust. IMM*, November 1987, pp. 31–37.

Barnes, P. B. (1980) *Computer-Assisted Mineral Appraisal and Feasibility*, Society of Mining Engineers, American Institution of Mining, Metallurgy and Petroleum Engineers. New York, 167 pp.

Boakye, E. B. (1989) Statistical and Geostatistical Evaluation of an Alluvial Gold Deposit in the Central Region of Ghana, Unpublished PhD Thesis, University of Wales, Cardiff, 334 pp.

Camisani-Calzolari, F. A. G. M. (1983) Pre-check for payability. *Nuc. Active*, **29**, 33–36.

Camisani-Calzolari, F. A. G. M., De Klerk, W. J. and Van der Merwe, P. J. (1985) Assessment of South African uranium resources: method and results. *Trans. Geol. Soc. South Afr.*, **88**, 83–97.

Carras, S. N. (1984) Comparative Ore Reserve Methodologies for Gold Mine Evaluation, in *Aust. IMM, Perth and Kalgoorlie Branches, Regional Conference, Gold Mining, Metallurgy and Geology*.

Diehl, P. and David, M. (1982) Classification of ore reserves/resources based on geostatistical methods. *CIM Bull.*, **75**, (838) 127–135.

Howe, A. C. A. and McCarthy, J. (1987) Canadian and American Guidelines on Ore Reserve Reporting, and a Comparison with those Proposed for Australia, in *Aust. IMM Sydney Branch, Resources and Reserves Symposium*, November 1987, pp. 23–25.

Koch, G. S. Jr and Link, R. F. (1970) *Statistical Analysis of Geological Data*, Wiley, New York.

Lafleur, P-J. (1986) Inplementation of computerized ore-reserves estimation of the Mobrun deposit, Rouyn, Quebec, in *Ore Reserve Estimation: Methods, Models and Reality. Proceedings of the CIMM Symposium, Montreal.*

Lane, K. F. (1988) *The Economic Definition of Ore: Cut-off Grades in Theory and Practice*, Mining Journal Books, London, 149 pp.

Lepeltier, C. (1969) A simplified statistical treatment of geochemical data by graphical representation. *Econ Geol.*, **64**, 538–550.

McKelvey, V. E. (1972) Mineral resource estimates and public policy. *Am. Sci.*, **60**, 32–40.

Peters, W. C. (1978) *Exploration and Mining Geology*, John Wiley, New York, 696 pp.

Rendu, J. M. (1981) *An Introduction to Geostatistical Methods of Mineral Evaluation*, South Afr. IMM, Johannesburg, 84 pp.

Smith, I. H. (1987) The Geology, Exploration and Evaluation of the Gold Deposits of Suriname, Unpublished PhD Thesis, University of Wales, Cardiff.

US Bureau of Mines and US Geological Survey (1980) *Resource/Reserve Classification System* (US Geol. Surv. Circular C831, Washington DC), June, 5 pp.

Wainstein, B. M. (1975) An extension of lognormal theory and its application to risk analysis models for new mining ventures. *J. South Afr. IMM*, **75**, 221–238.

Watermeyer, G. A. (1919) Application of the theory of probability in the determination of ore reserves. *J. Chem. Metall. Min. Soc. South Afr.*, **20**, 97–107.

Wellmer, E. W. (1989) *Economic Evaluations in Exploration*, Springer, Berlin, 180 pp.

BIBLIOGRAPHY

Carras, S. N. (1986) Concepts for Calculating Recoverable Reserves for Selective Mining in Open Pit Gold Operations, in *Aust. IMM Perth Branch. Selective Open Pit Gold Mining Seminar)*, 32 pp.

David, M., Froidevaux, R., Sinclair, A. J. and Vallée, M. (1986) Ore Reserve Estimation Methods, Models and Reality, in *Proceedings of CIMM Symposium, Montreal)*, May.

Hazen, S. W. Jr (1968) Ore reserve calculations. *Ore Reserve Estimation and Grade Control, CIMM*, **9**, (Spec. Vol.), 11–31.

Krige, D. G. (1966) Two dimensional weighting moving average trend surfaces for ore evaluation, in

Symposium on Mathematical Statistics and Computer Applications in Ore Evaluation, South Afr. IMM, Johannesburg.

McKinstry, H. E. (1948) Sampling ore and calculating tonnage, in *Mining Geology*, Prentice Hall, New York, pp. 35–39 (680 pp.).

O'Brian, D. T. and Weiss, E. (1968) Practical aspects of computer methods in ore reserve analysis. *Ore Reserve Estimation and Grade Control*, **9**, (CIMM Spec. Vol.), 109–113.

Parks, R. D. (1949) *Sampling, Examination and Valuation of Mineral Property*, 3rd edn, Addison-Wesley, Cambridge, MA.

Popoff, C. C. (1966) *Computing Reserves of Mineral Deposits: Principles and Conventional Methods* (Information Circular 8283), US Bureau of Mines, Washington, 113 pp.

Raymond, G. (1979) Ore estimation problems in an erratically mineralized orebody. *CIM Bull.* **72** (806), 90–8.

Sichel, H. S. (1951) New methods in the statistical evaluation of mine sampling data. *Trans. IMM*, **61**, 261–288.

Sichel, H. S. (1966) The estimation of means and associated confidence limits for small samples from lognormal populations. Symposium on Mathematical Statistics and Computer Applications in Ore Valuation. *J. South Afr. IMM*, (Spec. Iss.), 106–123.

Storrar, C. D. (1977) *South African Mine Valuation*, Chamber of Mines of South Africa, Johannesburg, 472 pp.

Taylor, H. K. (1972) General background theory of cut-off grades. *Trans. A. IMM*, **81**, A160–A179.

United Nations (1979) International classification of mineral resources. *Mining Mag.*, June, 535–536.

APPENDIX 3.1

USBM/USGS Classification of Resources and Reserves

TOTAL RESOURCES

These include all minerals having present or future value and comprised of identified minerals together with those presumed to exist from geological evidence. Total resources consist of two components, i.e. reserves and resources as defined below.

RESOURCE

This is defined as a concentration of naturally occurring solid, liquid or gaseous materials in, or on, the Earth's crust in such a form that economic extraction of a commodity is currently or potentially feasible. Resources can be subdivided into two main groups, i.e. identified-subeconomic resources and undiscovered resources. These are shown in the summary diagram (Table 3.2).

Identified-subeconomic resources are materials that cannot be classified as reserves but which may become so as a result of changes in the economic and legal environment. Paramarginal identified-subeconomic resources are those which are close to becoming economic at current metal prices or mining/processing costs or which are downgraded because of legal or political conditions. Submarginal identified-subeconomic resources require a substantially higher price, or a major cost reducing advance in technology, for their upgrading to the paramarginal or even into the reserve category.

Undiscovered resources are subdivided into hypothetical and speculative resources. The former are undiscovered materials that might reasonably be expected to exist in a known mining district under known geological conditions. Speculative resources are undiscovered materials that may occur either in known types of deposits in a favourable geological setting where no discoveries have been made, or in as yet unknown types of deposits that remain to be recognized. In both cases, exploration that confirms their existence and reveals quantity and quality will permit their reclassification as reserves or identified subeconomic resources.

RESERVE

The following classification is taken verbatim from the USBM/USGS recommendation.

Measured ore (proved) is ore for which tonnage is computed from dimensions revealed in outcrops, trenches, workings and drill-holes and for which the grade is computed from the results of detailed sampling. The sites for inspection, sampling and measurement are so closely spaced and the geological character is so well defined that the size, shape and mineral content are well established. The computed tonnage and grade are judged to be accurate within limits which are stated, and no such limit is judged to differ from the computed tonnage or grade by more than 20%.

Indicated ore (probable) is ore for which tonnage and grade are computed partly from specific measurements, samples or production data and partly from projection for a reasonable distance on geological evidence. The sites available for inspection, measurement and sampling are too widely or otherwise inappropriately spaced to outline the ore completely or to establish its grade throughout.

Inferred ore (possible) is ore for which quantitative estimates are based largely on broad knowledge of the geological character of the deposit and for which there are few, if any, samples or measurements. The estimates are based on an assumed continuity or repetition for which there is geological evidence; this evidence may include comparison with deposits of similar type. Bodies that are completely concealed may be included if there is specific geological evidence of their presence. Estimates of inferred ore should include a statement of the special limits within which the inferred ore may lie.

APPENDIX 3.2

APEO Classification of Reserves

The Association of Professional Engineers of the Province of Ontario (APEO) (1976–1985) have proposed the following definitions for reserves:

Mineral inventory: 'Uncategorized reserves calculated within stated minimum grade or quality and width or thickness parameters, without regard to internal sample spacing, for the purpose of quantifying the mineral content of a deposit.'
Ore: 'A mineral or metal bearing rock that is currently being mined (and processed) or that which could be mined and sold at a profit according to reasonable current expectations as set in a formal feasibility study. Ore should be defined in a formal reserve statement by a professional geologist.'
Ore-reserve statement: 'Ore-reserves statements should be accompanied by details of sampling grid, sampling and assay methods, calculation methods and confidence levels for estimates, cut-off grades and working methods'. An annual balance sheet should be included showing the changes in mineral inventories during the year. These include:

(1) Gains or losses due to additional drilling and development work;
(2) Gains or losses due to economic and/or technical changes which affect COG.

Reserve: 'A tonnage or volume of rock whose grade limits, mineralogy and other characters are known with a qualified and explicit degree of knowledge, in relation to a defined sampling grid and other related information and tests.'
Ore reserve: 'The reserves estimated for an ongoing mining operation, or related to a plan for a proposed mining operation, set out in a formal feasibility study when technical and economic feasibility is established.'
Geological reserves: '*In situ* material calculated from sampling data prior to any allowance for either tonnage or grade dilution.'
***In situ* reserves**: 'Reserves of mineralized rock that could be mined under specific economic or technical circumstances. They are not yet under development in an operating mine.'
Mineable reserves: 'That material demonstrated on the basis of a mining plan to be recoverable or extractable from the proven and probable reserves and deliverable to the mine exit.'

APPENDIX 3.3

AIMM/AMIC Classification of Resources and Reserves

RESOURCES

A resource is an *in situ* mineral occurence quantified on the basis of geological data and a geological cut-off grade only. Categories recognized are as follows:

Inferred. This is an estimate, inferred from geoscientific evidence, drill-holes, underground openings, or other sampling procedures and before testing and sampling information is sufficient to allow a more reliable and systematic estimation.

Indicated. These are sampled by drill-holes, underground openings, or other sampling procedures at locations too widely spaced to ensure continuity but close enough to give a reasonable indication of continuity and where geoscientific data are known with a reasonable level of reliability.

Measured. These are intersected and tested by drill-holes, underground openings, or other sampling and procedures at locations which are spaced closely enough to confirm continuity and where geoscientific data are reliably known.

RESERVES

The term 'ore-reserve' means that part of a 'measured' or 'indicated' resource, which could be mined including dilution and from which valuable or useful minerals could be recovered economically under conditions realistically assumed at the time of reporting.

Probable. These reserves are stated in terms of mineable tonnes/volumes and grades where the conditions are such that ore will probably be confirmed but where the *in situ* identified resource has been categorized as 'indicated' and has not been defined with the precision necessary for the 'measured' category. Probable ore-reserves include ore that has been sampled on a pattern too widely spaced to ensure continuity.

Proved. These are reserves stated in terms of mineable tonnes/volumes and grades in which the identified *in situ* resource has been defined in three dimensions by excavation or drilling, and should include additional minor extensions beyond actual openings and drill-holes, where the geological factors that limit the orebody are known with sufficient confidence, that it is categorized as a 'measured resource'.

APPENDIX 3.4

Coal Resources and Reserves

The following classification, which is specific for coal, is based on the 'Code for Calculating Resources and Reserves', Standing Committee on Coalfield Geology of New South Wales (June, 1984), and has been accepted by the Australian Minerals and Energy Council (AMEC). The following definitions have been taken verbatim from this Australian Code for Reporting Identified Coal Resources and Reserves (1986).

DEFINITIONS

(1) Coal resources are all of the potentially useable coal known in a defined area, and are based on points of observation and extrapolations from these points. Potentially useable coal is defined as coal which has, or could be beneficiated to give a quality acceptable for commercial usage in the foreseeable future and excludes minor coal occurrences.

(2) Coal reserves are those parts of the coal resources for which sufficient information is available to enable detailed or conceptual mine planning and for which such planning has been undertaken.

(3) A point of observation is an intersection, at a known location, of coal-bearing strata, which provides information about the strata by one or more of the following methods:

- Observation, measurement and testing of surface or underground exposure.
- Observation, measurement and testing of bore core.
- Observation and testing of cuttings, and use of down-hole geophysical logs of non-cored boreholes.

Geophysical techniques such as seismic surveys are not direct points of observation but may increase confidence in the continuity of seams between points of observation, especially in the broader resource categories.

The distances between points of observation and extrapolations from points of observation quoted for each resource category are normally the maximum under favourable geological conditions. More closely spaced points of observation will be required in areas where faulting, intrusion, seam splitting and other breaks in seam continuity are known to occur, or where the seam is subject to significant variation in thickness or quality.

CATEGORIES OF RESOURCES

(1) Measured resources are those for which the density and quality of points of observation are sufficient to allow a reliable estimate of the

coal thickness, quality, depth and *in situ* tonnage. Points of observation should provide a level of confidence sufficient to allow detailed planning, costing of extraction and specification of a marketable product. The points of observation generally should not be more than 1 km apart. Where geological conditions are favourable it may be possible to extrapolate known trends a maximum distance of 0.5 km from points of observation.

(2) Indicated resources are those for which the density and quality of points of observation are sufficient to allow a realistic estimate of the coal thickness, quality, depth and *in situ* tonnage and for which there is reasonable expectation that the estimate of resources will not vary significantly with more detailed exploration. Points of observations should provide a level of confidence sufficient to enable conceptual planning of extraction and to determine the likely quality of the product coal. Points of observation generally should not be more than 2 km apart. Where geological conditions are favourable, it may be possible to extrapolate known trends a maximum distance of 1 km from points of observation.

(3) Inferred resources are those for which the points of information are widely spaced and as a result, assessment of this type of resource may be unreliable. Points of observation should allow the presence of coal to be unambiguously determined.

Class 1 are those resources for which the points of observation allow an estimate of the coal thickness and general coal quality to be made, and the geological conditions indicate continuity of seams between the points of observation. Points of observation generally should not be more than 4 km apart. Extrapolations of trends should not extend more than 2 km from the points of observation.

Class 2 are those resources for which there is limited information and, as a result, the assessment of this type of resource may be unreliable. Provided the coal thickness can be determined, the order of magnitude of inferred resources Class 2 may be expressed within the following ranges:

1–10 million tonnes
10–100 million tonnes
100–500 million tonnes
500–1000 million tonnes
> 1000 million tonnes

TYPES OF RESERVES

(1) Mineable *in situ* reserves are the tonnages of *in situ* coal contained in seams or sections of seams for which sufficient information is available to enable detailed or conceptual mine planning and for which such planning has been undertaken. Mineable *in situ* reserves may be calculated only from measured and indicated resources. Measured resources are required for detailed mine planning, and are the preferred basis for mineable *in situ* reserves. Indicated resources may be used for conceptual mine planning. In general, further exploration will be required prior to commencement of mining operations. Mineable *in situ* reserves should be quoted separately for surface and underground mines and an outline of the proposed mining method(s) should be provided.

(2) Recoverable reserves are the tonnages of mineable *in situ* reserves that are expected to be recovered, i.e. that proportion of the seam(s) which will be extracted. If dilution is added to the recoverable reserves tonnage, the total equates to the 'run-of-mine' tonnage. If allowance is made for dilution it should be stated. In calculating recoverable reserves, a mining recovery factor must be applied to the mineable *in situ* reserves. This factor will depend on the mining method to be used. Unless a specific factor has been determined for conceptual studies, the historically proven mining recovery factor should be used. If information is not available, a mining recovery factor of 50% for underground reserves and 90% for surface reserves may be applied. An outline of the proposed mining method should accompany any statement of recoverable reserves.

(3) Marketable reserves are the tonnages of coal that will be available for sale. If the coal is to be marketed raw, the marketable reserves will be the same as the recoverable reserves plus dilution, i.e. the 'run-of-mine' tonnage. If the coal is to be beneficiated, marketable reserves are calculated by applying the predicted yield to the recoverable reserves. The basis of the predicted yield should be stated, e.g. 200 mm cores, slim cores, pretreated cores.

REPORTING OF RESOURCES AND RESERVES

All factors used to limit resources and reserves and necessary to verify the calculations (including the types of observations, e.g. cored hole, outcrop) must be stated explicitly. The relative density value adopted in calculating the coal tonnage should be noted, together with the evidence on which it is based. Tonnage estimates always should be rounded, commensurate with the accuracy of estimation.

Resource and reserves should be stated:

(1) For each seam.

(2) On a depth basis, in regular depth increments if sufficient information is available.

(3) On a seam thickness basis, the minimum thickness used should be stated and separate tonnages should be quoted for seams less than 1.5 m thick and seams equal to or greater than 1.5 m (this limit may be greater for brown coal, e.g. 3 m). The maximum thickness of any included non-coal bands should be stated. Normally where seams contain a non-coal band thicker than 0.3 m the two coal splits should be considered as separate seams and tonnages should be reported for each (the limit for non-coal bands may be greater for brown coal sequences, e.g. 1 m).

(4) On a quality basis, maximum raw coal ash should be stated and only that coal which can be used or beneficiated at an acceptable yield (to be stated) should be included in the estimate. Other raw coal quality parameters, particularly those which affect utilization behaviour, should be stated and further subdivision of the resources made if significant variations occur, e.g. heat-affected coal, oxidized coal.

In addition, for reporting of reserves the following information is required, as a minimum:

(1) An outline of the proposed mining method.

(2) Physical criteria limiting mining such as maximum and minimum working section, thickness, minimum separation of seams, maximum dip, geological structure, areas of prohibition.

(3) Quality criteria limiting mining such as ash content, volatile matter, yield, etc.

(4) For recoverable reserves, the mining recovery factor used.

(5) For marketable reserves, the predicted yield if the coal is to be beneficiated and the quality specification of the product coal.

(6) The overburden ratio expressed as bank cubic metres of overburden to tonnes of coal *in situ* for reserves amenable to surface mining.

(7) The depth of planning mining.

(8) The percentage of the resources which are the mineable *in situ* reserves within the area(s) proposed to be mined.

APPENDIX 3.5

Ore-Reserve Calculation – Worked Example

This worked example is designed to illustrate the way in which individual intersections can be assessed against specific economic criteria and then combined to produce a reserve estimate of a block of ore. Figure 3.43 shows a gold-bearing vein which dips northwards at 52° and which has been channel sampled at regular intervals along the back of an exploration drive. The grade (g/t) and the thickness (cm) of the vein at each sampling position are listed on the diagram. Given that the minimum mining grade is 5 g/tonne over a stoping width of 1.5 m (footwall to hangingwall), we have to determine whether the ore could be worked at a profit given the following information:

Estimated mining cost:	$23 per tonne ore
Estimated milling/concentration cost:	$12 per tonne ore
Administrative overheads:	$4 per tonne ore
Metallurgical recovery:	93.2%
Extraction ratio:	0.8
Unintentional dilution due to hangingwall instability:	approx. 10%
Gold price:	$457.60 (June 1988)
Tonnage factor:	2.68 tonnes per m^2

Proposed mining levels at 60 m vertical depth spacing.

CALCULATION

The first assumption that must be made is how far the information presented by the drive can be extrapolated upwards and downwards. At this stage, this will have to be half the distance to the next level both upwards and downwards, i.e. 30 m + 30 m = 60 m. As all the samples have been taken horizontally in the back, the MSW must be converted to its horizontal equivalent as follows:

$$\text{Horiz. MSW} = 1.5/\sin 52° = 1.90 \text{ m}$$

This step is also necessary because the ore-reserves will be calculated using the vertical height of the ore-block and not the dip-length.

Knowing that the MMG is 5 g/t, we can now calculate the minimum horizontal metal accumulation for an economic intersection; i.e. 190 cm × 5 g/t = 950 cm g/t. Next, the metal accumulation is calculated for each channel in turn and the result compared with the above value, as in Table A3.1.

The assay cut-offs are located as shown in the table; the western limit lying half-way between positive and negative channels, and the eastern limit at the fault. The strike length is then measured from Figure 3.43 as 69 m. The next step is to determine the average thickness of **ore** along this strike length which, because the samples are uniformly spaced, is the arithmetic mean of the stope widths. Where the vein is less than 190 cm wide, this value is substituted for the vein width

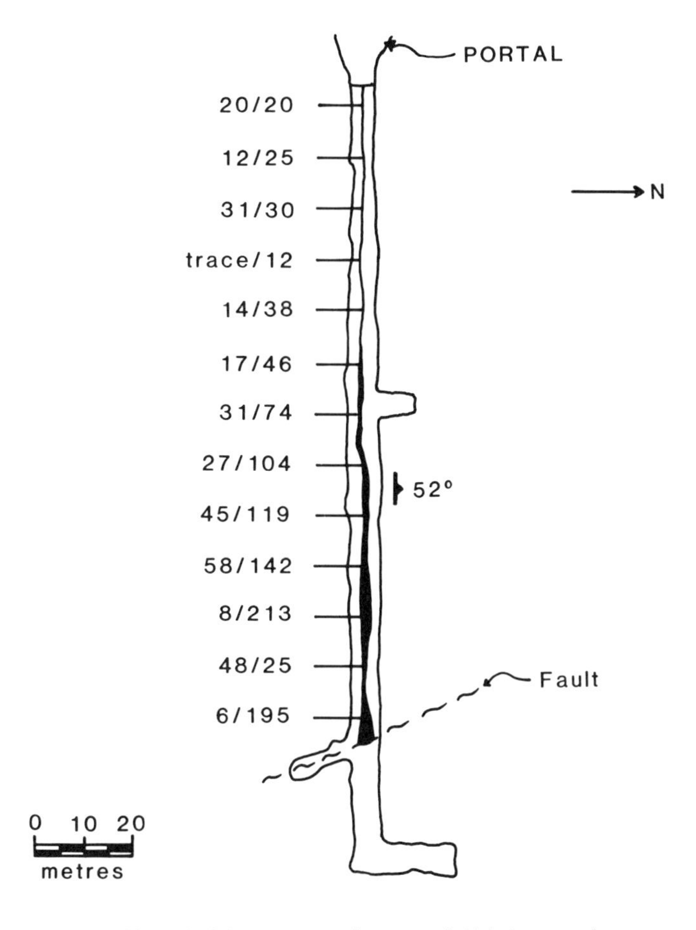

Fig. 3.43 Channel sample assay data for an auriferous quartz vein in an exploration adit.

to determine the stope width. In this case, five of the seven values are less than 190 cm.

$$\Sigma \text{ stope widths} = 1358 \text{ cm}$$

$$\text{Therefore mean width} = 1358/7 = 194 \text{ cm}$$

From this, the volume of ore can be computed as follows:

$$\text{Volume} = (1.94 \times 69 \times 60)\ \text{m}^3 = 8031.6\ \text{m}^3$$

Note. This calculation ignores ore removed during the development of the drive.

$$\text{Therefore tonnage} = (8031.6 \times 2.68) \text{ tonnes} = 21524.69 \text{ tonnes}$$

$$\text{Grade of ore} = \Sigma \text{ horizontal metal accumulation}/\Sigma \text{ stope widths}$$

$$= 22767/1358 = 16.77 \text{ g/t}$$

Note. This is the 'Geologist's Reserve' and makes no allowance for unintentional dilution and an extraction ratio.

Table A3.1 Channel values from gold vein (worked example B)

Channel values	Horiz. metal accumulation	
20/20	400	
12/25	300	
31/30	930	
Tr/12	—	
14/38	532	
17/46	782	
31/74	2294	Ore zone
27/104	2808	Ore zone
45/119	535	Ore zone
58/142	8236	Ore zone
8/213	1704	Ore zone
48/25	1200	Ore zone
6/195	1170	Ore zone

Mineable reserve:

(1) Allowing for extraction ratio:
$0.8 \times 21\,524.69 = 17219.75$ t

(2) Allowing for 10% dilution:
$1.1 \times 17\,219.75 = 18941.73$ t

Note. Dilution only applied to ore extractable.

If we assume that the 80% of the ore extracted has a grade equal to the overall *in situ* grade (i.e. 16.77 g/t), we can calculate the diluted grade from:

$$\text{Old tonnage} \times \text{old grade} = \text{new tonnage} \times \text{new grade}$$

This is because the contained metal remains unchanged assuming the dilution is by barren rock. Hence:

$$\text{Diluted grade} = 17\,219.75 \times 16.77/18\,941.73 = 15.25 \text{ g/t}$$

The mineable reserve is thus **18 942 t @ 15.25 g/t**

Total contained metal = 288 865.5 g
Recoverable metal = 269 222.6 g
or = 269 222.6/31.1 troy oz.
= 8656.7 troy oz.

Value = 8656.7×457.60
= $3961 306

Total mining cost per tonne of ore = 23 + 12 + 14 = $39
Therefore total mining cost = $39 \times 18\,942$
= $738 738

Profit margin = 3961 306 − 738 738
= $3222 568

Clearly very profitable!

APPENDIX 3.6
Program Listing for SG ORE

```
REM PROGRAM SGORE
REM COPYRIGHT A.E. ANNELS 1990
REM UNIVERSITY OF WALES, CARDIFF
REM THIS PROGRAM CALCULATES THE
  WEIGHT % OF ORE MINERALS
REM (CP, BN, CC, PO, PY, MAL, SPH,
  GAL, CARR) FROM SAMPLE ASSAY
  DATA
REM AND CALCULATES A WEIGHTED
  SPECIFIC GRAVITY

LABEL sgcalc
STREAM #1 : CLS #1: WINDOW #1 OPEN:
  WINDOW #1 FULL
ts = 0
wpcp = 0
wpbn = 0
wpcc = 0
wppy = 0
wppo = 0
wpcarr = 0
wpgal = 0
wpsph = 0
wpmal = 0
PRINT "DATA INPUT FOR SGORE"
PRINT " ~~~~~~~~~~~~~~~~ "
PRINT
INPUT "SAMPLE NO = ";sn$
INPUT "HOST ROCK SPECIFIC GRAVITY
  = ";hr
INPUT "INPUT THE GRADES OF
  CURRENT SAMPLE IN THE ORDER
  TCU,ASCU,CO,ZN,PB,S - WHERE
  VALUES DO NOT EXIST TYPE 0 ";tcu,
  ascu,co,zn,pb,s
IF tcu = 0 GOTO zinc
INPUT "ESTIMATE RELATIVE
  PERCENTAGES OF CP,BN,CC IN THIS
  ORDER";vcp,vbn,vcc
REM ESTIMATES ARE IN VOLUME % -
  NOW CONVERTED TO WT%
total = vcp*4.2 + vbn*4.9 + vcc*5.6
cp = vcp*4.2*100/total
bn = vbn*4.9*100/total
cc = vcc*5.6*100/total
wpcarr = co*2.355: carcu = wpcarr*0.229
scu = tcu - ascu - carcu
IF cp = 0 THEN GOTO nocp
wpcp = scu/(0.3464 + (bn*0.6332/cp) +
  (cc*0.7986/cp))
wpbn = bn*wpcp/cp
wpcc = cc*wpcp/cp
LABEL nocp
IF cp > 0 THEN GOTO scalc
wpcp = 0
wpbn = scu/(0.6322 + cc*0.7986/bn)
wpcc = wpbn*cc/bn
LABEL scalc
cps = wpcp*0.3493
bns = wpbn*0.2555
```

```
ccs = wpcc*0.2014
cars = wpcarr*0.3464
ts = ts + cps + bns + ccs + cars

LABEL zinc
IF zn = 0 GOTO lead
wpsph = zn*1.490
ts = ts + (wpsp*0.3290)

LABEL lead
IF pb = 0 GOTO oxide
wpgal = pb*1.155
ts = ts + (wpgal*0.134)

LABEL oxide
IF ascu = 0 GOTO sulphur
wpmal = ascu*1.745

LABEL sulphur
rs = s - ts
IF s = 0 GOTO listing
INPUT "WHAT IS THE RELATIVE % OF
  PY AND PO";py,po
ntot = py*5.0+po*4.6
py = py*5.0*100/ntot
po = po*4.6*100/ntot
wppy = rs/(0.534 + (po*0.4178/py))
wppo = po*wppy/py

LABEL listing
LPRINT TAB(26) "SAMPLE NO: ";sn$
LPRINT TAB(26) "*******************"
LPRINT
LPRINT TAB(27) "WT% CP =
  ";ROUND(wpcp,2)
LPRINT TAB(27) "WT% BN =
  ";ROUND(wpbn,2)
LPRINT TAB(27) "WT% CC =
  ";ROUND(wpcc,2)
LPRINT TAB(27) "WT% CAR =
  ";ROUND(wpcarr,2)
LPRINT TAB(27) "WT% MAL =
  ";ROUND(wpmal,2)
LPRINT TAB(27) "WT% GAL =
  ";ROUND(wpgal,2)
LPRINT TAB(27) "WT% SPH =
  ";ROUND(wpsph,2)
LPRINT TAB(27) "WT% PY =
  ";ROUND(wppy,2)
LPRINT TAB(27) "WT% PO =
  ";ROUND(wppo,2)
LPRINT

REM CALCULATION OF SG OF ORE
wphr = 100 - (wpcp + wpbn + wpcc + wpmal
  + wpcarr + wppy + wppo + wpgal +
  wpsph)
sgore = 100/(wpcp/4.2 + wpbn/4.9 + wpcc/5.6
  + wpmal/3.9 + wpcarr/4.9 + wppy/5.0 +
  wppo/4.5 + wpgal/7.5 + wpsph/4.1 + wphr/
  hr)
sgore = ROUND(sgore,3)
LPRINT "WEIGHTED SG FOR SAMPLE
  ";sn$; " = ";sgore
LPRINT " ~~~~~~~~~~~~~~~~~~
  ~~~~~~~~~~"
INPUT "DO YOU WISH TO CALCULATE
  ANOTHER SAMPLE? (Y,N)";as$
IF as$ = "Y" GOTO sgcalc
END


REM PROGRAM SGORE
REM COPYRIGHT A.E. ANNELS 1990
REM UNIVERSITY OF WALES, CARDIFF
REM THIS PROGRAM CALCULATES THE
  WEIGHT % OF ORE MINERALS
REM (CP, BN, CC, PO, PY, MAL, SPH,
  GAL, CARR) FROM SAMPLE ASSAY
  DATA
REM AND CALCULATES A WEIGHTED
  SPECIFIC GRAVITY

LABEL sgcalc
STREAM £1 : CLS £1: WINDOW £1 OPEN:
  WINDOW £1 FULL
ts = 0
wpcp = 0
wpbn = 0
wpcc = 0
wppy = 0
wppo = 0
wpcarr = 0
wpgal = 0
wpsph = 0
wpmal = 0
PRINT "DATA INPUT FOR SGORE"
```

```
PRINT " ~~~~~~~~~~~~~~~~~ "
PRINT
INPUT "SAMPLE NO = ";sn$
INPUT "HOST ROCK SPECIFIC GRAVITY
  = ";hr
INPUT "INPUT THE GRADES OF
  CURRENT SAMPLE IN THE ORDER
  TCU,ASCU,CO,ZN,PB,S – WHERE
  VALUES DO NOT EXIST TYPE 0 ";tcu,
  ascu,co,zn,pb,s
IF tcu = 0 GOTO zinc
INPUT "ESTIMATE RELATIVE
  PERCENTAGES OF CP,BN,CC IN THIS
  ORDER";vcp,vbn,vcc
REM ESTIMATES ARE IN VOLUME % –
  NOW CONVERTED TO WT%
total = vcp*4.2 + vbn*4.9 + vcc*5.6
cp = vcp*4.2*100/total
bn = vbn*4.9*100/total
cc = vcc*5.6*100/total
wpcarr = co*2.355: carcu = wpcarr*0.229
scu = tcu – ascu – carcu
IF cp = 0 THEN GOTO nocp
wpcp = scu/(0.3464 + (bn*0.6332/cp) +
  (cc*0.7986/cp))
wpbn = bn*wpcp/cp
wpcc = cc*wpcp/cp
LABEL nocp
IF cp > 0 THEN GOTO scalc
wpcp = 0
wpbn = scu/(0.6322 + cc*0.7986/bn)
wpcc = wpbn*cc/bn
LABEL scalc
cps = wpcp*0.3493
bns = wpbn*0.2555
ccs = wpcc*0.2014
cars = wpcarr*0.3464
ts = ts + cps + bns + ccs + cars

LABEL zinc
IF zn = 0 GOTO lead
wpsph = zn*1.490
ts = ts + (wpsp*0.3290)

LABEL lead
IF pb = 0 GOTO oxide
wpgal = pb*1.155
ts = ts + (wpgal*0.134)

LABEL oxide
IF ascu = 0 GOTO sulphur
wpmal = ascu*1.745

LABEL sulphur
rs = s – ts
IF s = 0 GOTO listing
INPUT "WHAT IS THE RELATIVE % OF
  PY AND PO";py,po
ntot = py*5.0+po*4.6
py = py*5.0*100/ntot
po = po*4.6*100/ntot
wppy = rs/(0.534 + (po*0.4178/py))
wppo = po*wppy/py

LABEL listing
LPRINT TAB(26) "SAMPLE NO: ";sn$
LPRINT TAB(26) "******************"
LPRINT
LPRINT TAB(27) "WT% CP =
  ";ROUND(wpcp,2)
LPRINT TAB(27) "WT% BN =
  ";ROUND(wpbn,2)
LPRINT TAB(27) "WT% CC =
  ";ROUND(wpcc,2)
LPRINT TAB(27) "WT% CAR =
  ";ROUND(wpcarr,2)
LPRINT TAB(27) "WT% MAL =
  ";ROUND(wpmal,2)
LPRINT TAB(27) "WT% GAL =
  ";ROUND(wpgal,2)
LPRINT TAB(27) "WT% SPH =
  ";ROUND(wpsph,2)
LPRINT TAB(27) "WT% PY =
  ";ROUND(wppy,2)
LPRINT TAB(27) "WT% PO =
  ";ROUND(wppo,2)
LPRINT

REM CALCULATION OF SG OF ORE
wphr = 100 – (wpcp + wpbn + wpcc + wpmal
  + wpcarr + wppy + wppo + wpgal +
  wpsph)
sgore = 100/(wpcp/4.2 + wpbn/4.9 + wpcc/5.6
  + wpmal/3.9 + wpcarr/4.9 + wppy/5.0 +
```

```
  wppo/4.5 + wpgal/7.5 + wpsph/4.1 + wphr/
  hr)
sgore = ROUND(sgore,3)
LPRINT "WEIGHTED SG FOR SAMPLE
  ";sn$;" = ";sgore
LPRINT " ~~~~~~~~~~~~~~~~~ "
  ~~~~~~~~~~
INPUT "DO YOU WISH TO CALCULATE
  ANOTHER SAMPLE? (Y,N)";as$
IF as$ = "Y" GOTO sgcalc END
```

4

Geostatistical Ore-reserve Estimation

4.1 INTRODUCTION

The basic theory behind geostatistical methods will be presented in this chapter in such a way as to minimize the use of mathematical expressions and notations. The intention is to give the mining geologist and engineer sufficient theory to be able to apply geostatistics in an intelligent and informed way. Considerable attention is thus given to the production and modelling of the semi-variogram, the basic tool of the geostatistician. This aspect is frequently underplayed which is a pity, for the whole process is totally dependent on the interpretation of the semi-variogram. The various stages of a geostatistical ore-reserve estimation procedure are explained, leading to the production of block (local) reserves and overall (global) reserves. The final stage of the whole exercise is the construction of a grade–tonnage curve.

4.2 THE APPLICATION OF GEOSTATISTICS

Geostatistical methods, and in particular the examination of the semi-variogram, can be used to determine:

(1) The optimum sample size.
(2) The optimun sample pattern.
(3) The optimum sample density.
(4) The area of influence of each sample which may be circular, elliptical, spherical or ellipsoidal.
(5) The nature of the mineralization, i.e. its characterization. Information gained from the semi-variogram allows us to determine the uniformity of the mineralization or the degree to which it has been concentrated by various processes during precipitation of the ore minerals, or remobilized during later metamorphism or secondary enrichment. An insight is gained into the relative importance of spatial and random influences operating during the mineralization process. Spatial controls may include such factors as distance from an igneous contact, a fault or a palaeo-shoreline, while random controls include fracture infillings or metamorphic lateral secretion veins.
(6) Predictability of grades, thickness, etc.

Other advantages include:

(1) The avoidance of arbitrary weighting methods such as those employed by inverse square distance.
(2) Reduction of the regression effect which results in high grades being overvalued, low grades undervalued.
(3) The application of an unbiased estimator of the grade of a deposit. Estimators include the

statistical mean, the log mean (geometric mean), the median, Sichel's *t* estimator, or the kriged mean (absolute or log). The optimum estimator is that which produces the best precision (the lowest variance).

(4) If the available data base is adequate, the method allows the determination of the best possible unbiased estimate of grade and tonnage. This is important when an operation is working close to its economic break-even point.

(5) Assignment of confidence limits and precision to estimates of tonnage and grade so that an assessment can be made as to whether they are acceptable, whether the sampling strategy should be modified and whether the improvement warrants the extra cost incurred.

4.2.1 Errors incurred in ore-reserve estimation

Many potential sources of error exist which may seriously affect the accuracy of an ore-reserve estimate. These combine to enhance the random component of the data variability (variance) and thus contribute to what is referred to as high nugget variance. In this section, we will examine some of these problem areas.

It is a pointless exercise quoting a **tonnage** to three or four significant figures when there are so many possibilities for error in its calculation. Errors can be introduced during the determination of:

(1) **Thicknesses**. This could be incorrect due to the misjudgement of intersected thickness, when core recoveries are poor, or to the use of an incorrect intersection angle. Variability of thickness tends to be limited over a range of a few metres (e.g. 0–30 m).

(2) **Area**. This is only an estimate based on extrapolation of ore beyond the limit of drilling, or between drill-holes, on a linear or arbitrary basis and could thus be incorrect.

(3) **Specific gravity/tonnage factor**. Few mines consider the effect of variations of specific gravity on the tonnage estimate and most apply a constant factor irrespective of changes in rock type (specific gravity range 2.4–3.2 t/m^3) or in grade of contained mineralization (specific gravity range 4.0–9.0 g/cm^3). A serious error may thus be incurred here. Computer estimation of the specific gravity is possible on the basis of rock type and assay data as explained in Chapter 3, section 3.4.3.

(4) **Core recovery**. This is usually measured but is rarely taken into account in ore-reserve calculation.

The problem is more serious in the case of grade which can fluctuate from a few parts per million to thousands of parts per million between samples. This is extremely large and thus estimation of grade is much less precise than for the tonnage of ore. The economics of an operation are thus disproportionately sensitive to grade. The wrong impression may be given if the grade of a deposit is quoted as 2.43% Cu if the confidence limits are very wide (implying a low confidence) or indeed are not assigned. The use of two places of decimals implies a high precision which might be acceptable in a stratiform deposit but which is unlikely in a porphyry copper deposit.

It is also a pointless exercise to try and reconcile the realized/extracted grade and the calculated grade, if the precision of the latter is poor. It is a matter of experience that the realized grade is usually less than the predicted grade by up to 10%, or even more in certain cases. This of course reflects both the computational methods used and abnormal dilution during efforts to maximize production. Often too little quality control is exercised in a desire to meet production targets which has a serious effect on grade, the most important indicator of the economic viability of a deposit.

The decision we need to make is how good our grade estimate needs to be having determined its precision. Many mines may accept that 10–20% is adequate, but recent calculations by the author have shown that, in the case of gold, the precision of grade estimates are often in the range 50–80%. Such levels are unacceptably high yet these mines are quoting grades to two places of decimals. At Nchanga Division of ZCCM Ltd in the Zambian Copperbelt, it is considered that, at the final development stage, the reserves should be proved to ± 20% but this is based on 90% confidence limits with precision = (1.645 × σ × 100/mean grade)%. Recent geostatistical studies on the underground section of the Lower Orebody at Nchanga have shown that this is the case for a drill-hole spacing of 30 × 30 m. However, in the so-called 'Thin Rich' ore which has highly variable but high grades, this criterion is not met and grade precisions are of the order of 45%.

We must remember that the 'true grade' can never be predicted accurately and that even the 'realized grade' is itself an estimate and thus incurs an error. The geologist is thus comparing his estimate with another estimate, both of which are in error.

Poor estimates can be due to:

(1) Lack of geological control because insufficient information is available to the geologist;
(2) Inadequate or inappropriate computational techniques;
(3) Variation in drilling density so that the confidence level in the reserves varies from area to area in the orebody;
(4) Errors in the basic assay data.

In the case of (4) this can be due to:

(1) Natural variability of the ore so that the sample is not representative of the area to which it is assigned. This is related to the nature of the host rock, the geological processes which formed the economic mineral phases, the amount of metal available, and the physiochemical conditions at the time.

(2) Poor core recovery. This is a major factor in reliability which is rarely taken into account. Even if 40% of the core is missing it is still assigned a grade based on that recorded from the recovered section of core. This is, however, a very difficult error to assess and more consideration should be given to this problem.

(3) The assay data set contains information from different sample types or sizes, e.g. AXT holes are combined with NX surface holes or chip samples are combined with drill samples. This ignores the different sample volumes which give different averaging effects in the case of patchy mineralization. Different sample types should be treated separately.

(4) Low intersection angles. Again these are not taken into account by most ore-reserve techniques. In the case of tabular deposits, the grade quoted will be less representative of a point down the dip of an orebody as the intersection angle decreases (Figure 4.1).

Also, if a system of subparallel veinlets dips at an angle close to that of the drill-hole there will be inadequate penetration of the mineralization to define low and high grade zones.

(5) Loss of a valuable component during drilling or handling, especially in the case of flaky or friable phases such as molybdenite.

(6) Rich nuggets may, or may not, be intersected (e.g. gold).

(7) Incomplete intersection of the ore-zone. Various methods may have been used to estimate the grade of the missing section, which include the use of the average value of the recovered portion, the assumption that it is barren, or setting it equal to the last value obtained in the drill-hole. In other cases, the core assay data are combined with chip sample data to complete the intersection. Ideally, incomplete intersections should be deleted from the data set before geostatistical analysis is attempted.

(8) Inadequate sampling for short-term financial reasons. Premature termination of a drilling programme at the exploration stage has often led to regret later. Even in large base-metal orebodies, a 30–60 m drill spacing may be inadequate to establish grade and continuity of ore.

(9) Biased sampling of core due to:

(a) contamination;
(b) poor handling procedures;
(c) effect of variable rock hardness in the sample length;

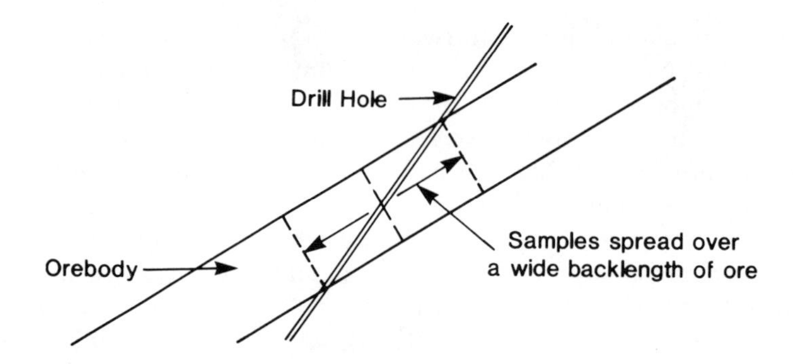

Fig. 4.1 The problem incurred by a low angle intersection of a tabular ore deposit.

(d) poor core splitting techniques, e.g. use of hand splitting rather than diamond sawing;

(e) too many sample reduction stages during the crushing and grinding of the original sample. A small sample is obtained for analysis which is a minute fraction of the original sample and which may no longer be representative due to accumulative errors. The original sample may also have been a poor representation of the block of ground in which it lay.

(10) Poor assaying due to:

(a) poor laboratory procedures, incomplete dissolution or contamination;

(b) poor standards and instrumental drift;

(c) variability between laboratories employed and the use of different analytical techniques.

4.2.2 Problems in implementing geostatistics

Many companies have faced considerable difficulties in their attempts to implement geostatistical techniques and some of the reasons are summarized below.

(1) The 'know-how' required.

(2) Inadequate program development backup and the need for mine-specific software.

(3) Managerial attitude and resistance to change.

(4) In some cases, the time needed to create computer files of an enormous backlog of assay data.

(5) The lack of continuity in manpower, for many mines suffer a rapid turnover often related to lack of a career structure.

(6) Lack of financial commitment to improve sampling procedures.

(7) The time needed to compute the initial semi-variograms and to devise a method which can be applied effectively in a particular operation. Much effort is often required to tailor standard geostatistical techniques to take into account the local geological conditions.

(8) The inability to produce a reliable semi-variogram because of the nature of the mineralization or the magnitude of the sampling/analytical errors contributing to a dominant nugget variance.

(9) Incorrect modelling of the semi-variogram.

(10) Lack of adequate computing facilities.

4.3 THE THEORY OF REGIONALIZED VARIABLES

The 'theory of regionalized variables' was largely developed by G. Matheron (1971) and is now

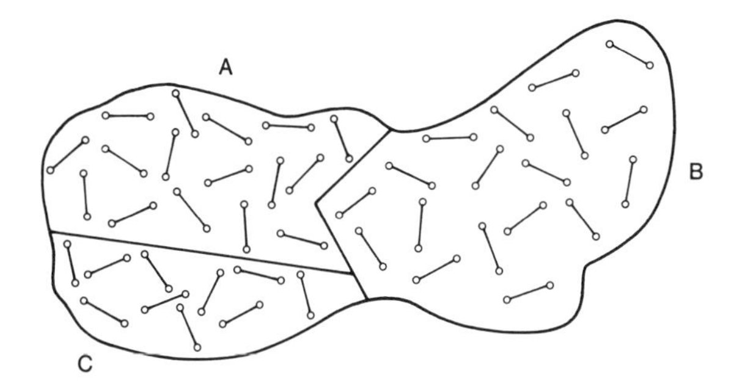

Fig. 4.2 Sample pairs within subareas of an ore deposit.

simply referred to as 'Geostatistics'. However, the first published work on this subject was by Matern (1960) who studied tree growth on behalf of the Swedish Forestry Commission in 1948 whilst at the University of Upsala. Geostatistics assumes that the statistical distribution of the difference in grade between pairs of point samples is similar throughout the entire deposit, or within separate subareas of the deposit, and that it is dependent upon the distance between, and the orientation of, pairs of samples. This is the concept of stationarity.

For each subarea (A, B, C in Figure 4.2) of the deposit, we can take every pair of samples at a distance h apart (e.g. 20 m) and calculate the difference in grade between these samples. We thus have three data populations which, once they have been normalized to a Gaussian distribution, can be tested with the F and t test to determine whether, at 95% confidence, they are statistically similar. If they are, then stationarity can be assumed to exist in the deposit. This can be repeated for different values of h (40, 60, 80, 100 m, etc.). If stationarity does not exist, then the subareas will have to be evaluated separately. These subareas may have been defined initially on the basis of some aspect(s) of their local geology.

Classical statistics considers only the magnitudes of the data and does not consider any evidence of a spatial control or trend. Geostatistics considers not only the value at a point, but also the position of that point within the orebody and in relation to the other samples.

4.3.1 Regionalized variables

A regionalized variable (RV) is the value of a point within an orebody/deposit whose magnitude is partly dependent on its position as indicated above, and also partly on its support.

Support

The support of an RV is its volume, shape and orientation; the shape could be cylindrical or prismatic, while the orientation is the attitude of the sample relative to the plane of the orebody. The value of an RV at a point will be different for a diamond drill core, a channel sample or a grab sample taken from the muckpile at the face and taken to represent one blast.

Typical examples of RVs that may be employed in a geostatistical study include:

1. Grade

% Cu, g/t Au, kg/m^3 SnO_2, lb U_3O_8/ton, carats/100 tons, % Ash, % clay (e.g. kaolin), metal units, metal equivalent values.

2. Physical parameters

% Porosity, % oil saturation, moisture content, volatile content, bleached brightness of clay, rock strength.

3. Metal accumulation

m% Cu (TT, VT, HT), cm g/t, kg/m^2.

4. Specific gravity

This is usually expressed as a tonnage factor (e.g. 2.8 t/m^3).

5. Service variables

Metal accumulation × specific gravity – This service variable is used to allow for large changes in SG in response to variation in grade or mineralogy. % Clay × bleached brightness.

6. Thickness

Horizontal, vertical or true components of thickness.

7. Geological structure

Perhaps represented by the depth above, or below, a datum plane of the median plane or the footwall of an orebody. It may also be possible to use orebody dip as an RV.

8. Monetary value

\$/t – This RV is especially useful for polymetallic orebodies. It reflects the current value of the metal contained within a specific volume.

Net smelter return – This reflects the profit that can be made from a block of ground after allowance is made for the cost of mining, processing, transport and overheads.

9. Environmental parameters

These change with time and not distance and include total dissolved solids in groundwater; radioactive contamination; nitrates; pH of water; metal content of vegetation, fish livers; etc.

10. Mill feed

The metal grade or content of valuable or deleterious trace/minor elements in mill feed which also change with time.

All these RVs could be expressed as absolute 'raw' values or as their two- or three-parameter log-transformed equivalent values, depending on the results of a statistical study.

Continuity

Some features of an orebody, e.g. width, have a high degree of continuity, i.e. they change slowly and regularly along strike or down-dip. Other features, e.g. grades across an orebody width, show very rapid changes indicating low continuity.

Anisotropy

The continuity of an RV may be greater in the plane of an orebody than across it (i.e. HW to FW), especially in the case of tabular sedimentary or lenticular orebodies. Even within the plane of the orebody, there may be differences in the continuity of an RV between the strike and down-dip directions, or parallel to a palaeo-shoreline and off-shore direction, or along the current direction and transverse to it. When this occurs, we have an anisotropic condition as opposed to an isotropic condition. In the latter case, no differences in continuity can be detected with direction.

Data correlation

As mentioned earlier, classical statistics does not take into account the spatial characteristics of a data set. For example, if we consider the following two sets of assay data, representing samples taken across the back of a drive at uniform intervals:

Set (a)	**Set (b)**
3, 5, 7, 9, 8, 6, 4, 2	9, 3, 8, 2, 6, 5, 7, 4.

we can see that they have the same mean and variance but clearly have markedly different spatial characteristics which can be expressed by the absolute differences between adjacent assays:

Set (a)	**Set (b)**
2, 2, 2, 1, 2, 2, 2	6, 5, 6, 4, 1, 2, 3.

These increments can thus be used to convey a picture of the spatial variability of the RV. Where the zones of influence of samples are large and are thus overlapping, only small increments or decrements exist between samples. Widely spaced samples or samples with small zones of influence, have greater differences between values.

4.3.2 Geostatistical methods

Transitive

In these, there is no assumption of stationarity and the covariogram is used in which covariance is plotted against distance between sample pairs.

Covariance is a measure of the interdependence between adjacent samples and its value thus drops to zero once the sample spacing is large enough for samples to be considered totally independent variables. The covariogram is produced by calculating the covariance of pairs of samples spaced at a distance equal to the minimum sampling interval (plus or minus a tolerance margin) and setting this value equal to $Cov(h)$ where h = lag 1. This process is then repeated for lag 2 (twice the first sample spacing) and so on, until $Cov(h)$ oscillates about zero. Note:

$$Cov = \Sigma_{i=1}^{n} [(X(x)_i - X_i)(X(x + h)_i - X_2]/n$$

where $X(x)_i$ and $X(x + h)_i$ represent the first and second samples respectively and X_1 and X_2 are the mean values of all the first samples and all the second samples respectively. More detailed explanations can be obtained from Davis (1973).

Intrinsic

These are based on an assumption of stationarity but this is not applied too rigidly as there are some models in which variance continues to increase as the samples are taken from larger and larger areas. We thus apply the concept of quasi-stationarity. These methods use the semi-variogram, whose computation is described later in section 4.5.

4.3.3 Requirements of a thorough geostatistical study

Each geostatistical study involves a series of clearly identifiable stages which are summarized below.

1. Data review

(a) An examination of sample type and size to ensure that a homogeneous sample/data set exists – reject incompatible data.

(b) An examination of the analytical precision of data produced by different analytical techniques or different laboratories – reject suspect analyses.

(c) An examination of the quality of sampling exercised during different sampling campaigns on the same deposit (see also 2. 'Statistical analysis' below).

(d) An examination to determine whether all samples intersected the entire width of the orebody – reject if incomplete intersections.

(e) If high grades were cut to an arbitrary level then the original assay values should be reinstated initially (it may be necessary to filter these out later).

(f) An examination of the sampling coverage – is it adequate and of uniform density and is additional sampling required?

(g) Verification of the data file for input errors or the determination of sample coordinates, etc.

(h) Subdivision of the data on the basis of host rock type/material type or style of mineralization.

2. Statistical analysis

(a) Calculation of the mean and variance of the data sets.

(b) Production of histograms of raw and log-transformed data.

(c) Examination for normality using:

- (i) the Chi square test;
- (ii) log-probability plots;
- (iii) skewness, kurtosis (see section 3.5) and coefficient of variation (σ/X – should be ≤ 1);
- (iv) data splitting in the case of bimodal populations.

(d) Three-parameter log-transformation, if necessary.

(e) Rejection of outliers – filtering.

(f) Calculation of the covariance of replicate assays to determine the error incurred during analysis.

3. Geological analysis

In order to be able to apply geostatistics effectively, a thorough understanding of the nature, structure and geological controls of the mineralization is necessary. Any feature which might adversely affect the geostatistical evaluation of the deposit must be appreciated at an early stage.

4. Structural analysis

(a) Production of experimental semi-variograms (variography) for individual areas, rock types, etc., based on the results of the statistical and geological analyses.

(b) Fitting of mathematical models to the semi-variograms.

(c) Determination of semi-variogram parameters using a process called cross-validation.

5. Local estimation

(a) Evaluation of individual blocks of ground.

(b) Assignment of kriging variance (i.e. error) to each RV determined.

6. Global estimation

Evaluation of the entire deposit based on results of the above.

7. Grade–tonnage curve

(a) Production of the experimental and theoretical grade–tonnage curve.

(b) Production of kriging variance–tonnage curves.

4.4 REGULARIZATION AND OREBODY SUBDIVISION

4.4.1 Tabular orebodies

The reduction of data to point values, amenable to plotting on a single plan or vertical longitudinal projection (VLP), is referred to as regularization. Thus the grades and thicknesses of all samples through an orebody intersection can be converted to a total thickness, a weighted grade and a metal accumulation. Flat dipping orebodies would thus have their grades plotted on a plan (i.e. projected vertically up to the plan level) along with **vertical** thickness and **vertical** thickness metal accumulations. Similarly, steep dipping orebodies have values projected horizontally on to VLPs – in this case the weighted grade is accompanied by **horizontal** thickness and **horizontal** thickness metal accumulation. Problems arise with steep orebodies which are also folded so that limbs of the folds (e.g. L1, L2 and L3 in Figure 4.3) overlap on the VLP (between x and y) and where the use of plans is not suitable because the data are too tightly compressed. This can only be overcome by the use of palinspastic maps (Chapter 1) on which all drill-hole intersections of the median plane of the orebody are

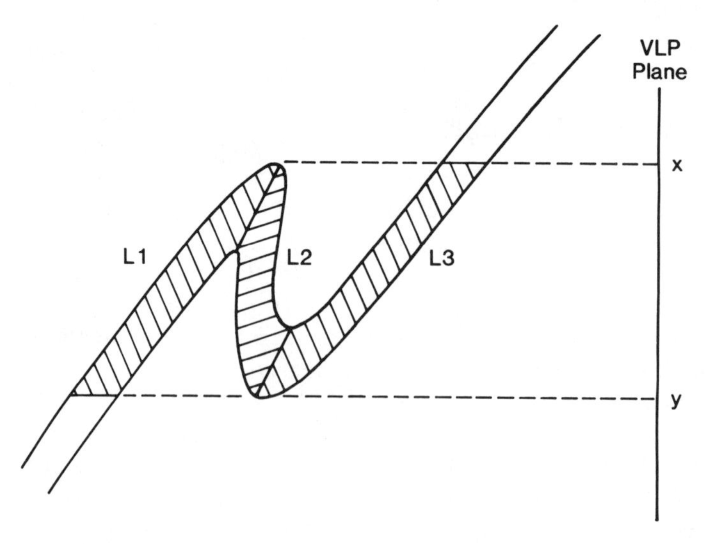

Fig. 4.3 A folded tabular deposit showing the superimposition of fold limbs on a vertical longitudinal projection.

plotted, along with anticlinal and synclinal axes and mine levels. On these maps weighted grades are accompanied by **true** thicknesses and **true** thickness metal accumulations. Although a distorted picture is obtained of each mining block, the global reserves of the orebody can be easily calculated, together with the reserves between each mine level and on each fold limb.

It should be noted that regularization is only possible if a complete intersection of the orebody, or material type under study, is made.

An alternative technique which can be applied to tabular deposits (stratiform and stratabound) and which overcomes the need to regularize the data, is referred to as stratigraphic slicing. In this the stratigraphic section, all or part of which may host the mineralization, is subdivided into a fixed number of intervals. Thus in each drill-hole, each interval represents the same stratigraphic unit as the equivalent interval in adjacent holes, although their intersected lengths may differ. Sample grades are thus composited to determine the intersection grades of each stratigraphic slice. Obviously, it is preferable if the initial sampling could be controlled by stratigraphy so that compositing is rendered unnecessary. Data sets can then be created, and semi-variograms produced, for each slice in turn. Stratigraphic slicing has been applied successfully at Navan Pb-Zn Mine in Eire where the Carboniferous 'Pale Beds', which host the sulphides, are subdivided into 45 intervals, each approximately 3 m thick. The subdivision is based on the recognition of internal marker horizons and begins at the contact between the underlying 'Muddy Limestone' and the 'Pale Beds' (Chapter 8, Case history in section 8.10).

It is clearly evident that, where the mineralization shows a strong sedimentary control, it should be subdivided stratigraphically so that the grades that are being compared reflect the metal content of laterally equivalent sedimentary horizons. This is particularly important where marked changes in lithology and permeability of the host sediment occur through the mineralized stratigraphic section. Such a situation exists at Mufulira in Zambia, where there are three superincumbent orebodies (A, B and C) separated by interorebody horizons which may be barren, low grade or even of economic grade. Copper sulphides thus occur over a stratigraphic thickness of approximately 65 m containing interbedded dolomites, feldspathic or argillaceous or carbonaceous quartzites, arkosic grits, argillites, dolomitic argillites and sandy argillites.

Before constructing a semi-variogram for an orebody, or part thereof, it is essential that the data be drawn from a homogeneous population. If the thickness values in one area of the orebody change suddenly so that they oscillate about a different mean, then this area should be treated separately. Similarly, if the grade or metal accumulation data show a similar change, then this should also be taken into account. A sudden change in the orebody dip will also require that the data be split at this point as marked changes in the vertical and horizontal components of thickness will occur. Failure to take such changes into account could make the resulting semi-variogram difficult to interpret and model. At Mufulira Mine it was found necessary to treat internal pyritic areas and fringe areas in the 'C' orebody separately. At Nchanga, the assay data were subdivided to allow for the sudden change of the Lower Orebody (in the Lower Banded Shale and underlying Transition) from an average thickness of 21 m at 6.7% total copper to a 'Thin Rich' equivalent in which grades of 10% occur over thicknesses often less than 2 m, but averaging 2.5 m.

4.4.2 Non-tabular orebodies

Regularization of deposits such as alluvial gold and tin deposits, lateritic gold and nickel (eluvial) deposits, bauxitic clays, and supergene iron ores (pisolitic valley fill deposits and those derived from banded iron formations), is best accomplished by the use of horizontal slicing. As most of these deposits will be mined in open-pit operations, the slices should conform with the proposed benches in the open-pit. A series of slice plans are thus produced with the composite grades for the slice interval plotted for each borehole. No thickness or metal accumulation data are now needed for, as these deposits are usually drilled with vertical holes, the support of each composite grade is constant (the bench height). Semi-variograms will thus be produced for each slice separately, although some may be combined at a later stage.

Horizontal slicing must be used with care, however, for allowance must be made for changes in geology, mineraology, grade or material type which may take place laterally within each slice/bench. Where this has not been done, as was initially the case in RTM's Cerro Colorado porphyry copper (J. O'Leary, personal communication) and in the Royal Hill eluvial gold deposit in Suriname (Smith, 1987), the semi-variograms produced were very erratic or showed 'pure nugget effect' and no mathematical model could be fitted. In the former example, the orebody was subdivided into zones in which grades were fairly homogeneous. The grade limits for each zone were determined by plotting cumulative frequency curves and observing breaks in the data population distribution. On examination of cross-sections, it was evident that these grade zones corresponded to zones of leaching, supergene (secondary) enrichment, and hypogene (primary) mineralization associated with typical porphyry alteration zones. Semi-variograms produced for each zone separately could now be modelled successfully. Similarly, at Royal Hill, once the data had been split on the basis of material type, i.e. lateritic clay versus kaolinitic clay and also on the basis of bedrock geology, i.e. metasediments versus metavolcanics, good semi-variograms were produced from which geostatistical ore-reserve estimates could be made.

4.5 PRODUCTION OF THE SEMI-VARIOGRAM

The method by which the experimental semi-variogram is produced depends on the distribution of the data, i.e. whether it is uniformly spaced on lines or totally erratic, or somewhere in between these two situations.

4.5.1 Regular linear data

Ideally, we need a data set of in excess of 200 values, each representing a sample of constant dimension in the direction of the proposed semi-variogram. Samples of differing volumes, e.g. NX and AX cores, are not acceptable for increases in volume decrease the nugget effect referred to earlier. The nugget effect for NX drill core sampling of an orebody will be different to that produced by AX sampling of the same orebody (i.e. lower) and will produce a different semi-variogram. Volume is therefore important, but shape and orientation must also be taken into account – all samples in the data set must have the same shape and orientation. The latter situation is, however, difficult to achieve due to the variable dip of boreholes and the ore-zone, and to the use of fan drilling underground. The samples should also span the entire thickness of the orebody.

The **experimental semi-variogram** $\gamma(h)$ values are obtained by application of the following formula:

$$\gamma^{\star}(h) = \Sigma(f(x + h) - f(x))^2/(2^{\star}N)$$

where N = total number of data pairs counted at each lag, and $f(x + h)$ and $f(x)$ are the values of RVs separated by a distance h.

Lag is purely the spacing at which the squared differences of sample values are obtained. Lag 1, for example, involves adjacent samples (i.e. A and B, B and C, etc., in Figure 4.4). The distance represented by lag 1 is thus the minimum sampling interval. Lag 2 requires that we calculate the squared differences of alternate samples (A and C, B and D, etc., in Figure 4.4). If the lag distance is 10 m, then the first $\gamma^{\star}(h)$ value calculated will consider all samples spaced 10 m apart (lag number = 1), the second value of $\gamma(h)$ will

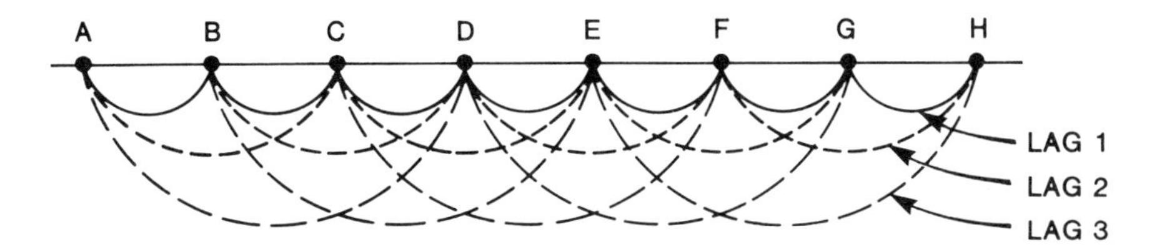

Fig. 4.4 Calculation of $\gamma(h)$ at different lags – regular linear data.

be obtained from a spacing of 20 m (lag 2), the third from a spacing of 30 m (lag 3), and so on. Normally the maximum lag number should not exceed half the sampling length, e.g. if a level has been channel sampled at 10 m intervals for a distance of 200 m, the highest lag calculated would be lag 10 or 100 m.

If a break occurs in a line of samples caused by a missing sample, then the variogram calculation should recognize this. If, for example, sample E is missing in Figure 4.4, then $N = 5$ instead of 7 at lag 1, $N = 4$ instead of 6 at lag 2, etc.

When several lines of samples are available, as when several levels or raises have been developed and sampled in a tabular orebody, the semi-variogram values $\gamma^\star(h)$ are produced for each and then the weighted average determined on the basis of number of differences squared on each level. In other words:

$$\gamma^\star(h) = \Sigma_i^n \, [(N)_i \times \gamma^\star h_i]/\Sigma(N)_i$$

where n = number of levels and $i = 1, 2, 3 \rightarrow n$.

The more lines the better!

The values of $\gamma^\star(h)$ are plotted against the corresponding lag to produce the semi-variogram. The rate of increase of $\gamma^\star(h)$ with lag is a reflection of the rate at which the influence of a sample decreases with distance and gives a precise meaning to the term zone of influence. The distance at which the semi-variance $\gamma^\star(h)$ remains constant corresponds to the point at which the covariance $Cov(h)$ between adjacent samples has decreased to zero. This defines the limit of the zone of influence of a sample (section 4.6.4 and Figure 4.13(a),(b)).

4.5.2 Irregular linear data

A. Irregular spacing along lines

Method 1

The lag is set equal to the minimum sampling distance – say 6 m.

A 11.8 m B 18.5 m C 7.5 m D 6.8 m E 20 m F

Dist AB	= 11.8/6 ≈ 2 lags	DE	= 6.8/6 ≈ 1 lag
BC	= 18.5/6 ≈ 3 lags	EF	= 20/6 ≈ 3 lags
CD	= 7.5/6 ≈ 1 lag	AC	= 30.6/6 ≈ 5 lags
BD	= 26/6 ≈ 4 lags		
CE	= 14.6/6 ≈ 2 lags		
DF	= 26.8/6 ≈ 4 lags		

etc.

This process can be dealt with in a computer program using:

$$\text{Lagno} = \text{INT(dist/lagdist} + 0.5)$$

for INT(. . . .) rounds down to the nearest integer below the value calculated in the brackets. If the expression in parentheses yields a value < 1.0 then 'lagno' is set to zero which may present problems in some programs. To avoid this, add 1.5 instead of 0.5 and then, when the $\gamma(h)$ values at each lag are printed to hardcopy or disk file, the lag value is reduced by one. Every possible combination of values is thus compared in this way, with the squared difference of the values in each case being assigned to the relevant lag (or 'lagno'), as calculated above.

Method 2

This method is similar to that described above.

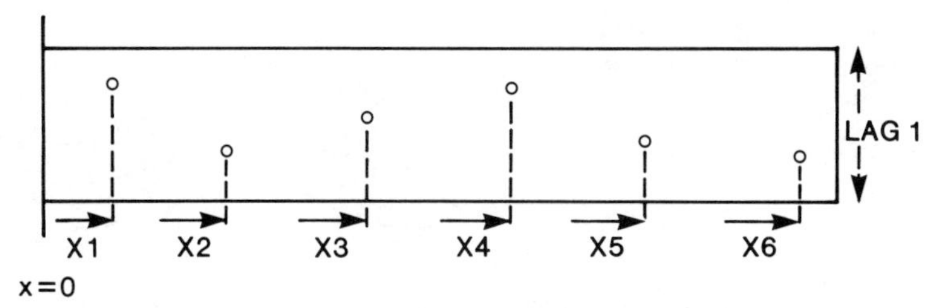

Fig. 4.5 Calculation of $\gamma(h)$ when the data points are slightly offset from a line at irregular spacings.

(1) Tabulate distance apart of each possible combination of pairs of samples.

(2) Tabulate $(\text{difference})^2$ equivalent to each of above distances.

(3) Group data into classes based on distance.

Note: If sampled length = 200 m then divide by 2 to obtain maximum value of h, i.e. 100 m.

For 10 lags we thus have a class interval of 10 m, i.e. lag 1 – 0–10 m, lag 2 – 10–20 m, etc.

All squared differences for each lag number are accumulated and then divided by two times the number of differences calculated.

B. Irregular spacing along lines and samples slightly offset to either side of line

Divide the orebody plan (or VLP) up into strips parallel to the direction of the proposed semi-variogram. Strips have widths equal to lag 1 for the direction at right angles to the strip. Determine the x (or y) increment in the direction of the strip.

The differences in x increment, e.g. X2–X1 in Figure 4.5, give the distance which is then treated as in Method 1 or 2 above. The squared differences of the values at each sample site are then calculated, using every possible combination of samples up to the maximum lag value selected. This can be repeated for strips elongated parallel to the Y axis and also in any other direction required.

4.5.3 Irregular data in both dimensions

When an orebody is drilled from surface or underground, the intersection pattern may be rather irregular but still retains some linearity. Intersections are, in other words, slightly off the drilling section because of unintentional drift of the drill-holes. In this situation we move a grid over the data until most of the sample points lie on, or close to, grid intersections or alternatively, to the centre of each grid square – a random stratified grid (RSG). It may be necessary to change the grid dimensions in order to achieve this. We can tolerate some squares with two samples so long as the majority only have one. The value nearest to each intersection is then accepted as the intersection value, if none is present, then a zero is used. A data matrix is produced, as in Figure 4.6, from which four directional semi-variograms can be calculated (A to D). The lag distances can be determined from the RGS cell dimensions, or the cell diagonals as necessary, and also recognizes all zero values as data voids.

4.5.4 Totally random data

Programs, such as MARVGM (A.G. Royle, University of Leeds) or SECTOR (A.E. Annels, University of Wales, Cardiff), are designed to treat irregular or clustered data which have been regularized on to a plan or VLP. A sector search routine is employed.

A. Input requirements for MARVGM/ SECTOR

(1) Number of points in the data file.

(2) X and Y coordinates on plan or VLP for each drill-hole/sample together with assay/ thickness data.

(3) RV for which variogram is to be drawn.

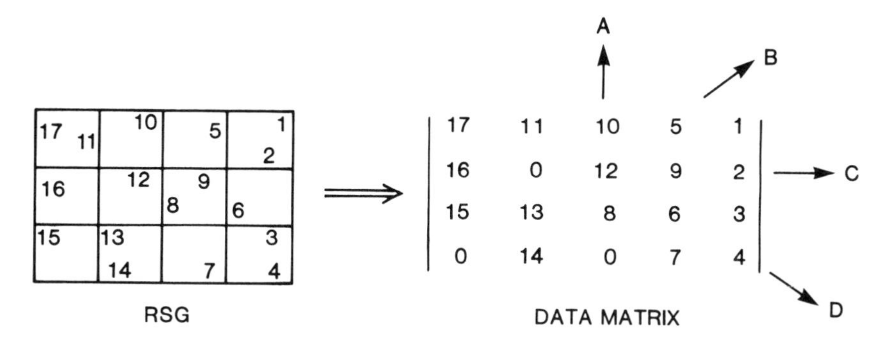

Fig. 4.6 Production of a data matrix after fitting a random stratified grid.

(4) Lag distance.

(5) Maximum lag number.

(6) Upper limit above which data values are filtered out (MARVGM only).

(7) Angular regularization (ALF for MARVGM, AR for SECTOR) which is the apex angle of the search sector.

(8) Direction in which the variogram is to be calculated BET (e.g. 0, 45, 90, 135 for SECTOR or the anticlockwise angle from an easterly bearing for MARVGM).

(9) For isotropic variograms, the angular regularization is set to 180° and BET to 90°.

B. Program details – MARVGM/SECTOR

Commencing at the first drill-hole in the data set, these programs initially set h = lag 1 and locate all drill-holes that fall within an area defined by the azimuth of the semi-variogram, by the angle of regularization and by the lag distance. This is done by calculating the distance to each drill-hole from the apex hole and then calculating the bearing to each hole. The latter is then compared to BET ± ALF/2 (or BET ± AR/2). If the borehole is shown to lie in the shaded area of Figure 4.7, then the squared difference of the values at the apex hole and this hole is calculated and accumulated for the relevant lag number. This exercise is repeated for each drill hole and for all values of 'h' up to the maximum lag number. The search cone apex is then moved to the next drill-hole in the data set and the whole exercise repeated. Once all holes have been covered in this way, the sum of all the squared differences at each lag are divided by two times the number of differences calculated at each lag, thus giving $\gamma^\star(h)$.

Note. ALF (or AR) is kept as small as possible to cut down computing time (typical values 5–10°), but this is only possible if the data are numerous enough to allow this.

Conical 3D search volume (program SVCONE)

This program operates in a similar way to MARVGM and SECTOR but can be used for non-regularized data from drill-holes intersecting a porphyry copper deposit or other non-tabular deposits not amenable to regularization except by compositing by bench. A conical search volume (Figure 4.8) is used and all samples falling within it, at a distance equivalent to the specified lag number (h) ± half the lag distance (LD) from the apex, are included in the calculation. The apex angle of the cone defines the tolerance, i.e. the angular regularization. The axis of the cone can be pointed down-dip or along strike.

4.6 SEMI-VARIOGRAM MODELS

Four main types of semi-variogram model may be recognized:

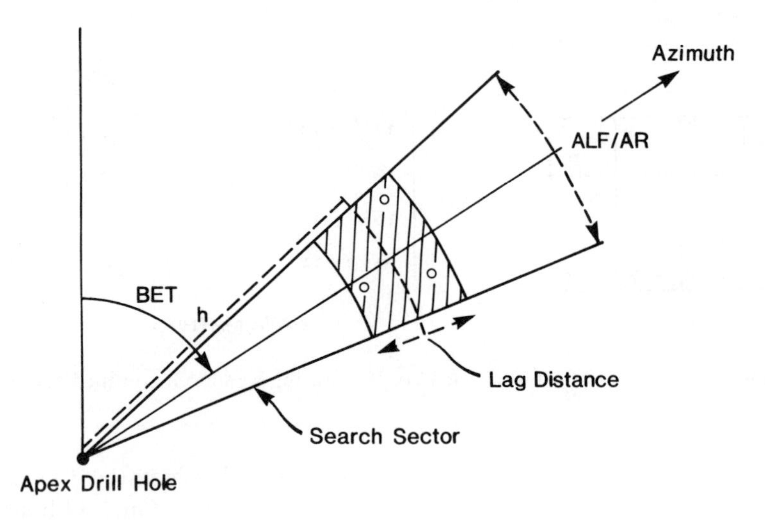

Fig. 4.7 Sector search method for irregularly distributed data points. (Note that in MARVGM the angle BET is measured anticlockwise from the east to west direction.)

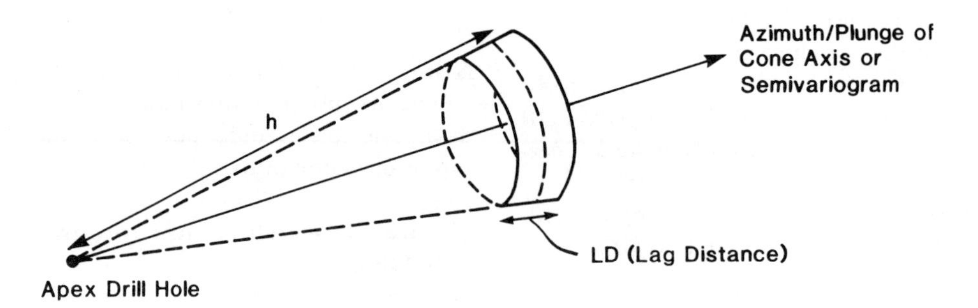

Fig. 4.8 A conical search volume for three-dimensional semi-variogram calculations.

(1) linear
(2) de Wijsian
(3) exponential
(4) spherical or Matheron

4.6.1 Linear scheme models

If, when $\gamma^*(h)$ is plotted against lag, the graph is a straight line, as shown in Figure 4.9, then the model can be written:

$$\gamma(h) = ph + k$$

where p is the slope, h the lag and k the intercept on the $\gamma(h)$ axis.

The variogram should go through the point ($h = L/3$, $\gamma(h) = \sigma^2$) where L is the length of the line sampled and for which the semi-variogram is being computed. The variance of the samples in the length of the data field, σ^2, is thus $p(L/3) + k$.

Where a gentle curve exists (Figure 4.10) a generalized linear model could be chosen of the form:

$$\gamma(h) = p \cdot h^{\alpha}$$

where α is < 2.

Linear models may exist in some iron ore deposits while an example of such a model is

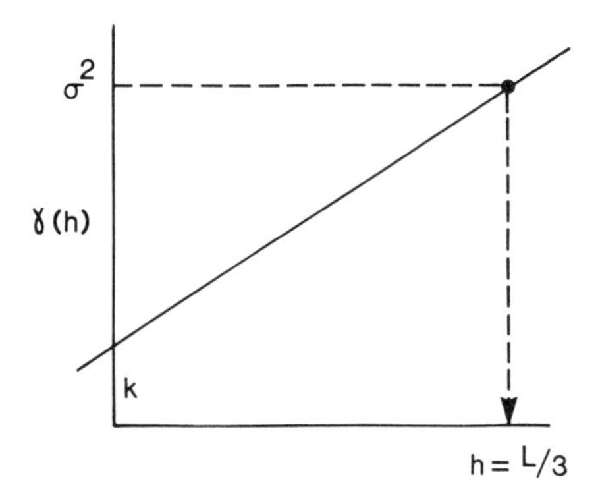

Fig. 4.9 A linear scheme semi-variogram.

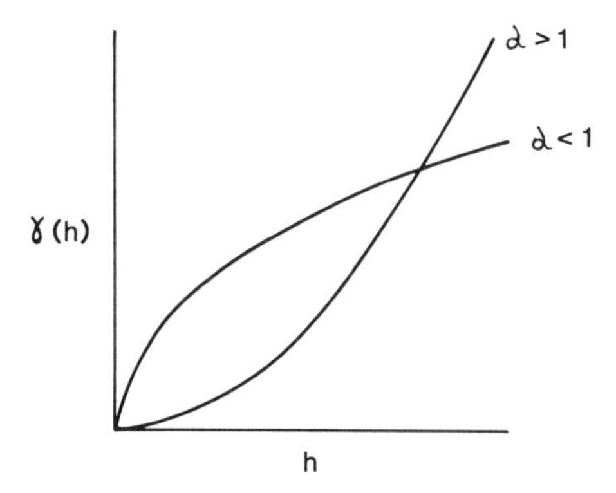

Fig. 4.10 Generalized linear scheme models, $\gamma(h) = p{\cdot}h\alpha$.

given in section 4.8 for the thickness of the Tarkwa AVS Main Reef in Ghana.

4.6.2 De Wijsian scheme models

In this scheme $\gamma(h)$ continues to increase beyond the value of σ^2, the variance of the data. At first sight it may appear to be of a generalized linear type but, if the $\gamma(h)$ values are plotted against the log of h, then a straight line is obtained. This is the de Wijsian model which, for parallel samples (e.g. channel samples) has the form:

$$\gamma(h) = 3\,\alpha[\ln h/L + 3/2]$$

where α is the coefficient of absolute dispersion, a measure of the spatial variation, and L is referred to as the equivalent thickness. Both of these can be determined by reading off the values of $\gamma(h)$ at two lags, e.g. lag 1 and lag 10, so that two simultaneous equations can be produced with L and α as the unknowns.

In this scheme the support of the sample or ore block must be defined and this is done on the basis of the linear equivalent (LE).

Rectangle: $LE = a + b$.

Parallelogram: $LE = (a^2 + b^2 + 2ab \sin\theta)^{1/2}$.

Triangle: $LE = [(a^2 + b^2 + c^2)/3 + 2S]^{1/2}$ where S = surface area.

Prism or cyliner: $LE = a + b + 0.7c$ where $a \geqslant b \geqslant c$.

Some examples of linear equivalents are listed in Figure 4.11.

If a panel of ore, and a sample cut into it, have linear equivalents of LE_p and LE_s respectively, then:

$$\sigma^2_{(S/P)} = 3\alpha \ln\,[LE_p/LE_s]$$

If the volumes of the panel and the sample are geometrically similar, then:

$$\sigma^2_{(S/P)} = \alpha \ln\,[\text{volume of panel/volume of sample}]$$

Such models are not common although they have been found to apply to some hydrothermal vein deposits, particularly tin, and to orebody thicknesses. An example of the fitting of a de Wijsian scheme model and the calculation of estimation variances, is provided in Appendix 4.1.

4.6.3 Exponential scheme models

If the semi-variogram rises slowly towards a sill but fails to flatten out to a constant value, then an exponential model is probable. There are two possible exponential schemes, Formery and Gaussian (Figure 4.12).

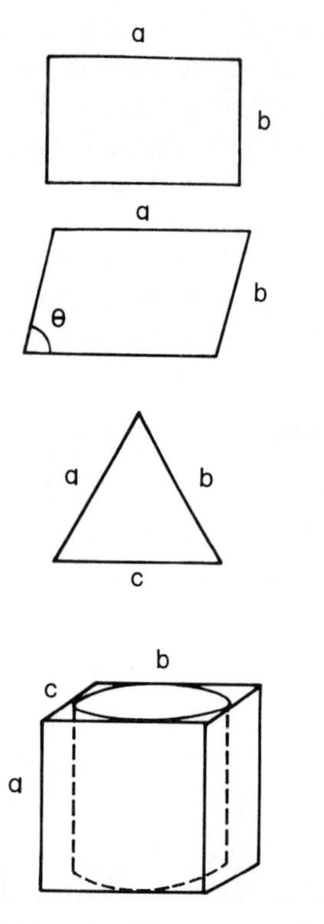

Fig. 4.11 Calculation of linear equivalents in the de Wijsian scheme.

A. Formery scheme
This has the form:

$$\gamma(h) = C[1 - \exp(-h/a)] + Co$$

In this case the tangent at the origin intersects the sill at $a/3$.

B. Gaussian scheme
This has the form:

$$\gamma(h) = C[1 - \exp(-h^2/a^2)] + Co$$

Here the tangent at the origin intersects the sill at $a/\sqrt{3}$.

4.6.4 Spherical scheme models

The spherical, or Matheron scheme, model is the most commonly found model for ore deposits. Its form is illustrated in Figure 4.13(a). The semi-variogram curve rises rapidly at low lags before gradually flattening out to a constant sill value at higher lags. A tangent to the curve, drawn through the first two or three points, defines values Co and C on the $\gamma^*(h)$ axis and it intersects the sill level ($Co + C$) at a point equal to $2a/3$ where 'a' represents the point at which the curve reaches the sill. The distance between the curve and the sill level at lags less than 'a' represents the covariance between samples at these lags, hence we can define the relationship between semivariance and covariance as:

$$\gamma(h) + Cov(h) = (Co + C)$$
or the statistical variance σ^2

Beyond 'a', therefore, the covariance drops to zero, i.e. there is no longer any correlation between the values of such widely spaced samples. The relationship between sample spacing and covariance is also illustrated in Figure 4.13(b) which shows how the overlap between the zones

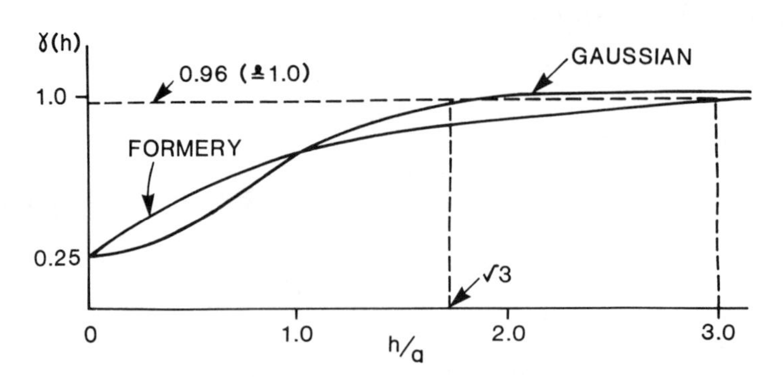

Fig. 4.12 Formery and Gaussian scheme semi-variograms.

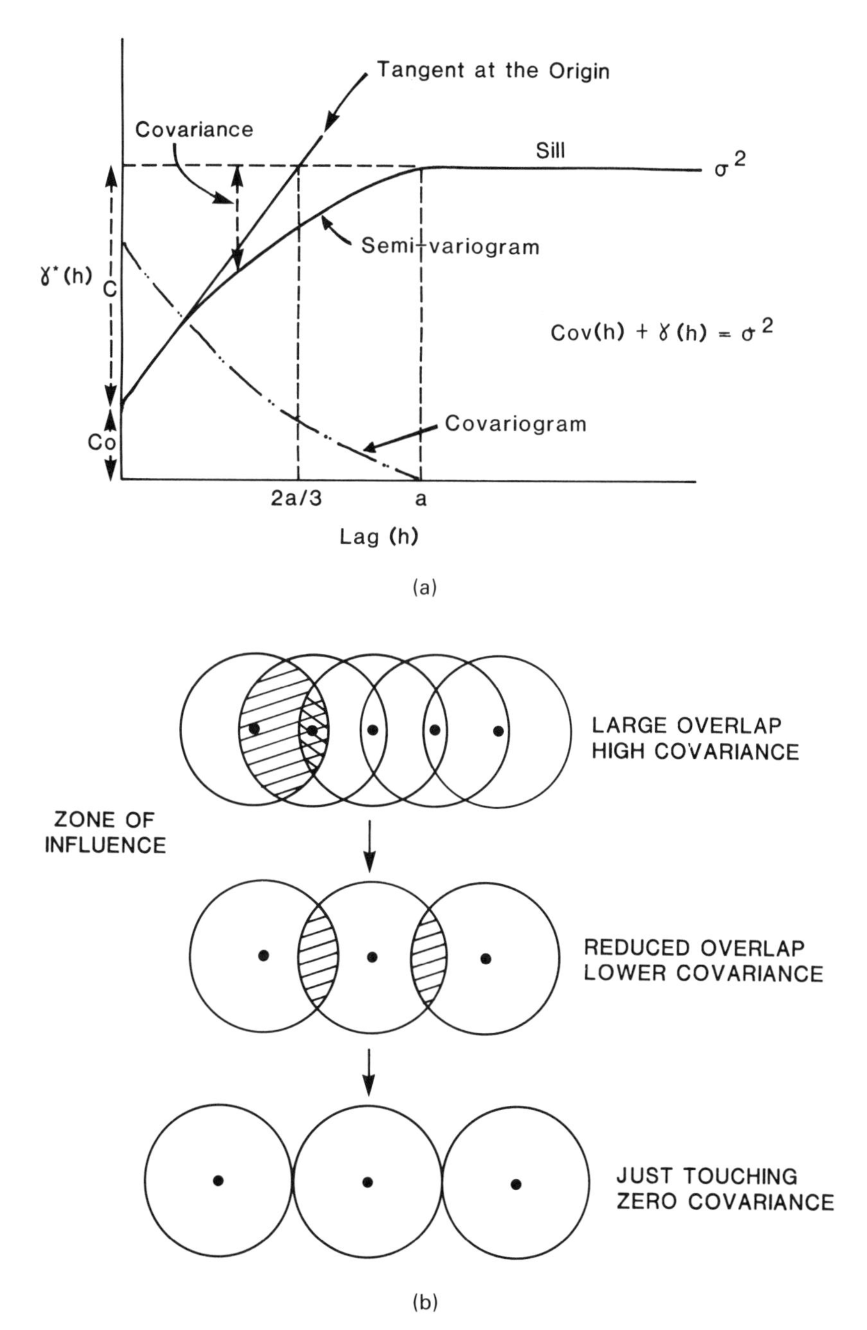

Fig. 4.13 (a) Diagram showing the relationship between the covariogram (dashed line) and the spherical scheme semi-variogram. (b) Basis of the semi-variogram and the explanation of the concept of zone of influence.

of influence of samples gradually decreases as the sample spacing increases. Where they are just touching, the sample spacing is exactly equal to '*a*', the geostatistical range.

The mathematical model for the curve is:

$$\gamma(h) = Co + C\,[1.5\,(h/a) - 0.5\,(h/a)^3] \quad \text{for } h < a$$

$$= Co + C \text{ for } h \geq a$$

where *Co*, *C* and '*a*' are the model semi-variogram parameters as defined below, and *h* is the lag distance.

Co is referred to as the nugget variance and represents the random portion of the variability of the regionalized variable. *Co* is the product of assaying and sampling errors plus microstructures at a scale smaller than the sampling interval.

C is the remaining variance which is the spatial component of the regionalization. $Co/C = \varepsilon =$ the nugget effect and is a measure of the importance of the random factor in an orebody. Stratiform sedimentary/diagenetic sulphide deposits may have a very low value of ε because of the uniform and gradational nature of the grade values perhaps strongly controlled by some regional process during formation of the deposit. However, metamorphic lateral secretional processes may cause local enrichments in grade causing ε to increase. This value of ε then could be used to measure the extent of metamorphic redistribution.

The value $Co + C$ is the constant value of $\gamma(h)$ which occurs beyond a value of *h* referred to as the range (*a*). This value '*a*' represents the zone of influence of the samples. This is an important value for it allows us to determine how far a particular sample value should be extrapolated and whether enough samples have been collected. The value $Co + C$ is also the geostatistical equivalent of the sample variance σ^2 of the data set. If the semi-variogram shows random fluctuations about a horizontal line then we have an orebody showing pure nugget effect (Figure 4.14). In this case, the best method to use for its evaluation would be one based on classical statistics. At an early stage of an exploration programme, the drill-hole spacing may be greater than the range and no structure can thus be detected at low lags. In this case, we will also have to resort to classical statistics to determine average grades, etc.

Although most semi-variograms are plotted on a distance base, a time base could have been used if we were considering changes in some variable with time. One example would be the variation of mill feed grade or the content of deleterious elements, such as As, Hg or Bi, over a monthly or quarterly basis. Similarly, hydrogeologists or environmental engineers may be interested in the variation of contaminants, nitrates or total dissolved solids in water from rivers, springs or boreholes.

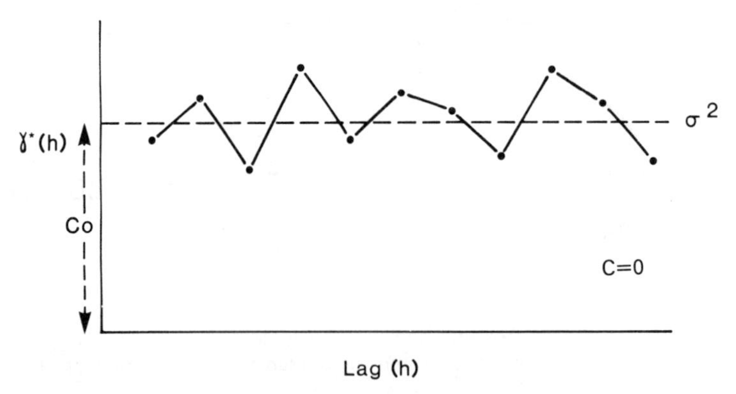

Fig. 4.14 Pure nugget effect.

4.7 SEMI-VARIOGRAM PHENOMENA IN THE SPHERICAL SCHEME

4.7.1 Proportionality in the spherical scheme

If a data set conforms to a log normal distribution then the variance of the data and hence the variogram sill are proportional to the square of the mean of the samples.

$$\text{Variance} = \Sigma_{i=1}^{n} (\text{mean of log values} - \text{log value}_i)^2/(n-1)$$

There will be no problem so long as data sets from different parts of an orebody have the same mean. If a deposit is very large then it is perhaps unrealistic to assume constant spatial variation and it is necessary to divide the deposit up into subareas or horizontal slices (levels) provided that there are still enough samples in each. Each subarea or level will have a semi-variogram with a different sill (Figure 4.15). This is the proportional effect.

This effect can be demonstrated if the variance σ^2, or $C + Co$, or just C (as Co is constant for subareas) is plotted against the (assay mean)^2 for these subareas as in Figure 4.16.

If semi-variograms are plotted on the basis of untransformed data is is possible to obtain two or more different sill levels but these may only be

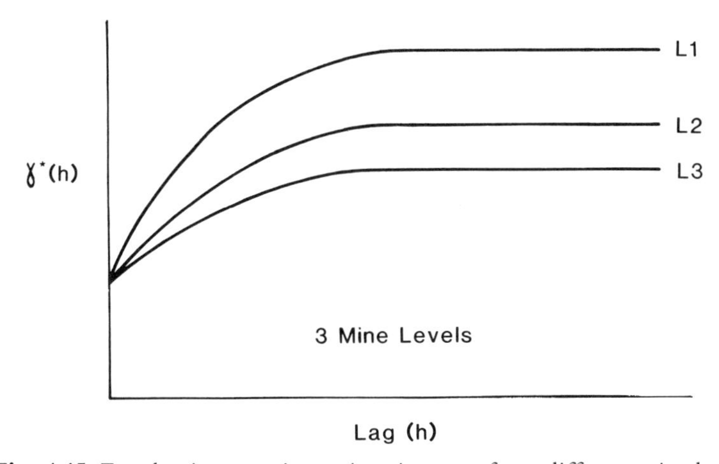

Fig. 4.15 Zonal anisotropy in semi-variograms from different mine levels.

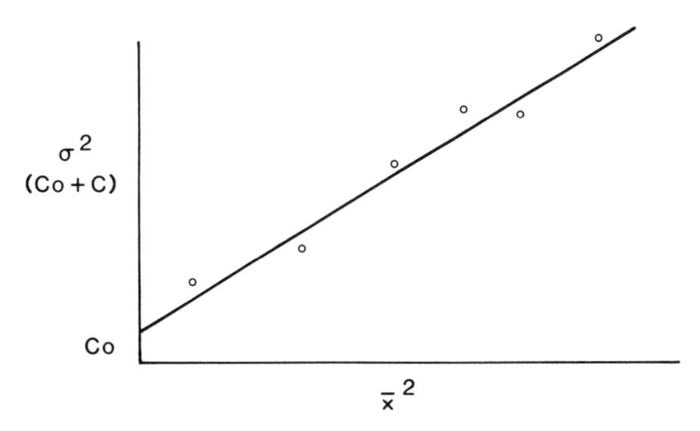

Fig. 4.16 Proportional effect demonstrated.

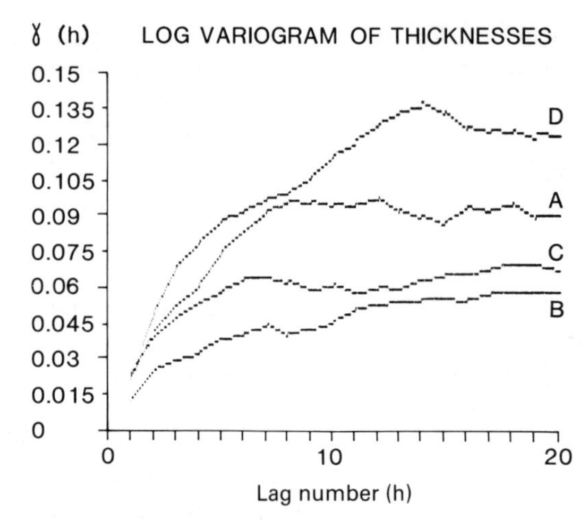

Fig. 4.17 Zonal anisotropy (proportional effect) in the Offin River placer, Ghana.

the product of apparent proportional effect. When the values are log-transformed this anisotropy may disappear. If it still remains, as was the case for the Offin River placer gold deposit (Figure 4.17), then a proportional effect must be accepted. Each subarea or slice of the deposit will thus have to be assigned a local sill value. A best fit regression line through the points will then enable the slope of this line (ϕ) to be determined so that C for each subarea can be calculated from:

$$C = \phi \times (\text{subarea mean value})^2 = \phi\,(X^2)$$

The theoretical semi-variogram for the subareas or slices will then be:

$$\gamma(h) = Co + \phi(X^2)\,[3h/2a - (h/a)^3/2] \quad \text{for } h < a \text{ or}$$

$$\gamma(h) = Co + \phi(X^2) \quad \text{for } h \geqslant a$$

The production of relative semi-variograms, as opposed to absolute semi-variograms, corrects for this proportionality. In this case, the $\gamma(h)$ values for each lag are divided by the $(\text{mean})^2$ value of the raw data on each level. A different mean will thus be used for each semi-variogram. A. G. Royle (personal communication, 1990) suggests that a proportional effect can be removed by calculating the relative $\gamma(h)$ by use of the following formula for each pair of samples at each lag value:

$$\gamma(h) = [(\text{value1} - \text{value2})/(0.5 \times (\text{value1} + \text{value2}))]^2$$

Half the mean of these $\gamma(h)$ values for each lag h gives the relative $\gamma(h)$, i.e.:

$$\gamma(h) = 0.5 \times \Sigma\,\gamma(h)/N(h)$$

where $N(h)$ is the number of pairs at lag h.

4.7.2 Drift

An assumption made in geostatistics is that no significant statistical trend occurs within the deposit which would cause a breakdown in stationarity. If such a statistical trend does occur, then a semi-variogram such as that illustrated in Figure 4.18 would result.

The drift in this case is, however, at distances beyond the range and it thus will not interfere with local estimation of deposits where the dimensions of the search area used to evaluate each ore-block are smaller than the distance represented by the point at which drift occurs. Where drift completely dominates the semi-variogram, it is perhaps best to fit a trend surface to the data and then to krige the residuals from this surface. It may be that this phenomenon only occurs on one or two lines in a deposit and all other semi-variograms based on single lines of data conform to a spherical scheme. In this case, there is no need to propose a generalized model of the form $\gamma(h) = ph^\alpha$ where $1 < \alpha < 2$. It is usually better to exclude such lines from the data set and produce a new semi-variogram for the deposit as a whole.

Parabolic behaviour of this type on all lines indicates the presence of strong drift, i.e. non-stationarity. In this instance we should use a technique called Universal Kriging (see Journel and Huijbregts, 1978) rather than Simple Kriging which is only relevant when quasi-stationarity exists.

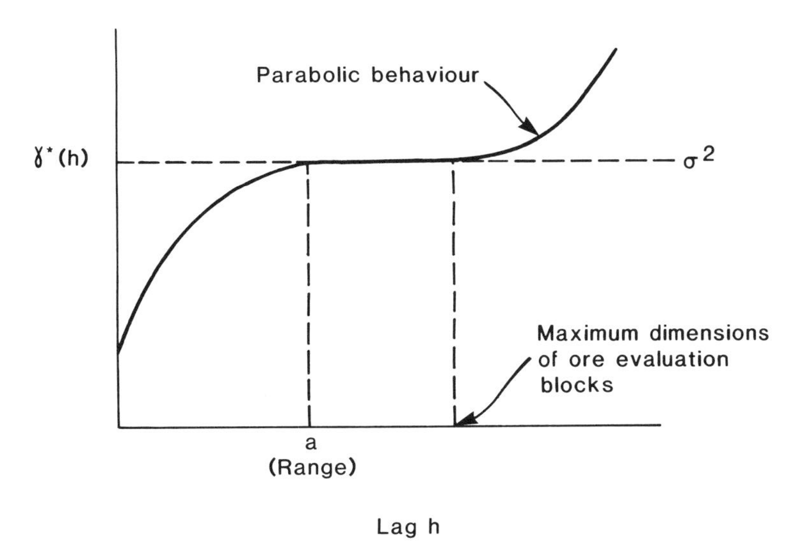

Fig. 4.18 Drift on a spherical scheme semi-variogram.

4.7.3 Directional anisotropism

Irrespective of the type of model we eventually fit to the data, we should check whether or not we have directional anisotropism. This occurs when different semi-variograms are obtained for different directions in an orebody. This means that, instead of having a circular zone of influence as in the isotropic case, we have an elliptical zone of influence. Ideally, the semi-variograms should be calculated in five different directions, four of which lie in the plane of the orebody, the other being across its dip (Figure 4.19). In order to more precisely define the long axis of the ellipse of anisotropism, as many as eight directional semi-variograms may have to be calculated. The $\gamma(h)$ values for each direction can be plotted on a polar plot, i.e. on lines radiating out from a central point, so that a better impression can be obtained of the form and orientation of the ellipse.

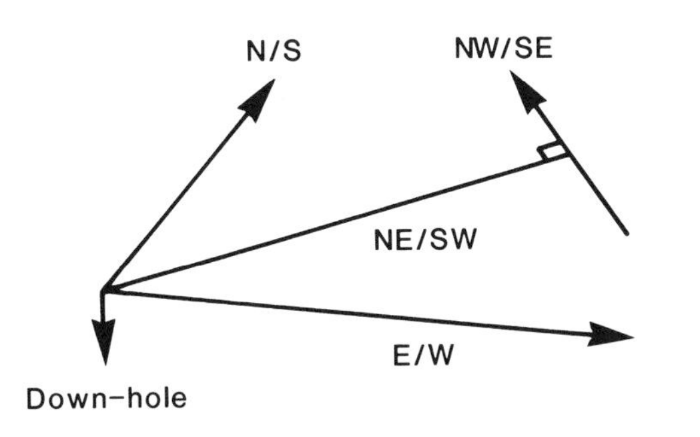

Fig. 4.19 Orthogonal semi-variogram directions in a mineral deposit.

Anisotropism is especially marked in alluvial deposits where the range across the deposit is short compared to that parallel to its length (i.e. downstream). This is illustrated in Figure 4.20. Boakye (1989), in his study of the Offin River placer deposit in Ghana, found that transverse ranges for raw thickness and log-transformed gold accumulations were 285 m and 145 m respectively, whereas in the down-stream direction the equivalent values were 400 m and 480 m. The implication of this is that sampling intervals should be shorter across the deposit than along it. The ratio, a_1/a_2, is referred to as the directional anisotropism coefficient which is usually in the range 1.0–1.5 (in Boakye's case, however, the ratios were 1.4 and 3.3).

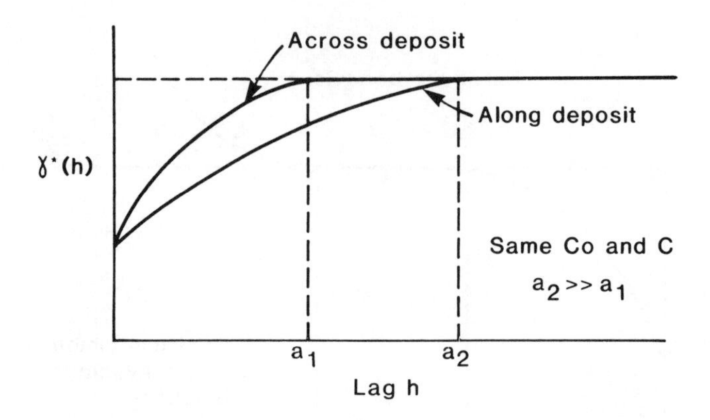

Fig. 4.20 Directional anisotropism in an alluvial deposit.

4.7.4 Hole effect

This effect may be recognized when areas of high grade mineralization alternate with areas containing low values. The result is a pseudo-periodicity which is reflected by an oscillation of the semi-variogram about the apparent sill level. An excellent example of this phenomenon is shown in Figure 4.21 from the Offin River placer, Ghana. This effect can be easily confused with the usual erratic oscillation of the semi-variogram about the sill value for lag values greater than the range.

4.8 MODEL FITTING IN THE SPHERICAL SCHEME

4.8.1 Simple spherical models

In practice, the experimental semi-variogram is more irregular than indicated above, especially beyond the range '*a*' where random fluctuations occur about the sill. Once the experimental semi-variogram has been produced, an attempt can be made to fit a simple spherical model. In the case of the Tarkwa semi-variogram for gold accumulations (Table 4.1 and Figure 4.22) we can deduce that, for channel sampling at 1.5 m intervals along drives in the reef:

$$Co = 2.20 \times 10^5 \text{ (cm g/t)}^2$$
$$C = 6.514 \times 10^5 \text{ (cm g/t)}^2$$

and

$$a = \text{lag } 4.5 \text{ or } 6.9 \text{ m.}$$

The model for the semi-variogram is thus:

$$\gamma(h) = 2.20 \times 10^5 + 6.514 \times 10^5[3h/(2 \times 6.9) - (h/6.9)^3/2] \text{ when } h < 6.9 \text{ m} \quad (1)$$

and

$$\gamma(h) = (2.20 + 6.514) \times 10^5 = 8.714 \times 10^5 \text{ when } h > 6.9 \text{ m} \quad (2)$$

We can produce the model curve by substituting various values of h in equations (1) and (2). Minor modifications of the variogram parameters can be made to gain a better fit.

In the case of the thickness semi-variogram, a straight line could be fitted to the data indicating either:

(1) A linear model with formula:

$$\gamma(h) = m(h) + Co = 7.93 \times 10^{-3}(h) + 0.012$$

(2) A spherical model with extreme drift.

Note. If the sampling spacing is closer in one direction than another, we can partly overcome the inadequacy by determining the semi-variogram for the closely sampled direction and then using the results to help model the other (they will have the same $Co + C$ and Co values as

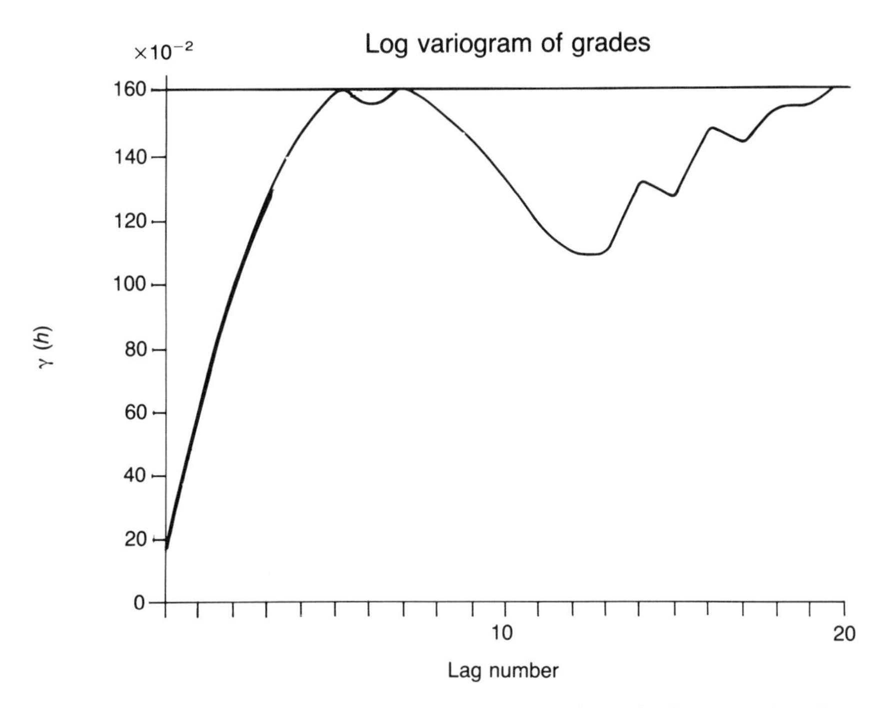

Fig. 4.21 Hole effect – an example of periodicity in a spherical scheme semi-variogram.

they are in the same orebody but will have different ranges if anisotropy exists).

4.8.2 Compound or nested semi-variograms

If a line drawn through the first two or three points of the semi-variogram meets the sill at a point whose value gives a range clearly less than that given by the point at which the semi-variogram flattens out at the sill, then a mixture of two spherical schemes is suspected.

In Figure 4.23, a semi-variogram for ln% Ni, the tangent through the first two to three points cuts the $\gamma(h)$ axis at 0.4 $(\ln\%)^2$ while the intersection with main sill at 2.55 $(\ln\%)^2$ occurs at a lag distance of 13 m, indicating that 'a' should be 20 m. However, the main curve does not approach the sill until the lag distance is 50 m. Therefore, a compound semi-variogram is assumed to exist consisting of a semi-variogram with a sill at 1.95 and another with a sill at 2.55 and a range of 50 m. The parameters can thus be listed as follows:

1st SV $Co = 0.40\ (\ln\%)^2$
$a_1 = 14\text{ m (as } 2a_1/3 = 9)$
$C_1 = 1.95 - 0.4 = 1.55\ (\ln\%)^2$

2nd SV $Co = 0.40$
$a_2 = 50\text{ m}$
$C_2 = 0.6\ (\ln\%)^2$

The compound model now has the form:

$$\gamma(h) = Co + C_1[3h/2a_1 - (h/a_1)^3/2] + C_2[3h/2a_2 - (h/a_2)^3/2]$$

For $h < 14$ m

$$\gamma(h) = 0.4 + 1.55[3h/28 - (h/14)^3/2] + 0.60(3h/100 - (h/50)^3/2]$$

For h between 14 m and 50 m

$$\gamma(h) = 0.4 + 1.55 + 0.60[3h/100 - (h/50)^3/2]$$

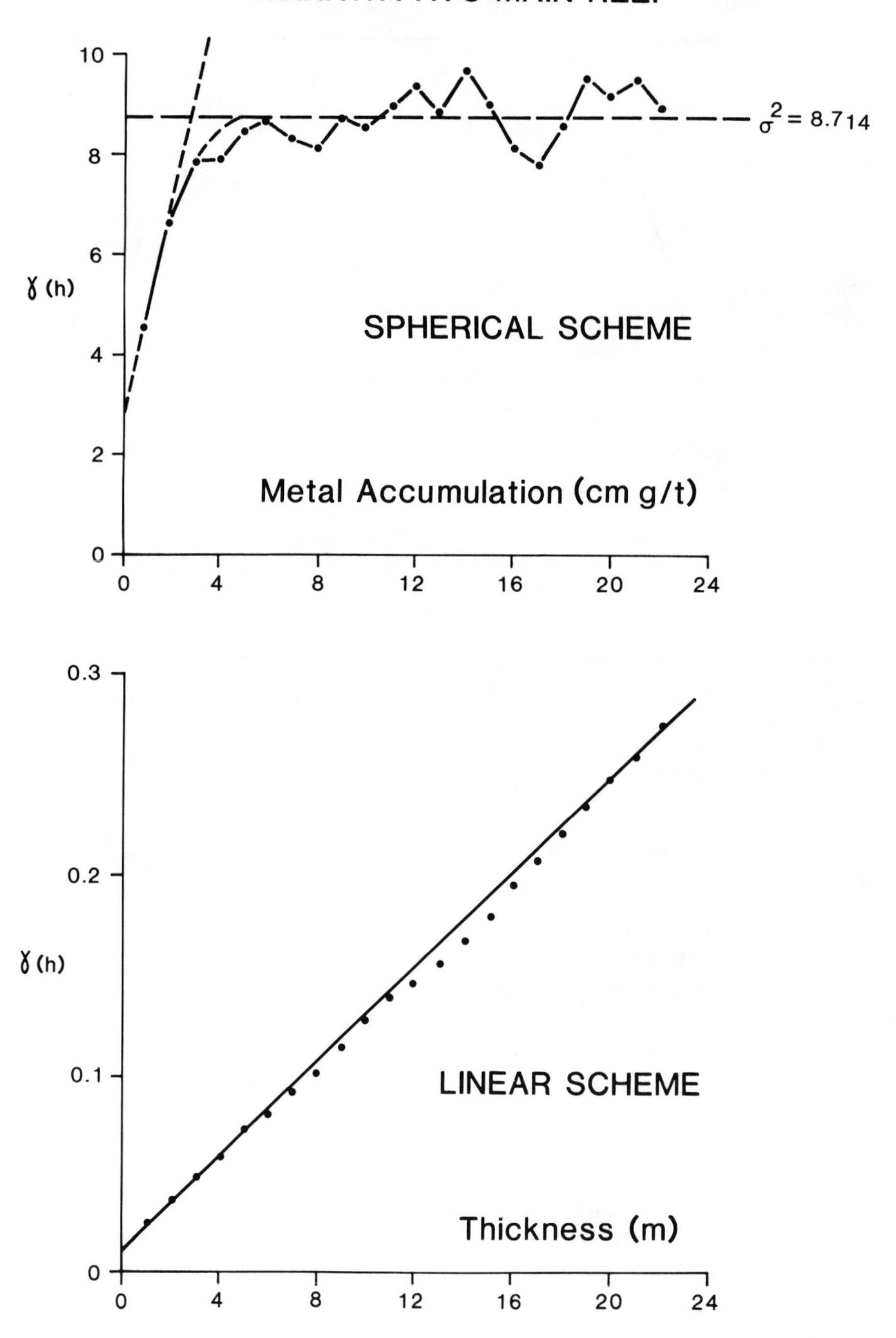

Fig. 4.22 Spherical scheme semi-variogram for metal accumulation and linear scheme for thickness from the AVS Main Reef, Tarkwa, Ghana.

For $h > 50$ m

$$\gamma(h) = 0.4 + 1.55 + 0.60 = 2.55$$

We can thus calculate and plot the theoretical semi-variogram by systematically varying h from 5 m to 80 m, taking into account the three portions of the curve outlined above. If this model does not give a good fit to the experimental data, we can modify C_1, C_2 and a_1, a_2 until this is achieved.

Nested structures of this type may be caused by the presence of richer ore-shoots within a sparsely mineralized quartz vein. Figure 4.24 represents a VLP of such a vein showing the main elements of the mine development. Semi-variograms drawn along the levels would thus be nested. The first range (small, e.g. 15–20 m) would represent the width of the small structures, i.e. the ore-shoots, whereas the second range would represent the wider regionalization of mineral in the vein. Semi-variograms drawn on the raises would, however, be simple spherical as the ore-shoots are steeply plunging.

Compound spherical models are common in alluvial gold deposits where the shorter range reflects individual channels and the longer range the full width of the pay zone. It is also possible

Table 4.1 $\gamma^*(h)$ values AVS Main Reef, Tarkwa, Ghana

Lag	Diffs. (no.)	Accumulation (cm g/t)^2 (× 10^5)	Thickness (m^2)
1	351	4.495	0.026
2	349	6.574	0.037
3	347	7.843	0.048
4	345	7.898	0.060
5	343	8.463	0.073
6	341	8.739	0.081
7	339	8.299	0.094
8	337	8.110	0.104
9	335	8.703	0.115
10	333	8.572	0.128
11	331	8.979	0.139
12	329	9.349	0.146
13	327	8.814	0.156
14	325	9.699	0.168
15	323	9.006	0.180
16	321	8.122	0.197
17	319	7.809	0.208
18	317	8.572	0.221
19	318	9.505	0.235
20	313	9.184	0.249
21	311	9.519	0.258
22	309	8.984	0.275

Lag distance = 1.5 m.

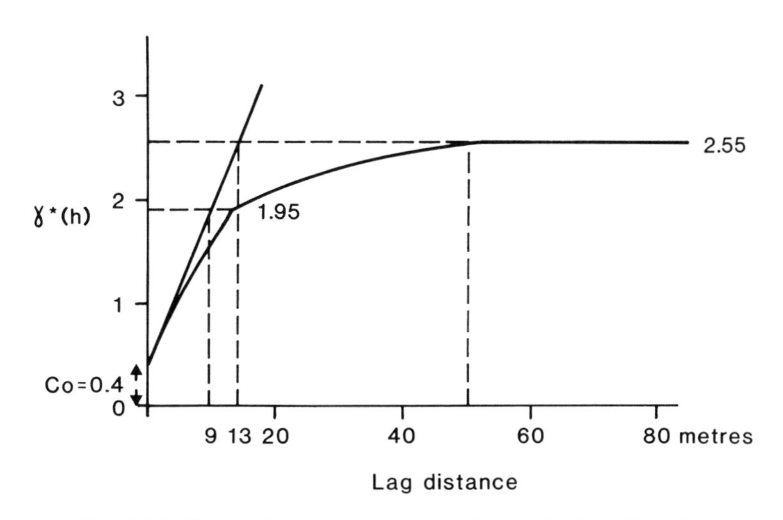

Fig. 4.23 Composite semi-variogram for grade (log$_e$ % Ni).

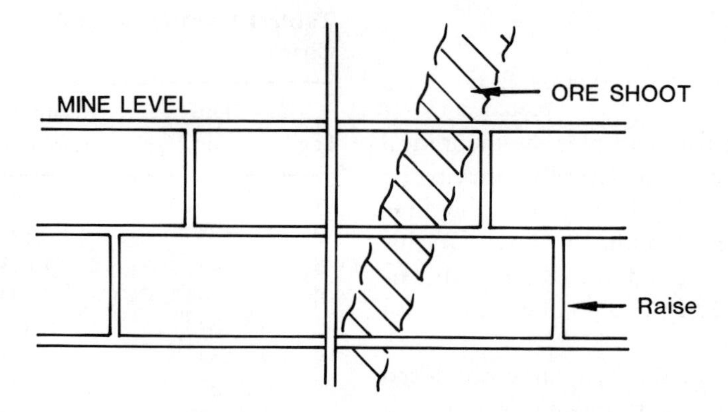

Fig. 4.24 Ore-shoot in a quartz vein plus associated mine development (VLP).

to have a third range reflecting the lower grade material in adjacent alluvial flats which have also been sampled during the main programme. Examples of complex models were obtained by Boakye (1989) for the semi-variograms for the Offin River placer, Ghana (Figure 4.25).

4.8.3 Two-stage spherical scheme models

These are usually the product of combining unrelated data sets unintentionally, e.g. two unrelated veins or two phases of mineralization

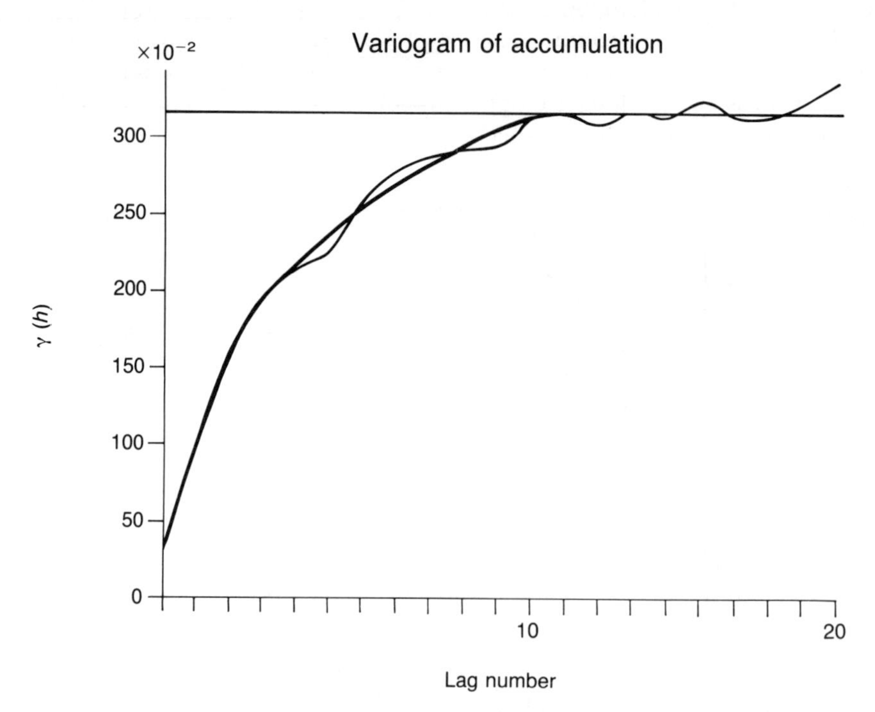

Fig. 4.25 Composite semi-variogram of gold accumulation, Offin River placer, Ghana.

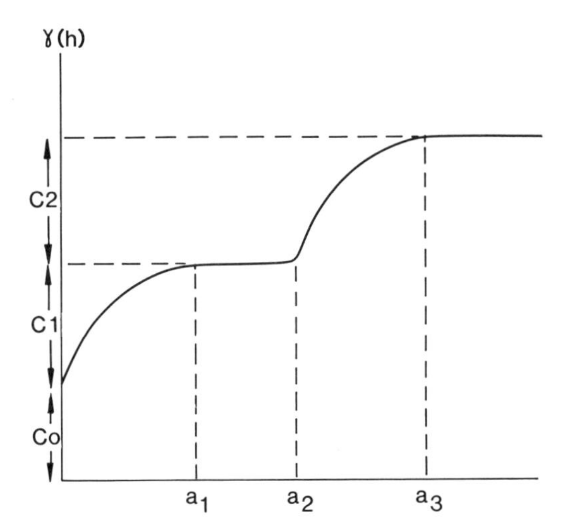

Fig. 4.26 Typical two-stage semi-variogram.

which have different characteristics. The typical form of such semi-variograms is illustrated in Figure 4.26 while the mathematical model applied in this case is:

$$\begin{aligned}\gamma(h) &= Co + C_1\,[3h/2a_1 - (h/a_1)^3/2] && \text{for } 0 < h < a_1\\ &= Co + C_1 && \text{for } a_1 \leqslant h < a_2\\ &= Co + C_1 + C_2(3(h - a_2)/2(a_3 - a_2) - (h - a_2)^3/2\\ &\quad (a_3 - a_2)^3) \text{ for } a_2 \leqslant h < a_3\\ &= Co + C_1 + C_2 \text{ for } h > a_3\end{aligned}$$

4.8.4 Point kriging for semi-variogram modelling (cross-validation)

Point kriging is a technique which allows the determination of the best linear unbiased estimate of a point using a series of weighting factors (K_i) to weight all values of an RV falling within a specified search area. The method works in a similar way to inverse distance weighting. The sum of all the weighting factors ($\Sigma\ K_i$) is set to one while these factors themselves are calculated from a series of simultaneous equations (Appendix 4.2). The other values for these equations are obtained from the chosen semi-variogram model equation.

Point kriging can be used to test the chosen semi-variogram model – this is a process called cross validation or jacknifing. The program estimates the value at each drill-hole or sample location, after removing the observed value, by kriging all those adjacent values which fall in the search area around this point. If the data are on a regular 15 × 15 m grid, as in Figure 4.27(a), then by removing the central point, a total of eight points will be kriged if the X max and Y max values (referred to as the semi-axis lengths) are both set at 25 m. The search area semi-axis lengths should be less than, or equal to, the

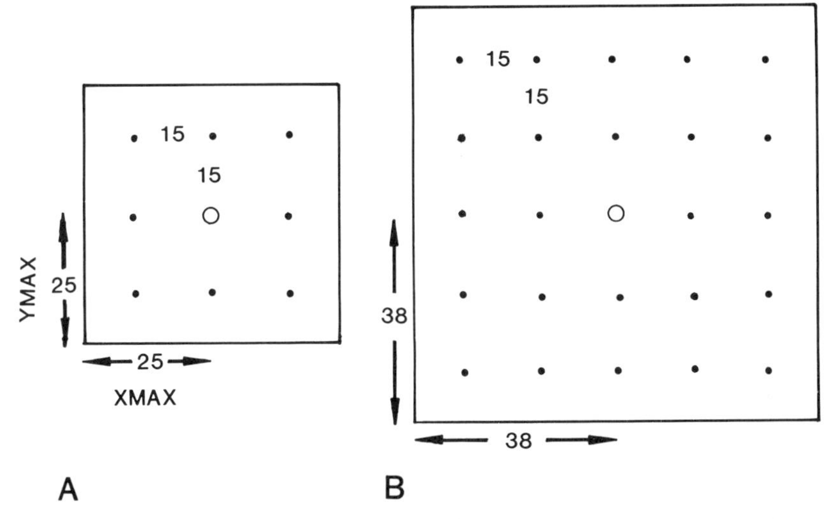

Fig. 4.27 Point kriging search areas showing points captured around central point.

range. If the range of the semi-variogram (isotropic) is 40 m then another possible search area would be one with a semi-axis value of 38 m (Figure 4.27(b)). A total of 24 points would then be kriged. The recommended minimum number of points is six.

Each drill-hole thus has an observed value Z_i and also a kriged estimate $Z_i^\star$ for the regionalized variable at this point. Most point kriging programs then output the X and Y coordinates of each point together with Z_i, $Z^\star_i$ and $(Z_i - Z^\star_i)^2$. The final outputs are listed below:

(1) Mean algebraic error:

$$\left(\Sigma_{i=1}^{N} (Z_i - Z_i^\star)\right)/N$$

Where Z_i is the actual value at each point and N is the number of points. This calculation takes into account the sign of $Z_i - Z^\star_i$.

(2) Mean absolute error:

$$\left(\Sigma_{i=1}^{N} \mid Z_i - Z_i^\star \mid\right)/N$$

This is the mean of the differences but this time the sign is ignored.

(3) Mean kriging variance:

$$\left(\Sigma_{i=1}^{N} \sigma_K^2\right)/N$$

(4) Mean square error of estimation:

$$\left(\Sigma_{i=1}^{N} (Z_i - Z_i^\star)^2\right)/N$$

(5) Number of points valued by point kriging.

If the model allows an accurate estimation of the data population then, the value of 1 approaches zero and is not more than 1% of Z (the mean of all the Z_i values); 4 should be almost equal to 3; 5 should be as large as possible.

A significant difference between 4 and 3 may be due to outliers, i.e. abnormally high or low values in a data set, which greatly increase the (difference)^2 values between these and adjacent points. Removal of these outliers may allow the mean squared differences value to approach the mean point kriging variance. Another way to test the semi-variogram model is to plot Z_i against $Z_i^\star$ and if the values are uniformly distributed about a best fit regression line whose slope is 45°, then we can say that conditional unbiasedness has been achieved.

4.9 1D-REGULARIZATION (SPHERICAL SCHEME)

After regularization, a process which effectively converts a 3D orebody into a 2D plan, the sample dimensions relative to the ranges in the plane of the orebody are usually very small. A point sample is thus one with no significant size. If, however, the sample size is large in the plane of the orebody, or the range in one or more directions is small, then the experimental semi-variogram must be converted into a point semi-variogram to remove the effect of sample support. The problem is particularly acute where a 'down-the-hole' semi-variogram is to be calculated and where the sample length may be 1.5 m in a direction where the range is likely to be small, e.g. 5–10 m.

The true range 'a' of a point semi-variogram = ($a^\star$ − the sample length in the direction of the semi-variogram) where $a^\star$ is the range from the experimental semi-variogram. The point variogram values of Co and C are then determined from the following equations and the 1D-regularization graphs (Figure 4.28).

$$\gamma_1 = Co + C \cdot X_1 \quad \text{(for lag 1)} \qquad (1)$$
$$\gamma_\infty = Co + C \cdot X_2 \quad \text{(value at range } a^\star\text{)} \qquad (2)$$

For lag 1 and lag ∞ we can thus read off from the graphs the values of X_1 and X_2 for the corresponding values of h/a and the value of l/a where h = lag distance and l = sample length. The simultaneous equations can then be solved for Co and C. These are then substituted in the equation for the Matheron scheme semi-variogram, as has been done in the following example, to obtain the regularized model $\gamma(h)$ values.

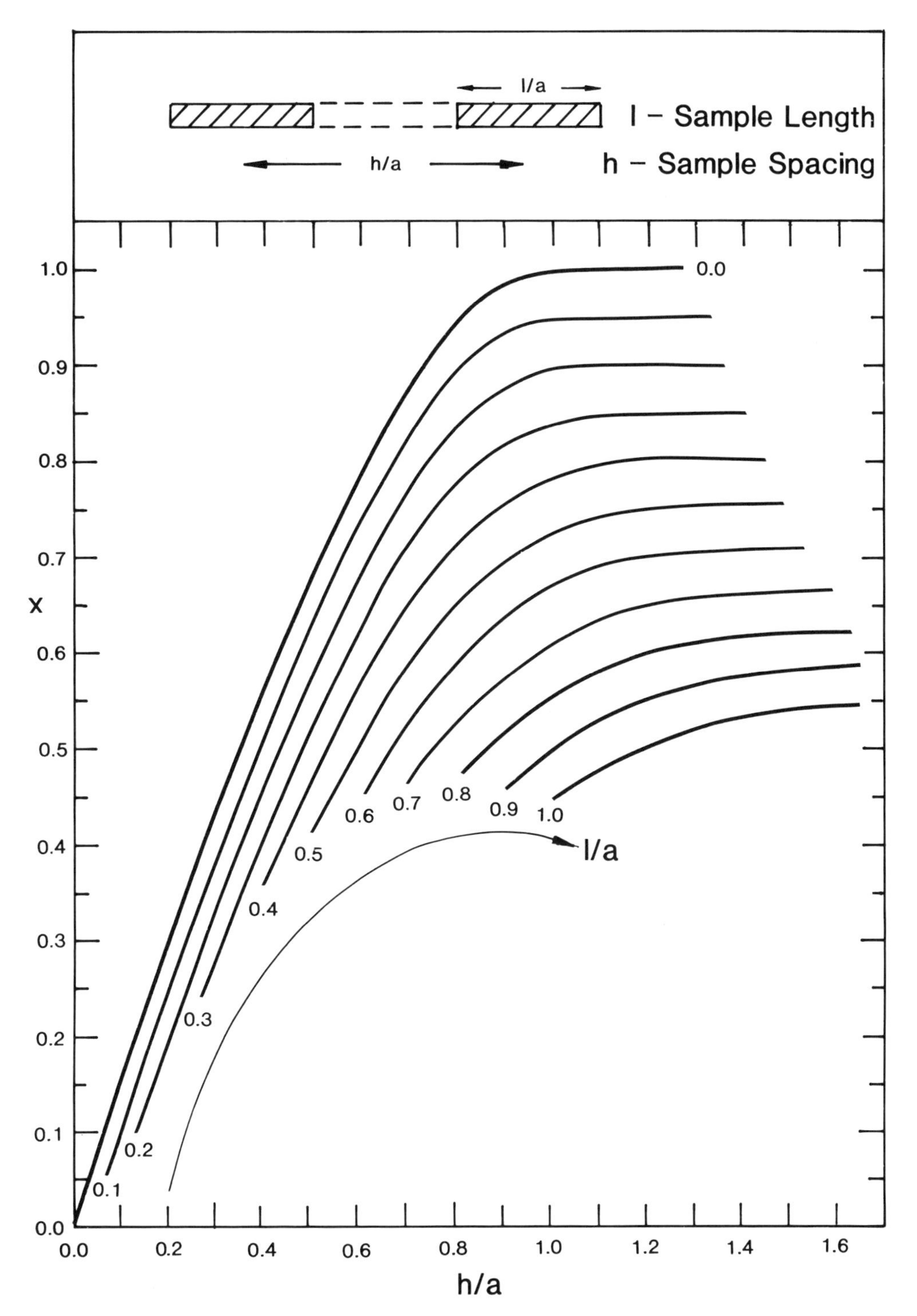

Fig. 4.28 1D-regularization graphs.

4.9.1 Example of the application of 1D-regularization

Alluvial diamond deposit

An alluvial diamond deposit has been sampled by cross trenching. Each sample is representative of the gravel removed from 20 m sections of this 1 m wide trench. The regionalized variable used to compute the reserves of the deposit was the weight (in carats) of diamonds found in each 1 × 20 m section of the trench. The dimensions of the RV were thus $M{\cdot}L^{-2}$, corresponding to the accumulation. The $\gamma(h)$ values of accumulation are as follows:

lag 1 (20 m)	0.0147
lag 2 (40 m)	0.0213
lag 3 (60 m)	0.0226

after which the variogram oscillated about a sill value of 0.0226. The apparent range $a^\star = 60$ m, therefore the true range $a = 60 - 20 = 40$ m. Hence $L/a = 20/40 = 0.5$. We can now determine the point variogram values of Co and C using the 1D-regularization graphs.

For lag 1 $h/a = 20/40 = 0.5$, therefore $X_1 = 0.415$
For lag ∞ $h/a = 60/40 = 1.5$, therefore $X_2 = 0.760$

We can thus solve the simultaneous equations (1) and (2) quoted earlier:

$$\gamma(20) = 0.0147 = Co + C(0.415)$$
$$\gamma(\infty = 60) = 0.0226 = Co + C(0.760)$$

Hence $Co = 0.0052$ and $C = 0.0229$.

Applying the mathematical model for the spherical scheme semi-variogram:

$$\gamma(h) = Co + C[3h/2a - (h/a)^3/2] \quad \text{for } h < a$$
$$\text{and } \gamma(\infty) = Co + C \quad \text{for } h \geqslant a$$

the point semi-variogram model $\gamma(h)$ values are:

$\gamma(20) = 0.0210$
$\gamma(30) = 0.0263$
$\gamma(40) = 0.0281$ $(\gamma(\infty) = Co + C = 0.0052 + 0.0229)$

4.10 BLOCK RESERVE ESTIMATES BY KRIGING

The evaluation of individual blocks or RSG panels of ore can be undertaken by a technique called simple kriging. This method thus replaces such techniques as inverse distance weighting (linear, square, cubic) or grid superimposition on contour maps of assay grades, thicknesses and accumulation.

A major criticism of methods which use weighting by the inverse of distance, or the inverse square of distance to block centres, is that they give the same result irrespective of block size. Figure 4.29 shows two co-axial blocks of markedly different size which would be assigned the same value – not a satisfactory situation. Also they do not allow any estimate to be made of the error involved in the grade evaluation of the block. Kriging is able to overcome both of these problems.

The kriging process is referred to as BLUE in that it gives the Best Linear Unbiased Estimate for a block. It aims to minimize estimation variance of a block which thus makes it an optimum estimator.

A search area (Figure 4.30) is established around the ore-block whose dimensions are equal to, or just less than, twice the range in each direction (the smallest range in the case of nested semi-variograms). Where an isotropic case exists, a circular search area can be used with a

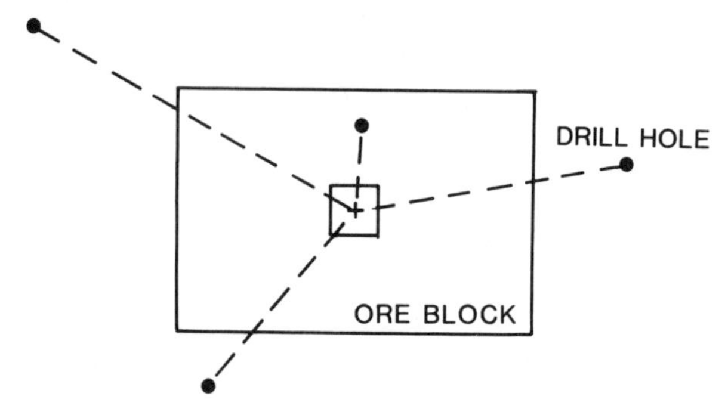

Fig. 4.29 Ore blocks showing samples to be used for block kriging.

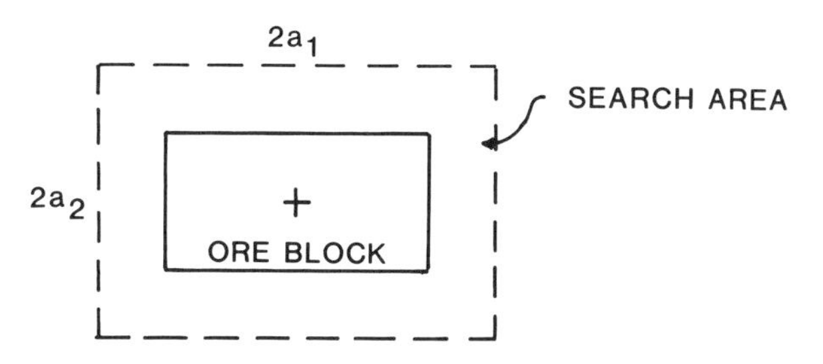

Fig. 4.30 Kriging search area.

radius just less than the range. In the case of the Nchanga Lower Orebody, isotropic semi-variograms with a range of 150 m were obtained and thus a search radius of 120 m was applied.

All drill-holes located within the search area are allocated an optimum weighting coefficient – the kriging coefficient K so that:

$$\Sigma_{i=1}^{n} (K_i) = 1$$

These coefficients are derived from the mathematical model of the semi-variogram for the RV under study (usually thickness and metal accumulation). The block value is then derived in the usual way, viz:

$$\Sigma_{i=1}^{n} (Z_i \times K_i)$$

where Z_i represents the value of the RV at each hole in the search area. Ideally there should be 15 or 16 samples in the search area and if the number drops below 4 then the block is not estimated. At Nchanga a 'throw-out angle concept' has been used to select 12 drill-holes for kriging within a circular search area. In this method, all holes located in the search area are listed in decreasing order of distance from the centre. The angle (θ) subtended by the centre of the block to be evaluated and the first and second holes in the list is then calculated followed by the angle between the first and the third hole, and so on to the end of the list. If any of these are less than a specified tolerance (e.g. 3°), then the first hole in the pair (i.e. that farthest from the block centre) is eliminated. This is repeated with the next hole in the list (which could now be the first hole) and so on until all holes in the search area have been compared with all the others. If there are still more than 12 holes, then the exercise is repeated with progressively larger tolerances until this figure is reached. This effectively reduces the effect of duplication by boreholes with the same relative position with respect to the block centre (Figure 4.31). It also significantly reduces the running time for the kriging program.

Simple kriging is a robust estimator and can deal with a certain amount of skewness in the data distribution. When log-transformed RVs are used, however, the method is less robust and is very distribution dependent. Computer programs are available using both methods and an understanding of the mathematical basis is not absolutely essential for the mine geologist. However, for those who wish to delve deeper into the subject, more information is provided in Appendix 4.4. Basically the method involves the solving of a series of kriging equations (simultaneous equations) which relate the covariance between each sample value in turn, with all the other sample values in the search area plus the

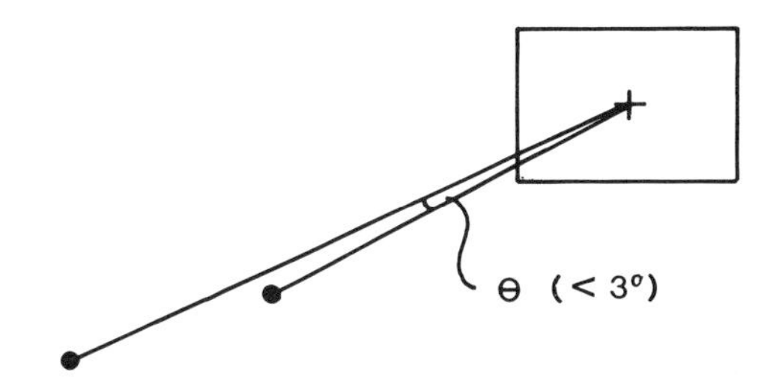

Fig. 4.31 Illustration of the throw-out angle technique.

covariance between this sample value and that of the block. Each sample–sample covariance is weighted by the relevant kriging coefficient, K_i. Thus, if there are 12 samples in the search area then 12 equations are solved for the 12 coefficients together with a thirteenth which states that $K_1 + K_2 + K_3 + \ldots + K_{12} = 1$.

Only values nearest the block have any significant effect on the estimate for the block as distant values are effectively filtered out – the screen effect. The nugget effect ε is very important in kriging, for, as Co/C increases, the screen effect begins to break down and as a result values farther away begin to influence the block estimate. For a given value of ε, larger values of the range 'a' decrease the kriging coefficients given to nearby samples. Thus for low values of ε and short ranges, only samples close to the block play any role in evaluating the block. Like inverse weighting, kriging is a smoothing technique.

Simple kriging programs produce kriged estimates of grade, thickness and metal accumulation, along with the relevant kriging variances ($\sigma_K{}^2$), for each ore-block in turn. The kriging variance is a reflection of the error incurred in using simple kriging to evaluate each block. The process has, however, minimized this error during the calculation of the kriging coefficients. The variables used may be either raw or log-transformed.

4.11 GLOBAL RESERVE EVALUATION BY KRIGING

There are four main stages in the estimation of the global reserves of a deposit from the local kriged estimates of blocks:

(1) Estimation of tonnage and its error variance.
(2) Estimation of overall grade and its error variance.
(3) Estimation of the quantity of contained metal and the associated variance.
(4) Determination of the precisions of estimates.

4.11.1 Relative estimation variance of tonnage

In order to calculate the tonnage of an orebody we need:

(1) The area of the orebody ($S^\star$);
(2) The thickness estimate for the orebody – the average kriged thickness (Th_K) of all blocks above the cut-off grade;
(3) The density (d).

Hence:

$$\text{tonnage} = S^\star \times Th_K \times d$$

Area estimates
If we have a drilling grid in which n holes are in ore and the remainder are barren, then the estimated area of the orebody is:

$$S^\star = n \cdot a_1 \cdot a_2$$

where a_1 and a_2 are the dimensions of the grid and also of the ore-block centred on each hole.

We need to know how good an estimate this is, i.e. what is the estimation variance of $S^\star$? The accuracy of the estimate depends on a_1, a_2 and the irregularity of the orebody fringe.

The problem lies in the fact that our estimate of the location of the orebody's boundary is based on the point data provided, in 2D, by drill-holes, as in Figure 4.32. Its actual position could

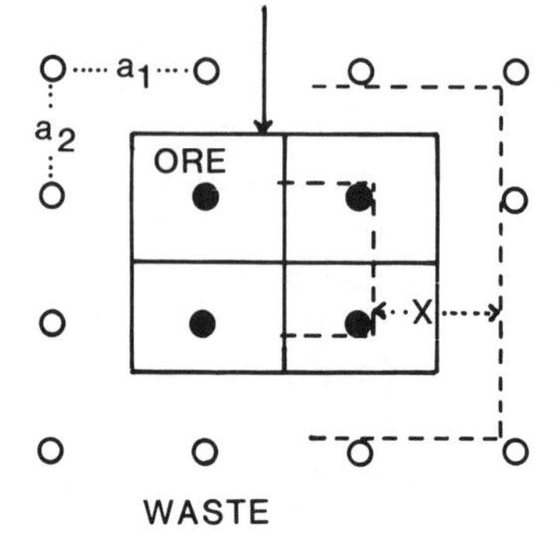

Fig. 4.32 Delineation of an orebody fringe on the basis of RSG panels.

lie anywhere between the extremes shown, which are a distance 'x' apart. The maximum possible error is thus ± half the drill-hole spacing in this direction. The irregularity of the orebody fringe is measured by its semi-diametral variation D.

Semi-diametral variation of RSG panels

The easiest method for determining estimation variances for areas of orebodies represented by groups of positive (i.e. ore grade) RSG panels is based on the formula for relative estimation variance (REV) which is:

$$\sigma^{\star 2}/S^{\star 2} = [D_\alpha/6 + 0.0609(D_\beta)^2 D_\alpha]/n^2$$

where n represents the number of blocks and where D_α and D_β are the semi-diametral variations determined by counting the total number of RSG block sides lying between +ve and −ve blocks in the directions α and β respectively and dividing by two. The direction β is chosen so that D_β is greater than D_α.

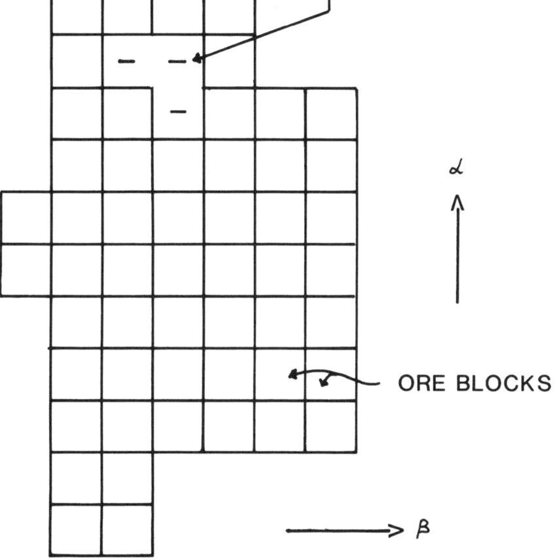

Fig. 4.33 A grouping of RSG panels (ore and waste) from Navan Mine, Eire.

In the example in Figure 4.33, $D_\alpha = 18/2 = 9$ and $D_\beta = 26/2 = 13$.

This grouping of RSG panels, each 60 × 60 m, represents part of the 2–5 lens at Navan in Eire. It was produced by:

(1) constructing an unrolled plan of the drill-hole intersections in this lens;
(2) regularizing drill-hole grades;
(3) plotting true thickness, m% (combined Pb + Zn) and grade values;
(4) fitting an RSG grid to the drill-holes;
(5) outlining those RSG panels with grades greater than the cut-off grade which, in this instance, totalled 53.

The total area estimate $S^\star$ for the 2–5 lens = $53 \times 60^2 = 1.908 \times 10^5 \text{m}^2$ and the relative estimation variance of this estimate is:

$$\sigma_{S\star}{}^2/S^{\star 2} = [1.5 + 0.0609(13)^2/9]/53^2$$

Hence $\sigma_{S\star}{}^2/(1.908 \times 10^5)^2 = 0.0009411$ and $\sigma_{S\star} = 5853.24$.

Hence at 95% confidence our area estimate is:

$$1.908 \times 10^5 \pm 0.117 \times 10^5 \text{m}^2 \qquad (\pm 2\,\sigma_{S\star})$$

Relative estimation variance of thickness

Once the limits of the orebody have been defined on the basis of kriged grades, or kriged accumulations, and also on the basis of mining practicality, a histogram of the kriged thicknesses for these ore-blocks should be produced. If normality is apparent, then the mean of the thickness values can be taken to represent the 'average' thickness of the deposit. Where a positive skew exists, log transformation may be necessary before using the log mean as the estimator for the deposit. The value of the true log mean can be determined as described in section 3.5. Unfortunately this approach cannot be used to determine the overall estimation variance from the block kriging variances. Journel and Huijbregts (1978) suggest that the global estimation variance for the kriged blocks is best approximated by using extension variances calculated from auxiliary functions. This method will be discussed in detail later in section 4.14 and an example of the calculations involved presented in section 4.14.4. The relevant formula for the global estimation variance is, however, as follows:

$$\sigma_{Th}^2 = (Co + C(\sigma_e^2))/N$$

where N is the number of ore-blocks containing samples, σ_e^2 is the extension variance found in geostatistical tables for either a regular centred sampling grid or an RSG, as the case may be. From this we can calculate the REV of thickness as follows:

$$REV_{Th} = \sigma_{Th}^2/(Th_K)^2$$

Relative estimation variance of density

Only rarely can this element be calculated as it requires that for each sample a density is determined (or estimated from the assay data) and then semi-variograms produced in the usual way. From this, kriged density and associated variance for each block can be determined. The mean kriged density for all the blocks can then be calculated. The problem still remains as to how to compute the overall estimation variance of the density estimate. A.G. Royle (personal communication 1990) suggests that the simplest approximation would be to use the statistical variance of the densities as the estimation variance. This approximate value is higher but gives a safer result.

Relative estimation variance of tonnage

The error incurred in estimating areas, thicknesses and density are combined in the form of REVs to produce the total error which is expressed in the following way:

$$\sigma^2_{\text{ton}}/\text{tons}^2 = \sigma_{S\star}^2/S\star^2 + \sigma_{Th}^2/Th_K^2 + \sigma_d^2/d^2$$

Knowing the tonnage estimate, we can now determine the standard deviation of this estimate and hence the 95% confidence limits or precision. It will be noted that the formula for REV of tonnage includes the term $1/(Th_K)^2$ hence, where the ore is thin, as at the fringes of an orebody, this markedly increases the REV of the thickness and hence of the tonnage.

4.11.2 Relative estimation variance of grade

Where the support of a weighted grade value is variable, it is necessary to apply the following formula to determine the overall grade of a deposit:

$$G = \Sigma \text{ metal accumulation}/\ \Sigma \text{ thickness}$$

Thus, we need to determine the REVs for both accumulation and thickness. However, as these regionalized variables are related, it is necessary to take into account the covariance between them. The REV for grade is thus:

$$\sigma_g^2/g^2 = \sigma_{acc}^2/Acc_K^2 + \sigma_{Th}^2/Th_K^2 - 2\cdot\text{relative } Cov_{(acc,\ Th)}$$

But the correlation coefficient, $r = Cov_{(x,y)}/(\sigma_x^2\cdot\sigma_y^2)^{1/2}$. Hence relative covariance $= r(\sigma_{acc}\cdot\sigma_{Th})/(Acc_K\cdot Th_K)$. The REV equation can thus be rewritten as follows:

$$\sigma_g^2/g^2 = \sigma_{acc}^2/Acc_K^2 + \sigma_{Th}^2/Th_K^2 - 2\cdot r\ (\sigma_{acc}\cdot\sigma_{Th})/(Acc_K\cdot Th_K)$$

Exactly the same method is used for the determination of the mean kriged accumultion (Acc_K), the global estimation variance of accumulation (σ_{acc}^2), and the REV of accumulation, as was used for thickness in the previous section. All that is needed to complete the calculation is the correlation coefficient, r, between the accumulations and the thicknesses which is then multiplied by two times the product of the relative standard deviations for accumulation and thickness.

Blocks at the edge of the orebody may have very high kriging variances due to the lack of drill-holes. Once again, the REV for grade involves the term $1/Th_K^2$ in both the REVs for accumulation and thickness and hence the inclusion of thin sections of the orebody will significantly increase the error in the grade estimation. It may be advisable to delete such blocks from the reserve calculation. To reduce the high kriging variances on orebody fringes, it is important that we increase the density of drilling in these areas and avoid the natural tendency to underdrill economically less interesting areas. We thus need a higher density of drilling to determine whether marginal blocks should be declared economic or subeconomic. Our estimates for such blocks need to be more precise or we may be assigning marginal ore to waste and vice versa.

There is a counter-argument to the above approach, however, which states that less attention should be paid to these fringe blocks in an effort to ensure that the central area is thoroughly evaluated with a high level of confidence. It is this core of a deposit which should determine the viability of an operation. Once this has been established then attention should be given to the assignment of the marginal blocks to ore or waste categories.

4.11.3 Relative estimation variance of quantity of metal

The quantity of contained metal is calculated from:

$$Q = G_K \times t/100 \quad \text{(if grade is quoted in \%)}$$

or

$Q = G_K \times t$ (for those metals quoted in g/t or kg/t).

where: G_K = mean of the kriged block grades in the orebody and t = tonnage.

The associated estimation error can be calculated using the equation for REV for the various components, i.e.:

$$\sigma_Q^2/(Q)^2 = \sigma_G^2/(G_K)^2 + \sigma_{\text{ton}}^2/(\text{ton})^2$$

From this equation we can thus determine σ_Q^2.

Alternatively, the quantity of metal may be determined from:

$$Q = S \times Th_K \times G_K \times TF/100$$

where:

S = the area estimate
Th_K = the mean kriged thickness
G_K = the mean kriged grade
TF = tonnage factor

This formula can be modified to:

$$Q = S \times Acc_K \times TF/100$$

where Acc_K is the kriged metal accumulation.

Hence, the relative estimation variance equation becomes:

$$\sigma_Q^2/Q^2 = \sigma_S^2/S^2 + \sigma_{acc}^2/Acc^2 + (\sigma_{TF}^2/TF^2)$$

The element in brackets is generally ignored for the information for its calculation is rarely available.

4.11.4 Precision

The precision of the various estimates can be calculated (as a percentage) and the figures obtained compared to what one would consider to be an acceptable level. If the value is too high, then additional sampling is necessary to increase the data density.

Precision is calculated using the following equation:

$$P = (2 \times \sigma/m) \times 100$$

where:

P = precision (%)
σ = standard deviation of the variable being considered
m = mean of the variable being considered

The standard deviations required for this equation are determined by multiplying the REVs for grade, tonnage and quantity of metal by the square of the estimates for each and then taking the square root. The 95% confidence limits for each estimate can also be quoted as $\pm 2{\cdot}\sigma$ although the procedure involved for the confidence limits for log-transformed data is more complex (Appendix 4.1).

4.12 GRADE–TONNAGE CURVE

This is a useful tool for the mining engineer as it tells him at a glance exactly how much ore exists above certain cut-off grades. It allows him to change his cut-off at will and know what effect this action will have on his reserves.

The kriged grades within the orebody are grouped into 15–20 classes and % cumulative frequency for the kriged thicknesses associated with each of the grades in each class calculated. The accumulation is undertaken from the **highest** grade class down to the lowest as we are, in this instance, interested in the values above certain cut-off limits. The cumulative frequency data for thickness are then plotted on the **lower** class limits for each grade class. As the block area and tonnage factor are constant, the tonnage is directly proportional to thickness. Thus for a given COG value the percentage of the total tonnage above this grade can be read from the graph. This graph is, however, only an initial estimate of the true grade–tonnage curve which can be determined from:

$$\sigma_B^2 = \sigma_{B\star}^2 + \sigma_{mk}^2$$

where:

σ_B^2 = variance of the distribution of 'true' block grades
$\sigma_{B\star}^2$ = variance of the kriged block estimates
σ_{mk}^2 = mean of the kriging variances.

A new curve can now be drawn (Figure 4.34) using this new data variance but with the same mean kriged grade. The area under this curve can be determined using the tabulated values for areas under the standard normal curve and the 'Z' values where:

$$Z = \text{(lower class limit} - \text{mean kriged grade)} / \text{s.d. of kriged grades}$$

Similar curves can be produced for grade kriging variances against % tonnage and also against mean grade. From these, we can thus estimate the tonnages and mean grades over specific

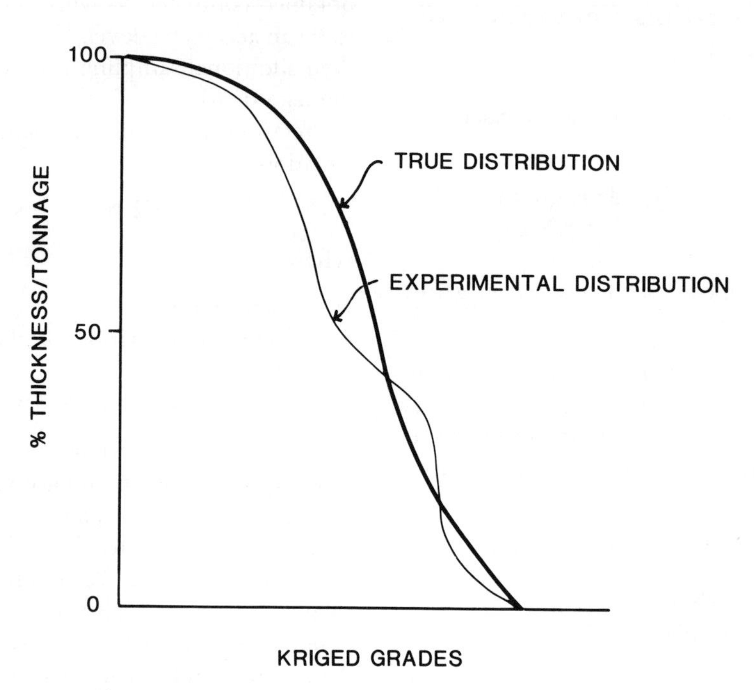

Fig. 4.34 Grade–tonnage curve – experimental and model.

ranges of σ_k^2. Ranges with low σ_k^2 are well-proven ore and those with high σ_k^2 are poorly known reserves. This method can thus form the basis of a properly quantified ore-reserve classification.

4.13 KRIGING VARIANCES AND ORE-RESERVE CLASSIFICATION

The kriging variances of ore-reserve blocks can be used as a basis for the establishment of an ore-reserve classification for a particular deposit. The procedure involves the production of a frequency curve of all the kriging variances within the orebody limits. The curve is then examined to see whether evidence exists for a complex population representing perhaps three superimposed populations. Such a situation is illustrated in Figure 4.35. On the basis of this curve, three ranges of variance can be defined which are then assigned to the three main reserve categories as below:

Kriging variance	**Category**	
0–0.0075	'Probable reserves'	(1)
0.0075–0.0135	'Possible reserves'	(2)
> 0.0135	'Inferred/indicated reserves'	(3)

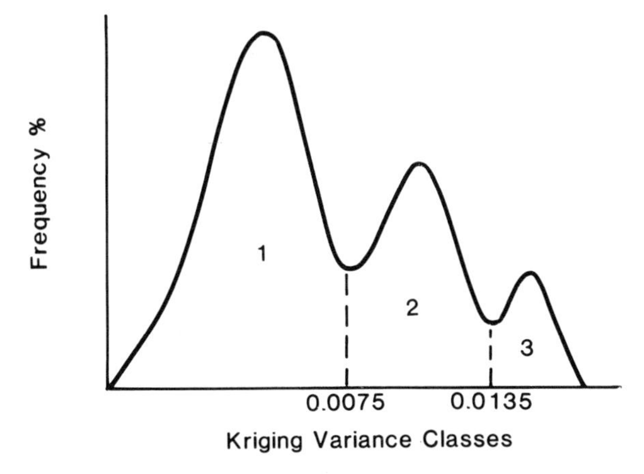

Fig. 4.35 Frequency curve for kriging variance (exaggerated).

The kriging variance determined for each ore-block is very much dependent on the number of samples that were available within the defined search area and their spacing. It is thus possible to examine the blocks in each of the three categories in turn to establish the number of samples used and their spacing. This study could reveal, for example, that category (1) blocks were all evaluated by in excess of eight holes; all category (2) grades were based on four to seven holes and finally, all category (3) block grades were based on three or less holes. We thus have a means of relating our reserve categories directly to density of drilling. It must be pointed out, however, that this is not a universal method which would allow comparison between different mines and ore-bodies. It is single deposit specific. Also, it may not be possible to recognize clear-cut subdivisions of the kriging variance population. The method also assumes that some portions of the deposit are sufficiently well drilled to be classified as proven. This may not be the case – the whole deposit may be under-drilled.

Royle (1977b) suggests that a rational system of classifying ore-reserves should be based on both the kriged estimates and associated kriging variance. He states that for each block within the orebody limits, the probability that its 'true' value lies above the cut-off grade (COG) can be estimated. The first step in this process is to list the block grade estimates in order of increasing kriging variance. For each, the value:

$$D = (\text{kriged block value} - \text{COG})/\sigma_K$$

is calculated, where D is the deviate expressed as a ratio of the kriging standard deviation. If COG = 3 g/t and a kriged block grade is 3.125 g/t and its kriging variance is 0.04 (g/t)^2, then:

$$\begin{aligned} D &= (3.125 - 3.0)/\surd(0.04) \\ &= 0.125/0.2 = 0.625 \end{aligned}$$

Referring to normal probability tables, the probability that the actual block value is less than COG is 0.2660. Hence, the probability that it is greater than the COG is 1 − 0.266 = 0.734 (or 73.4%).

Table 4.2 Ore reserve classification

Identified				Undiscovered	
Demonstrated					
Measured		Indicated			
Proved	probable	(possible)	Inferred	Hypothetical	Speculative
± 10%*	±20%	±40%	±60%		
>80%†	60–80%	40–60%	20–40%	10–20%	<10%
Economically significant resources				Resource base	

After Diehl and David (1982).
*Error tolerance.
†Assurance.

Proven reserves would thus be defined on the basis that all blocks in this category would have probabilities that their true grade exceeds COG, above a set percentage level. Most of these are likely to lie in the more densely sampled internal areas of orebodies. Those blocks with very high kriging variances and with a probability that their true grades exceeds the COG falls below a second set level, would be placed in the possible or indicated category. These will occur in sparsely sampled areas and especially those close to the margins of the deposit where an edge effect is present. Those reserves lying between these two levels would then be considered as probable reserves. Royle suggests that the calculation be done for a range of COGs as the COG may change from area to area in the deposit.

Diehl and David (1982) have proposed a more rigid basis for reserve classification (Table 4.2) than that provided by current classifications (e.g. USBM/USGS, section 3.2). This defines levels of uncertainty ('assurance') and precisions ('error tolerance') in each case. The latter is based on the relative kriging standard deviation $\sigma_K/Z^\star_K$. Hence, for the interface between proven and possible reserves,

$$\text{Precision} = 10 = (\sigma_K \times 100 \times \mu_{0.8})/Z^\star_K$$

where $\mu_{0.8} = 1.282$ for the standard normal distribution (probability, $P = 0.80$). Hence, we have:

$$\sigma_K/Z^\star_K = 10/(100 \times 1.282) = 0.078$$

Had an assurance of 95% been required, the relative kriging standard deviation would have had to have been as low as 0.051 for ore to be classified as proved.

At present, there is no international agreement as to what levels of confidence and precision constitute the various classifications of ore-reserve and thus geostatistical definitions have not yet been accepted. The problem is exacerbated by the fact that only a relatively small proportion of mines have introduced geostatistical methods generally. Further discussion of the application of geostatistics to the problem of ore-reserve classification can be found in David (1988) and especially in Chapter 8 of this work.

4.14 EXTENSION VARIANCES IN THE SPHERICAL SCHEME

If a block of ore is evaluated by the arithmetic mean, X, of sample values around its margins, classical statistics tells us that the error of this mean is $\sigma/\sqrt{n}$, so that at the 95% confidence levels we have $X \pm 2\sigma/\sqrt{n}$. This implies that if we increase the number of samples the precision of our estimate gradually improves until we eventually have an extremely low precision. This of course cannot be correct for all we are obtaining is a better estimate of a thin skin around the block. What we need to know is the error incurred in extrapolating the edge values into the block. This is extension variance. How well marginal samples value a block is dependent on the relative magnitude of the range 'a' and the block dimensions. A small range in a large block will mean that the central area of the block is effectively unsampled. Increasing the number of samples at the margins does nothing to prove the situation.

Where we have evaluated a block by drill-holes either located at the centre, at the corners, or floating randomly within the block, we need to determine the variance of the error (extension variance) to see how well we have accomplished this task. The extension variance produced by a central drill-hole in a large block is greater than that for a smaller block.

4.14.1 Auxiliary functions

Extension variances can be calculated for a series of different sampling situations using what are called auxiliary functions. There are three linear functions denoted by the letters γ, F and X while there are four 2D functions, F, γ, X and Q. The former would be represented by $\gamma(L)$, $F(L)$ and $X(L)$, where L represents the length of the line (h), to which the sample is to be extended, expressed as a ratio to the geostatistical range 'a' in this direction, i.e.:

$$L = h/a$$

Similarly 2D functions would appear in the form $F(x,y)$ or $Q(x,y)$ etc., where x and y are the sides of the rectangular block expressed as a ratio to the range in the two directions as in Figure 4.36.

Tables exist for each of these functions to avoid the complex mathematics involved in their calculation. A sample may be extended into a line or into a 2D block and hence the two types of functions. Graphs also exist (Figures A4.4–A4.11 in Appendix 4.5) so that we can directly obtain the value of variance without having to look up the tables of values of several auxiliary functions. The values obtained from these graphs

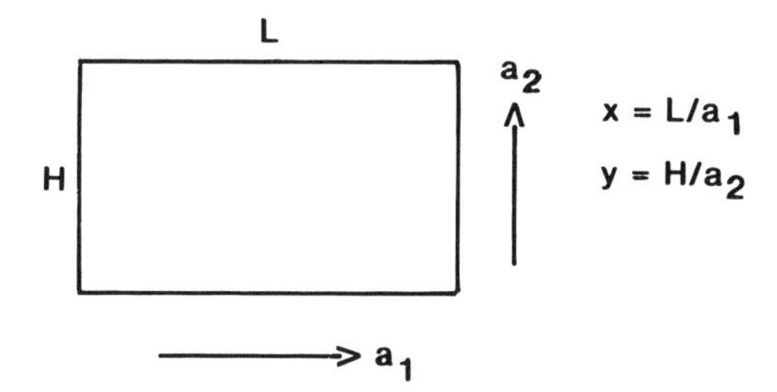

Fig. 4.36 Block dimensions expressed as a ratio of the relevant range 'a'.

and tables are σ_e^2/C where C is the regionalized variance from the relevant semi-variogram and σ_e^2 is the variance of error incurred by extending a value to a line or block. The total estimation variance σ_E^2 consists of two terms, the extension variance as determined above and the nugget variance Co, which are combined as follows:

$$\sigma_E^2 = Co/n + C(\sigma_e^2/C)$$

where n is the number of samples being extrapolated to the line or to the block.

4.14.2 Extension variance models

There are a variety of models which could be applied, depending on the type and pattern of sampling, which are summarized below. They all require that the sample be a point sample, i.e. one without significant size relative to the range.

A. A point sample (*ps*) at the centre of a line of length *L* (= *h*/*a*)

ps

--- L ----

hence $\quad \sigma_e^2 = 2\cdot X(L/2) - F(L)$

and $\quad \sigma_E^2 = Co + C(\sigma_e^2/C)$

Use Figure A4.4 for *F* and *X* functions (or Figure A4.5).

B. A point sample at any point on a line of length *L* (*h*/*a*)

ps

---- L ----

hence $\quad \sigma_e^2 = F(L)$

and $\quad \sigma_E^2 = Co + C(\sigma_e^2/C)$

Use Figure A4.4.

C. Two point samples at ends of a line of length *L* (*h*/*a*)

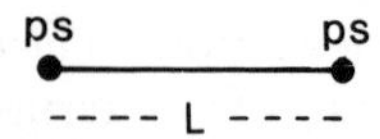

hence $\quad \sigma_e^2 = 2\cdot X(L) - F(L) - 1/2\cdot\gamma(L)$

and $\quad \sigma_E^2 = Co/2 + C(\sigma_e^2/C)$

Use Figure A4.4 for *F*, *X* and γ functions (or Figure A4.5).

D. A point sample 'floating' at random within a rectangle of side *x* and *y* where *x* = *L*/*a* and *y* = *H*/*a*

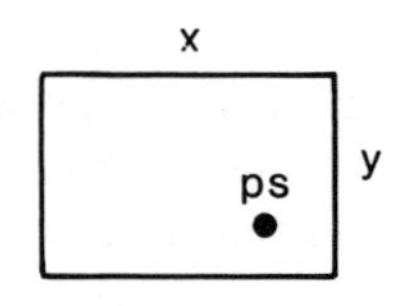

hence $\quad \sigma_e^2 = F(x,y)$

and $\quad \sigma_E^2 = Co + C(\sigma_e^2/C)$

Use Figure A4.7.

E. A point sample at the centre of a rectangle

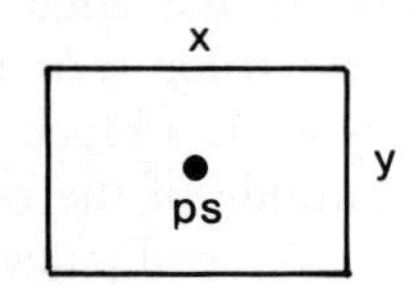

hence $\quad \sigma_e^2 = 2\cdot Q(x/2, y/2) - F(x,y)$

and $\quad \sigma_E^2 = Co + C(\sigma_e^2/C)$

Use Figure A4.8 (or Figure A4.6) if $x = y$ (i.e. curve $\sigma_{E_1}{}^2$).

F. Point samples at the four corners of a rectangle

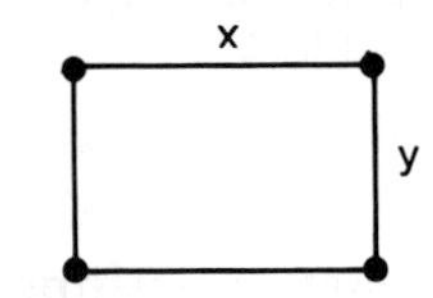

hence $\quad \sigma_e^2 = 2\cdot Q(x,y) - F(x,y) - \tfrac{1}{4}\gamma(x) + \gamma(y) + \gamma((x^2 + y^2)^{1/2})$

and $\quad \sigma_E^2 = Co/4 + C(\sigma_e^2/C)$

Use Figure A4.9 (or Figure A4.6) if $x = y$ (i.e. curve $\sigma_{E_2}{}^2$).

G. Central continuous sample

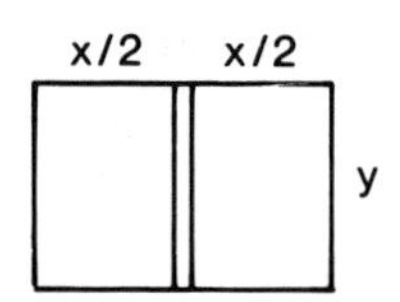

hence $\quad \sigma_e^2 = 2\cdot X(y/2, x) - F(x,y) - F(y)$

and $\sigma_E^2 = Co/\infty + C(\sigma_e^2/C)$

Use Figure A4.10.

H. Two continuous samples at block ends

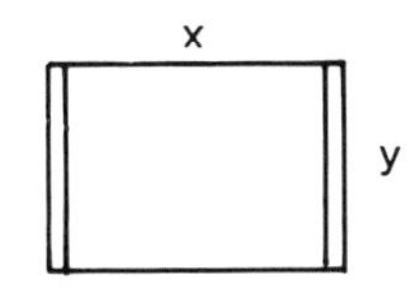

hence $\sigma_e^2 = 2 \cdot X(x,y) - F(x,y) - \frac{1}{2}F(y) - \frac{1}{2}\gamma(x,y)$

and $\sigma_E^2 = Co/\infty + C(\sigma_e^2/C)$

No figures exist for this model and thus tables are required (see Appendix 4.5, Table A4.1.

I. Continuous sample at one end

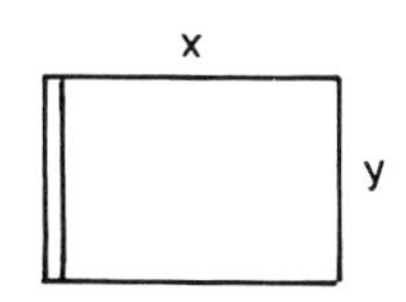

hence $\sigma_e^2 = 2 \cdot X(x,y) - F(x,y) - F(y)$

and $\sigma_E^2 = Co/\infty + C(\sigma_e^2/C)$

As above, use tables.

J. Continuous sampling in levels and raises

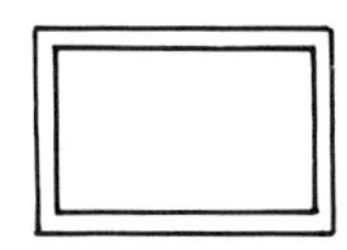

$$\sigma_E^2 = Co/\infty + C(\sigma_e^2/C)$$

Tables are available for this model in Appendix 4.5, Table A4.2.

K. Continuous sampling in central raise and sub-level

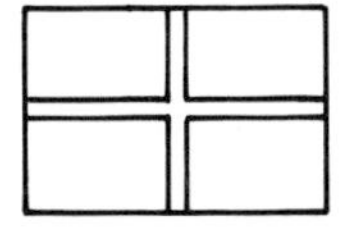

$$\sigma_E^2 = Co/\infty + C(\sigma_e^2/C)$$

Tables are available for this model in Appendix 4.5, Table A4.3.

4.14.3 Effect of sample type and location on extension variance

Worked example

Assume a block of length 200 ft and height 250 ft is to be evaluated in an orebody for which the following semi-variogram parameters have been determined:

1. Isotropic
2. $Co = 0.2$
3. $C = 2.0$
4. $a = 500$ ft

The block dimensions are converted to ratios of range as below:

$$h/a = 250/500 = 0.5$$

$$l/a = 200/500 = 0.4$$

Sampling method D: $\sigma_E^2 = 0.2 + 2(0.340) = \mathbf{0.840}$

Sampling method E: $\sigma_E^2 = 0.2 + 2(0.169) = \mathbf{0.538}$

Sampling method F: $\sigma_E^2 = 0.2/4 + 2(0.115) = \mathbf{0.280}$

Sampling method G: $\sigma_E^2 = 0 + 2(0.049) = \mathbf{0.098}$

Sampling method H: $\sigma_E^2 = 0 + 2(0.071) = \mathbf{0.142}$

Sampling method I: $\sigma_e^2 = 2X(0.4,0.5) - F(0.4,0.5) - F(0.4)$
$= 2.(0.434) - 0.342 - 0.197$
$= 0.329$

Hence $\sigma_E^2 = 0 + 2(0.329) = \mathbf{0.658}$

Sampling method J: $\sigma_E^2 = 0 + 2(0.033) = \mathbf{0.066}$

Sampling method K: $\sigma_E^2 = 0 + 2(0.020) = \mathbf{0.040}$

Conclusions

(1) Four samples at the corners of the block almost halve the extension variance compared to one at the centre.

(2) RSG sampling gives higher variances than regular sampling.

(3) Continuous samples* give much lower variances.

(4) Continuous samples* positioned either E–W or N–S through the centre of the block give the best results, followed by sampling all round

the block margins, then at two ends and finally at one end only.

*Note. This type of sample is rarely taken in a mine (only if drives/raises are in the orebody and a shift production grade is determined). These can, however, be obtained by calculating the weighted average of each channel sample across the orebody and then the weighted or arithmetic average of these 'point samples'. We must then calculate the extension variance involved in extending a central point sample to its line of influence along the level or raise. What we have now created is the geostatistical equivalent of a continuous line of samples. This line term of extension variance is then added to the extension variances determined in models (G) onwards.

4.14.4 Application of extension variance at Navan Pb/Zn Mine, Eire

It is possible to calculate the global reserves and associated variances of a deposit without having produced kriged estimates and kriging variances. The following example demonstrates this technique at Navan.

Thickness estimate and estimation variance

The arithmetic average value of thickness for a portion of this orebody is 37.68 m. The four directional semi-variograms for thickness give $Co = 0$ and $C = 348$ (m^2) but the range varies from 245 m (E–W) to 140 m (N–S). The 53 RSG panels fitted to this orebody are 60 × 60 m.

Considering one panel initially:

$$H/a = 60/140 = 0.429$$

$$L/a = 60/245 = 0.245$$

In the case of a 'floating' drill-hole the extension variance σ_e^2 is obtained from Figure A4.7, in Appendix 4.5.

If we use one RSG panel to evaluate the orebody, then

$$\sigma_E^2 = Co + 348\,[0.26] = 0 + 90.48$$

However, we have 53 panels to evaluate the orebody, therefore

$$\sigma_{Th}^2 = 90.48/53 = 1.7072$$

Relative estimation variance $\sigma_{Th}^2/Th^2 = 1.7072/(37.68)^2 = 1.20 \times 10^{-3}$.

In this instance we do not have information for the density and hence the REV for density will have to be ignored. Had the individual drill-hole values for density been available, semi-variograms would have been produced in the usual way and σ_d^2/d^2 calculated. The REV for the area estimate was determined earlier in section 4.11.1.

$$REV \text{ tonnage} = \sigma_{\text{ton}}^2/\text{ton}^2 = \sigma_{Th}^2/Th^2 + \sigma_S^2/S^{\star 2}$$

$$= 1.20 \times 10^{-3} + 0.94 \times 10^{-3}$$

$$= 2.14 \times 10^{-3}$$

$$\text{But tonnage} = 1.908 \times 10^5 \times 37.68 \times 2.894 \quad (SG = 2.894)$$

$$= 20.806 \times 10^6 \text{ tonnes}$$

$$\text{Therefore } \sigma_{\text{ton}}^2 = (20.806 \times 10^6)^2 \times 2.14 \times 10^{-3}$$

$$\text{and } \sigma_{\text{ton}} = 20.806 \times 10^6 \times 0.0463 = 0.9633 \times 10^6$$

Thus at 95% confidence the tonnage is $20.806 \times 10^6 \pm 1.927 \times 10^6$ tonnes and precision = 9.26%.

Extension variance of grade estimate

If metal accumulations (m%) and thickness (m) are produced for each RSG panel and the global grade determined as below:

$$G = \Sigma(\text{m\%})/\Sigma(\text{m})$$

then the *REV* is:

$$\sigma_g^2/g^2 = \sigma_m^2/m\%^2 + \sigma_{Th}^2/Th^2 - 2 \cdot r(\sigma_{m\%} \cdot \sigma_{Th})/(m\% \cdot Th)$$

In this equation σ_{Th}^2/Th^2 has already been calculated. The semivariogram parameters for the metal accumulation of combined lead and zinc are as follows:

$$Co = 4645\ (\text{m}\%^2)$$
$$C = 88\,250\ (\text{m}\%^2)$$
$$a_{EW} = 205\ \text{m}$$
$$a_{NS} = 115\ \text{m}$$

The arithmetic mean of the metal accumulations = 422.2 m%. For a floating drill-hole in a rectangle we use Figure A4.7, Appendix 4.5 again. As

$$L/a = 60/205 = 0.293 \text{ and } H/a = 60/115 = 0.522,$$

then

$$\sigma_E^2 = 4645 + 88\,250\ (0.315) = 32\,443.75$$

However, there are 53 blocks in the orebody, therefore:

$$\sigma_E^2 = 32\,443.75/53 = 612.146$$

and

$$\sigma_E = 24.74$$

Hence, the metal accumulation, and its 95% confidence limits, are 422.2 ± 49.5 m% and the precision = 11.72%.

If all the drill-holes had been at the centres of the RSG panels then we would have used Figure A4.8, Appendix 4.5 and thus:

$$\sigma_E^2 = 4645 + 88\,250\ (0.155) = 18\,323.75$$

For 53 blocks

$$\sigma_E^2 = 18\,323.75/53 = 345.73\ (\text{m}\%^2)$$
$$\sigma_E = 18.594$$

Hence the metal accumulation estimate is now:

$$422.2 \pm 37.19\ \text{m}\%$$

and the precision = 8.8%, i.e. a significant improvement.

The REV for metal accumulation:

$$= 345.73/(422/2)^2$$
$$= 1.94 \times 10^{-3}$$

This can now be combined with the REV for thickness, together with the correction for covariance, m% versus m, to determine the REV for grade. This value can then be multiplied by the grade estimate squared (10.305^2) to give the estimation variance of grade from which the 95% confidence limits and the precision can be calculated as before.

Royle (1977a) provides another useful example of the use of extension variances in ore-reserve estimation to which the reader is referred.

4.14.5 Principle of composition of line and block terms

So far, we have dealt with internal drill-hole samples in a block. What do we do with channel samples taken on levels or sub-levels? Channel samples along a line can be considered as point samples after their weighted averages are determined (regularization). These values then have to be extended to a line, i.e. each has to be extended over a distance equivalent to the channel interval. The error incurred is the line term of variance. This line grade then has to be extended into a panel of ore and the error is the block term of variance. The total estimation variance is then the sum of these terms.

Example – AVS Main Reef, Tarkwa Goldfield, Ghana

In section 4.8.1 we saw that for this deposit cm g/t values conform to a spherical model where a = 6.9 m (isotropic), $Co = 2.2 \times 10^5$, $C = 6.514 \times 10^5$ and $Co/C = 0.33$. The block dimensions are 30 m × 25 m (Figure 4.37).

A. Line term

Use model (A) described earlier and Figure A4.5 in Appendix 4.5.

ps

h = 1.5 m

The extension variance of one channel sample into its 1.5 m zone of influence (i.e. the sample

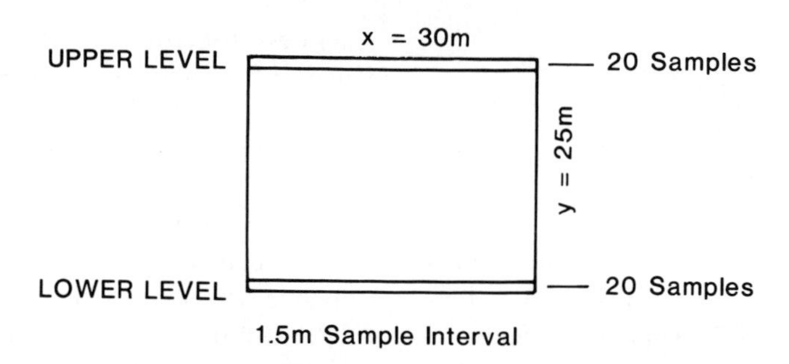

Fig. 4.37 Channel sampling along levels at Tarkwa, Ghana.

interval) can be calculated as follows:

$$L/a = 1.5/6.9 = 0.22$$
$$\sigma_E^2 = Co + C(\sigma_e^2/C)$$
$$= 2.2 \times 10^5 + 6.514 \times 10^5\ (0.056)$$
$$= 256\,478.4$$

But we have 40 samples to evaluate the levels. Therefore $\sigma_E^2 = 256\,478.4/40 = 6412$.

B. Block term

The channel sample data can now be effectively considered as a continuous line of data (infinite number of point samples) which has to be extended into the block.

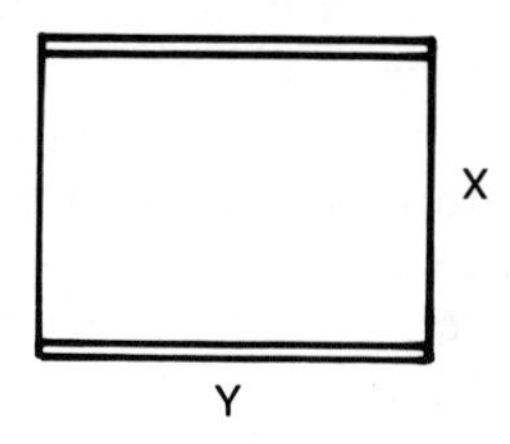

Using Model (A) and geostatistical tables in Appendix 4.5:

$y = H/a = 30/6.9 = 4.348$
and $x = L/a = 25/6.9 = 3.623$

Therefore $\sigma_E^2 = Co/\infty + 6.514 \times 10^5\ (0.079) = 51\,460.6$.

Total estimation variance $= 51\,460.6 + 6412 = 57\,872.6$.

Therefore $\sigma_E = 240.57$.

The block term is much greater, as might be expected, because the line is sampled every 1.5 m but lines have to be extended over a width of 25 m.

The cm g/t value of this block (arithmetic average, as no kriged estimate is available) = 836.

Therefore 95% confidence limits ($\pm\ 2\sigma_E$) = 481.13.

As can be seen below, the very small range (6.9 m) and the large nugget effect (0.33) result in a poor precision for the estimate of the block gold grade.

$$\text{Precision} = 2{\cdot}\sigma/X \times 100$$
$$= 481.13/836 \times 100 = 57.55\%$$

This block is inadequately sampled and it is thus necessary to either resort to (a) diamond drilling to obtain internal samples, or (b) internal development work for the same purpose. It should be noted that closer spaced channel sampling is not warranted as the line variance is already low – it is the block term that must be reduced.

4.15 VOLUME–VARIANCE RELATIONSHIP

So far we have discussed the effect of sampling pattern and density on the variance of our reserve estimates. Another factor that must be considered is the effect of sample size on the estima-

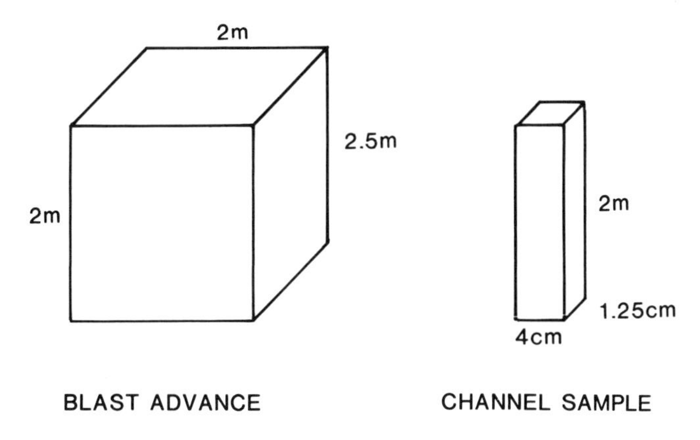

Fig. 4.38 Geometries of samples compared.

tion variance. Is it worth increasing the sample size? Is the extra cost warranted?

Classical statistics tells us that if sets of samples of random variables numbering n_1, n_2 etc. are taken, the standard error of their mean is given by $\sqrt{\sigma^2/n_1}$, $\sqrt{\sigma^2/n_2}$ etc., where σ^2 is the variance of the parent population. Thus the standard error decreases as the number of samples increases. Eventually, with a large enough set of samples the precision of our estimate would approach zero. Taking a larger sized sample can also be likened to taking a larger number of samples. It contains a larger number of point samples than the smaller volume. Again classical statistics would tell us that, the larger the sample volume, the better the estimate. The degree of this apparent improvement can be demonstrated (Figure 4.38) by considering the use of a sample representing face advance per shift as opposed to a channel sample.

$$\text{Volume of advance} = 1 \times 10^7\ \text{cm}^3$$

$$\text{Volume of channel} = 1 \times 10^3\ \text{cm}^3.$$

On this basis, the blast sample variance would be $1/10^4$ of that for the channel, giving remarkably narrow confidence limits to our estimate. Unfortunately this happy state of affairs does not hold with regionalized variables as it applies only to the random component *Co*.

4.15.1 Volume–variance in the spherical scheme

Krige's formula gives a general expression for the volume–variance relationship. It is as follows:

$$\sigma^2_{ps/Block} = \sigma^2_{ps/Sample} + \sigma^2_{Sample/Block}$$

or

$$\begin{aligned}\sigma^2_{Sample/Block} &= \sigma^2_{ps/Block} - \sigma^2_{ps/Sample}\\ &= F(x,y)_{Block} - F(x,y)_{Sample}\end{aligned}$$

The total variance ($\sigma^2_{ps/v}$) of a point sample in a larger volume is, as we have seen before, the sum of two terms, i.e. *Co* and the variance of a floating sample in the block (i.e. the 2D F function expressed as $F(L/a, H/a)$). The value of *Co* thus has an important effect on the total variance and it can be minimized by reducing the sampling and analytical errors. If the sample volume is doubled then, theoretically, the nugget variance *Co* is halved. This reduction is due to the 'averaging out' effect of a larger volume. However, this is only one component of the total variance and thus doubling the sample volume will not reduce this by half. The effect of different values of range '*a*' and sample size is shown in the example below.

Comparative sample variance

If we assume that a sample is taken at random from an ore-block B whose dimensions are 300 m × 300 m, we can examine the volume–variance relationship in two different situations.

Case I – range = 200 m (isotropic)

If a sample S_1 (15 cm × 10 cm) is taken from the ore-block (Figure 4.39) then:

L/a and H/a

= 15/20 000 and 10/20 000 respectively.

Both are very small, therefore $F(S_1)$ approaches zero.

For the block B, however,

$$F(\mathrm{B}) = F(300/200, 300/200)$$

which, from Figure A4.7, Appendix 4.5, of 2D F functions (rectangle + contained sample) = 0.81 C.

If the sample is increased to 30 × 30 cm, $F(S_2)$ is still very small, therefore little change occurs in variance.

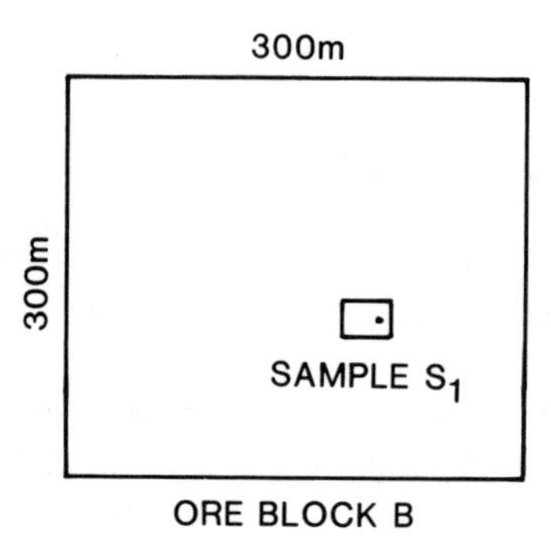

Fig. 4.39 A sample S_1 taken within an ore block B.

Case II – range = 1 m (isotropic)

For the first sample size

$$F(S_1) = F(15/100, 10/100) = F(0.15, 0.1) = 0.10\,C$$

For very high values of L/a, H/a, as is now the case, F(B) approaches 1C, therefore using Krige's equation:

$$\sigma^2_{(S_1/B)} = F(\mathrm{B}) - F(S_1) = C(1 - 0.10) = 0.9\,C \quad (1)$$

For the second sample size (30 × 30 cm):

$$F(S_2) = F(30/100, 30/100) = 0.23\,C$$

Therefore

$$\sigma^2_{(S2/B)} = C(1 - 0.23) = 0.77\,C \quad (2)$$

We see that a large range 'a' means that an increase in sample size has little effect, however, when the range decreases the variance begins to show a significant drop (compare equations (1) and (2)). In the example, the effect on Co is ignored – only C, 'a' and sample size are considered. The total extension variance in Case II for the two sample sizes should really be:

$$\sigma^2_{S_1/B} = Co_{S_1} + C_{S_1}(0.9)$$
$$\sigma^2_{S_2/B} = Co_{S_2} + C_{S_2}(0.77)$$

The semi-variograms, and hence the semi-variogram parameters, for the two different sample sizes will be different and thus, to be able to determine the full effect of the increase in sample size, we have two options:

1. Run two separate sampling campaigns so that experimental and then point model semi-variograms can be produced for S_1 and S_2.
2. Estimate the values for the second sample size from the actual semivariogram parameters for the first. In this instance, as the sample volume ratio S_1/S_2 = 150/900, $Co_{S_2} = Co_{S_1} \times$ 150/900. Generally it is found that C is little changed so we could let $C_{S_1} = C_{S_2}$.

We thus see that the impact of changing sample size is not only dependent on 'a' but also on the relative magnitudes of Co and C in the two cases.

4.15.2 Volume–variance in the de Wijsian scheme

De Wijsian scheme models were discussed in section 4.6.2 and a worked example is presented in Appendix 4.2. It was seen that the volume of an orebody or sample is expressed in terms of linear equivalents. We can use these to investigate the effects of increasing sample size for a given orebody. Let us assume that a block of ground, 30 m × 50 m, has been sampled using first a T46 core barrel and then a T66 barrel; in both cases a 1 m sample length was used. The average thickness of the ore in this block proved to be 3.21 m. The question that we need to answer is whether the increased drilling cost was warranted.

Volume T46 core sample
$$= \pi(1.6/100)^2 \times 1 = 8.04 \times 10^{-4}\text{m}^3$$

Volume T66 core sample
$$= \pi(2.6/100)^2 \times 1 = 2.12 \times 10^{-3}\text{m}^3$$

The volume ratio (Vol_{66}/Vol_{46}) is thus 2.64. Classical statistics would suggest that this increase in volume would result a decrease in the variance to almost one-third of the value for the T46 drilling, thus implying a marked improvement in precision.

The geostatistical approach to this problem would be first to calculate the linear equivalents of the block and the two samples ($LE = a + b + 0.7\,c$) and then to determine the variance when the two sample types are used to evaluate the block:

$$LE_{46} = 1.0 + 0.032 + 0.7(0.032) = 1.0544$$
$$LE_{66} = 1.0 + 0.052 + 0.7(0.052) = 1.0926$$
$$LE_{Block} = 50 + 30 + 0.7(3.21) = 82.247$$

If the samples and the block are assumed to be geometrically similar, then:

$$\sigma_{Sample/Block} = \alpha \log\text{(sample volume/block volume)}$$
$$= 3\alpha \log_e(LE_{Sample}/LE_{Block})$$

Hence

$$\sigma^2_{46/Block} = 3\alpha_{46}[\ln 82.25 - \ln 1.05]$$

and

$$\sigma^2_{66/Block} = 3\alpha_{66}[\ln 82.25 - \ln 1.09]$$

The value of 3α would be obtained from the semi-variograms for the two sample sets. The values in brackets are little changed and hence the impact of changing core size is minimal. A much greater impact will be gained by changing from core sampling to sampling each blast muckpile at the stope-face.

4.16 INDICATOR KRIGING (IK)

4.16.1 Ore/waste estimation

Indicator kriging is an enhancement of the simple kriging technique, the difference being that it does not calculate a grade or metal accumulation value but the proportion of the block which can be expected to contain values above a given cut-off. The technique is particularly applicable where strict ore/waste boundaries exist within given blocks, e.g. large copper porphyries where grade zoning is the major control, and in low grade deposits where the cut-off value is of major concern. Simple kriging estimates the expected grade or metal accumulation of a given block and, as such, is a true reflection of its value providing that the block is mined as a complete unit. It does not, however, allow the user to determine whether the assigned value of the block is heavily biased by a single high value or if it contains areas of both ore and waste material.

We can examine the problem further by considering a block of mineralized ground which has been evaluated by drill holes at its four corners. The grade assigned to the block will, in this instance, be assumed to be the arithmetic mean of the four holes (i.e. the thickness is constant). In the case of block A in Figure 4.40, the average grade is 5.75 g/t and, given a cut-off grade of 5 g/t, it would thus be considered as ore. Although block B has an identical grade and would also be classified as ore, this decision is highly dependent on one value, a fact which is not reflected directly in the result.

Suppose now that the type of material used to value the block is taken into account so that ore material is indicated by the value 1 and waste by

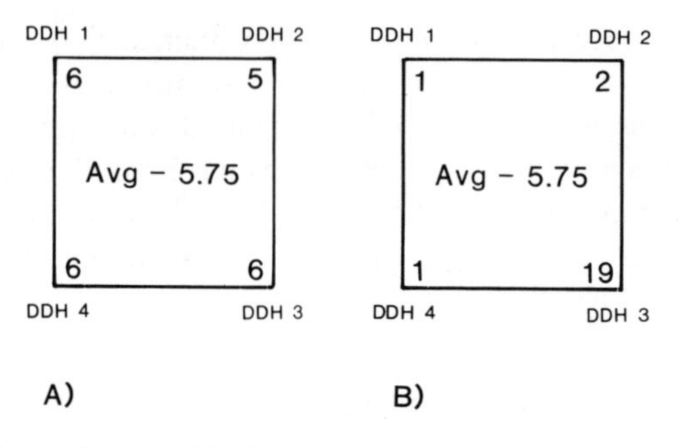

Fig. 4.40 Two blocks showing gold grades of corner boreholes. Block A has a mean indicator value of 1.0 and Block B a value of 0.25, for a cut-off grade of 5 g/t.

the value of 0. Then, in the first example, the sample values would be categorized, according to a cut-off grade of 5 g/t, as follows:

Drill hole	Grade	Category	Indicator value
1	6 g/t	ore	1
2	5 g/t	ore	1
3	6 g/t	ore	1
4	6 g/t	ore	1

The average grade of the block suggests that it is ore and the indicator values suggest that it can be expected to contain 100% ore-grade material. If we now consider the second example, we have:

Drillhole	Grade	Category	Indicator value
1	1 g/t	waste	0
2	2 g/t	waste	0
3	1 g/t	waste	0
4	19 g/t	ore	1

Although the average grade of the block suggests that the block is ore, the mean indicator value is 0.25 suggesting that only 25% is of ore-grade material. This fact must be considered during the construction of the final grade/tonnage curve to avoid an overestimation of the ore tonnage.

The application of the above technique in geostatistical ore-reserve estimation procedures was first proposed by Journel (1983) and further refined by Lemmer (1984). The mathematical expression for an indicator variable is $i(x;z)$ which is based on the grade $z(x)$ of a sample point x and on the cut-off grade z. Hence:

$$i(x;z) = \begin{cases} 1, & \text{if } z(x) > z \\ 0, & \text{if } z(x) \leqslant z \end{cases}$$

Once all the grade (or accumulation) data have been transformed in this way, experimental semi-variograms are generated and mathematical models fitted as described in sections 4.5 and 4.8. Where a spherical scheme model is deemed applicable, the model equation is:

$$\gamma_I(h;z) = I_o + I[1.5(h/a) - 0.5(h/a)^3] \text{ for } h < a \quad (1)$$
$$= I_o + I \text{ for } h \geqslant a \quad (2)$$

where I_o and I are equivalent to Co and C in a grade semi-variogram. These indicator semi-variogram parameters are then used to produce block indicator values using simple kriging techniques (section 4.10), the methodology being referred to as indicator kriging (IK). This kriged indicator value for a block thus represents the recovery function for that block at a specified cut-off. The above procedure can be repeated for a range of cut-off values.

The final product of the exercise will thus be a block plan of kriged grades and indicator values which is of considerable benefit to an operation where selective mining is possible within the confines of ore-evaluation blocks. Revised block tonnages can thus be computed and used to produce a more realistic grade–tonnage curve (the tonnages being determined after repeated use of IK for a range of cut-off grades).

4.16.2 Semivariogram modelling with IK

Lemmer (1986) states that indicator semi-variograms are much more robust with respect to anomalous outliers than grade or accumulation semi-variograms. Indicator semi-variograms can thus be modelled with more confidence. Also, when the nugget effect is not too small, as in the case of Witwatersrand gold accumulations (cm g/t), then an approximate linear relationship exists between $\gamma_I(h;z)$ and $\gamma(h)$. As a result, the modelling of indicator semi-variograms can be used to provide a more reliable model for the associated grade or accumulation semi-variogram. The method involves the plotting of $\gamma_I(h;z)$ against $\gamma(h)$ for successive lags (for a specific cut-off). A least-squares best-fit line is then plotted through the points. The equation of this line is thus:

$$\gamma_I(h;z) \approx A(z) + B(z)\gamma(h) \qquad (3)$$

where $A(z)$ is the intercept on the $\gamma_I(h;z)$ axis and $B(z)$ is the gradient of the line.

Co and C can now be predicted by using I_o and I from the indicator semi-variogram and by setting $h = a$ and $h = 0$ in equations (3) and (1) respectively, as below:

When

$$h = a,\ \gamma_I(h;z) = I_o + I \text{ and } \gamma(h) = Co + C$$

therefore

$$I_o + I = A(z) + B(z)(Co + C)$$

Hence

$$Co + C = [I_o + I - A(z)]/B(z)$$

When

$$h = 0,\ \gamma_I(h;z) = I_o$$

and

$$I_o = A(z) + B(z)(Co)$$

hence

$$Co = [I_o - A(z)]/B(z)$$

The range is assumed to be the same in both semi-variograms. This predicted model can now be superimposed on the experimental grade/metal accumulation semi-variogram to test whether a good fit has been achieved.

REFERENCES

Boakye, E. B. (1989) Statistical and geostatistical evaluation of an alluvial gold deposit in the central region of Ghana. Unpublished PhD thesis, University of Wales (Cardiff), 334 pp.

David, M. (1988) *Handbook of Applied Advanced Geostatistical Ore Reserve Estimation*, Elsevier (*Developments in Geomathematics 6*), Elsevier, Amsterdam, 216 pp.

Davis, J. C. (1973) *Statistics and Data Analysis in Geology*, John Wiley, New York, 550 pp.

Diehl, P. and David, M. (1982) Classification of ore reserves/resources based on geostatistical methods. *CIM Bull.*, **75**, (838), 127–135.

Journel, A. G. and Huijbregts, Ch.J. (1978) *Mining Geostatistics*, Academic, London, 600 pp.

Journel, A. G. (1983) Non-parametrics estimation of spatial distributions. International Association for Mathematical Geology Journal, **15**, 445–468.

Lemmer, I. C. (1984) Estimating local recoverable reserves via IK, in *Geostatistics for Natural Resources Characterization* (eds G. Verly *et al.*), Reidal, Dordrecht, pp. 349–364.

Lemmer, I. C. (1986) Grade-indicator Plots for Gold Reef Data: Their Use in the Determination of Variogram Parameters, in *Extended Abstracts, GeoCongress 86*, Geol. Soc. South Afr., Johannesburg, pp. 237–240.

Matern, B. (1960) *Spatial Variation*, Almaenna Foerlaget, Stockholm, 144 pp.

Matheron, G. (1971) *The Theory of Regionalized Variables and its Applications*, Centre de Morphologie Mathematique, Fontainbleau, 211 pp.

Royle, A. G. (1977a) Global estimates of ore reserves. *Trans. A IMM*, **86**, A9–A17.

Royle, A. G. (1977b) How to use geostatistics for ore reserve classification. *World Mining*, February, 52–56.

Smith, I. H. (1987) Geology, exploration and evaluation of the gold deposits of Suriname. Unpublished PhD thesis, University of Wales (Cardiff), 639 pp.

BIBLIOGRAPHY

Armstrong, M. (1984) Improving the estimation and modelling of the Variogram, in *Geostatistics for Natural Resources Characterization*, Vol. 1 (eds G. Verly *et al.*), Reidal, Dordrecht, pp. 1–20.

Blais, R. A. and Carlier, P. A. (1968) Applications of geostatistics in ore evaluation. *CIMM*, **9**, (Spec. Vol. Ore Reserve Estimation and Grade Control), 48–61.

Brooker, P. I. (1975) Optimal Block Estimation by Kriging, in *Aust. IMM, Proceedings*, No. 253.

Brooker, P. I. (1976) Block Estimation at Various Stages of Deposit Development, in *Proceedings of the 14th International APCOM Symposium, Penn. State University*, pp. 995–1003.

Clarke, I. (1979) *Practical Geostatistics*, Applied Science, New York, 129 pp.

Clarke, I. and White, B. (1976) Geostatistical Modelling of an Orebody as an Aid to Mine Planning, in *Proceedings of the 14th International Symposium, Penn. State University*, pp. 1004–1012.

Crozel, D. and David, D. (1985) Global estimation variance formulae and calculation. *Math. Geol.*, **17**(8), 785–796.

Dagbert, M. and David, M. (1983) New developments in the categorization of ore reserves by geostatistical methods, in *AIME Fall Meeting, Salt Lake City*.

David, M. (1970) Geostatistical ore estimation – a step by step case study, *CIMME* , **12**, (Spec. Vol. Decision Making in the Mineral Industry) 185–191.

David, M. (1972) Grade tonnage curve: use and misuse in ore reserve estimation. *Trans. A IMM*, **81**, 129–132.

David, M. (1977) *Geostatistical Ore Reserve Estimation*, Elsevier, Amsterdam, 364 pp.

Davis, B. (1984) Indicator kriging as applied to an alluvial gold deposit, in *Geostatistics for Natural Resources Characterization*, Part 1 (eds G. Verly *et al.*), Reidal, Dordrecht, pp. 337–348.

Dowd, P. A. (1984) The variogram and kriging: robust and resistant estimators, in *Geostatistics for Natural Resources Characterization*, Part 1 (eds G. Verly *et al.*), Reidal, Dordrecht, pp. 91–106.

Journel, A. G. and Sans, H. (1974) Ore grade control in subhorizontal deposits. *Trans. IMM Bull.*, **83**, A74–84.

Journel, A. G. (1984) Recoverable reserve estimation: the geostatistical approach. *Mine Engineering*, **37**, 563–568.

Krige, D. G. (1951) A statistical approach to some mine valuation problems on the Witwatersrand. *J. Chem. Metall. Min. Soc. South Afr.*, **52**, 119–139.

Krige, D. G. (1976) Some basic considerations in the application of geostatistics to gold ore valuation. *J. South Afr. IMM*, **76**, 383–91.

Marachal, A. (1975) Geostatistique et applications minieres. *Ann. Mines*, Nov., 1–12.

Matheron, G. (1955) Application des methods statistiques a l'estimation des gisements. *Ann. Mines*, Dec., 50–75.

Matheron, G. (1963) Principles of geostatistics. *Econ. Geol.*, **58**, 1246–1266.

Matheron, G. (1965) *Les Variables Regionalisees et Leur Estimation*, Masson et Cie, Paris, 305 pp.

Newton, M. J. (1973) The application of geostatistics to mine sampling patterns, in *Proceedings 11th International APCOM Symposium, University of Arizona, Tucson*, April, pp. 44–58.

Norrish, N. I. and Blackwell, G. H. (1987) A mine operator's implementation of geostatistics. *CIMM Bull.*, **80**, 103–12.

Parker, H. M. (1979) Volume variance relationship: a useful tool for mine planning. *Eng. Mining J.*, **180**, 106–23.

Raymond, G. F. (1979) Ore estimation problems in an erratically mineralized orebody. *CIMM Bull.*, **72** (806), 90–8.

Rendu, J. M. (1970) Geostatistical approach to ore reserve calculation. *Eng. Mining J.*, **171**, 112–18.

Royle, A. G. (1975) A Practical Introduction to Geostatistics, unpublished course notes, Department of Mining and Mineral Sciences, Leeds University.

Royle, A. G. (1979) Estimating small blocks of ore, how to do it with confidence. *World Mining*, **32**, 55–57.

Royle, A. G., Newton, M. J. and Sarin, H. K. (1972) Geostatistical factors in design of mine sampling programmes. *Trans. IMM*, **81**, A81–A88.

Royle, A. G. and Hosgit, E. (1974) Local estimation of sand and gravel reserves by geostatistical methods. *Trans. IMM*, **83**, A53–A62.

Royle, A. G. *et al.* (1980) *Geostatistics*, McGraw Hill, New York, 168 pp.

Sichel, H. S. (1951–2) New methods in the statistical evaluation of mine sampling data. *Trans. IMM*, **61**, 261–88.

Sinclair, A. J. and Deraisme, J. A. (1974) Geostatistical study of the Eagle Copper Vein, N. British Columbia, *CIMM Bull.*, **67**, 131–42.

APPENDIX 4.1

Determination of Confidence Limits for Log-transformed Data

If kriging has been applied to log-transformed data, it will be absolutely necessary to convert the log estimates into absolute (i.e. untransformed) estimates. The necessary equation is as follows:

$$Z^\star = e^{y - \mu + 1/2\sigma^2}$$

where:

$Z^\star$ = absolute estimate
y = log estimate
μ = Lagrange multiplier
σ^2 = logarithmic kriging variance

To convert a logarithmic kriging variance into an absolute kriging variance, the necessary equation is:

$$\sigma_K{}^2 = X^2 \cdot e^{\sigma e(B,B)}[1 + e^{\mu - \sigma^2}(e^{\mu} - 2)]$$

where:

$\sigma_K{}^2$ = kriging variance for the estimate X_B
X = average value of orebody
$\sigma e(B,B)$ = variance of X_B
X_B = log value of block B
μ = Lagrange multiplier
σ^2 = logarithmic kriging variance

APPENDIX 4.2

Worked Example – de Wijsian Scheme

(Taken from Royle, 1975)

A gold-quartz reef has been channel sampled at 5 ft intervals over a strike length of 270 ft in two levels and in two intervening sub-levels so that a block of ore 270 ft × 150 ft is defined whose average thickness is 4.35 ft. A weighted semi-variogram of metal accumulation (inch dwt/ton) has been calculated for these levels with $\gamma(h)$ values as follows:

Lag	$\boldsymbol{\gamma(h)(\times 10^3)}$
1	633
2	686
3	731
4	833
5	768
6	650
7	755
8	780
9	902
10	895
11	921
12	1061
13	683
14	746
15	803

The statistical variance of the accumulations = 839 000 (inch dwt/ton)^2.

A plot of the semi-variogram is produced on cm-log paper in Figure A4.1 and a best-fit straight line plotted through the data points nearest to the $\gamma(h)$ axis. From this:

$$\gamma(1) = 0.633 \times 10^6$$
$$\gamma(10) = 0.825 \times 10^6$$

Expanding the formula for the de Wijsian Scheme we have:

$$\gamma(h) = 3\alpha \ln h - 3\alpha \ln l + 9\alpha/2$$

Hence

$$0.825 \times 10^6 = 3\alpha \ln 50 - 3\alpha \ln l + 9\alpha/2 \qquad (1)$$
$$0.633 \times 10^6 = 3\alpha \ln 5 - 3\alpha \ln l + 9\alpha/2 \qquad (2)$$

Subtracting (1) and (2) we have:

$$0.192 \times 10^6 = 3\alpha (\ln 50 - \ln 5) = 3\alpha \ln 10$$

therefore $3\alpha = 0.192 \times 10^6/2.3026 = 83\,384$.

Substituting in (1) and dividing both sides by 3α we have:

$$(0.825 \times 10^6)/83384 = \ln 50 - \ln l + 3/2$$
$$\ln l = \ln 50 + (3/2) - 9.894$$
$$= 3.912 + 1.5 - 9.894 = -4.482$$
$$\text{therefore } l = \mathbf{0.01131} \text{ (equivalent thickness)}$$

Hence the model can be written:

$$\gamma(h) = 83\,384\ ([\ln (h/0.01131)] + (3/2))$$

The variance produced by using the samples to

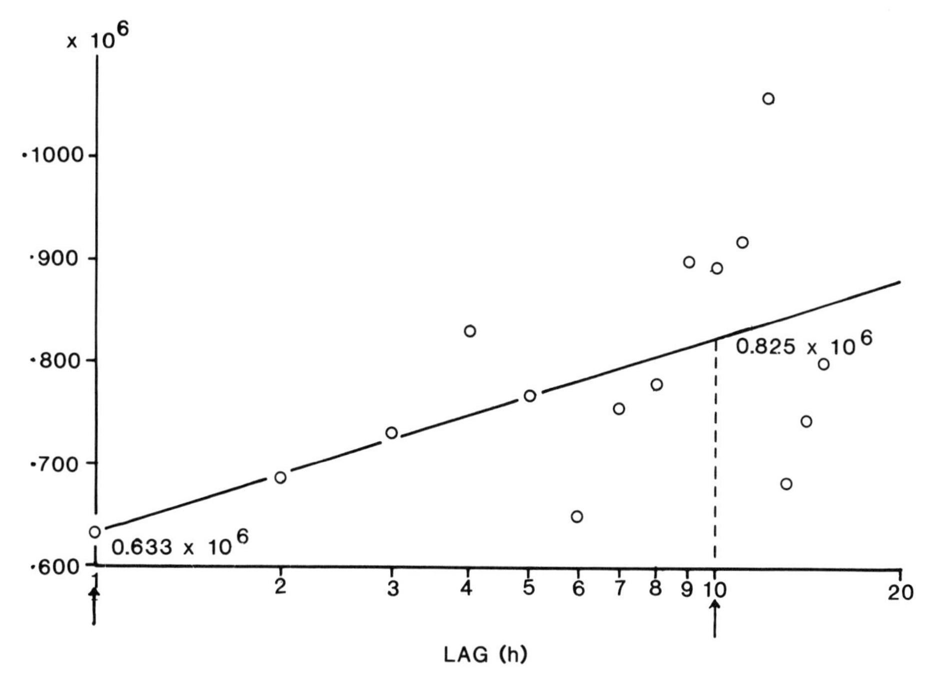

Fig. A4.1 De Wijsian scheme semi-variogram of a gold quartz vein where lag 1 is 5 ft.

evaluate the block is:

$$\begin{aligned}\sigma^2_{S/B} &= 3\alpha \ln\,(LE_{Block}/\textit{Equivalent thickness of sample})\\ &= 83\,384 \ln\,[(270 + 150 + 0.7 \times 4.35/0.01131]\\ &= 83\,384 \ln\,(423/0.01131) = \mathbf{877\,996}\end{aligned}$$

This value is close to the statistical variance quoted earlier.

The average value of σ^2 for the lines of samples (calculated individually and then averaged) along the levels and sub-levels is 765 340. The theoretical value for these samples is, however:

$$83\,384 \ln\,(270/0.01131) = \mathbf{840\,552}$$

Thus the model gives a reasonable approximation to the observed value.

APPENDIX 4.3

Mathematical Basis of Point Kriging

In Figure A4.2, 'O' represents the point to be evaluated using the four surrounding data points. The first of the kriging equations that must be solved to obtain the weighting factor K_i is thus:

$$K_1\,\sigma_{1,1} + K_2\,\sigma_{1,2} + K_3\,\sigma_{1,3} + K_4\,\sigma_{1,4} + \mu = \sigma_{0,1}$$

where $\sigma_{1,1}$ is the covariance between sample 1 and itself, $\sigma_{1,2}$ is the covariance between sample 1 and 2, etc., μ is the Lagrange multiplier and $\sigma_{0,1}$ is the covariance between the point to be evaluated and sample 1. Three other equations of this type can also be written out comparing samples 2, 3 and 4 in turn with each of the other points. The $(n + 1)$th equation is then:

$$K_1 + K_2 + K_3 + K_4 = 1$$

(n, the number of points used = 4).

In section 4.6.4 it was shown that the covariance at a specific lag distance h was $(Co + C) - \gamma(h)$. Thus, in the n kriging equations we can substitute $\gamma(h)$ for covariance without affecting the values of K_1 obtained. The $\gamma(h)$ values used are calculated from the mathematical model for the spherical scheme:

$$\gamma(h) = Co + C\,(3h/2a - 0.5\,h^3/a^3) \text{ for } h < a$$

or

$$\gamma(h) = Co + C \quad \text{for } h \geqslant a$$

The 'h' values inserted into these equations represent the distances between each pair of samples, e.g. 1 and 2, thus giving $\gamma_{1,2}$. The full set of kriging equations is thus:

$$\begin{aligned}
K_1\gamma_{1,1} + K_2\gamma_{1,2} + K_3\gamma_{1,3} + K_4\gamma_{1,4} + \mu &= \gamma_{0,1}\\
K_1\gamma_{2,1} + K_2\gamma_{2,2} + K_3\gamma_{2,3} + K_4\gamma_{2,4} + \mu &= \gamma_{0,2}\\
K_1\gamma_{3,1} + K_2\gamma_{3,2} + K_3\gamma_{3,3} + K_4\gamma_{3,4} + \mu &= \gamma_{0,3}\\
K_1\gamma_{4,1} + K_2\gamma_{4,2} + K_3\gamma_{4,3} + K_4\gamma_{4,4} + \mu &= \gamma_{0,4}\\
K_1 + K_2 + K_3 + K_4 &= 1
\end{aligned}$$

Once the K_i values have been calculated by solving these equations, the value at 'O' is determined by summing the products of each data value and its weighting factor. Barnes (1980, pp. 57–9)* presents a useful worked example of point kriging in the case of a vein-type silver deposit. The kriging estimation variance, i.e. the error incurred in estimating the point sample value, is calculated as described in Appendix 4.4.

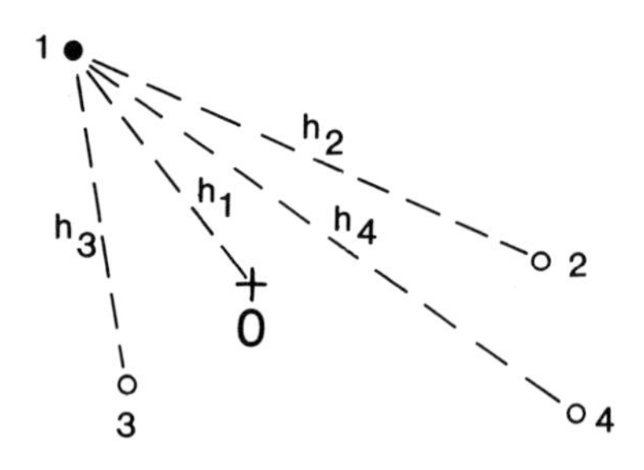

Fig. A4.2 Four samples used to evaluate the point 'O' by point kriging.

*Barnes, P. B. (1980) Computer-assisted mineral appraisal and feasibility, Society of Mining Engineers, American IMM and Petroleum Engineers, New York, 167 pp.

APPENDIX 4.4

Mathematical Basis of Block Kriging

Kriging estimates the value of a block from surrounding data in such a way as to minimize the expected squared error of estimation (kriging variance). The kriged estimate of a block ($Z^{\star}_B$) is calculated as follows:

$$Z^{\star}_B = K_1Z_1 + K_2Z_2 + \ldots + K_nZ_n = \Sigma_{i=1}^{n} K_iZ_i$$

where K_i are the weighting factors and where $K_1 + K_2 + \ldots + K_n = 1$.

It is thus the weighted average of the sample values, Z_i. The estimation variance, σ^2_K, is expressed mathematically as

$$E(Z^{\star}_B - Z_B)^2$$

where Z_B represents the true value of the block. In order to determine the weighting factors which minimize the estimation variance, we take partial derivatives with respect to K_i, set these to zero and solve for K_i. As:

$$\Sigma_{i=1}^{n} K_i = 1$$

a side condition must also be added in which μ is a Lagrange parameter:

$$F = E(Z^{\star}_B - Z_B)^2 + 2\mu\,(\Sigma_{i=1}^{n} K_i - 1)$$

The partial derivative for weight K_i is:

$$\frac{\delta F}{\delta K_i} = \frac{\delta}{\delta K_i} E(Z^{\star}_B - Z_B)^2 + 2\mu(\Sigma_{i=1}^{n} K_i - 1)$$

$$= \frac{\delta}{\delta K_i}(E(K_1Z_1 + K_2Z_2 + \ldots + K_nZ_n$$

$$- Z_B)^2 + 2\mu(\Sigma_{i=1}^{n} K_i - 1)$$

$$= 2[K_1E(Z_iZ_1) + K_2E(Z_iZ_2) + \ldots$$

$$+ K_nE(Z_iZ_n) - E(Z_iZ_B)] + 2\mu$$

The partial derivative for the Lagrange parameter is:

$$\frac{\delta F}{\delta \mu} = \Sigma_{i=1}^{n} K_i - 1$$

To minimize F, the partial derivatives are set to zero, thus yielding a set of $n + 1$ equations with $n + 1$ unknowns:

$$\begin{array}{ccccccccc}
K_1E(Z_1Z_1) & + & K_2E(Z_1Z_2) & + \ldots + & K_nE(Z_1Z_n) & + & \mu & = & E(Z_1Z_B) \\
K_1E(Z_2Z_1) & + & K_2E(Z_2Z_2) & + \ldots + & K_nE(Z_2Z_n) & + & \mu & = & E(Z_2Z_B) \\
\vdots & & \vdots & \vdots & \vdots & & & & \vdots \\
K_1E(Z_nZ_1) & + & K_2E(Z_nZ_2) & + \ldots + & K_nE(Z_nZ_n) & + & \mu & = & E(Z_nZ_B) \\
K_1 & + & K_2 & + \ldots + & K_n & + & \mu & = & 1
\end{array}$$

By subtracting m^2 from both sides of the first n equations, it is possible to convert the terms within parentheses to covariances.

For example for the ith equation:

$$K_1E(Z_1Z_1) - K_1m^2 + K_2E(Z_iZ_2) - K_2m^2 + \ldots + K_nE(Z_iZ_n) - K_nm^2 + \mu = E(Z_iZ_B) - m^2$$

This is possible since the K_i must sum to 1. Since $E(Z_i) = m$ when stationarity exists, we can rewrite the equations as follows:

$$\begin{aligned}
K_1\sigma_{1,1} + K_2\sigma_{1,2} + \ldots + K_n\sigma_{1,n} + \mu &= \sigma_{1,B} \\
K_1\sigma_{2,1} + K_2\sigma_{2,2} + \ldots + K_n\sigma_{2,n} + \mu &= \sigma_{2,B} \\
\vdots \quad \vdots \quad \vdots \quad \vdots \quad \vdots \quad & \vdots \\
K_1\sigma_{n,1} + K_2\sigma_{n,2} + \ldots + K_n\sigma_{n,n} + \mu &= \sigma_{n,B} \\
K_1 + K_2 + \ldots + K_n &= 1
\end{aligned}$$

Note $\sigma_{1,1}$ represents the covariance between a sample and itself while $\sigma_{1,2}$ represents the covariance between sample 1 and sample 2, etc.

These equations are solved to produce the weighting factors used for the kriged estimates.

The covariance between two samples $\sigma_{i,j}$ is determined from the semi-variogram as:

$$\sigma_{i,j} = (Co + C) - \gamma_{i,j}$$

In the above kriging equations we can use $\gamma_{i,j}$ instead of covariance – it gives the same result. The distance h between samples i and j is calculated on the basis of their X and Y coordinates and the result substituted in:

$\gamma(h) = Co + C\,[3h/2a - (h/a)^3/2]$ for $h < a$

or

$\gamma(h) = Co + C$ if $h \geq a$

In the isotropic case

$h = [(x_1 - x_2)^2 + (y_1 - y_2)^2]^{1/2}$.

However, in the anisotropic case

$$h' = [(x_1 - x_2)^2 + A^2(y_1 - y_2)^2]^{1/2}$$

where A = the anisotropy coefficient = range in the x direction divided by the range in the y direction.

To determine the covariance between a sample and a block ($\sigma_{i,B}$), the block is considered to be represented by a grid of j points (Figure A4.3).

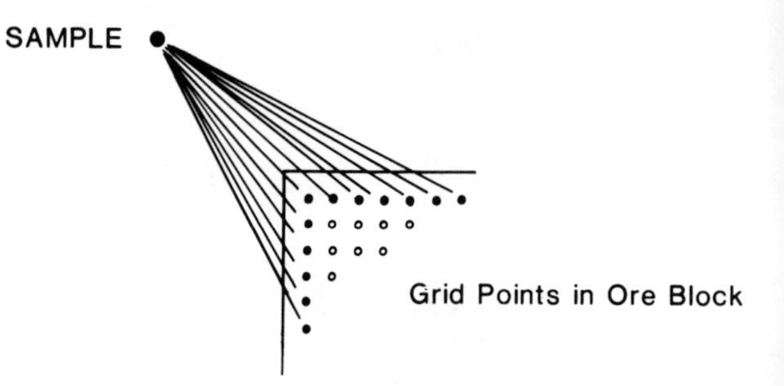

Fig. A4.3 Block kriging using a mesh of points within the block.

The covariance between each of these and the sample is determined and the average computed. The grid size could be 10 × 10, therefore the value of $\sigma_{i,B}$ would be the mean of 100 values of $\sigma_{i,j}$. The distance between each grid point and the sample i is calculated and then input into the semi-variogram equation to determine $\gamma_{i,j}$.

KRIGING ESTIMATION VARIANCE

The estimation variance:

$$\begin{aligned}
\sigma^2_e &= E(Z^\star_B - Z_B)^2 \\
&= E(Z^\star_B)^2 - 2E(Z^\star_BZ_B) + E(Z_B)^2
\end{aligned}$$

As before, we can add and subtract the mean:

$$\begin{aligned}
\sigma^2_e &= E(Z^\star_B)^2 - m^2 - 2E(Z^\star_BZ_B) + 2m^2 + E(Z_B)^2 - m^2 \\
&= \sigma^2_{Z\star B} - 2\sigma_{Z\star B,ZB} + \sigma^2_{ZB}
\end{aligned}$$

$$\sigma^2_e = \Sigma^n_{i=1}\ \Sigma^n_{j=1}\ K_iK_j\sigma_{i,j} - 2\ \Sigma^n_{i=1}\ K_i\sigma_{i,B} + \sigma^2_{ZB}$$

This formula is valid for any weighted average estimator. In the case of a kriging estimator it can be simplified. If each of the kriging equations is multiplied by the ith weight:

$$K_iK_1\sigma_{i,1} + K_iK_2\sigma_{i,2} + \ldots + K_iK_n\sigma_{i,n} + K_i\mu = K_i\sigma_{i,B}$$

and then the results are summed we have:

$$\Sigma_{i=1}^{n} \Sigma_{j=1}^{n} K_i K_j \sigma_{i,j} + \mu \Sigma_{i=1}^{n} K_i = \Sigma_{i=1}^{n} K_i \sigma_{i,B}$$

or

$$\Sigma_{i=1}^{n} \Sigma_{i=1}^{n} K_i K_j \sigma_{i,j} + \mu = \Sigma_{i=1}^{n} K_i \sigma_{i,B} \text{ as } \Sigma_{i=1}^{n} K_i = 1$$

Kriging estimation variance can thus be expressed as:

$$\sigma^2_e = \sigma^2_{ZB} - \Sigma_{i=1}^{n} K_i \sigma_{i,B} + \mu$$

where σ^2_{ZB} is the variance of the block defined as the average difference in values of pairs of points within the block.

APPENDIX 4.5

Extension Variance Graphs and Tables for the Spherical Scheme

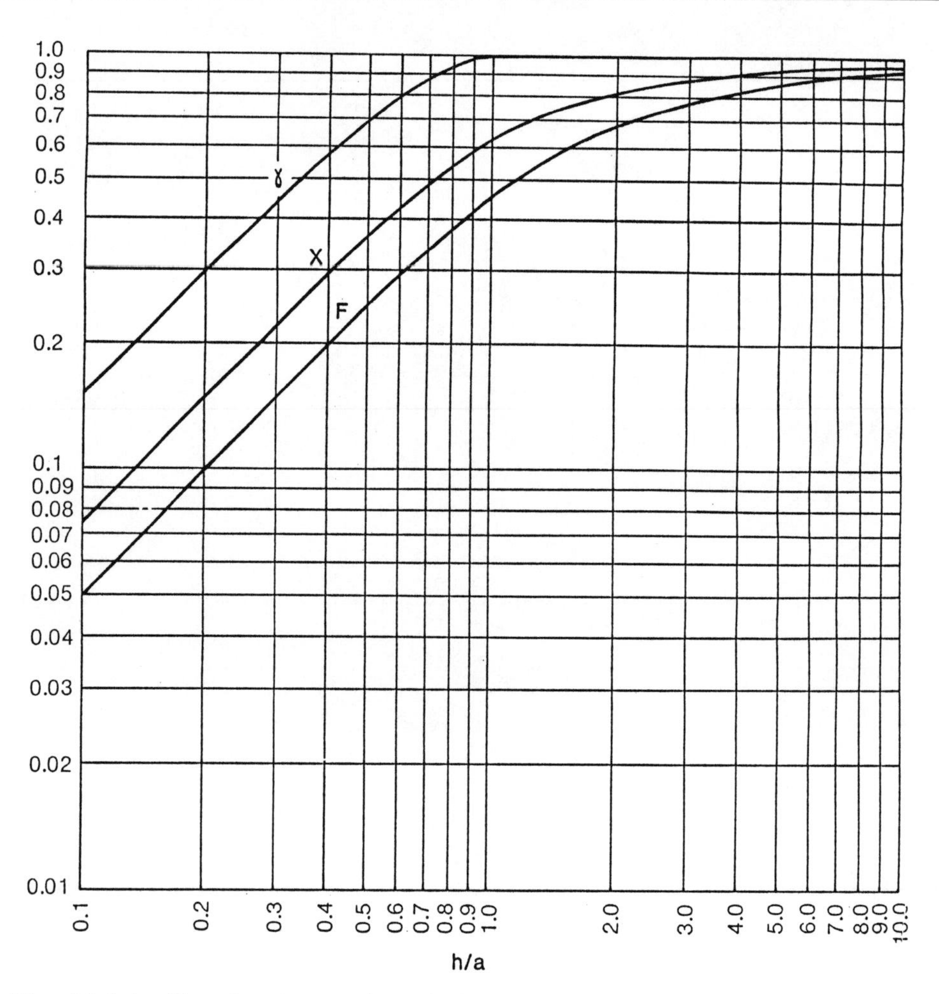

Fig. A4.4 Auxiliary functions – spherical scheme. One dimensional γ, X and F functions.

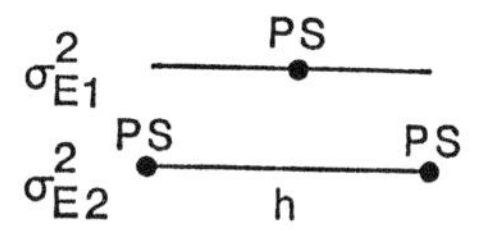

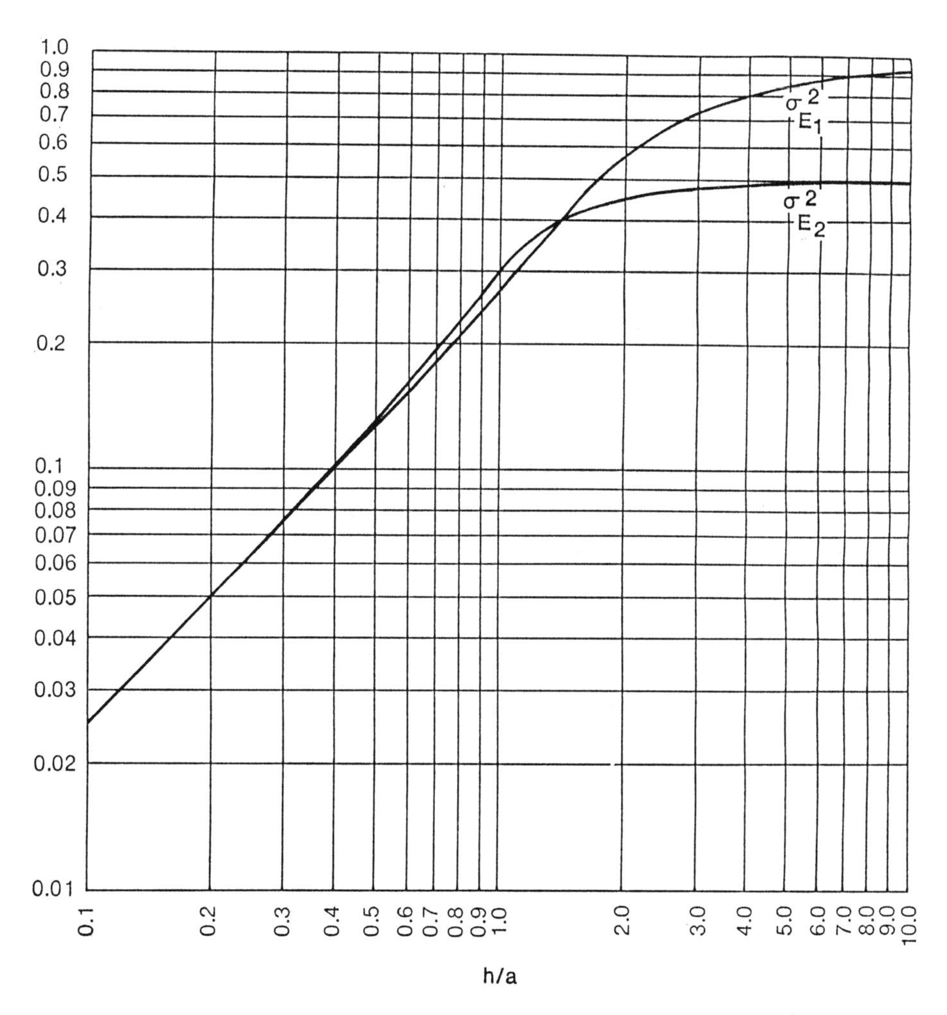

Fig. A4.5 Spherical scheme. One dimensional extension variances $\left(\frac{1}{c} \times \sigma_e^2\right)$.

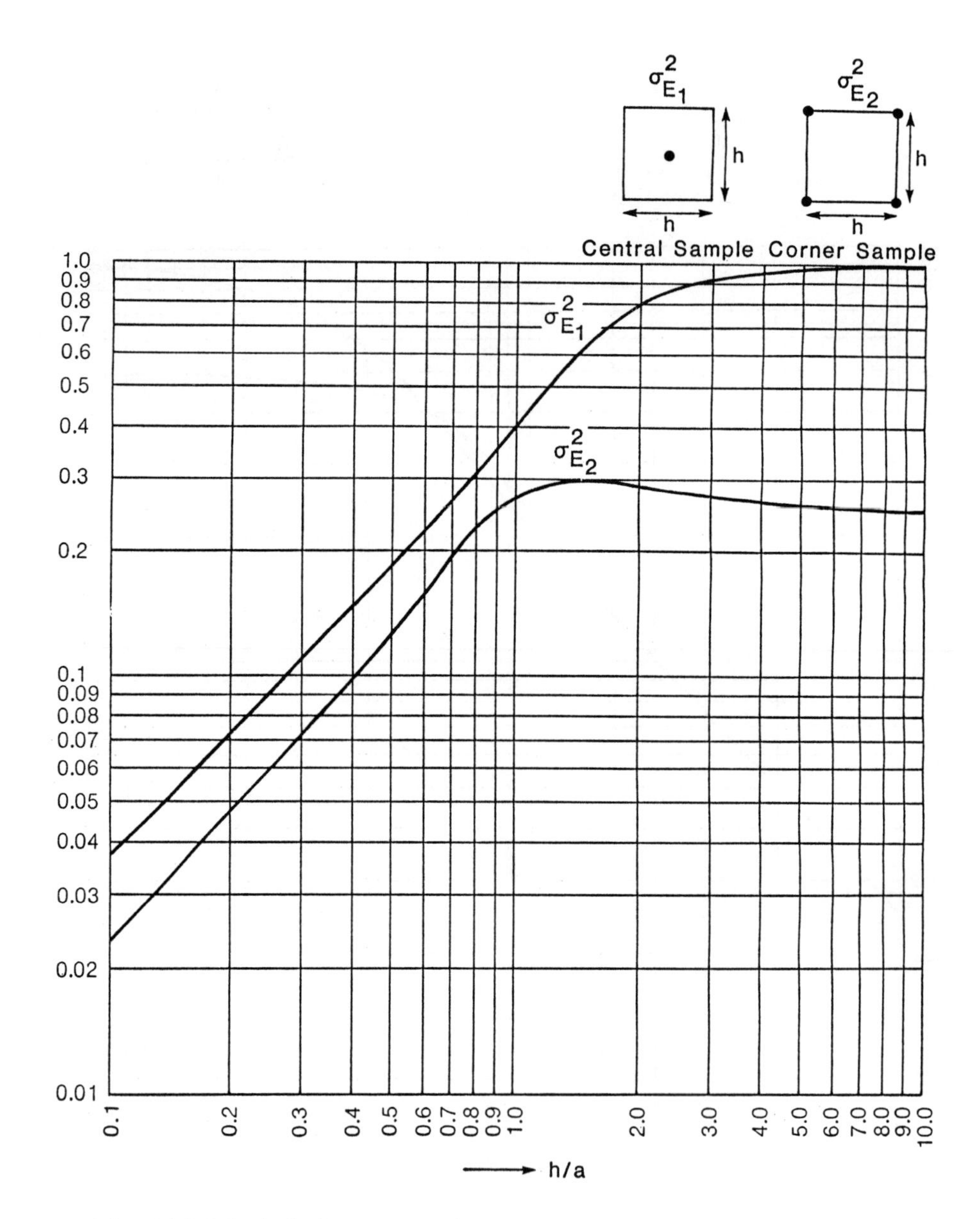

Fig. A4.6 Spherical scheme. Two dimensional extension variances (square grid).

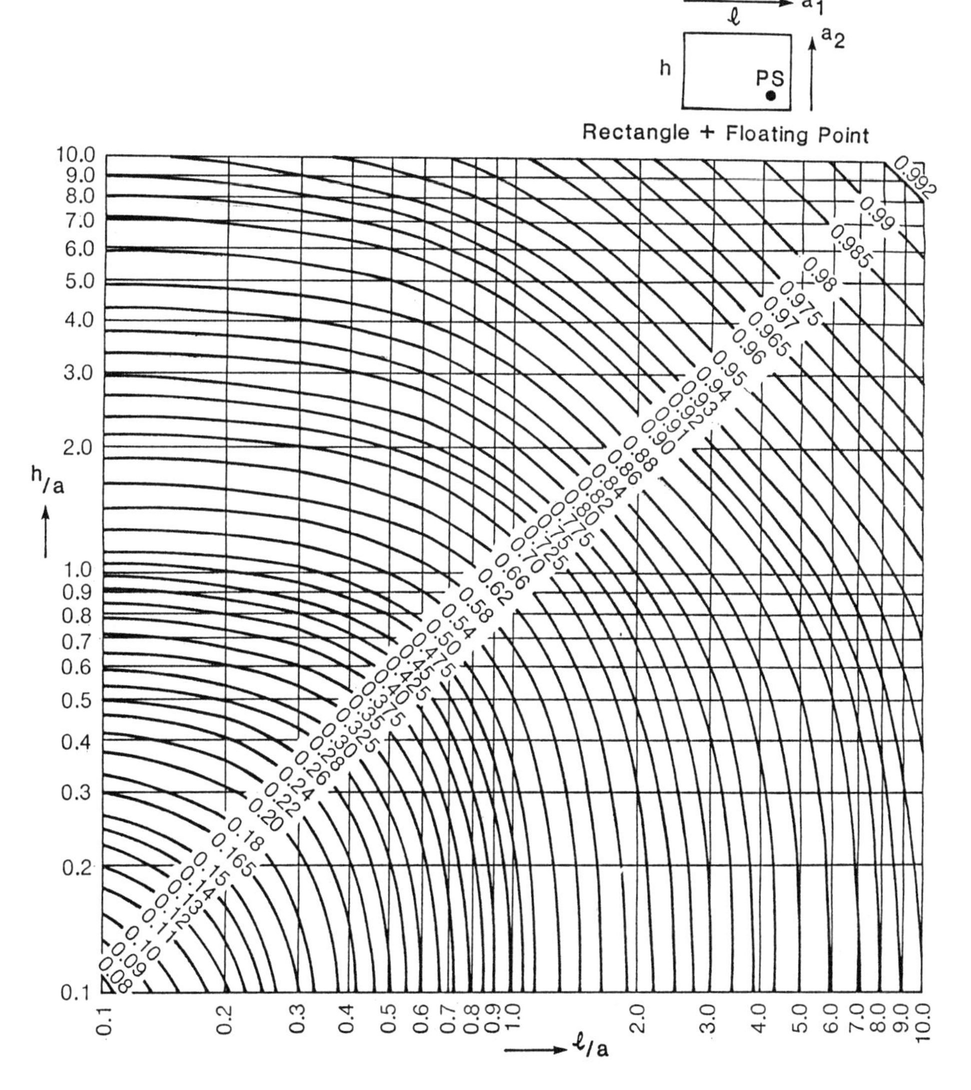

Fig. A4.7 Spherical scheme. Two dimensional auxiliary function 'F' in $\frac{1}{c}\left[F\left(\frac{l}{a_1}, \frac{h}{a_2}\right)\right]$.

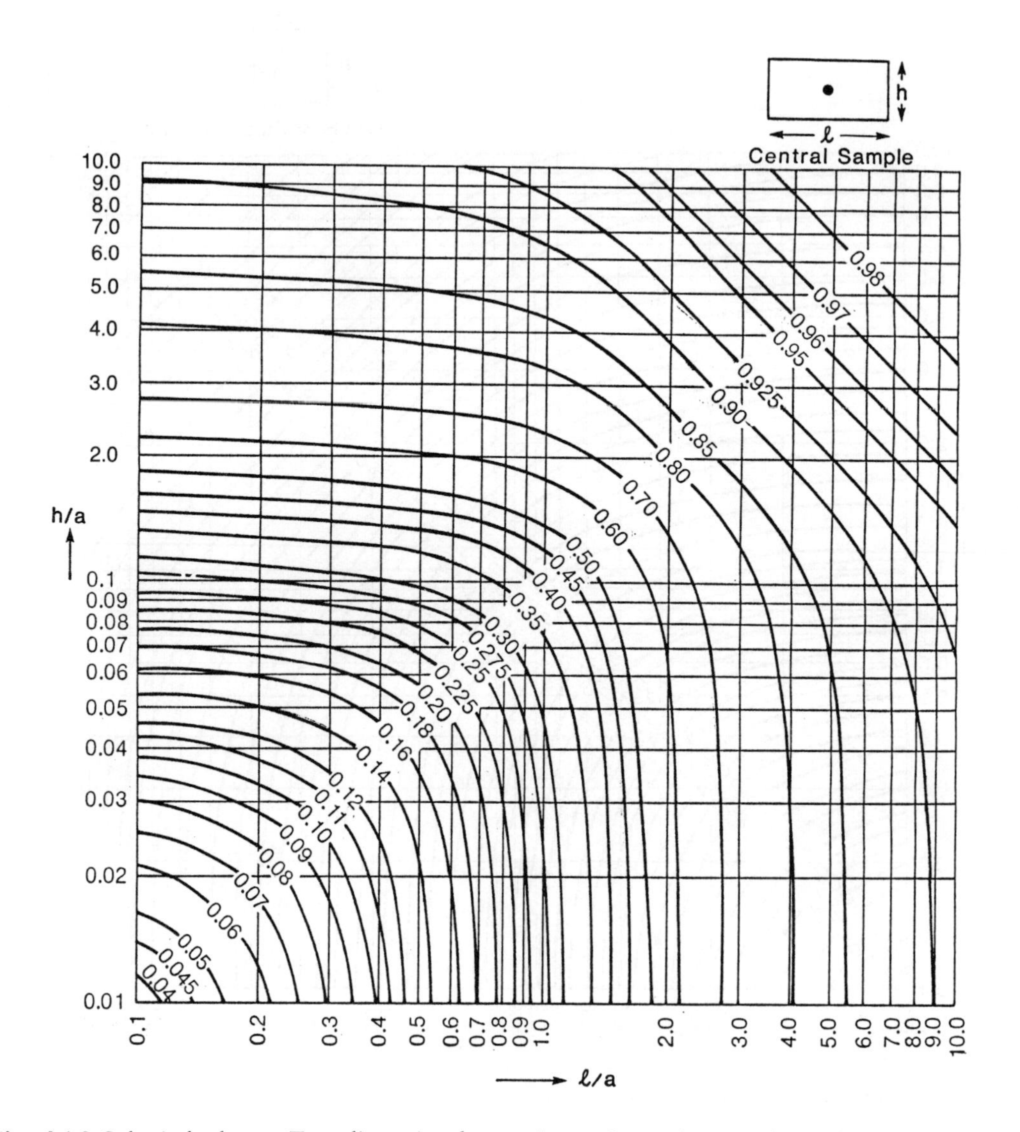

Fig. A4.8 Spherical scheme. Two dimensional extension variances (rectangular grid, central samples).

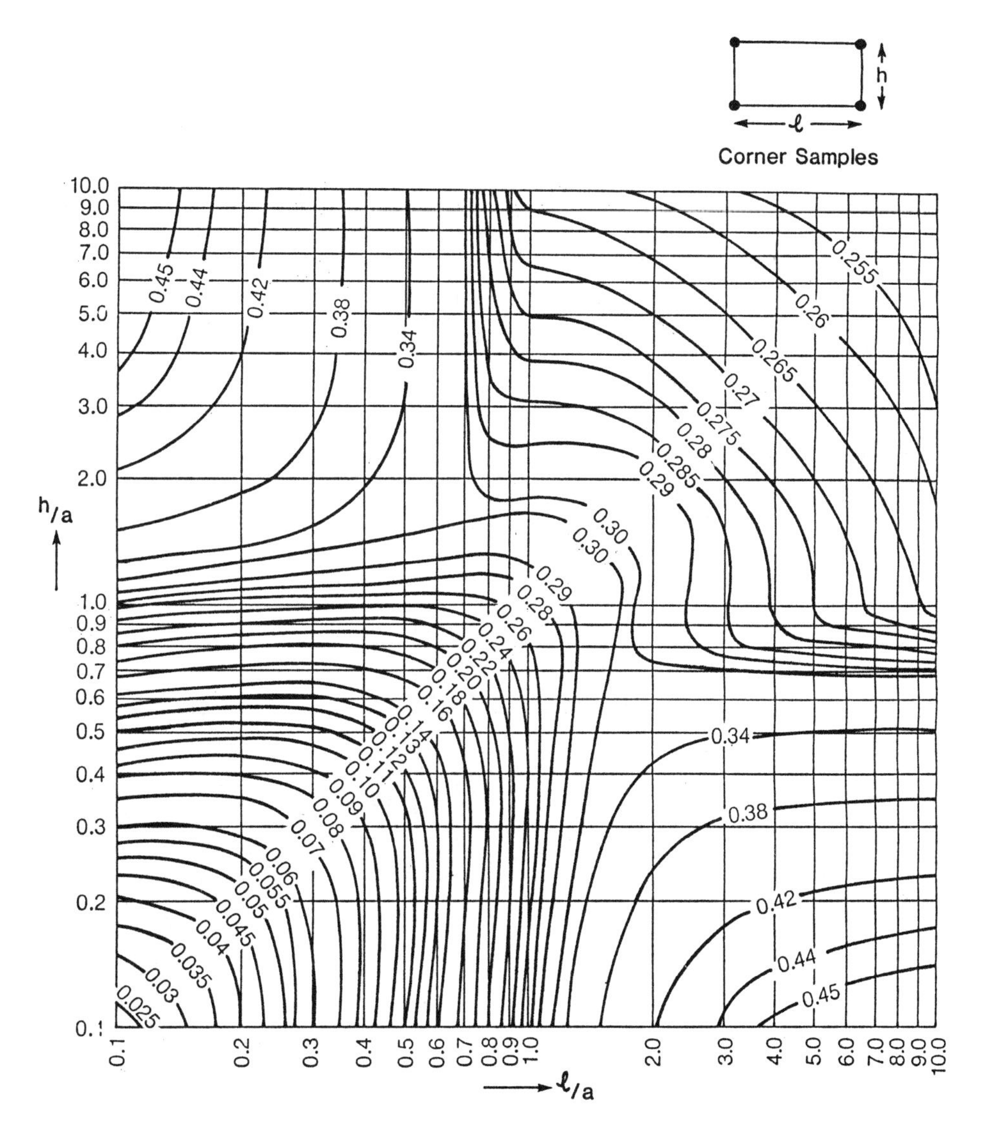

Fig. A4.9 Spherical scheme. Two dimensional extension variances (rectangular grid, corner samples).

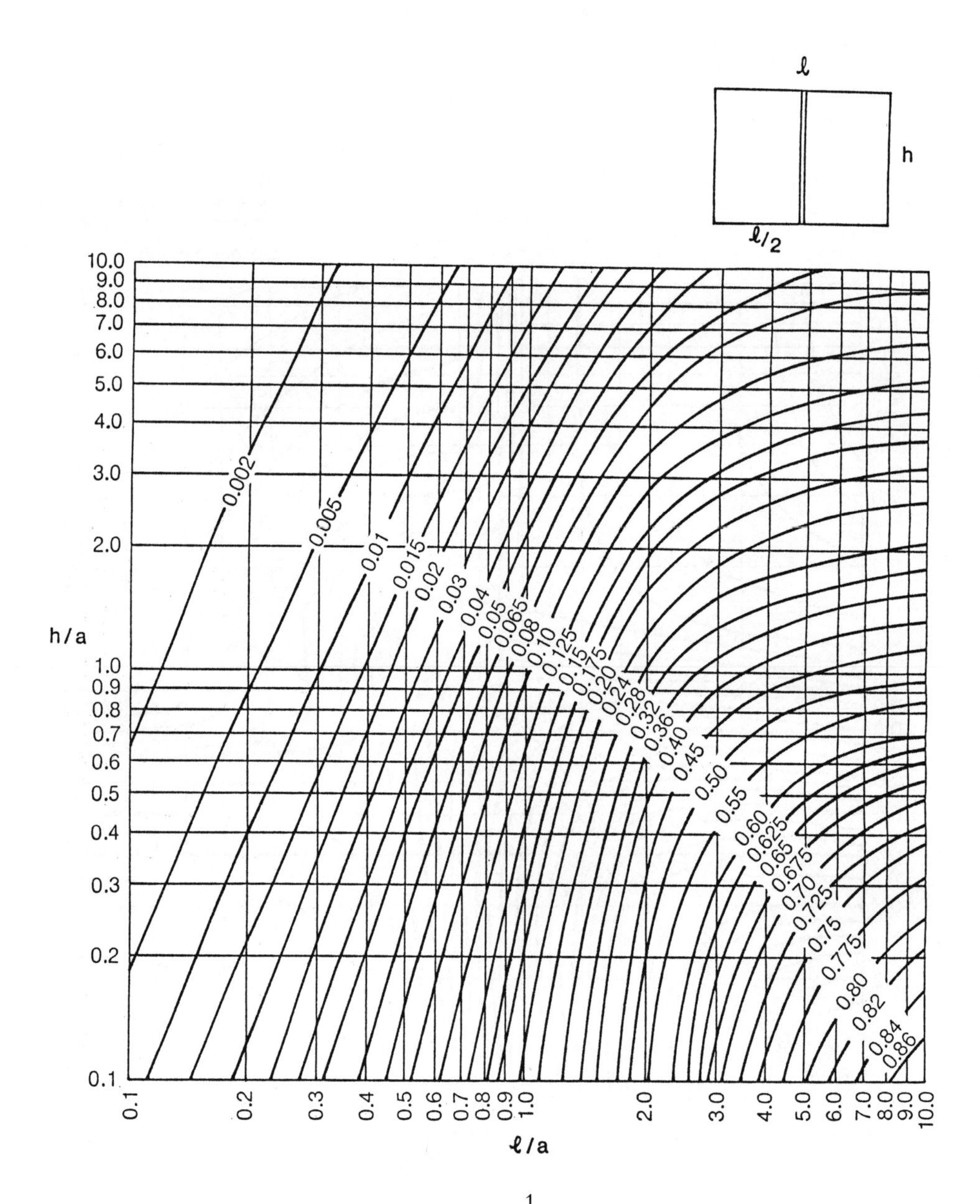

Fig. A4.10 Spherical scheme. Extension variance $\left(\frac{1}{c}\sigma_e^2\right)$, continuous sample in rectangular block $l \times h$.

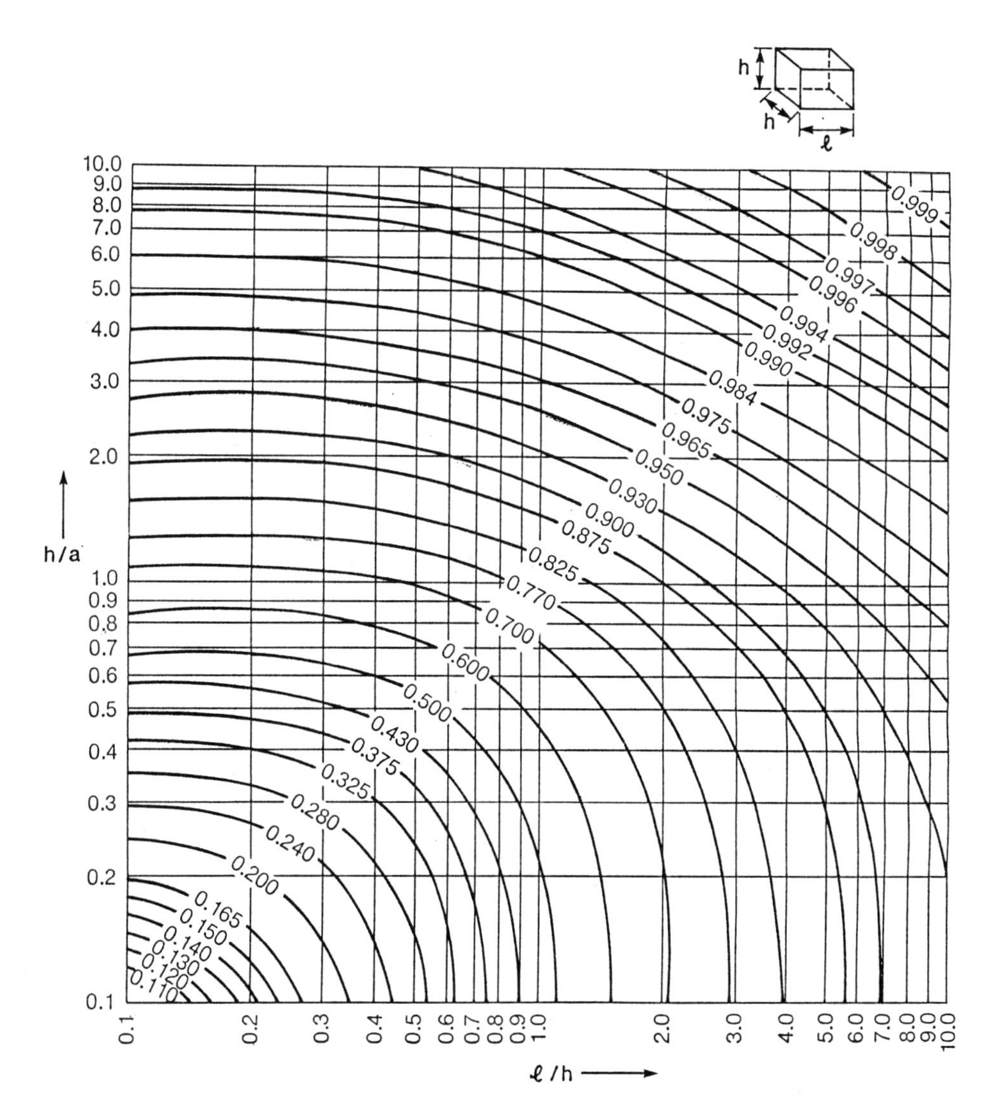

Fig. A4.11 Spherical scheme. Estimation variance (parallelopiped).

Table A4.1 Extension variance tables for model with two continuous samples at block ends

x = l/a

y = h/a

h/a	l/a 1.0	1.5	2.0	2.5	3.0	3.5	4.0	4.5	5.0
1.0	0.151	0.220	0.244	0.255	0.261	0.265	0.267	0.269	0.270
1.5	0.113	0.165	0.183	0.191	0.195	0.198	0.200		0.202
2.0	0.090	0.130	0.144	0.151	0.154	0.157	0.158		0.160
2.5	0.074	0.107	0.119	0.124	0.127	0.129	0.130		0.132
3.0	0.063	0.091	0.101	0.106	0.108	0.110	0.111		0.112
3.5	0.055	0.079	0.088	0.092	0.094	0.095	0.096		0.097
4.0	0.049	0.070	0.078	0.081	0.083	0.084	0.085		0.086
4.5	0.043	0.063	0.070	0.073	0.075	0.076	0.076	0.077	0.077
5.0	0.039	0.057	0.065	0.066	0.067	0.068	0.069	0.069	0.070
5.5	0.036	0.052	0.058	0.060	0.062	0.063	0.063	0.063	0.064
6.0	0.033	0.048	0.053	0.055	0.057	0.058	0.058	0.058	0.059
6.5	0.031	0.044	0.049	0.051	0.053	0.053	0.054	0.054	0.054
7.0	0.029	0.041	0.046	0.048	0.049	0.050	0.050	0.050	0.051
7.5	0.027	0.039	0.043	0.045	0.046	0.046	0.047		
8.0	0.025	0.036	0.040		0.043	0.044			
8.5	0.024	0.034	0.038		0.041	0.041			
9.0	0.022	0.032	0.036		0.038	0.039			
9.5	0.021	0.031	0.034		0.036	0.037			
10.0	0.020	0.029	0.032		0.035	0.035			

h/a	l/a 5.5	6.0	6.5	7.0	7.5
1.0	0.271	0.272	0.272	0.272	0.273
1.5	0.203	0.203	0.203	0.204	0.204
2.0	0.160	0.160	0.161	0.161	0.161
2.5	0.132	0.132	0.133	0.133	0.133
3.0	0.112	0.112	0.113	0.113	0.113
3.5	0.098	0.098	0.098	0.098	0.098
4.0	0.086	0.086	0.087	0.087	
4.5	0.077	0.077	0.078	0.078	
5.0	0.070	0.070	0.070	0.070	
5.5	0.064	0.064	0.064	0.064	
6.0	0.059	0.059	0.059	0.059	
6.5	0.055	0.055	0.055	0.055	
7.0	0.051	0.051	0.051	0.051	

Table A4.2 Extension variance tables for model with continuous sampling in levels and raises

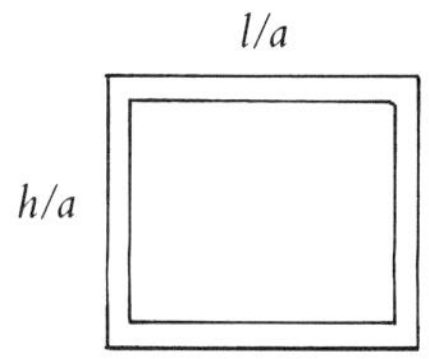

h/a	*l/a* 0.10	0.15	0.20	0.25	0.30	0.35	0.40	0.45	0.50
0.10	0.007	0.008	0.008	0.008	0.008	0.007	0.007	0.007	0.006
0.15	0.008	0.011	0.012	0.012	0.013	0.012	0.012	0.012	0.011
0.20	0.008	0.012	0.014	0.016	0.017	0.017	0.017	0.017	0.017
0.25	0.008	0.012	0.016	0.018	0.020	0.021	0.021	0.021	0.022
0.30	0.008	0.013	0.017	0.020	0.022	0.024	0.025	0.026	0.026
0.35	0.007	0.012	0.017	0.021	0.024	0.026	0.028	0.029	0.030
0.40	0.007	0.012	0.017	0.021	0.025	0.028	0.030	0.032	0.033
0.45	0.007	0.012	0.017	0.021	0.026	0.029	0.032	0.034	0.036
0.50	0.006	0.011	0.017	0.022	0.026	0.030	0.033	0.036	0.039
0.55	0.006	0.011	0.016	0.022	0.026	0.031	0.035	0.038	0.041
0.60	0.006	0.011	0.016	0.021	0.027	0.031	0.035	0.039	0.043
0.65	0.005	0.010	0.016	0.021	0.027	0.032	0.036	0.040	0.044
0.70	0.005	0.010	0.015	0.021	0.027	0.032	0.037	0.041	0.045
0.75	0.005	0.010	0.015	0.021	0.026	0.032	0.037	0.042	0.046
0.80	0.005	0.009	0.015	0.021	0.026	0.032	0.037	0.043	0.047
0.85	0.005	0.009	0.015	0.020	0.026	0.032	0.038	0.043	0.048
0.90	0.004	0.009	0.014	0.020	0.026	0.032	0.038	0.043	0.049
0.95	0.004	0.009	0.014	0.020	0.026	0.032	0.038	0.044	0.049
1.00	0.004	0.009	0.014	0.019	0.026	0.032	0.038	0.044	0.050

h/a	*l/a* 0.55	0.60	0.65	0.70	0.75	0.80	0.85	0.90	0.95	1.00
0.10	0.006	0.006	0.005	0.005	0.005	0.005	0.005	0.004	0.004	0.004
0.15	0.011	0.011	0.010	0.010	0.010	0.009	0.009	0.009	0.009	0.008
0.20	0.016	0.016	0.016	0.015	0.015	0.015	0.015	0.014	0.014	0.014
0.25	0.022	0.021	0.021	0.021	0.021	0.021	0.020	0.020	0.020	0.019
0.30	0.026	0.027	0.027	0.027	0.026	0.026	0.026	0.026	0.026	0.025
0.35	0.031	0.031	0.032	0.032	0.032	0.032	0.032	0.032	0.032	0.032
0.40	0.035	0.035	0.036	0.037	0.037	0.037	0.038	0.038	0.038	0.038
0.45	0.038	0.039	0.040	0.041	0.042	0.043	0.043	0.043	0.044	0.044
0.50	0.041	0.043	0.044	0.045	0.046	0.047	0.048	0.049	0.049	0.050
0.55	0.043	0.046	0.048	0.049	0.051	0.052	0.053	0.054	0.055	0.055
0.60	0.046	0.048	0.051	0.053	0.055	0.056	0.058	0.059	0.060	0.061
0.65	0.048	0.051	0.053	0.056	0.058	0.060	0.062	0.064	0.065	0.066
0.70	0.049	0.053	0.056	0.059	0.061	0.064	0.066	0.068	0.070	0.071
0.75	0.051	0.055	0.058	0.061	0.065	0.067	0.070	0.072	0.074	0.075
0.80	0.052	0.056	0.060	0.064	0.067	0.070	0.073	0.076	0.078	0.080
0.85	0.053	0.058	0.062	0.066	0.070	0.073	0.076	0.079	0.082	0.083
0.90	0.054	0.059	0.064	0.068	0.072	0.076	0.079	0.082	0.085	0.087
0.95	0.055	0.060	0.065	0.070	0.074	0.078	0.082	0.085	0.088	0.090
1.00	0.055	0.061	0.066	0.071	0.075	0.080	0.083	0.087	0.090	0.092

Table A4.2 continued

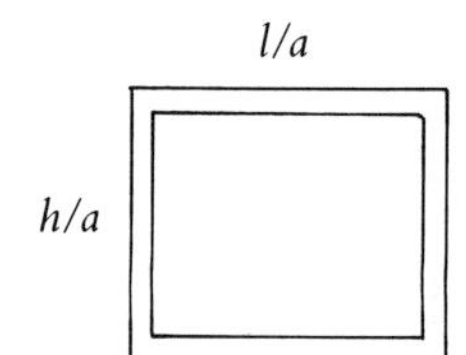

	l/a								
h/a	1.0	1.5	2.0	2.5	3.0	3.5	4.0	4.5	5.0
0.10	0.004	0.003	0.002	0.001	0.001	0.001	0.001	0.001	0.001
0.15	0.008	0.006	0.004	0.003	0.002	0.002			0.001
0.20	0.014	0.010	0.007	0.005	0.004	0.003	0.003	0.002	0.002
0.25	0.019	0.015	0.011	0.008	0.006	0.005			0.005
0.30	0.025	0.020	0.015	0.011	0.009	0.007	0.006	0.005	0.005
0.35	0.032	0.025	0.019	0.015	0.012	0.010			0.006
0.40	0.038	0.031	0.023	0.018	0.015	0.013	0.011	0.009	0.008
0.45	0.044	0.037	0.028	0.022	0.018	0.015			0.010
0.50	0.050	0.042	0.033	0.027	0.022	0.019	0.016	0.014	0.013
0.55	0.055	0.048	0.038	0.031	0.026	0.022			0.015
0.60	0.061	0.054	0.043		0.030	0.025	0.022	0.019	0.017
0.65	0.066	0.060	0.048		0.033	0.029			0.020
0.70	0.071	0.065	0.053		0.037	0.032	0.028	0.025	0.023
0.75	0.075	0.070	0.058		0.041	0.036			0.025
0.80	0.080	0.075	0.063		0.045	0.039	0.035	0.031	0.028
0.85	0.083	0.080	0.067		0.049	0.043			0.031
0.90	0.087	0.084	0.072		0.053	0.046	0.041	0.037	0.033
0.95	0.090	0.088	0.076		0.056	0.049			0.036
1.00	0.092	0.091	0.079	0.068	0.059	0.052	0.047	0.042	0.038
1.50	0.091	0.097	0.088	0.079	0.071	0.046	0.058		0.049
2.00	0.079	0.088	0.083	0.076	0.069	0.063	0.058		0.049
2.50	0.068	0.079	0.076	0.070	0.065	0.060	0.055		0.048
3.00	0.059	0.071	0.069	0.065	0.060	0.056	0.052		0.046
3.50	0.052	0.064	0.063	0.060	0.056	0.052	0.049		0.043
4.00	0.047	0.058	0.058	0.055	0.052	0.049	0.046		0.041
4.50	0.042	0.053	0.053		0.049	0.046	0.044	0.041	0.039
5.00	0.038	0.049	0.049		0.046	0.043	0.041		0.037
5.50	0.035	0.045	0.046		0.043	0.041	0.039		
6.00	0.033	0.042	0.043		0.041	0.039	0.037		
6.50	0.030	0.039	0.041		0.039	0.037	0.033		
7.00	0.028	0.037	0.038		0.037	0.035	0.034		
7.50	0.027	0.035	0.036		0.035	0.034	0.032		
8.00	0.025	0.033	0.034		0.033	0.032			
8.50	0.024	0.031	0.033		0.032	0.031			
9.00	0.022	0.030	0.031		0.030	0.030			
9.50	0.021	0.028	0.030		0.029	0.028			
10.00	0.020	0.027	0.029		0.028	0.027			

Table A4.2 continued

	l/a				
h/a	5.5	6.0	6.5	7.0	7.5
1.00	0.035	0.033	0.030	0.028	0.027
1.50	0.045	0.042	0.039	0.037	0.035
2.00	0.046	0.043	0.041	0.038	0.036
2.50	0.045	0.042	0.040	0.038	0.036
3.00	0.043	0.041	0.038	0.037	0.035
3.50	0.041	0.039	0.037	0.035	
4.00	0.039	0.037	0.035	0.034	
4.50	0.037	0.036	0.034	0.033	
5.00	0.036	0.034	0.033	0.031	
5.50	0.034	0.033	0.031	0.030	
6.00		0.031	0.030	0.029	
6.50			0.029	0.028	
7.00				0.027	

Table A4.3 Extension variance tables for model with continuous sampling in the central raise and sub-level

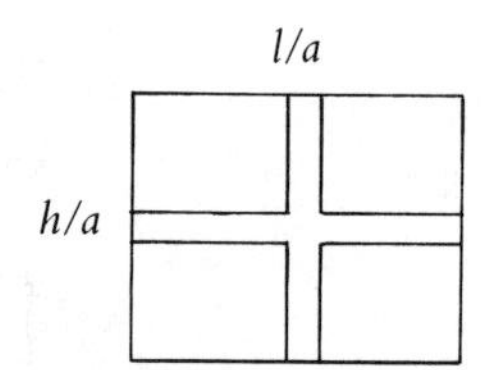

h/a	l/a 0.10	0.15	0.20	0.25	0.30	0.35	0.40	0.45	0.50
0.10	0.004	0.005	0.005	0.005	0.005	0.005	0.005	0.004	0.004
0.15	0.005	0.007	0.007	0.008	0.008	0.008	0.008	0.008	0.007
0.20	0.005	0.007	0.009	0.010	0.010	0.010	0.011	0.011	0.010
0.25	0.005	0.008	0.010	0.011	0.012	0.013	0.013	0.013	0.013
0.30	0.005	0.008	0.010	0.012	0.013	0.014	0.015	0.016	0.016
0.35	0.005	0.008	0.010	0.013	0.014	0.016	0.017	0.018	0.018
0.40	0.005	0.008	0.011	0.013	0.015	0.017	0.018	0.019	0.020
0.45	0.004	0.008	0.011	0.013	0.016	0.018	0.019	0.020	0.022
0.50	0.004	0.007	0.010	0.013	0.016	0.018	0.020	0.022	0.023
0.55	0.004	0.007	0.010	0.013	0.016	0.019	0.021	0.023	0.024
0.60	0.004	0.007	0.010	0.013	0.016	0.019	0.021	0.023	0.025
0.65	0.004	0.007	0.010	0.013	0.016	0.019	0.022	0.024	0.026
0.70	0.004	0.007	0.010	0.013	0.016	0.019	0.022	0.024	0.027
0.75	0.003	0.006	0.010	0.013	0.016	0.019	0.022	0.025	0.027
0.80	0.003	0.006	0.009	0.013	0.016	0.019	0.022	0.025	0.028
0.85	0.003	0.006	0.009	0.013	0.016	0.019	0.022	0.025	0.028
0.90	0.003	0.006	0.009	0.012	0.016	0.019	0.022	0.025	0.028
0.95	0.003	0.006	0.009	0.012	0.016	0.019	0.022	0.026	0.029
1.00	0.003	0.006	0.009	0.012	0.016	0.019	0.022	0.026	0.029

h/a	l/a 0.55	0.60	0.65	0.70	0.75	0.80	0.85	0.90	0.95	1.00
0.10	0.004	0.004	0.004	0.004	0.003	0.003	0.003	0.003	0.003	0.003
0.15	0.007	0.007	0.007	0.007	0.006	0.006	0.006	0.006	0.006	0.006
0.20	0.010	0.010	0.010	0.010	0.010	0.009	0.009	0.009	0.009	0.009
0.25	0.013	0.013	0.013	0.013	0.013	0.013	0.013	0.012	0.012	0.012
0.30	0.016	0.016	0.016	0.016	0.016	0.016	0.016	0.016	0.016	0.016
0.35	0.019	0.019	0.019	0.019	0.019	0.019	0.019	0.019	0.019	0.019
0.40	0.021	0.021	0.022	0.022	0.022	0.022	0.022	0.022	0.022	0.022
0.45	0.023	0.023	0.024	0.024	0.025	0.025	0.025	0.025	0.026	0.026
0.50	0.024	0.025	0.026	0.027	0.027	0.028	0.028	0.028	0.029	0.029
0.55	0.026	0.027	0.028	0.029	0.029	0.030	0.031	0.031	0.032	0.032
0.60	0.027	0.028	0.029	0.030	0.031	0.032	0.033	0.034	0.034	0.035
0.65	0.028	0.029	0.031	0.032	0.033	0.034	0.035	0.036	0.037	0.037
0.70	0.029	0.030	0.032	0.034	0.035	0.036	0.037	0.038	0.039	0.040
0.75	0.029	0.031	0.033	0.035	0.037	0.038	0.039	0.040	0.041	0.042
0.80	0.030	0.032	0.034	0.036	0.038	0.039	0.041	0.042	0.044	0.045
0.85	0.031	0.033	0.035	0.037	0.039	0.041	0.043	0.044	0.045	0.047
0.90	0.031	0.034	0.036	0.038	0.040	0.042	0.044	0.046	0.047	0.049
0.95	0.032	0.034	0.037	0.039	0.041	0.043	0.045	0.047	0.049	0.050
1.00	0.032	0.035	0.037	0.040	0.042	0.045	0.047	0.049	0.050	0.052

Table A4.3 continued

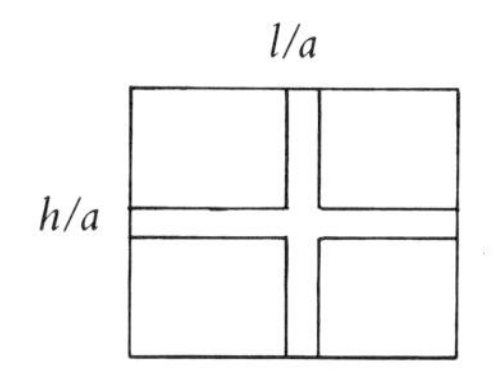

h/a	l/a 1.0	1.5	2.0	2.5	3.0	3.5	4.0	4.5	5.0
0.10	0.003	0.003	0.002	0.001	0.001	0.001	0.001	0.001	0.001
0.15	0.006	0.005	0.004	0.003	0.002	0.002	0.002	0.001	0.001
0.20	0.009	0.007	0.006	0.005	0.004	0.003	0.003	0.002	0.002
0.25	0.012	0.010	0.009	0.007	0.006	0.005	0.004	0.003	0.003
0.30	0.016	0.014	0.012	0.010	0.008	0.006	0.005	0.004	0.004
0.35	0.019	0.017	0.015	0.012	0.010	0.008	0.007	0.006	0.005
0.40	0.022	0.020	0.018	0.015	0.013	0.011	0.009	0.008	0.007
0.45	0.026	0.024	0.022	0.018	0.015	0.013	0.011	0.010	0.008
0.50	0.029	0.027	0.026	0.021	0.018	0.015	0.013	0.012	0.010
0.55	0.032	0.030	0.029	0.025	0.021	0.018	0.015	0.014	0.012
0.60	0.035	0.034	0.033	0.028	0.024	0.020	0.017	0.015	0.013
0.65	0.037	0.037	0.036	0.031	0.026	0.023	0.019	0.017	0.015
0.70	0.040	0.040	0.040	0.035	0.029	0.025	0.022	0.019	0.017
0.75	0.042	0.043	0.043	0.038	0.032	0.028	0.025	0.021	0.019
0.80	0.045	0.046	0.047	0.041	0.035	0.030	0.027	0.023	0.021
0.85	0.047	0.048	0.050	0.044	0.038	0.035	0.029	0.026	0.023
0.90	0.049	0.051	0.053	0.046	0.041	0.036	0.031	0.028	0.025
0.95	0.050	0.053	0.056	0.050	0.044	0.039	0.033	0.030	0.027
1.00	0.052	0.061	0.059	0.053	0.046	0.041	0.036	0.032	0.029
1.50	0.061	0.077	0.080	0.075	0.069	0.063	0.057		
2.00	0.059	0.080	0.087	0.085	0.080	0.074	0.069		
2.50	0.053	0.075	0.085	0.085	0.082	0.077	0.073		
3.00	0.046	0.069	0.080	0.082	0.080	0.076	0.072		
3.50	0.041	0.063	0.074	0.077	0.076	0.074	0.071		
4.00	0.036	0.057	0.069	0.073	0.072	0.071	0.068		
4.50	0.032	0.053	0.064		0.069	0.067	0.065	0.063	
5.00	0.029	0.048	0.060		0.065	0.064	0.063		
5.50	0.027	0.045	0.056		0.062	0.061	0.060		
6.00	0.025	0.042	0.053		0.059	0.058	0.057		
6.50	0.022	0.039	0.050		0.056	0.056	0.055		
7.00	0.021	0.036	0.047		0.053	0.053	0.053		
7.50	0.019	0.034	0.044		0.051	0.051	0.051		
8.00	0.018	0.032	0.042		0.048	0.049			
8.50	0.017	0.030	0.040		0.046	0.047			
9.00	0.016	0.029	0.038		0.045	0.045			
9.50	0.015	0.028	0.036		0.043	0.043			
10.00	0.014	0.026	0.035		0.041	0.042			

5

Design and Evaluation of Open-pit Operations

5.1 INTRODUCTION

The computation of reserves available for a potential open-pit operation must be undertaken in conjunction with a preliminary pit design, for it is this which defines the limits of the mineable mineralization and not the geological fringes of the deposit. Even at the exploration stage, the geologist must bear in mind the constraints imposed by open-pit mining and must continually reassess whether the deposit under evaluation can still be mined by this means and whether a profit can be made at present prices and costs. Future possible trends in these economic parameters must also be taken into account. It is a pointless exercise to continue expensive drilling on a deposit whose grade or tonnage or depth or morphology, or any combination of these, makes its exploitation no longer an attractive proposition.

This chapter thus aims to summarize some of the methods available for both the design and evaluation of open-pits. It will examine both manual and computerized techniques, including the economic optimization of the pit design, and will also consider the design in situations where surface limitations are placed on the pit perimeters and in others where none exist. Also considered are deposits which are either tabular, and thus have well-defined assay hangingwalls and footwalls, or which are large, irregular and have poorly defined limits. The latter must thus be evaluated on the basis of an envelope, which encloses both waste and high grade material, but whose overall grade is deemed to make it economically viable.

The methods employed will be illustrated by reference to gold deposits in British Columbia and Western Australia, amongst others, and to a hypothetical tabular deposit, a figment of the author's fertile imagination! A further example of a limestone quarry design can be found in Chapter 8 (section 8.8).

5.2 DESIGN OF OPEN-PIT OPERATIONS

5.2.1 Design parameters

Detailed surface topographic information
This information may be in the form of detailed contour plans, perhaps stored to computer file via a digitizer, or as computer files of survey spot heights, including drill-hole collars. In the latter case, this data base may be used to grid the surface elevation using trend surface fitting, inverse distance weighting or block kriging techniques. The grid fitted would be so designed as to

match that eventually used to compute bench or global reserves. Alternatively, the surface may be modelled from spot heights using digital terrain modelling (DTM) which effectively is a triangulation method. This is discussed in more detail in section 5.2.5.

Batter

Initial pit designs are usually made with an overall slope of 45° and are later modified on the basis of geotechnical information. Different faces (design sectors) may thus be allocated different slopes and, in individual design sectors, the slope may change with depth. Batters may be set to 30–35° in overburden, increased to 35–40° in weathered bedrock and increased to 55° in 'fresh' rock. To some extent, the overall pit batter should be dependent on the final anticipated depth of the pit. Roberts, Hoek and Fish (1972) recommend that the batter should not exceed 60° at depths of 65 m and at depths of 300 m it should be less than 40°.

Bench heights

These can vary from pit to pit, and also within pits, depending on the equipment used, the depth of the pit and on the local geology or degree of weathering. Benches in unconsolidated or weak overburden, or in weathered ground, may be relatively thin, e.g. 2–5 m, while those in hardrock may be considerably in excess of this. A recent (1988) survey by the *Canadian Mining Journal* showed that for a wide range of different orebodies, the benches varied from 6 m to 20 m in height. However, there was a tendency for the larger operations, mining in excess of 10 000 t/d, to operate with benches in excess of 9 m. At the Continental Pit, Butte, Montana, benches are increased from 12 m in alluvium to 24 m in competent rock. Smaller operations, mining less than 5000 t/d, generally use benches in the range 6–8 m. Where poor continuity in grades exists in the vertical dimension, and where selective mining is feasible, it is standard practice to subdivide benches into flitches, or horizontal slices, 2–3 m thick. This is the situation in the Australian open-pit gold mines (section 7.2.2, Case history VI). One advantage of using small bench heights is that a better fit can be obtained to the margins of the mineralized body, especially if each bench is also subdivided into small ore blocks. Mining dilution in each bench will also be reduced.

Bench faces

The slopes of bench faces are partly governed by whether they are active faces, in which case only short-term stability is required, or whether they are to be permanent benches and long-term stability is a major consideration. In the former case, much steeper slopes may be tolerated. The slope achieved is, however, largely controlled by the equipment used. Walton and Atkinson (1978) quote 60–80° for loading shovels, 45–90° for hydraulic shovel excavators, 30–90° for hydraulic backhoes and 30–80° for front-end loaders. Rarely, however, is the exposed or active face of a bench left standing vertically. Initial pit designs may assume this to be the case to simplify the exercise, however, final designs will usually allow for a 60–70° rock face. Maximum fragmentation (explosive energy) occurs when blastholes are inclined 45° into the pit or quarry, thus producing a free face with this inclination. However, as indicated above, a 60–70° inclination is used as a compromise to prevent a serious reduction in the overall gradient of the pit face (batter). A final pit or quarry design will thus contain contours representing both the crest and toe of each bench.

Where a significant thickness of overburden exists which has to be pre-stripped by ripping, scraping, draglines or even face shovels (where it is highly variable in thickness), the overburden should be stripped back so as to clear the crest of the first bedrock bench by a distance equivalent to the depth of the overburden and then given a considerably reduced gradient, e.g. 1 in 5 (i.e. approx 11°) in the case of clay (Figure 5.1).

Berm width

As each bench is mined, it will be cut back towards the limit of the overlying bench until it

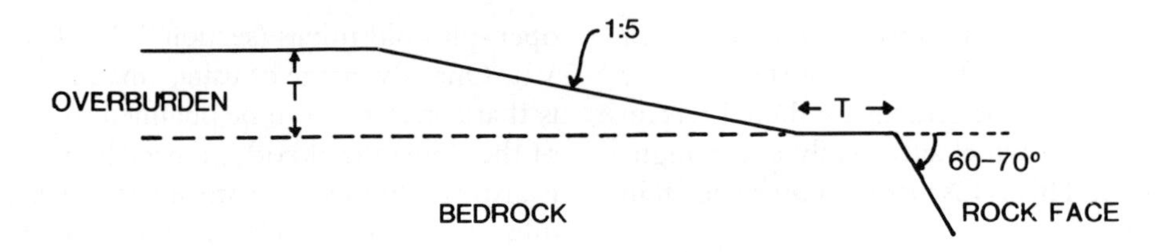

Fig. 5.1 Design of benches in unconsolidated overburden.

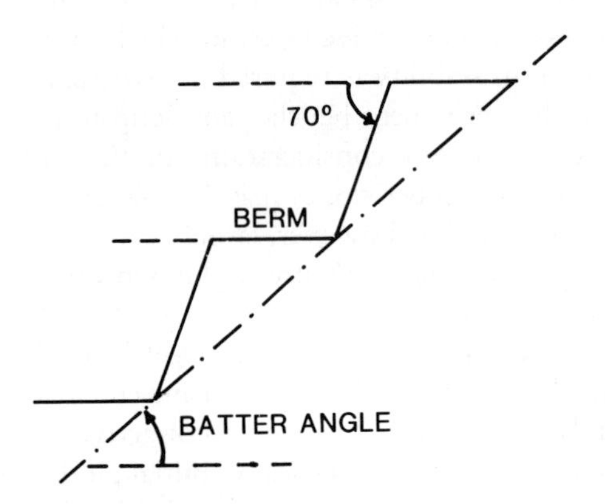

Fig. 5.2 Berms and benches in bedrock.

reaches a distance equivalent to the specified berm width (Figure 5.2). Generally, this width varies from 6 m to 35 m with a typical average close to 10 m. However, where benches are thinner, e.g. 5 m, the berm may be as low as 4 m. Where the anticipated depth of the pit is large, e.g. greater than 200 m, then a larger berm tends to be used, e.g. 35 m. The berm thus to some extent reflects the size of the orebody. Porphyry copper deposits tend to have large berms and small stockwork gold deposits smaller berms. The width of the haul road, which generally follows the berms, is dictated by the size of trucks used which, in turn, is related to the size of the orebody and anticipated production capacity. Typical haul road benches would be between 18 m and 30 m wide. Geotechnical considerations will also be important. Wide berms effectively reduce the pit batter and thus improve the stability of weak faces. They are also an important safety measure as they prevent falling debris reaching the operational benches lower in the pit.

Pit bottom depth

No fixed rule can be applied here for the depth that an orebody can be mined to is dependent on many factors such as changes in stripping ratio (section 5.3.2) with depth, increased costs of mining and haulage with depth, the value of the metal or mineral being mined, the size of the deposit and the production and mill capacity. The depth is thus dictated either by a design optimization procedure (e.g. Lerchs and Grossman – section 5.4.2) or by physical constraints on the pit perimeter at the surface. The deeper we mine, the larger the circumference of the pit. In this situation we will eventually impinge on a river, road, building, etc. A review of operations around the world reveals that few pits exceed 300 m depth (exceptions include Lornex, British Columbia at 350 m; Nchanga in Zambia at 335 m; Berkerley Pit, Butte, Montana at 550 m).

Haul roads

These are usually added to the pit design once the depth of the pit bottom has been established. It is commenced at the bottom bench and allowed to rise across the overlying benches at gradients of 8–12%. These ramps may be 'all cut' ramps in that they cut through the benches to the pit perimeter; they may be circular ramps spiralling upwards around the pit walls, or there may be of the switch-back variety, confined to one pit wall (perhaps due to greater rock competence or load-carrying capacity). The Berkerley pit which closed in 1982, contained 25 miles of haul road allowing transport of the copper ore from depths of approximately 550 m. Figure 5.3 shows this pit and the switch-back haul road on the far face.

Fig. 5.3 The Berkeley Pit, Butte Montana. (Courtesy of Larry Dodge, Big Sky Magic, Montana.)

Figure 5.4 shows a similar haul road in RTM's Atalya Pit in the Pyrite Belt of southern Spain.

In choosing which type of haul road to use, it is necessary to realize that spiral roads may result in long haulage distances in large pits, especially if bad sections have to be avoided. Zig-zag or switch-back haul roads are ideally suited to stratiform deposits where a permanent road is needed on the face following the orebody footwall. The problem with these roads is that, where the pit wall is steep, there may not be sufficient radius to allow bends to be developed to reach the next section of the road and thus additional cut backs into waste have to be made. Such considerations may make spiral roads more economical (Taylor, 1971).

Other factors

Geotechnical information

This includes details of rock strengths, discontinuities in the rock mass and their relationship to the orientation of the face in each design sector (the potential for simple plane or wedge failure). A kinematic analysis (section 5.3.5) might reveal variations in the mode of possible failure at different levels in the pit or in different faces (e.g. rotational failure in upper benches).

Hydrogeological information

This includes annual precipitation, the size of the catchment area, groundwater recharge rates, water-table depths and fluctuations thereof, piezometric pressures, hydraulic gradients, porosity/permeability/perviousness of the intersected strata, nature of the surface drainage, the existence of confined aquifers and aquicludes, the location of old flooded workings, etc.

Overburden

What is the depth of overburden to be stripped and how variable is it in thickness?

Fig. 5.4 RTM's Atalya Pit in southern Spain showing a switch-back haul road descending the pitwall (right-hand side of photograph).

Metallurgical zones

Can areas of different metallurgical properties be recognized, and will all be mined, or will some be excluded from the pit or taken as waste or stockpile feed?

Production capacity

What should the anticipated life of the mine be? This is dictated initially by the design capacity of the mill/concentrator and by the production capacity of the mine relative to the tonnage of ore that can be encompassed by the pit design. If the total tonnage of rock in the proposed operation is 7.0 million tonnes and an operating life of 10 years is anticipated and a stripping ratio of 2.5 : 1 (tonnage of waste : tonnage of ore) has been determined, then, for a 350-day operation, the pit production would be 2000 t/d and the mill capacity needed to match this would have to be in excess of 570 t/d (2000/3.5, as 2.5 t of waste has to be mined to allow the production of 1 t of ore). The *Canadian Mining Journal* (1988) review of Canadian mining operations reveals that typical mill capacities for gold operations range between 500 and 3100 t/d; for polymetallic sulphide operations, between 1000 and 13 500 t/d (most in the range 3000 and 8000 t/d); and for porphyry copper-molybdenum operations, between 8600 and 88 000 t/d.

Physical limitations

What geographical or man-made features limit the size of the pit?

Location of waste dumps and stockpiles

The need for baffle embankments

These may be required for environmental reasons, e.g. to reduce noise or to impede 'fly-rock' during blasting or merely to reduce the visual impact of the mine site.

Location of plant areas, mill and concentrator

Care should be taken here to avoid sterilization of potential or indicated resources which may, in the future, become viable reserves with an increase in metal price or an improvement in processing technology.

Mining selectivity

The degree of selectivity that can be achieved by the introduction of front-end loaders, etc.

Tonnage factors

Careful estimates of bulk densities (wet and dry) are needed from drill cores or surface trenches and pits. Different tonnage factors may be applied to overburden, weathered rock, different metallurgical types, and competent waste rock.

Ore and waste transportation system

This could include load-haul-dump trucks (with or without trolley assist) or conveyor belt systems.

5.2.2 Manual pit design – tabular deposits

This section introduces the basic ideas behind pit design for tabular deposits via a worked example. Each stage in the process is illustrated by text figures. It should be pointed out that the design is one which attempts to encompass all the potentially accessible mineralization. No attempt is made at this stage to optimize the design from an economic point of view, a process which would almost certainly result in a smaller/shallower pit.

The orebody, which is to be mined, is conceived as being a polymetallic sulphide deposit containing gold and silver, together with minor concentrations of lead and zinc. It has a concordant base which dips uniformly to the SSW at 38°. Its continuity is disrupted in the central area by a normal fault which dips at 68° to the NW and whose down-throw is in this direction. Two exposures of the fault have been located on the hillside, as shown in Figure 5.5. Although the bedrock is close to surface in the eastern portion of the area, the western area is overlain by alluvial deposits. Only in one locality is the mineralization (now a gossan) actually exposed and this is also shown on Figure 5.5. The deposit has been explored by vertical drilling (Figure 5.6) on an offset grid based on NNE trending lines, 100 m apart, and the holes are on 100 m centres. The data available are thus the vertical thickness of the mineralized zone and the gold equivalent grade, from which vertical cm g/t values can be calculated.

Stage I

The first step in the pit design process is the production of an accurate surface contour plan based on existing topographic maps of the area, the collar coordinates of drill-holes and other spot-height surveys. On to this plan are drawn any geographical features such as roads, forests, pylons, rivers, etc. that may have an impact on pit design. In this case (Figure 5.5), the only relevant feature is a highway around the south-western corner of the map. Any available surface geological information, such as the inferred fault line, dips and mineralized outcrop, is also included. A second map can then be produced on which the sampling and drilling baseline is marked, together with all drill section lines and drill collars (Figure 5.6). The elevation of the bedrock surface, beneath the overburden (alluvial sands and gravels), can now be extracted from drill-logs and contours of this surface produced (dashed lines in Figure 5.6), together with a line representing the feather-edge of the alluvium against the hillside. The initial location of this line is based on interpolation between drill-holes but this can be confirmed by study of aerial photographs, by field traverses and by trenching or machine augering.

Stage II

The next step is to determine the elevation of the orebody footwall from the drill-logs and produce a structure contour plan using a contour interval equal to the proposed bench height for the pit and with contour values identical to those chosen for these benches. In order to construct Figure 5.7, we have chosen a 20 m bench height commencing at the 1180 m level (the bench datum). We have also assumed a uniformly

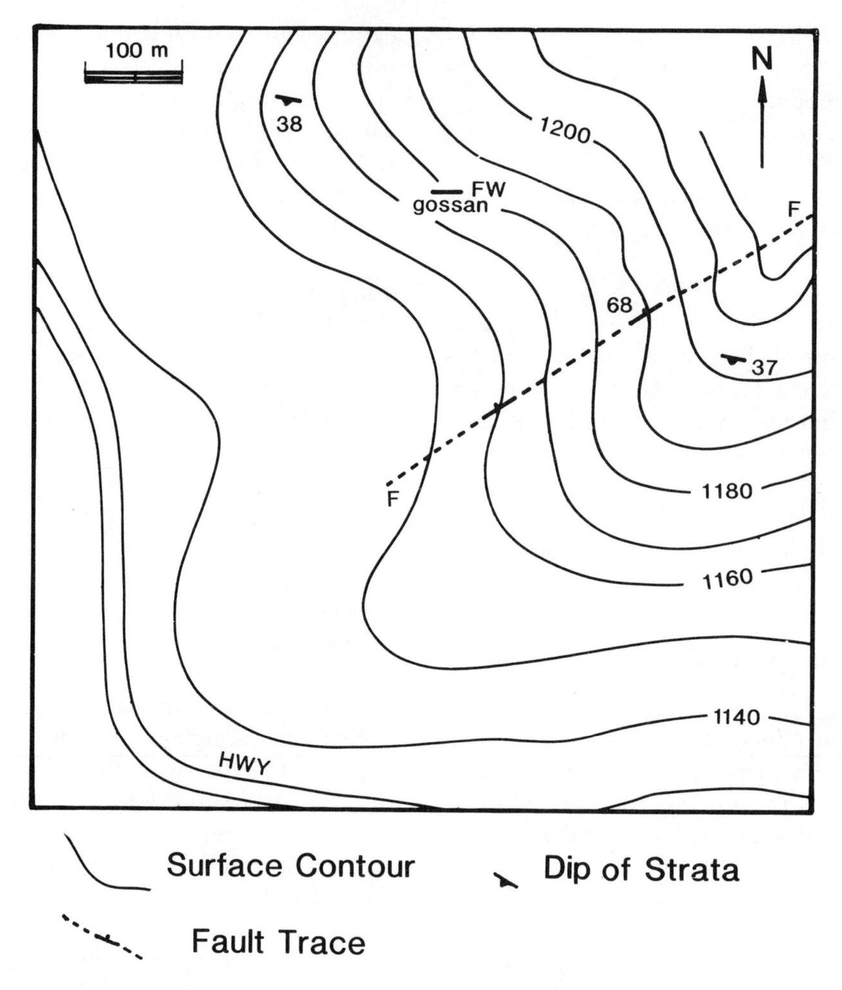

Fig. 5.5 Surface features within the site area for the proposed open-pit.

dipping orebody, with constant strike, in order to facilitate understanding and avoid the production of an excessively complex drawing at so small a scale. The contour interval, for example on a 1 : 5000 scale plan, would be 2.56 mm for a dip of 38°.

Although a 20 m bench was chosen, in reality we would probably be forced to use a lower height to reduce dilution of a tabular deposit whose thickness is generally less than 10 m. In Figure 5.8, we can see the impact of dilution if a 10 m (vertical thickness) tabular deposit is mined using 10 m and 20 m benches and if we assume that only rectangular prisms of rock could be mined by the available equipment and that 100% recovery of the ore is required. In the case of the 20 m benches, the end area of the prism is 20 × 38.4 = 768 m^2, whereas the end area of the orebody is 768–512 = 256 m^2, implying a 200% dilution of ore. In the case of the 10 m benches, the equivalent values are 256 m^2 and 128 m^2, implying a 100% dilution. It is evident that, if this dilution is to be reduced significantly, each bench will have to be mined via a series of flitches.

Using drill-hole information, combined with

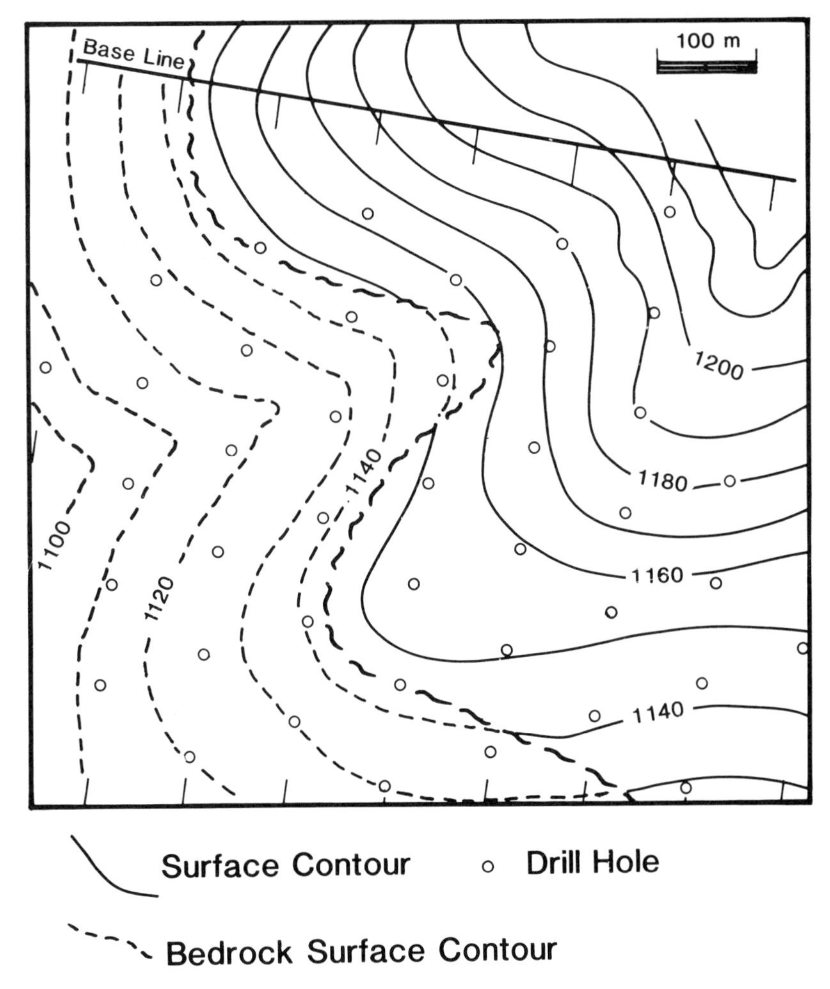

Fig. 5.6 Drill layout and bedrock topography.

the surface trace of the fault, we can now produce structure contours of the fault plane, again using the same contours chosen for the footwall structure map (Figure 5.7). In this case, as the fault dips at 68°, the contour interval will be 1.62 mm on a 1 : 5000 scale plan. It should be pointed out that, in a real life situation, a much larger plan would be produced, perhaps at a scale of 1 : 500. The intersection of the orebody footwall on either side of the fault plane can now be determined by locating the intersection of each footwall contour with the equivalent contour on the fault. The two intersection traces thus define a barren gap in plan view and provide truncation points for the orebody contours on either side of the fault.

Finally, in this stage, the outcrop traces of the orebody footwall (or its suboutcrop beneath the alluvium) can be located, again on the basis of the one outcrop marked and on the intersection of each topographical contour with the equivalent structure contour. The feather-edge of the orebody can be inserted, if required at this stage, by contouring the hangingwall of the orebody in the near surface area and repeating the above procedure to locate the hangingwall outcrop (or

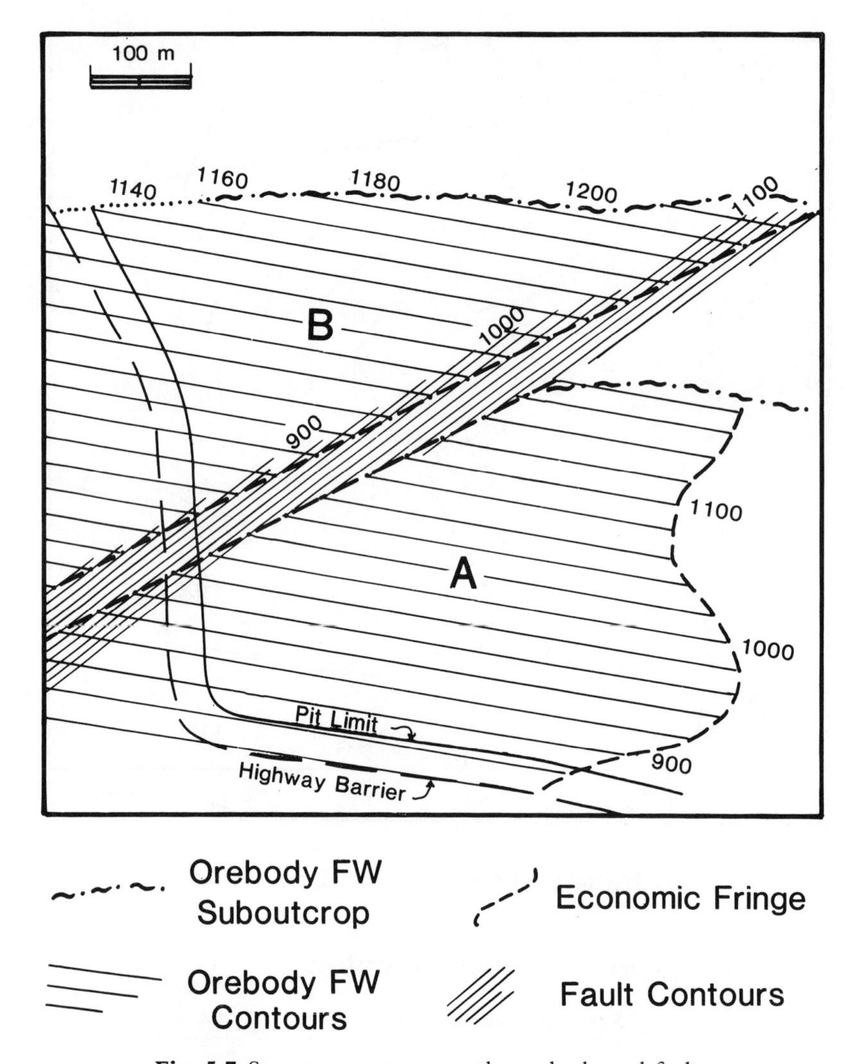

Fig. 5.7 Structure contours on the orebody and fault.

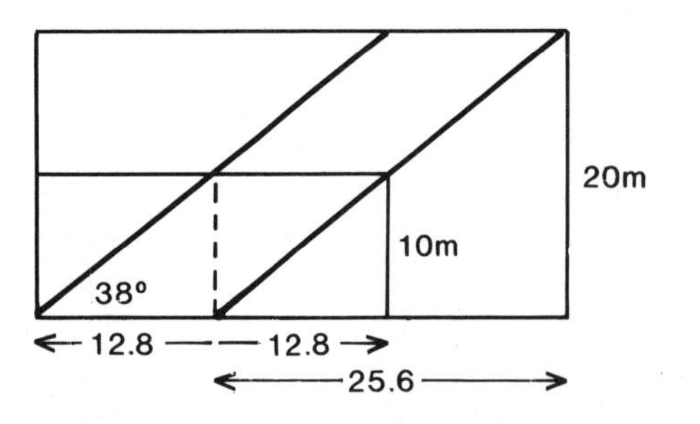

Fig. 5.8 Dilution of a tabular deposit in benches of different thickness.

sub-outcrop) trace. This is shown on both Figures 5.9 and 5.10.

Stage III

Using the outcrop traces and the fault truncation lines as limits, the vertical thickness of the mineralized zone can now be contoured (Figure 5.9), together with the metal accumulation, which, in this case, is the vertical cm g/t gold equivalent value (Figure 5.10). The latter plan can then be used to define the orebody fringe. If a limiting

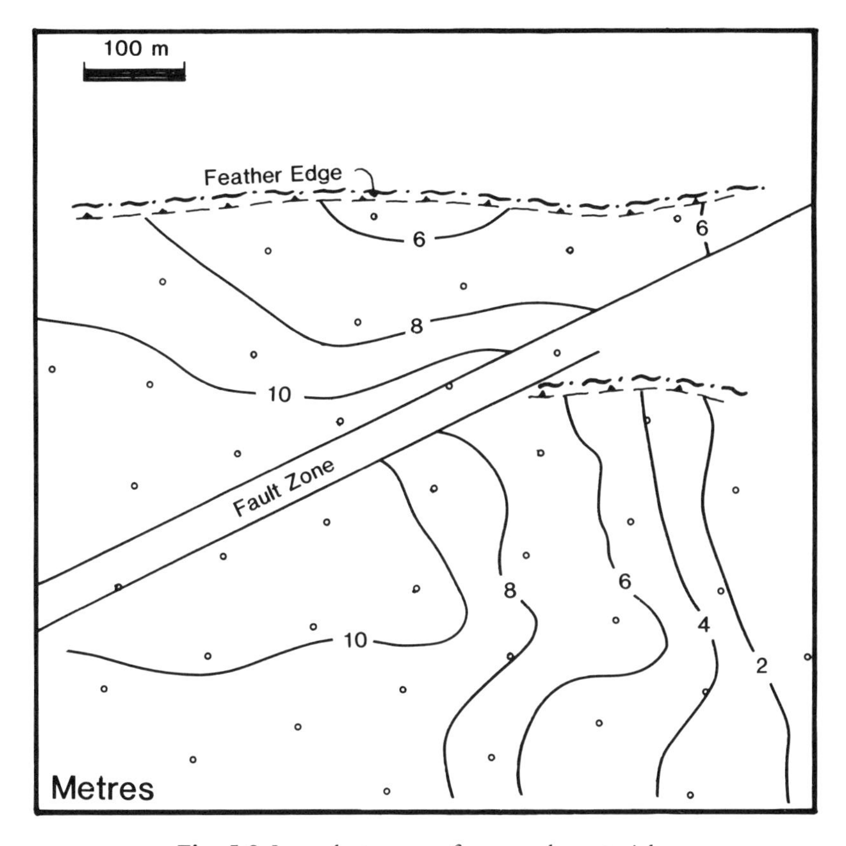

Fig. 5.9 Isopachyte map of ore-grade material.

value of 1000 cm g/t is chosen, then we can superimpose this limit on Figure 5.7 along with the geographical limits imposed by the highway in the W and S. In the latter case, a 30 m barrier has been established from the nearest edge of the highway, representing the limit of overburden stripping. As the thickness of the overburden rarely exceeds a 20 m bench in thickness, the outer limit of the pit at bedrock can also be delineated. If a 30° slope is accepted for the overburden, then this would be cut back for 35 m. We thus have a barrier which should be at least 65 m wide between the crest of the first bedrock bench and the edge of the highway. If a berm on the bedrock surface, equivalent in width to the thickness of the overburden, is used as suggested in section 5.2.1 ('Bench faces'), then this barrier will be significantly wider.

Examination of Figure 5.7 reveals that two separate areas have been defined (A and B). Thus the design exercise will initially treat these two areas separately so that each can be assessed independently. Later, the two will be integrated and the combined pit design reassessed.

Stage IV

As a 45° pit-wall has been accepted for the first initial pit design, and vertical faces assumed for each bench, then the berm width must be set at 20 m in order to maintain the required batter. A 45° line representing the pit-wall can now be extrapolated downwards from the pit perimeter limits on the down-dip side of the orebody (as defined earlier) until it meets the orebody footwall. In the case of Pit A (Figure 5.11), this occurs at the 980 m elevation. The pit bottom

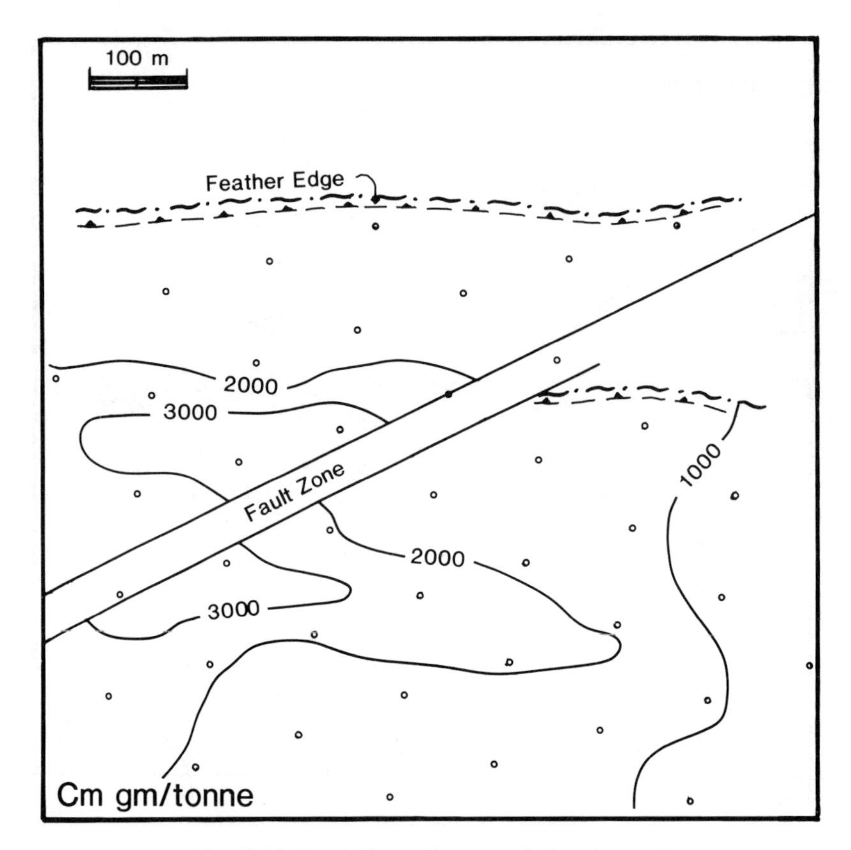

Fig. 5.10 Vertical metal accumulation (cm g/t).

perimeter is thus drawn in at the 1000 m elevation, as in Figure 5.12. This thus represents the crest of the lowest bench in this pit whose base would thus be at 980 m. The area outlined by this perimeter must, of course, be wide enough to allow for the turning circle of machines and the space needed for shovels to load the trucks. For example, a 5 m^3 shovel or a 45 t truck requires a minimum width of 35 m. Benched pit walls can now be drawn in at horizontal scale intervals equivalent to 20 m until they break the bedrock surface. In the case of the southern pit wall, this occurs at the 1120 m elevation. The pit will be essentially 'footwall-following' on its northern side so that the bench crests will coincide with the structure contours. The western wall of Pit A will then be extended downwards from the surface limits to determine the mining limit on the plane of the orebody (i.e. mineralization to the W of this limit will not be mined), whilst on the eastern fringe, as no geographical constrictions exist, the bench perimeters are built upwards at 45° from the economic fringe, as defined by the 1000 cm g/t contour, until the surface is reached. Each bench reaching the surface will thus be truncated at the equivalent surface contour. Careful linking of these benches will then produce our first attempt at a pit design.

Figure 5.12 also shows a possible access point for a ramp/haul road into the pit and thus consideration would have to be given to the possibility of a tunnel beneath the highway, roughly on the line of the bedrock valley shown in Figure 5.6. This would be cut in alluvium, and thus

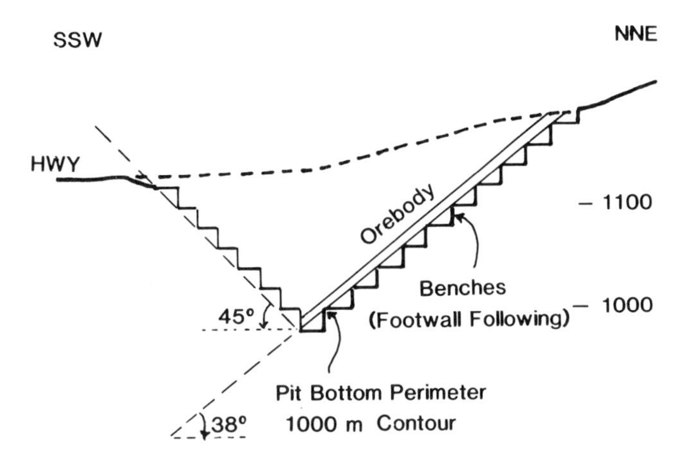

Fig. 5.11 Cross-section of Pit A showing location of benches and pit bottom.

require lining, but it would give access to a possible mill/plant site on the W side of the highway. Waste dumps could also be located in this area.

A similar procedure is used for the design of Pit B and the results are shown in Figure 5.13. Here the pit bench contours are controlled by the surface outcrop in the north, by the intersection of the orebody with the fault in the east, and by the highway barrier in the west.

The combined pit design is presented in Figure 5.14, from which it is apparent that only minor

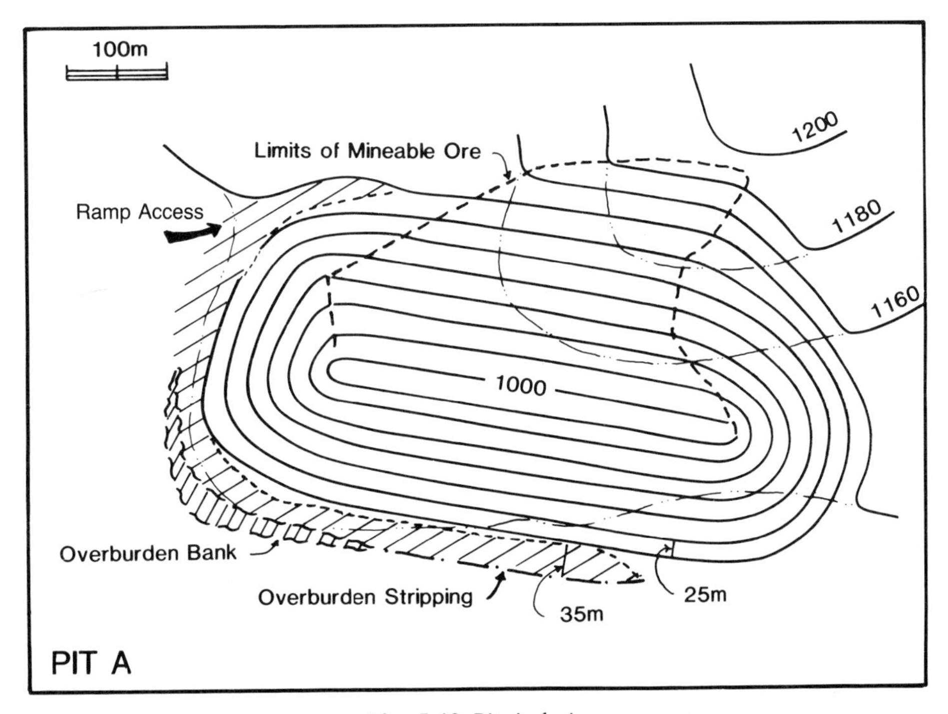

Fig. 5.12 Pit A design.

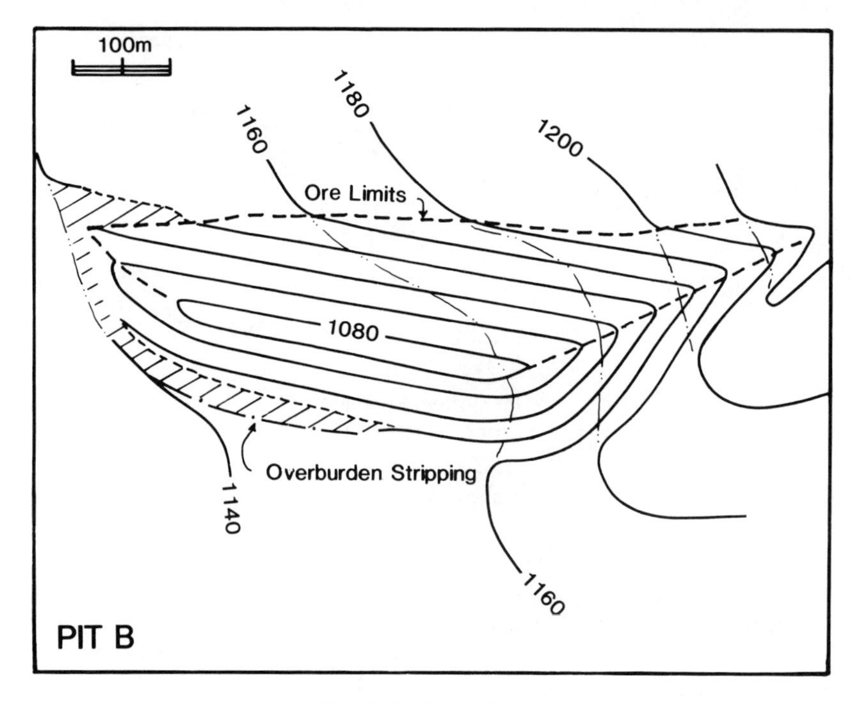

Fig. 5.13 Pit B design.

modifications were necessary to allow integration of the two. Very little overlap of the designs occurs and therefore little improvement has been achieved in the stripping ratio. In some situations, the impact can be considerably greater, for waste stripping from one pit could markedly improve the viability of an adjacent pit. This would occur because the pit wall of one pit intersects that of the other reducing the stripping ratio for this second pit.

Stage V

Once the pit designs are completed, it is necessary to compute the total volume of ore and waste (including overburden) to be removed. This can be done by planimetering each bench contour (Figure 5.15) to determine the enclosed area. Note, however, that the 1000 m contour represents the bench between the 980 m and 1000 m elevations. In the case of Pit A, planimetry would proceed up to the 1120 m contour but then the procedure would have to change at the 1140 m contour because the bench breaks surface at this level. Similarly, in Pit B, above the 1000 m contour, allowance will have to be made for the effect of surface topography. Figure 5.16 demonstrates the two methods that could be used. Here the 1160 m and 1180 m bench contour are seen intersecting the equivalent surface contours. The first method involves averaging two areas; the first of these is that enclosed by the 1180 m contour (i.e. area I in Figure 5.16) and that (area II) enclosed by the 1160 m and 1180 m bench contours and the lines joining the surface intersection points (WY and XZ). The result is then multiplied by 20 m, the bench height. Alternatively, a surface contour at the mid-bench elevation (1170 m) can be constructed and the area between this and the 1180 m bench contour planimetered. Although bench limits do not daylight on all sides of a hill in this case, had this occurred, then the volume between each pair of bench limits would have been determined by multiplying the average of the adjacent surface

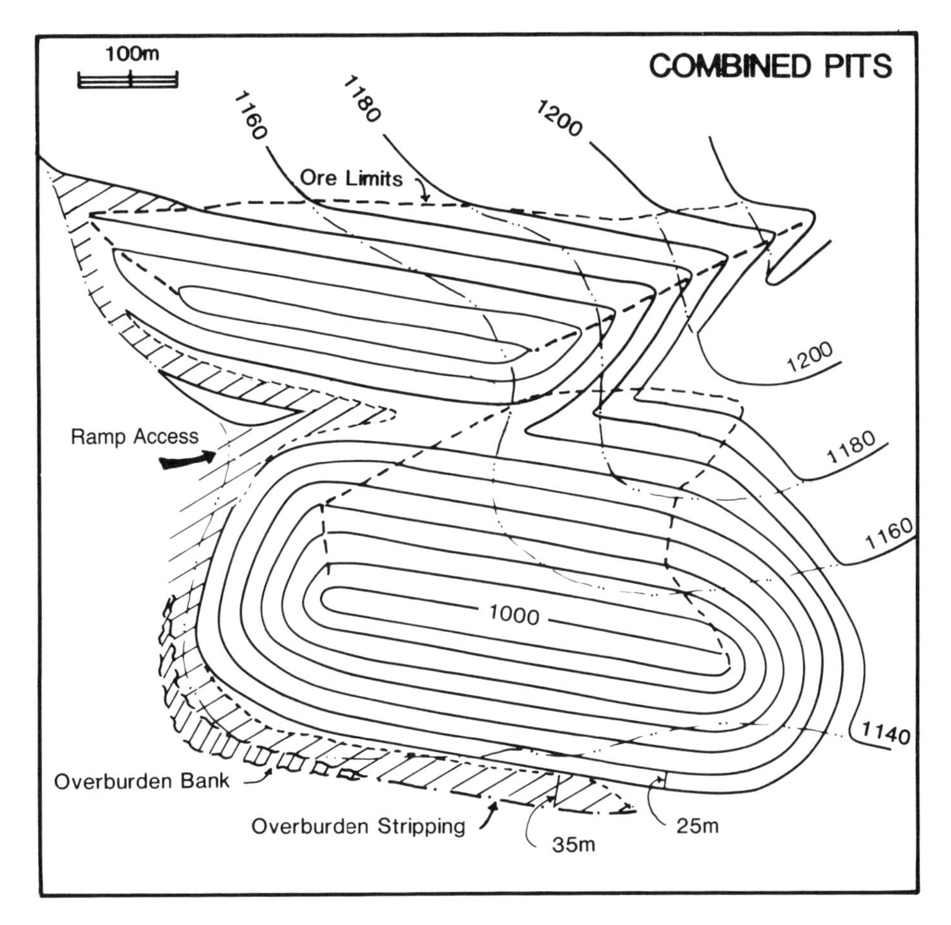

Fig. 5.14 Combined pit design.

contour areas by the bench height. Had the summit of a hill been truncated, its maximum elevation would have been estimated and the difference between this value and the bench bottom elevation calculated. The result would then have been multiplied by half the area contained by the bench bottom contour. The volume obtained would only be a rough approximation to the true value, but it would probably be adequate in most cases. A better result would be obtained by subdividing the bench into thin slices, perhaps 1 m high, and then by calculating the volume of each separately before combining into an overall volume.

The total pit volume calculated by planimetry, however, includes overburden whose specific gravity and mining cost will be different to those for ore and competent waste rock. To determine the volume of overburden, the areas enclosed by the feather-edge of the alluvials and the pit limits were covered by a matrix of 50 m square blocks. At the centre of each, or at the centre of gravity of portions of blocks lying within these limits, the surface and bedrock surface elevations were estimated from the relevant contours on Figure 5.6. A subtraction of each pair of values gave the vertical thickness of alluvials which was then multiplied by the relevant block fraction and by the block area. These products were accumulated giving the total volume of overburden. Table 5.1

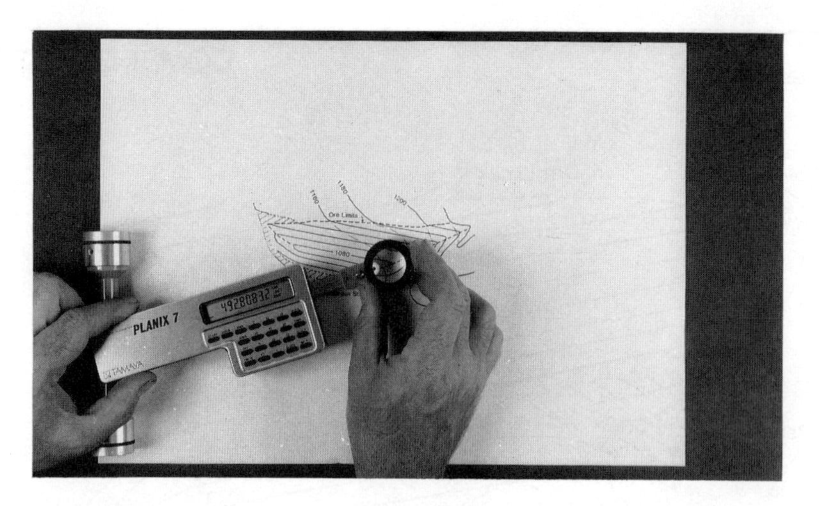

Fig. 5.15 Planimetry of the benches in Pit B to determine the total extraction volume.

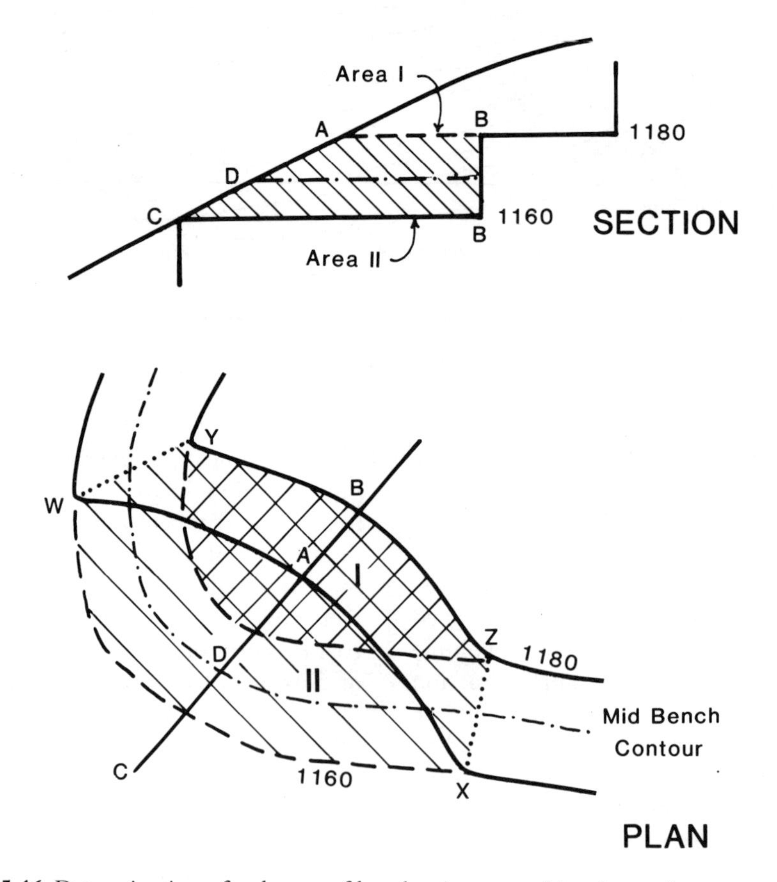

Fig. 5.16 Determination of volumes of benches intersected by the surface topography.

Table 5.1 Volume calculations for Pits A and B

Contour	Bench	Area
Pit A★		
1000	980	10 250
1020	1000	29 583
1040	1020	53 916
1060	1040	77 250
1080	1060	106 166
1100	1080	138 000
1120	1100	169 500
1140†	1120	188 250
1160	1140	95 000
1180	1160	30 250
Pit B‡		
1080	1060	7 800
1100	1080	28 200
1120	1100	54 625
1140†	1120	88 916
1160	1140	70 500
1180	1160	30 416
1200	1180	13 583
1220	1200	5 833

★Total volume = 898 165 × 20 = 17 963 300 m^3; volume of overburden = 754 750 m^3; volume of ore and waste = 17 208 550 m^3.
†Includes overburden.
‡Total volume = 299 873 × 20 = 5997 460 m^3; volume of overburden = 652 375 m^3; volume of waste rock and ore = 5345 085 m^3.

presents the results of the volume calculations for the two pits.

Stage VI

This stage involves the calculation of the reserves made available by the two pit designs, which are of course only a portion of the total resource identified. The simplest way to compute the reserves in the present case is to superimpose the ore limits on Figures 5.9 and 5.10 and to establish a matrix of ore-blocks parallel to the strike of the orebody. Figure 5.17 shows these ore limits and the 100 × 50 m block matrices. As contouring of both the thickness and metal accumulation data was possible, the grid superimposition method of ore-reserves was employed (section 3.10.1) but the block values could have been calculated using the moving window, inverse distance weighting or kriging methods, especially if the data had been more erratic in value and thus difficult to contour. The information used to calculate the reserves is listed in Table 5.2, whereas Table 5.3 presents a compilation of the results. Both were produced using Lotus 1-2-3 and a tonnage factor of 3.8 was assumed. As can be seen, the tonnages are small and the grades low. An economic appraisal of this pit will be considered later in section 5.3.8.

5.2.3 Manual pit design – irregular deposits

The manual design of pits to exploit highly irregular deposits is usually a trial and error process which is based on the modelling of the ore-deposit on sections or plans (benches) by the use of envelopes or blocks. It is a highly subjective process and is based on the geologist's intuition and experience. As in the previous example it does, however, allow an initial appraisal to be made of the potential economic viability of a deposit.

Ore-envelope method

The envelope method requires that the geologist is able to define the limits of the ore-zone on transverse sections (Figure 5.18(a)) in such a way as to enclose as much high grade ore as possible and also exclude low grade material. It will be necessary to include some waste areas but the amount of this material should be minimized. Various alternative ore outlines can be drawn in this way and each tested by determining the average grade of all assay samples (length weighted) falling within each outline. Once it proves impossible to obtain a higher grade without producing a highly irregular shape which would be impossible to mine, this outline will be accepted and the tonnage calculated (area × section spacing × tonnage factor). This exercise is

Table 5.2 Open-pit design – database for reserves

Block	V. thick	Met. acc.	Ore-fract.	VT × OF	MA × OF
Pit A					
3	4.5	1185	0.10	0.45	118.5
4	2.6	1090	0.30	0.78	327
7	7.0	1380	0.25	1.7375	345
8	5.6	1270	0.95	5.32	1206.5
9	3.0	1075	0.85	2.55	913.75
11	9.2	1625	0.15	1.38	143.75
12	7.9	1520	0.85	6.715	1292
13	6.4	1325	1.00	6.4	1325
14	4.0	1060	0.60	2.4	636
16	10.3	1770	0.75	7.725	1327.5
17	8.8	1690	1.00	8.8	1690
18	6.7	1480	1.00	6.7	1480
19	4.8	1070	0.45	2.16	481.5
21	11.2	1950	0.90	10.08	1755
22	9.3	1950	1.00	9.3	1950
23	7.1	1740	1.00	7.1	1740
24	5.3	1310	0.75	3.975	982.5
26	11.9	2810	0.85	10.115	2388.5
27	10.2	2680	1.00	10.2	2680
28	7.7	2260	1.00	7.7	2260
29	6.1	1600	1.00	6.1	1600
30	4.2	1075	0.25	1.05	268.75
Pit B					
37	6.2	1055	0.35	2.17	369.25
38	5.5	1025	0.70	3.85	717.5
41	6.5	1000	0.30	1.95	300
42	5.7	1010	0.75	4.275	757.5
43	5.8	1090	0.90	5.22	981
44	6.9	1120	1.00	6.9	1120
45	6.9	1110	0.90	6.21	999
46	5.9	1060	0.20	1.18	212
48	7.9	1300	0.85	6.715	1105
49	7.1	1400	1.00	7.1	1400
50	6.4	1500	1.00	6.4	1500
51	6.9	1550	1.00	6.9	1550
52	7.5	1460	1.00	7.5	1460
53	7.5	1290	0.35	2.625	451.5
56	8.2	1600	0.45	3.69	720
57	7.8	1690	1.00	7.8	1690
58	7.3	1750	1.00	7.3	1750
59	7.9	1800	1.00	7.9	1800
60	8.5	1675	0.55	4.675	921.25

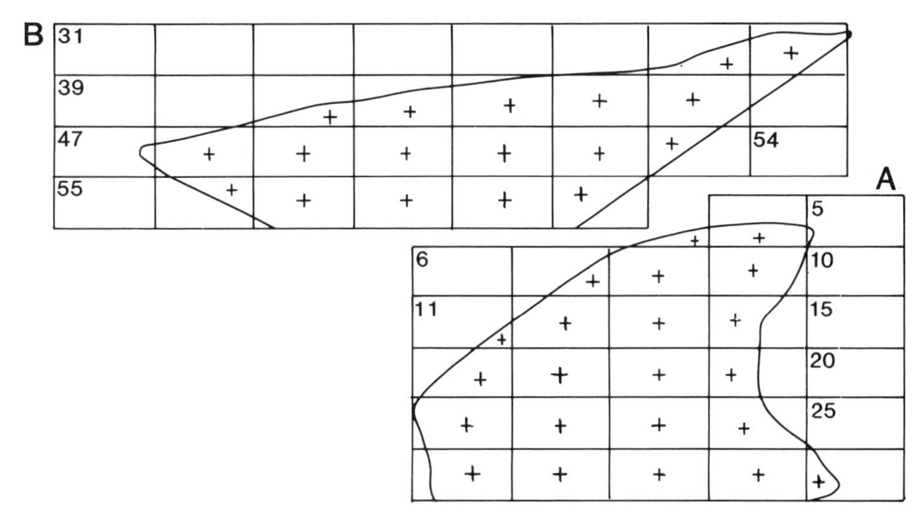

Fig. 5.17 Ore-reserve blocks and ore limits.

repeated for all sections and then the information is transferred to a series of horizontal plans drawn at the proposed bench-mid or bench limit locations. In this way, the strike termination of the deposit can also be represented (Figure 5.18(a)). A series of longitudinal sections can now be produced from the plans and transverse sections.

Having produced a 3D representation of the deposit via 2D plans and sections, the geologist can now take the transverse section on which the ore-envelope penetrates to the deepest level and

Table 5.3 Open-pit design – ore reserves. Global reserves for pits

Pit A		
Volume	593 687.5 m^3	
Tonnage	2256 012.5 t	
Grade	2.27 g/t	Gold = 164 999.2 oz
Pit B		
Volume	501 800 m^3	
Tonnage	1906 840 t	
Grade	1.97 g/t	Gold = 120 973.5 oz
Combined pit		
Volume	1095 487.5 m^3	
Tonnage	4162 852.5 t	
Grade	2.14 g/t	Gold = 285 970.9 oz
Value £68 009 759 or 106 095 224.1 US$		
Gold price 371.0 US$		
Exchange rate = 1.56 US$/lb		

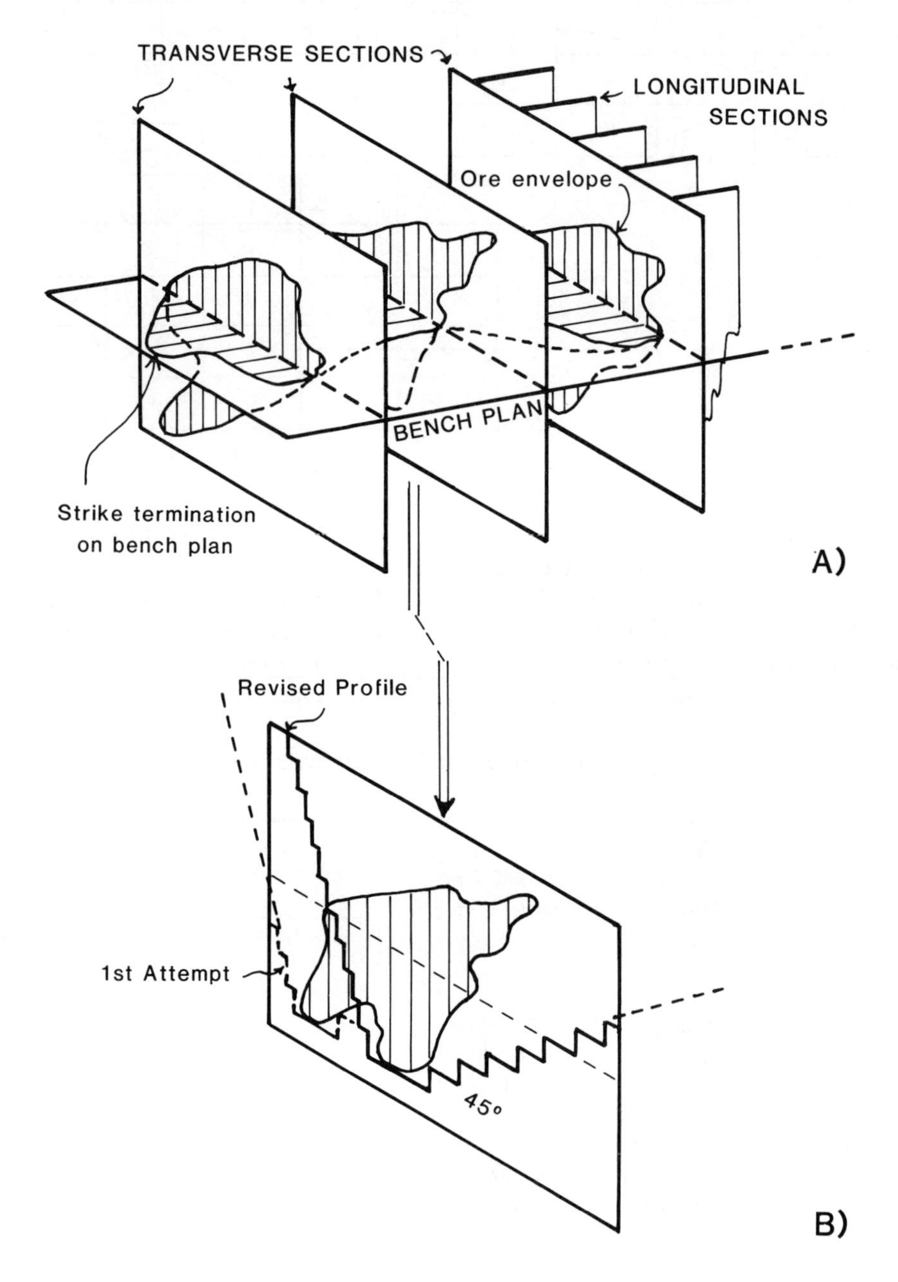

Fig. 5.18 (a) Ore envelopes defined on transverse sections. (b) Benched pit outline superimposed on section.

construct a pit bottom perimeter. From this perimeter, benched pit-walls are extended upwards at 45° or less if the shape of the ore-zone so dictates (Figure 5.15(b)). The intention is to enclose as much of the ore-zone within these pit walls as possible without also taking in too much waste in the process. The sections on either side are then benched in the same way. Rapid fluctuations in the position of the pit-walls can be avoided by superimposing groups of sections (e.g. three at a time) on a light table. It is thus useful if the sections are drawn on tracing paper/draughting film and horizontal lines representing the bench positions drawn in. The end pit-walls (along strike) can be produced in a similar manner but taking into account the positions of the pit-walls as located on the two extreme transverse sections. How successful the exercise has been can then be judged by drawing the bench perimeters on the bench plans. At this stage, it may be necessary to smooth out some of the irregularities in the outlines to facilitate mining. This may involve taking in more waste, or losing more ore than originally intended. Alternative pit designs can be made by selecting progressively shallower pit bottoms and also by omitting portions of the ore-zone which may be protruding from the main body and resulting in too much waste having to be mined on upper benches (Figure 5.15(b)). This is a very time-consuming exercise but eventually the pit can be refined and evaluated as described later in section 5.3. Further refinements, such as haul roads, sumps, etc., can be added if the results look favourable.

Block method

This method involves the creation of a 3D block matrix by inverse distance weighting methods (spherical or ellipsoidal search volumes – section 3.10) or by kriging. The block sizes should reflect the SMU (selective mining unit) to be used and should also be small enough to allow accurate definition of the shape of the orebody. Figure 5.19 shows an example of a section drawn across such a matrix showing block centres (marked with crosses) and ore grade blocks (shaded). If the ore-blocks have heights equal to the bench height, or to some exact fraction of them, then it is an easy matter to locate bench faces to enclose as much ore as possible. On this figure, an attempt has been made to follow the footwall

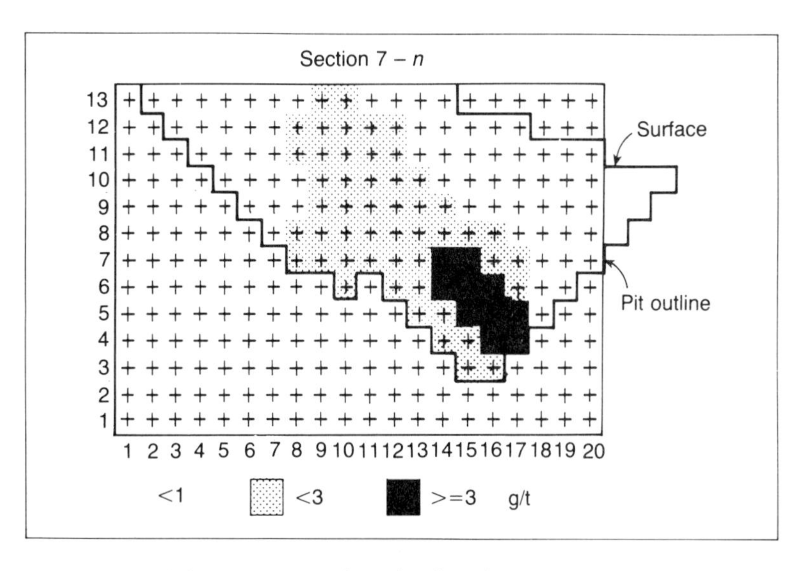

Fig. 5.19 Block section of a gold deposit in British Columbia showing grade zones and proposed pit outline.

of the ore-pod as closely as possible, before breaking away at 45° into waste. The truncation of the ore-zone on the right hand side of the diagram (due to a fault) necessitates a pit-wall inclined at 45° right to the point where it breaks the surface. In this case, the surface has been modelled by blocks equivalent in size to the ore-blocks to make computation of volumes easier. A more accurate representation of the surface profile would have been obtained by subdividing each block into sub-blocks. Other than the fact that the orebody is modelled by blocks, and thus has a stepped outline, the method used to produce a pit design is identical to that described for ore-envelopes.

5.2.4 Sectional modelling (string files)

The SURPAC software package (Surpac Mining Systems Ltd) has been produced to enable the

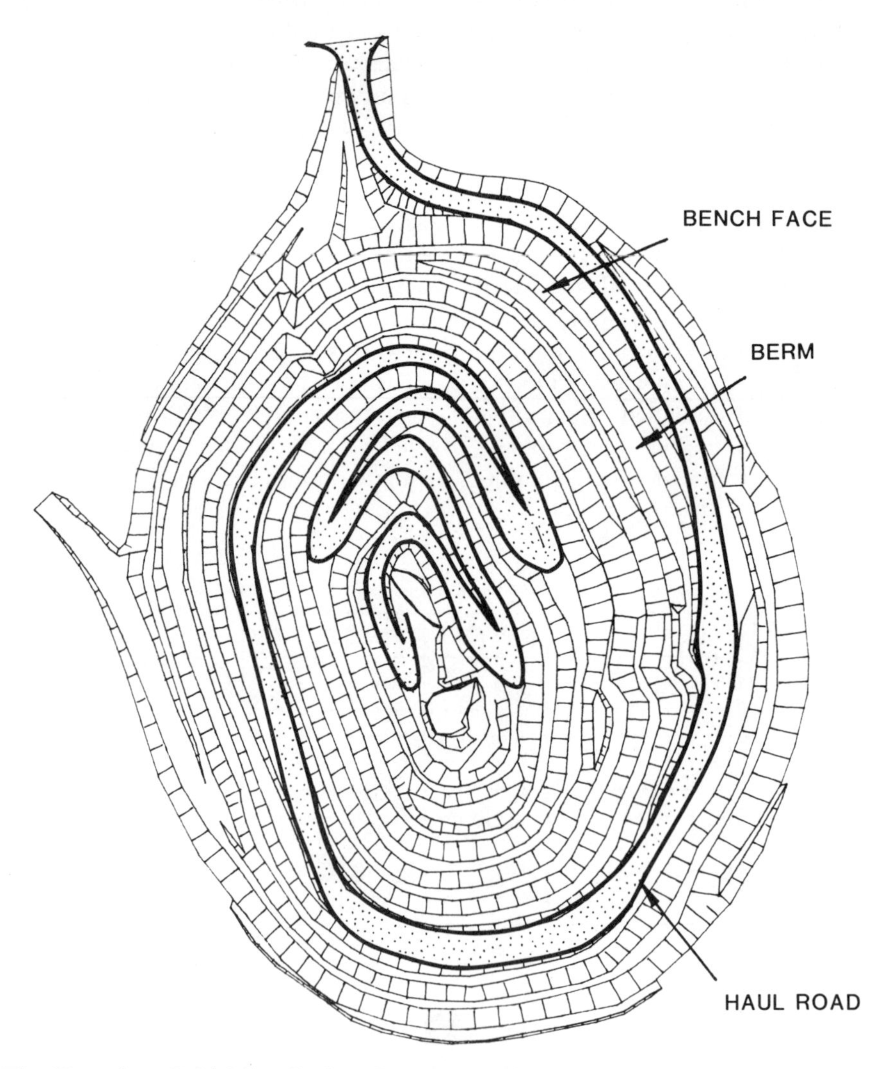

Fig. 5.20 The Horseshoe Gold Mine Project Open-pit produced from a survey string file (SURPAC). (File courtesy of Barrack Mine Management Pty Ltd.)

modelling of orebodies and the design of open-pits suitable for their exploitation. It was originally developed from a surveying package and works by manipulating data stored in files as 'strings'. A string of data is set of 3D coordinates and a point description code for each, which delineate and describe a single physical feature. Four main types of string exist:

(1) Open strings, such as a set of samples taken down a borehole.
(2) Closed strings, which can be used to represent pit crest lines, ore outlines on plan or section, or bench crests and toes (Figure 5.20 shows a survey string file produced by SURPAC of the Horseshoe Lights Open-Pit shown in Figures 5.21(a) and (b)).
(3) Closed strings with isolations, which could be used to represent several ore-zones within one bench perimeter.
(4) Random point strings, which could be used to represent a set of borehole collars.

SURPAC also has the capacity to produce digital terrain models (DTMs) in which a surface is represented by a series of non-overlapping triangles joining drill-hole collars or spot heights or both. These triangles can also be constructed so that they do not cross specified lines, referred to as breaklines, e.g. fault lines, pit contours or surface contours (stored as open or closed string files).

The first stage in the open-pit design process is the digitization of the surface contours from the site plan. This information is then combined with the drill-hole collar XYZ file and a DTM model is produced with the contours as break lines. Surface topographic sections are then produced from the DTM on lines corresponding with the drill-hole section lines. Borehole survey, assay and geological information is then extracted from the data base and joined with this topographical information to produce drill sections. The ore-zone(s) on each section is (are) then defined by the geologist either interactively on the screen or on hard copies produced by a plotter. An example is presented in Figure 5.22. Different zones can be defined on each section by assigning each a string range number, 1,2,3, etc. These may be separate ore-pods, areas of different metallurgical type or areas of different host rock lithology. These areas are digitized clockwise whilst internal waste zones are digitized in an anticlockwise direction (Figure 5.23(a)).

The next stage is to extract the assay string files for each section so that the program can calculate the length-weighted grades of all samples falling within each digitized area as defined by the location and the string range numbers. From this, a grade-tonnage report can be produced listing the geological reserves associated with each section and also for the global reserve.

The pit design process requires that the relative elevation of the base of the pit and the bench height be defined so that horizontal sections can be produced at each mid-bench level. On each of these the line intercepts of each ore-zone on each section are plotted. The geologist can then interpret or modify the orebody shape on each mid-bench plan and then digitize the outline back into SURPAC as a new closed string file. Again length-weighted grades can be determined for each zone as before. If a waste area has to be defined, then this is achieved by adding an anti-clockwise isolation as in Figures 5.23(b) and (c).

The pit plan can be divided into design sectors, i.e. areas with particular batters and berm widths as shown in Figure 5.23(d). A base of pit closed string is then produced by digitizing points clockwise so as to enclose the lowest bench-mid ore outline. At this stage, a ramp entry point could also be included to allow a clockwise ascent of the pit wall from the base (Figure 5.24(a)). The type of ramp required is selected (all-cut or spiral) together with its gradient and ramp berm width. A separate string file stores the coordinates of all the ramp entry points as each bench outline is generated during the pit design process. The two strings shown in Figure 5.24(a) are then joined and a crest line produced for this lowest bench. The next mid-bench ore string is then superimposed and the next toe and crest lines automatically generated on the basis of

Fig. 5.21 (a) Vertical aerial photograph of the Horseshoe Lights Open-pit near Meekathara, Western Australia, showing spiral ramp. A plan of this pit, based on a SURPAC string file, is presented in Fig. 5.20.

the specifications and design sectors provided earlier. The user can modify this new toe line to take in more ore, should this be necessary (Figure 5.24(b)). The impact of this change can be judged by checking the overlying bench-mid plan. This may reveal that the modification does not take in sufficient ore to warrant the extra waste that will have to be mined at the higher levels and thus that the original crest line should be restored. This procedure is repeated for all the benches until the pit intersects the surface. String files of the mid-bench surface contours would have been produced at an earlier stage for those benches likely to cut the surface. These allow the truncation of the pit design, at the higher levels, where two strings of equal altitude intersect. It should be noted that the parameters set for each design sector could be made specific for a particular grouping of benches. Higher benches could thus be given different batters by creating a second design sector/slope file and repeating the above procedure with a new pit bottom perimeter (equivalent to the top bench in the underlying group of benches) and bench height, if necessary. This allows weathered rocks or unconsolidated overburden to be given different slopes, berms and even bench heights.

Once all the benches have been designed in this

Fig. 5.21 (b) Oblique shot of the pit showing location of the ore and waste dumps, spiral ramps and benches. (Photographs courtesy of Barrack Mine Management Pty Ltd.)

way, all the toe and crest outlines for all the benches are displayed and the ramp string file appended. This model can now be examined on the basis of transverse or longitudinal sections or in perspective, from different view points, to provide a 3D image of the pit (Figure 5.25).

The program then creates an ore and waste model by intersecting the ore outline(s) with the pit outline for each bench, as demonstrated in Figures 5.24(c) and (d).

The grades and tonnages of material encompassed by the pit design can now be calculated on the basis of material type (up to 29 types possible). Material type is often just defined on the basis of grade ranges but it could be based on metallurgical zones as defined on the original cross-sections and on the level plans. Each material type can also be assigned its own specific tonnage factor. The results listing can be on the basis of a bench by bench analysis or by groups of benches or the whole pit. Volumes, tonnages and grade are listed for each material type class (which includes a waste category) in each bench. The design can now be assessed on the basis of economic and technical criteria and then modified to maximize the financial return over the proposed life of the pit.

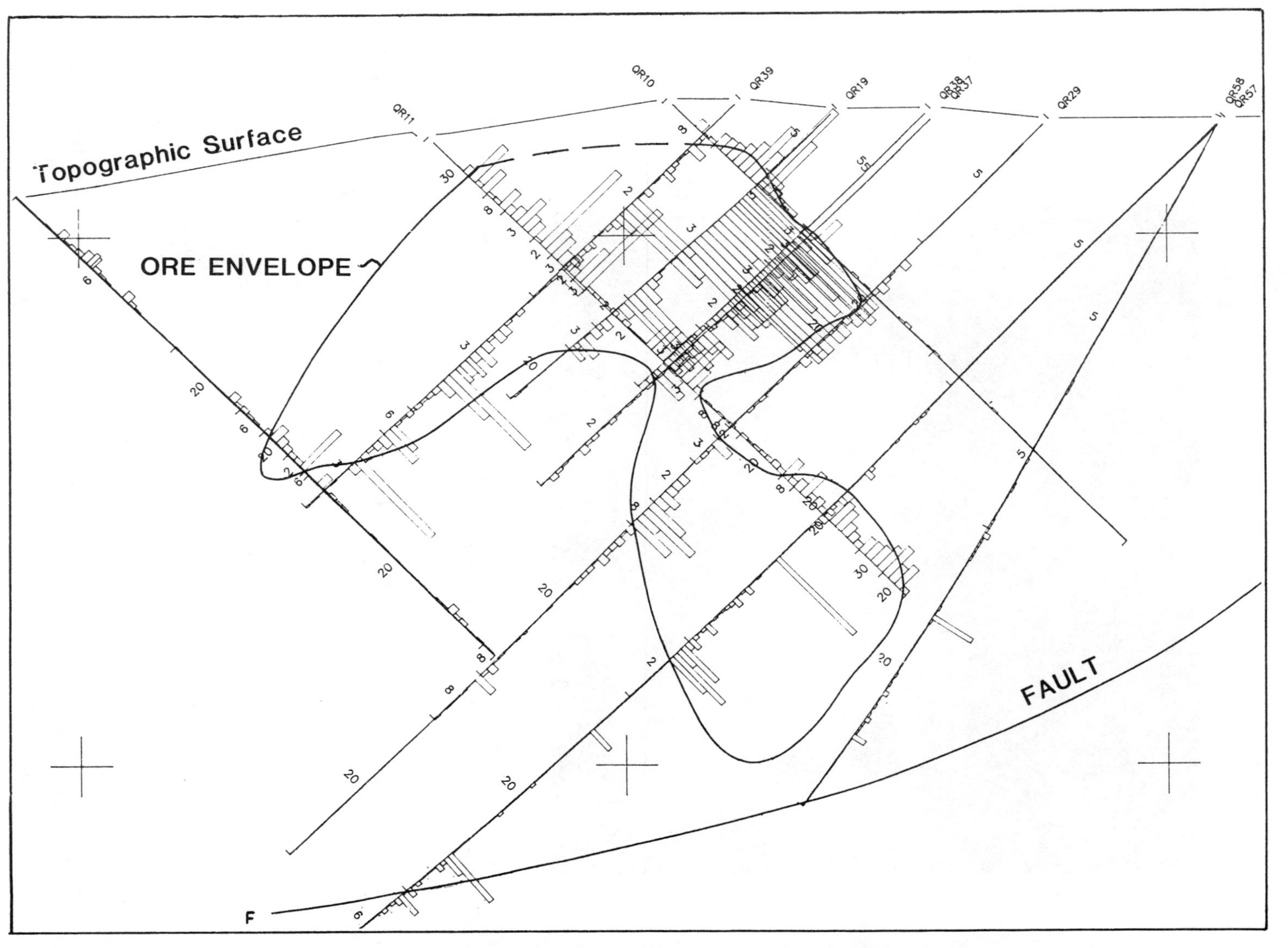

Fig. 5.22 An assay section produced by SURPAC on which the ore-envelope has been manually inserted prior to digitization. Codes on drill traces indicate rock or alteration types. The information is taken from a potential gold operation in British Columbia. (Courtesy of Fox Geological Consultants Ltd, Vancouver.)

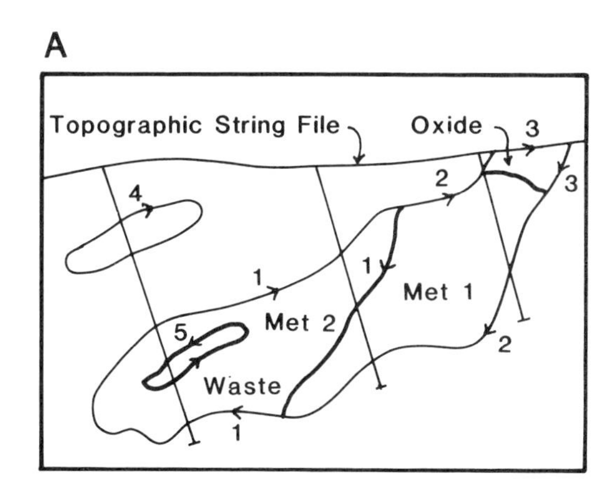

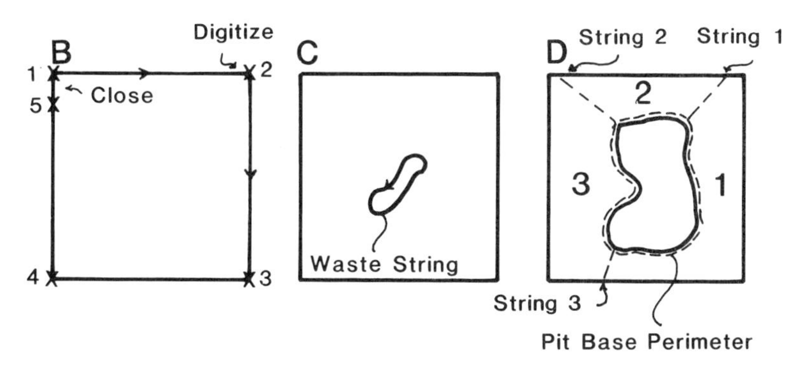

Fig. 5.23 (a) Outlining ore and metallurgical zones using SURPAC string files. (b), (c) Production of a waste string. (d) Definition of pit slope sectors.

5.2.5 Computerized 3D block modelling

The DATAMINE integrated mine planning system (Mineral Industries Computing Ltd/ DATAMINE International) has been discussed in general terms along with SURPAC in section 1.15. It is a modular system which includes the facility for the production of 3D block models of a deposit either directly, or via the construction of solid wireframe models, as described later. Each of these blocks may have values representing variables, such as grade. Grade can be calculated by a number of methods which include kriging, inverse distance weighting and polygonal weighting. Other facilities within the system include pit design, haul road design and the economic optimization of the pit design using the 3D Lerchs and Grossmann algorithm.

The DATAMINE block model is a 3D structure of cuboid cells each of which has a size and one or a number of values, such as grade, associated with it. The attractive feature of this structure is that the cells can be selectively split into sub-cells of different size and shape to obtain a better representation of a geological structure, ore-deposit or excavation (underground or surface). Because only the cells at the boundary of

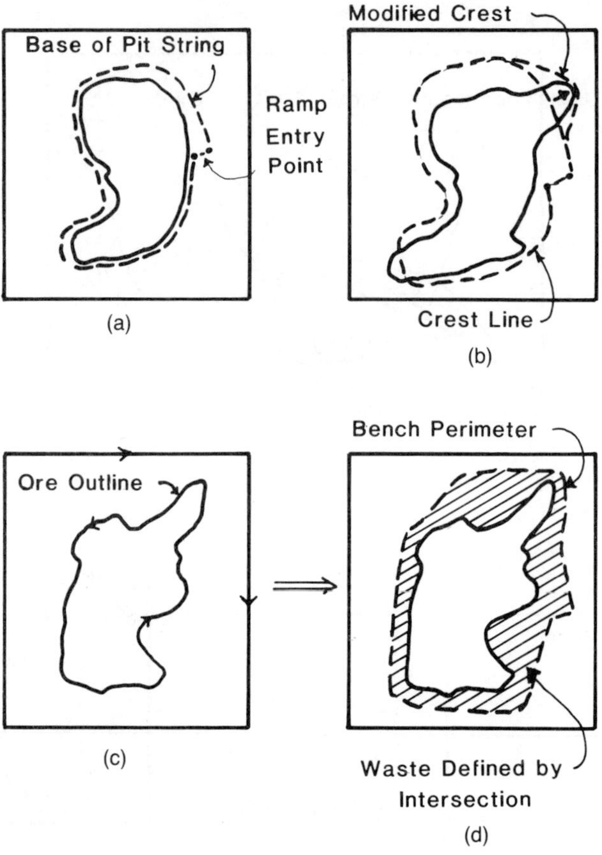

Fig. 5.24 (a) Base of pit string and ramp entry point digitized around ore outline. (b) Modification of second crest line to take in more ore. (c), (d) Creation of an ore and waste model for a bench.

the physical feature need to be subdivided, a considerable saving on computer time and media space is achieved. With an inflexible 3D block matrix, all blocks would have to be reduced in size to achieve the same degree of fit to a geological shape, thus increasing enormously the number of calculations needed and the space required to store the model. The blocks may be given dimensions which allow them to match the proposed bench heights for the pit or allow them to be subdivided so that two or more exactly span the height of the bench. To further reduce calculation time, the drill-hole assay grades may be composited over lengths equivalent to these bench heights. This reduces the coefficient of variation and variance of the assay population and produces a data set with constant geometric support amenable to geostatistical modelling.

The block model can be limited vertically by modelling the surface topography by interpolation between spot heights and then representing it by cells and sub-cells below the surface only. Alternatively, DATAMINE has a DTM facility which uses an irregular mesh of triangles to represent a surface, each constructed by joining adjacent surface spot heights with straight lines. The block model of cells and sub-cells can then be created underneath this surface.

Although such a block model facilitates the evaluation of individual blocks of ground, its shortcoming is that, even with the use of the cell-splitting option, it does not accurately represent

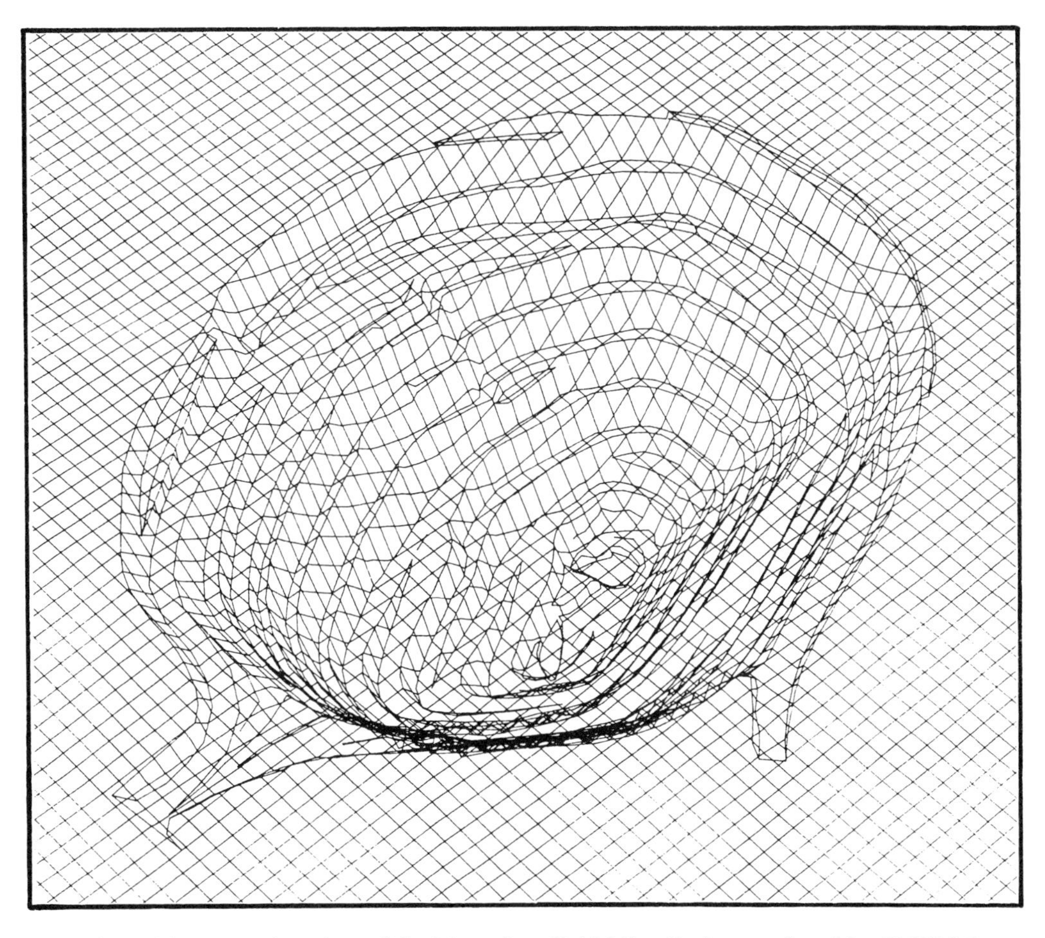

Fig. 5.25 Perspective view of the Horseshoe Gold Mine Project produced by SURPAC.

the shape of the deposit under evaluation. DATAMINE has overcome this by employing a technique used extensively in CAD systems, i.e. wireframe modelling. This is equivalent to the production of surface DTMs in that the shape of the orebody is defined by triangular facets produced by joining points on its surface using a semi-automatic Delaunay triangulation technique. In Figure 5.26, wireframe modelling has been used to produce an isometric projection of underground driveages and stopes. Excavation and orebody volumes can now be determined directly.

The geological representation and grade evaluation of an orebody can now be undertaken by generating a block model structure within the limits of the wireframe model. Splitting of cells intersected by the surface of the wireframe model allows a close fit to be achieved (Figures 5.27(a), (b), 5.28). Alternatively, an existing 3D block model can be intersected by a wireframe model so that only those blocks within its limits are evaluated. In this way, the amount of computer time spent in calculating the values of unwanted blocks can be reduced. The above assumes, of course, that the orebody limits are clearly defined at an early stage and can thus be modelled in this way. Henley and Wheeler (1988) provide further

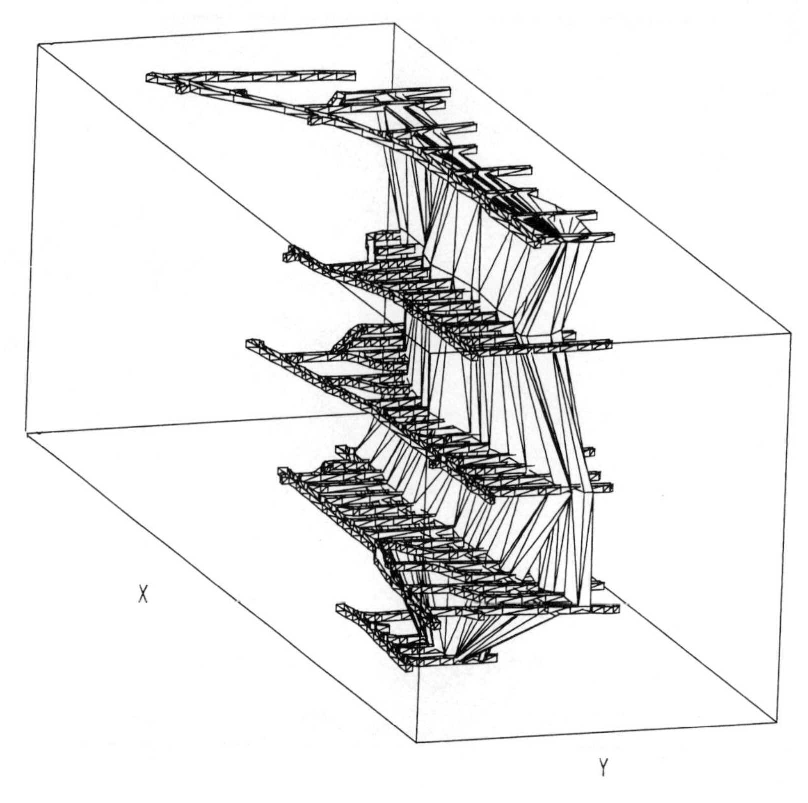

Fig. 5.26 Wireframe model (isometric projection) of underground driveages and stopes produced by DATA-MINE. (Courtesy of Datamine International.)

examples of the use of wireframe methods. Where the limits of potentially economic mineralization are ill defined, then an alternative approach is to form boundaries around groups of blocks which meet a certain cut-off criteria. This technique can be refined by allowing blocks cut by this perimeter to be reduced to sub-cells, redetermining grade for these sub-cells and then allowing the perimeter to be redrawn to exclude more low grade/waste material. Iteratively, a perimeter is produced which encompasses as much above cut-off grade material as possible (Figure 5.29). This is repeated for all sections and then the set of 'cut-off boundaries' is linked and a wireframe model of the orebody created. Combining all the blocks within the wireframe allows the global grade and tonnage of the orebody to be determined.

Pit design can be achieved manually on the graphics screen working through the block model from level to level, or it can be produced using the floating cone method as described in section 5.4.1. An enhanced open-pit design feature allows multiple pit design and evaluation using different bench heights and separate toe and crest perimeters for each bench. Haul roads can be inserted via an interactive graphics process and the pit design re-evaluated. Figure 5.30 shows a typical pit design produced by DATA-MINE in which optimization has been achieved using the Lerchs and Grossmann technique, described later in section 5.4.2. A case history of

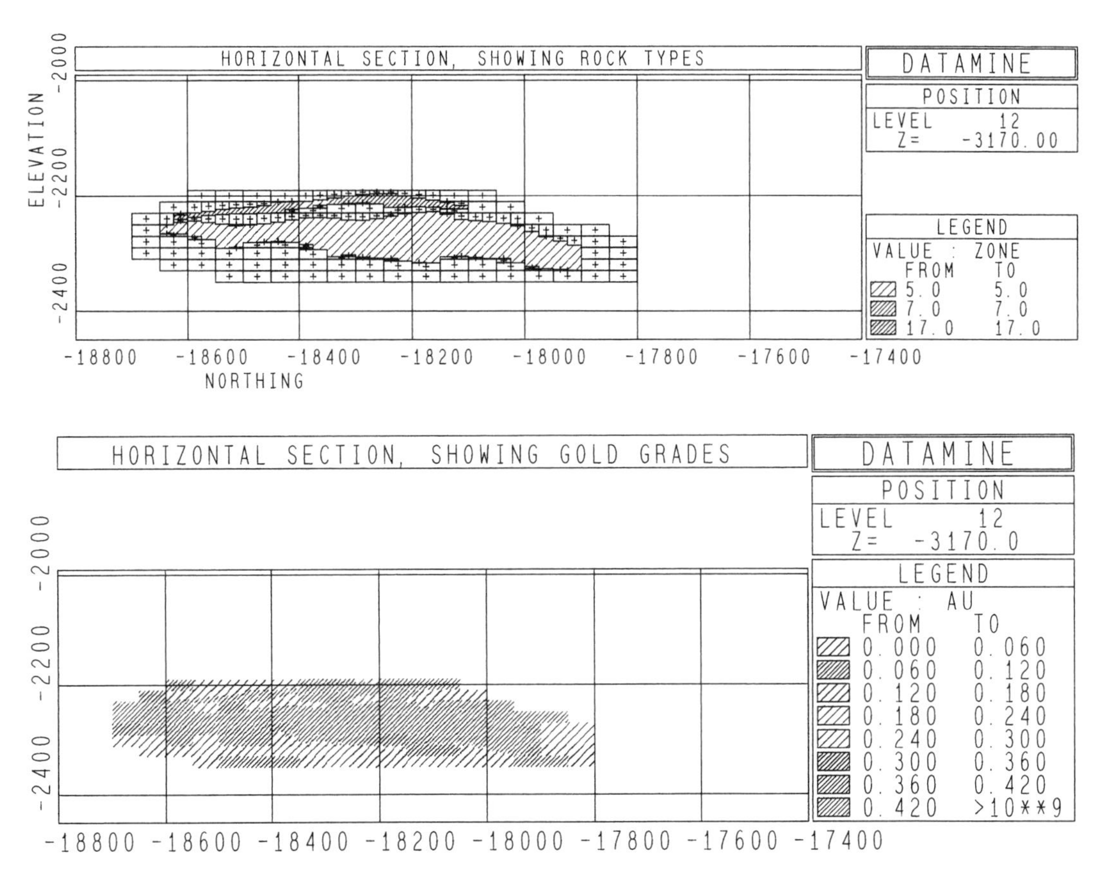

Fig. 5.27 Horizontal geological and grade sections produced from the wireframe model shown in Fig. 5.26. (Courtesy of Datamine International.)

the use of DATAMINE for the design of a limestone quarry is provided in Chapter 8 (section 8.8).

5.2.6 3D solids modelling

The LYNX Mining System (Lynx Geosystems Inc., Vancouver) allows the modelling of a mineral deposit via components or 3D solid, but irregularly shaped, blocks. These solids are created one at a time and each is centred on an interpretative section of the deposit produced at an earlier stage in the process. Tie-lines are inserted between adjacent sections using the mouse to correlate equivalent features on the upper and lower surfaces of the deposit. The system then auto-interpolates the profile of the deposit on to vertical planes located half way between the two adjacent sections: these are referred to as the back-plane and the fore-plane. These then become the end areas of a block centred on the section line whose upper and lower surfaces are defined by the tie-lines.

Block volumes can be determined by the solids of integration theory based on solid geometry and integral calculus. Grades can be determined for 3D cuboid matrices using inverse distance weighting (spherical, discoidal or ellipsoidal search volumes) or kriging methods. This cuboid model is then intersected with the 3D solid model created above to determine the overall grade, or by a mining cut model (e.g. for the

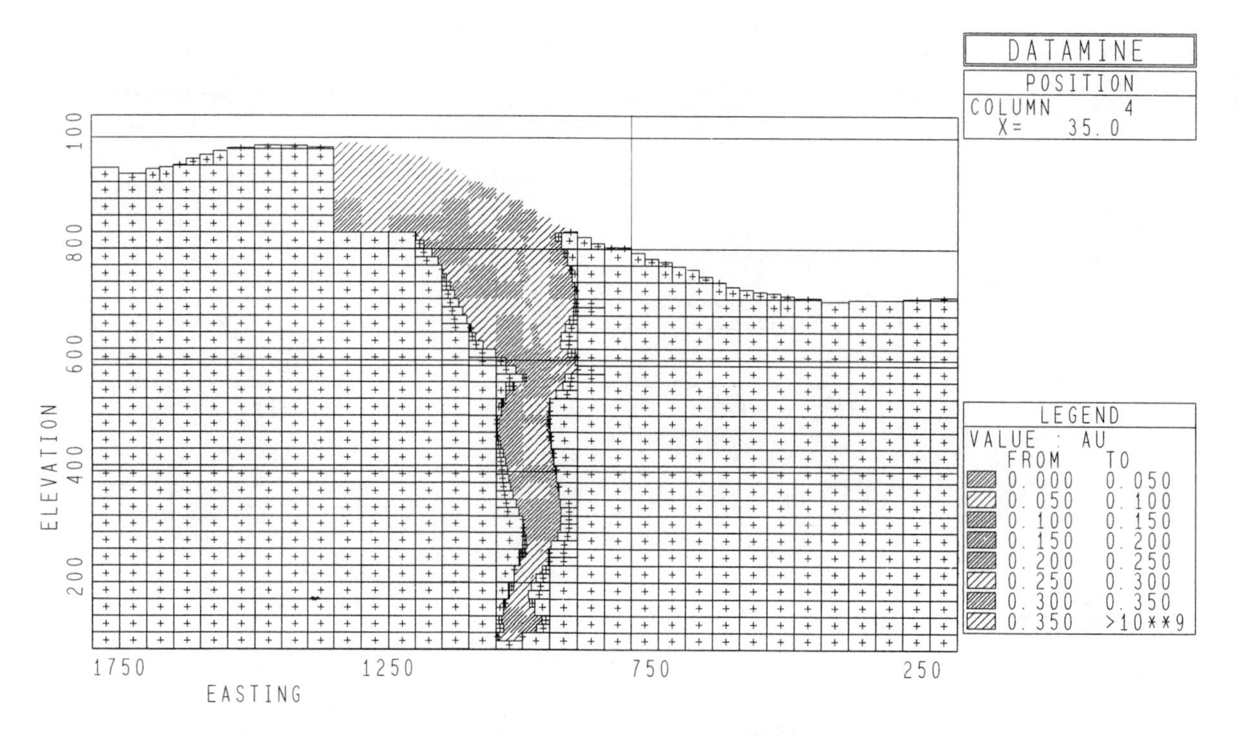

Fig. 5.28 Block modelling of an orebody via the wireframe technique. (Courtesy of Datamine International.)

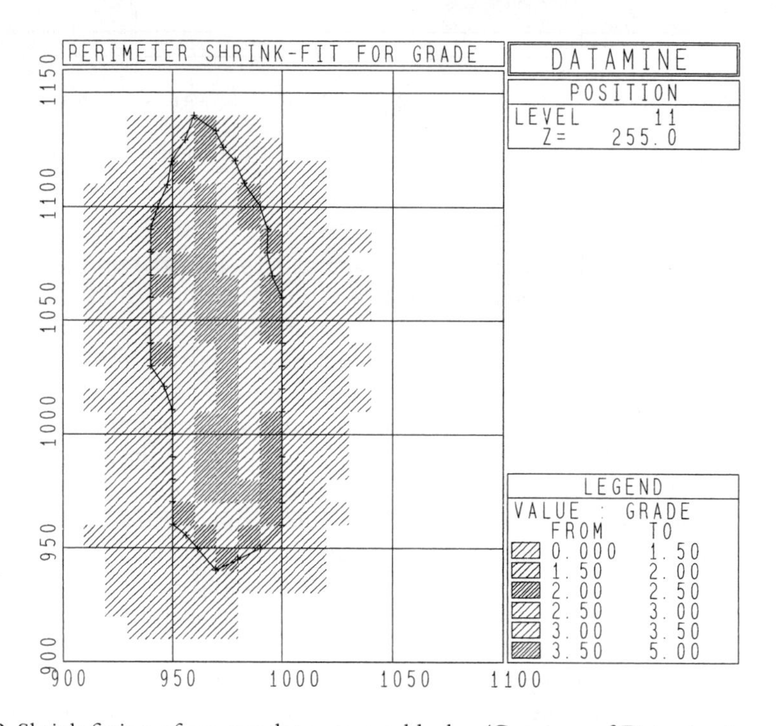

Fig. 5.29 Shrink fitting of ore-envelopes to ore-blocks. (Courtesy of Datamine International.)

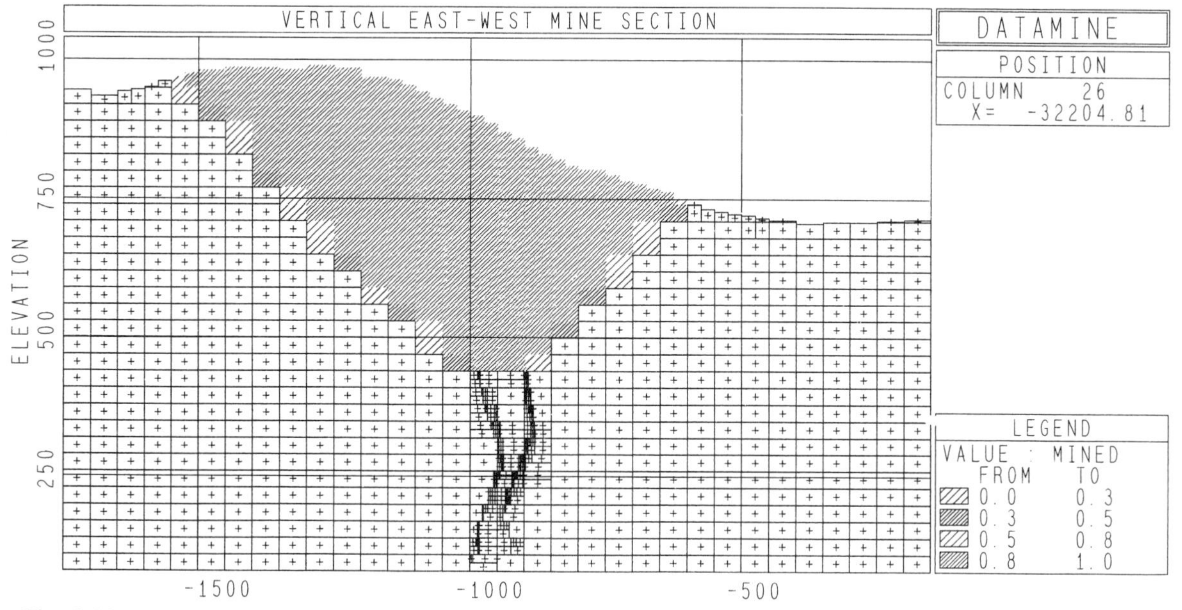

Fig. 5.30 Optimum pit design produced by DATAMINE using the Lerchs–Grossmann technique. (Courtesy of Datamine International.) The original block model is shown in Fig. 5.26.

first bench), to determine the grade at each stage in the mining process. The overall pit design is thus built up from a series of 3D solids, each representing a bench whose outline has been determined by examination of the geological sections through the deposit.

5.3 EVALUATION OF OPEN-PIT OPERATIONS

Once a preliminary design has been accomplished for the proposed open-pit, it is necessary to test the practical and economic viability of this design. Later, in section 5.4, we will discuss the various methods of optimization of this design on the basis of economic and technical criteria.

Although the factors which affect the economic viability of the operation can be listed separately, as below, they are multidimensional and interdependent. Changes in one can have considerable impact on the others. The critical factors include:

(1) metal/mineral price;
(2) market demand;
(3) finance available;
(4) tonnages and grades available at different cut-off grades;
(5) stripping ratios at different cut-off grades;
(6) sequence and method of mining;
(7) anticipated life of the operation;
(8) mining and processing costs;
(9) mining and processing capacity throughout the proposed life of the operation.

Space does not allow a detailed discussion of the factors which affect consumption, market demand and price. However, financing of mining operations will be discussed in Chapter 6. This section will thus concentrate on a review of the technical and economic parameters which need to be taken into account.

5.3.1 Sequence of mining

The order in which ore and waste blocks are mined in a pit will have an impact on the relative amounts of waste and ore mined at any one time and thus on the magnitude of feed to the mill. At the same time, the grade of ore will vary with the ore blocks mined and this will thus have an impact on the production, and revenue derived from, the valuable metal or mineral. The optimum mining plan will thus have to be determined by using different production rates and concentrator capacities as well as by changing the order in which ore and waste blocks are mined. The optimum plan will be that which maximizes the NPV (net present value) of the operation (section 6.5.3) on the basis of a realistic discount rate (e.g. 10–15%). The end product of the exercise will thus be an inventory of reserves which increments on a yearly basis. Thus, for each year of the mining plan, we will have the tonnage and grade of ore mined by bench, and by metallurgical zone in each bench, together with the tonnage (and sometimes grade) of waste.

The impact that the sequence of mining has on the waste to ore ratio can be illustrated by reference to Figure 5.31(a) and (b). If each bench is cut back to its ultimate design position before commencing the next, then it is obvious that in the early years of the operation we will be mining large amounts of waste in today's dollars or pounds. Much of the ore below, that is released for extraction, will not be mined for many years and hence, because of discounting, it may have a value which is less than that required to pay for the earlier waste removal. A mining method which calls for incremental push backs of the bench perimeters is thus considered a better policy. (Figure 5.31(b)). Alternatively, bench batters in waste will be given a lower angle at early stages in the operation and will later be cut back to their final positions once the lowest benches are being mined as in Figure 5.31(c).

5.3.2 Stripping ratio and strip index

Both of these parameters give a valuable indication as to the potential viability of an open-pit operation. Stripping ratio is defined as:

$$\text{Tonnes of waste/tonnes of ore} \qquad (1)$$

This definition is preferred to one which is based on volumes only, as it allows for the different tonnage factors of waste and ore. Also, within the waste category, we have the possibility of several different rock types with different factors together with weathered rock and residual or transported overburden with generally much lower tonnage factors than for competent rock. Frequently, the stripping ratio is calculated as follows:

$$\text{(Tonnage of rock in pit} - \text{tonnage of ore released)/tonnage of ore} \qquad (2)$$

Care must be exercised in using formula (2) for the tonnage of ore calculated (especially in the case of inclined tabular deposits) may be the *in situ* reserves based on the geologist's recognition of hangingwall and footwall assay cut-offs or an ore-envelope. It thus does not allow for bench dilution, as referred to in section 5.2.2, and as a result the tonnage of waste is less than calculated by the above formula by an amount equal to the dilution of the ore. Some of the waste is thus trucked to the mill, and not to the waste tip, and thus incurs an additional milling cost.

The *Canadian Mining Journal* (1988) review of open-pit operations in Canada reveals that stripping ratios in gold deposits are generally less than 8.0 : 1 whereas the ratios for copper-molybdenum porphyries are less than 2.1 : 1, with many less than 1 : 1. Worldwide, stripping ratios in excess of 10 : 1 are rare. On their own, these ratios are not particularly meaningful as they do not take into account the grade or value of the contained metal or mineral. A high grade or high value deposit could thus tolerate a higher stripping ratio than a low grade/value deposit. As a result the strip index is more useful, i.e.:

$$\text{Stripping ratio/grade} \qquad (3)$$

The lower the strip index, the greater the economic potential of the pit.

In order to assess the stripping ratio calculated for a given pit, we need to determine the break-even stripping ratio. This occurs when the profit produced equals zero or:

$$\text{Profit} = RV - (M + P) - \text{SR} \times W = 0 \qquad (4)$$

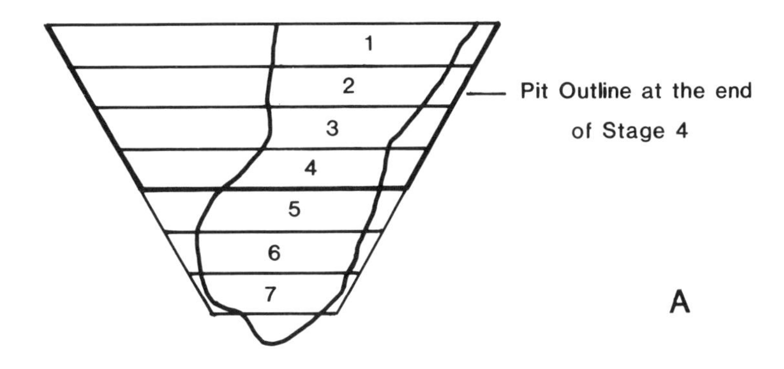

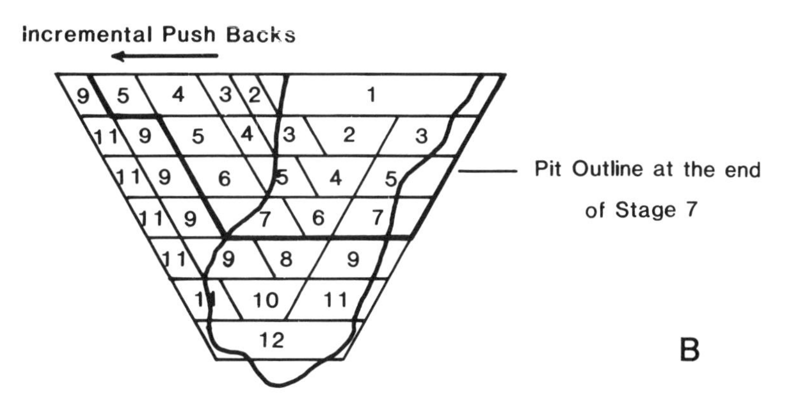

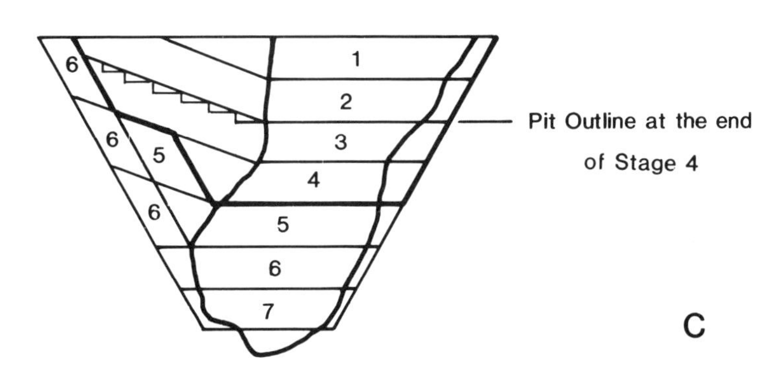

Fig. 5.31 (a) Mining sequence in which each bench is mined completely before proceeding to the next. (b) Mining sequence with incremental push backs of the pit wall to the ultimate position. (c) Use of lower slopes to reduce the amount of waste mined at early stages.

where:

RV is the recoverable value per tonne of ore calculated from the product of grade G (g/t or percentage ÷ 100), metallurgical recovery MR (expressed as a decimal) and the current or projected metal/mineral price (e.g. \$ per gram or £ per tonne). The grade in this case should be reduced to allow for mining dilution.
M is the combined mining and transport cost for the ore, plus an element of the administrative overheads.
P is the total cost of milling, concentrating, smelting and refining, again with an element of the administrative overheads.
SR is the stripping ratio.
W is the stripping cost per tonne of waste.

Hence,

$$\text{Breakeven stripping ratio } (BSR) = \{RV - (M + P)\}/W \quad (5)$$

An example of the calculation of breakeven stripping ratio can be found in section 5.3.8.

Breakeven stripping ratios can be used to gain an approximate estimate of the maximum depth to which a podiform or tabular orebody can be mined economically in an open-pit. It is assumed that mining will follow the footwall of the orebody and that the opposite face has a batter of 45°. This method is based on the requirement that the revenue generated from each block of ore meets the cost of mining the additional waste that must be excavated to expose this block. The ore-block in this case can be defined by a depth increment (DI), e.g. 5 m, and by the horizontal width of the orebody (HW) plus an additional amount ($DI/\tan\theta$) to allow for mining dilution (Figure 5.32). This is necessary as the calculation of RV in equation (5) involves the use of a diluted grade. For a mid-block depth BD the cross-sectional area of additional waste involved is approximately $2 \times BD \times DI - D/2$ where $D = DI^2/\tan\theta$. The error in approximation increases as the orebody dip θ deviates from 45° but generally is small. The end area of the ore block is $DI \times HW + D$, thus the stripping ratio associ-

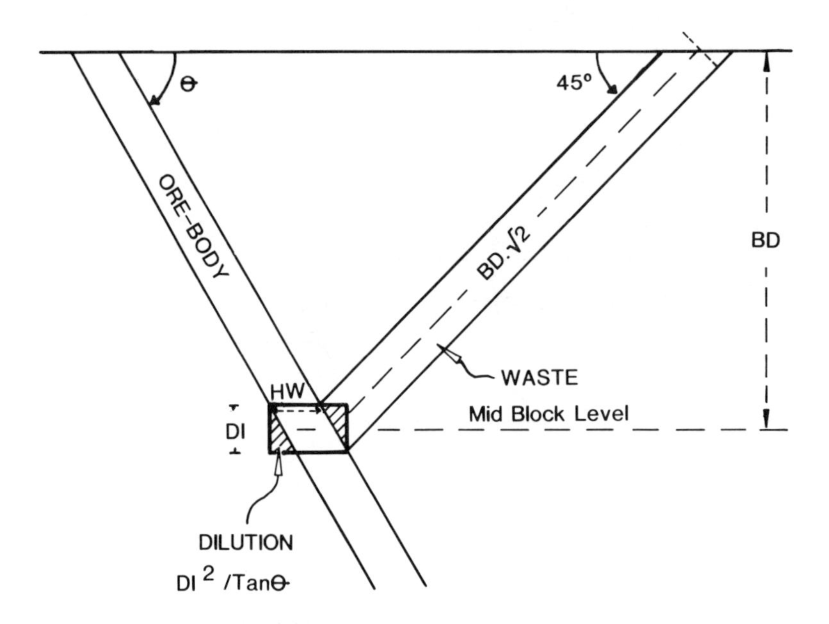

Fig. 5.32 Determination of maximum pit depth by applying breakeven stripping ratio to ore-blocks defined by a depth increment (DI).

ated with this block is:

$$\{2 \times BD \times DI - D/2\}/\{DI \times HW + D\}$$

At the breakeven point:

$$BSR = \{2 \times BD \times DI - D/2\}/\{DI \times HW + D\}$$

or

$$BD = \{BSR(DI \times HW + D) + D/2\}/(2 \times DI)$$

BD in this case is the maximum depth of the block-mid. The pit bottom will thus be at a depth of $BD + DI/2$. Had no bench dilution been taken into account then the equation for maximum depth could have been simplified to:

$$BD = (BSR \times HW)/2$$

If $\theta = 60°$, $DI = 5$ m, $HW = 10$ m and $BSR = 8$ then $D = 25/\tan 60°$ and:

$$\begin{aligned} BD &= \{8(5 \times 10 + 14.43) + 7.215\}/2 \times 5 \\ &= 52.2 \text{ m} \end{aligned}$$

The pit bottom is thus at 52.3 + 2.5 = 54.8 m and the ore dilution is 14.434 × 100/50 = 28.87%. It should be noted that the above calculations make no allowance for the extra waste incorporated in the pit by the batters at the along-strike limits of the pit.

5.3.3 Cut-off grade

Where the mineral deposit under evaluation is non-tabular, i.e. its limits have to be defined on the basis of an envelope enclosing high and low grade material and waste, the distinction between cut-off grade and minimum mining grade, as defined in section 3.3.1, is no longer relevant. An overall operating cut-off grade (OCOG) could be produced by rewriting equation (4) as follows:

$$RV = \text{profit} + (SR \times W) + M + P$$

i.e.

$$OCOG = \{\text{profit} + (SR \times W) + M + P\}/(MR \times \text{price}) \quad (6)$$

A specified profit margin per tonne of ore can thus be included, but if this is set to zero, we determine the breakeven cut-off grade. An example is given in section 5.3.8.

It should be stressed, however, that this is an overall value for the pit, based on current prices and costs, and discounting will be necessary over the life of the pit. Cut-off grade is a dynamic factor which will change during the life of the operation due to both economic and technical factors. It may be necessary to apply a higher OCOG at a later stage in the life of the mine as transport and pumping costs increase. However, it should be realized that if we set a higher OCOG, this requires a greater degree of selectivity which in turn requires flexible equipment scheduling and greater manpower (Leigh and Blake, 1971). If the OCOG is lowered, we need less flexibility in mining and can operate with larger equipment and a smaller labour force. Dilution factors are also generally lower and the combined result is to reduce operating and capital costs and metal loss.

The determination of the correct COG policy to apply to an operation is complex but it should produce the maximum NPV from the operating cash-flow (Lane, 1988). The NPV over the projected life of an operation is determined from:

$$NPV = [\Sigma\{NC_n \times (1 + i)^{-n}\}] - I$$

where NC_n is the net cash-flow for year n, i is the discounting factor and I is the initial investment including interest charges. If the interest rate on capital is 10% then the discounting factor should be 15% to allow for the risk involved in mining.

It is, however, also essential that contractural obligations to customers be met and hence the metal production profile, resulting from the chosen mining sequence and COG, must be carefully examined. A decline in production may cause problems from this point of view and a decrease in COG may be necessary to stabilize output levels towards the end of the life-span of the pit. NPV calculations may suggest that high grade areas should be mined at an early stage but marketing may dictate that blending should take place between high and low grade areas.

5.3.4 Haul roads

Initial pit designs may not include haul roads. However, Taylor (1971) points out that their incorporation can have a significant impact on tonnages, grades and stripping ratios in that additional waste may have to be mined, or ore sterilized, to allow for the increased berm width necessary to support the road. As haulage costs frequently constitute 30–50% of the mining costs, it is important that any haul road design minimizes transport costs, allows permanency by avoiding areas of poor slope stability and avoids congestion.

5.3.5 Geotechnical analysis

Once an initial design has been produced, it is essential that a kinematic analysis be undertaken to assess the mode of failure that might take place and to determine which faces, if any, are liable to fail by one or more of these modes.

Kinematic analysis

There are four possible modes of failure:

Rotational shear

In which sliding occurs on a curved surface and which is common in less competent rocks, e.g. soils, overburden and weathered rock. It is thus likely to be found in the upper benches of the pit.

Simple plane shear

In which failure takes place if a plane daylights into the pit and has a strike within 20° of the strike of the face in which it occurs. Factors which control whether or not failure will take place include the dip of the plane relative to the angle of stability for that surface and whether release surfaces (e.g. steep dipping joints) are available which define the boundaries of a block and which allow the block to move independently of the surrounding rock mass. In the latter case, a steep tensional crack approximately parallel to the strike of the face is essential. Additional factors influencing whether a failure is likely are the presence of water pressures in the joints, etc., and the natural cohesion of the surfaces. The various factors are combined mathematically in a modified version of Coulomb's equation:

$$\tau = C + \{(W \cos \alpha/A) - U\} \tan \phi \qquad (7)$$

where τ is the shear strength of the plane, W is the weight of a block with basal area A, α is the inclination of the plane, U is the water pressure in the plane of sliding counteracting the normal force on the plane due to the weight of the block, and ϕ and C are the angle of shear resistance and the natural cohesion of the plane respectively, as determined from shear box tests. If the driving force ($W \sin \alpha/A$), i.e. the component of the weight of the block in the down-slope direction, exceeds τ, then the block will move. The stability rating of a plane can be gauged on the basis of the safety factor, being the ratio of the resisting force τ to the driving force, i.e.:

$$[C + \{(W \cos \alpha/A) - U\} \tan \phi]/(W \sin \alpha/A) \qquad (8)$$

Generally a safety factor in excess of 3 is preferred. Values in the range 1–3 are considered metastable, while values < 1 represent unstable blocks.

The use of equation (8) can be demonstrated with a worked example. Let us assume that joints and bedding planes define blocks whose average dimensions are such that the weight of material on the slip plane (inclined at 25° into the pit) is 78 kN/m^2 (10 kN = 1 t). Shear box tests indicate that the natural cohesion C = 29 kN/m^2 and the angle of shear resistance ϕ = 8°. Water pressure on the slip plane is estimated to be 10 kN/m^2.

Hence, applying equation (8):

$$SF = [29 + (78 \cos 25 - 10) \tan 8]/78 \sin 25 = 1.14$$

This value lies in the stable field but is insufficiently high to enable the face to be declared safe. In the event of mining, those faces in which these sliding planes daylight, will have to be carefully monitored and remedial action taken, if necessary. If the apparent dip of the pit face, in the direction of sliding, is cut back to an angle less than the dip of the sliding plane, then stability is ensured. However, this might be accomplished at the cost of increasing the stripping ratio to uneconomic levels. The use of concave pit slopes may alleviate the problem for, if the radius of curvature of a concave slope is less than its height, then the slope angle can be 10° steeper than that suggested by conventional stability analysis. Where problems exist with simple plane failure, the faces affected should also be made as short as possible. Additional remedial action for metastable faces could include the use of inclined tensioned rock-bolts, rock-bolts and chains, metal strapping between rock-bolts, wire mesh and improved face drainage via low angle drill-holes in bench faces, drainage cross-cuts and drives, or even a peripheral curtain of large diameter drill-holes fitted with submersible pumps.

Whether or not a block lying on a plane of discontinuity will slide, topple, or slide and topple, is another consideration that should be taken into account. A block with a high height to length ratio will both slide and topple if the angle of inclination of the plane is high. As this ratio decreases, and the attitude of the plane decreases, sliding only will take place until the angle of stability is reached. Very high blocks lying on planes inclined at less than the angle of stability may topple without sliding.

Wedge failure

This takes place where two sets of discontinuities (e.g. two sets of joints or a joint set and bedding planes) intersect in such a way that their line of intersection (LOI) plunges into the pit at an angle exceeding the angle of friction/stability, but less than the apparent dip of the face in the direction of the LOI. Sliding of wedges will also be facilitated if a steep release plane is also present roughly parallel to the strike of the face. Whether various combinations of planes are likely to result in wedge failure can be tested by plotting the great circles representing the various planes on stereonets along with those representing the faces and a circle representing the angle of friction (dry or wet). Space is not available here to provide a more detailed description of the method and the interested reader is thus referred to Priest (1985) for more information.

Safety factors can also be calculated for those combinations of planes whose LOI daylights into the proposed pit. Calder (1970) has determined the necessary equation which is as follows:

$$F = [\sin \beta \cdot \cos \alpha \cdot \tan \phi \, (1/\tan w + 1/\sin w) + \cos \beta \cdot \tan \phi]/\sin \beta \cdot \sin \alpha \qquad (9)$$

where F is the required safety factor, β is the dip of one of the two planes defining the wedge, α is the pitch of the LOI on this plane, ϕ is the friction angle and w is the wedge angle. The wedge angle is the angle between the two arcs representing the discontinuities and is measured on a great circle when the LOI is rotated on to the E–W axis of the stereonet (Figure 5.33(a)). The great circle is displaced 90° from the point representing the LOI, on the opposite side of the net. From the calculation of the safety factor, a decision can be made as to whether to change the design of the pit or whether to monitor any metastable faces during extraction and take remedial action as and when necessary. Where particularly dangerous wedge configurations are evident, the geologist has three options, (a) to flatten the dip of the face, (b) to change the strike of the face or (c) to change both the dip and strike. All will have impacts on the stripping ratio and economics of the pit. These changes are all aimed at reducing the apparent dip of the face in the direction of the LOI of the wedge (Figure

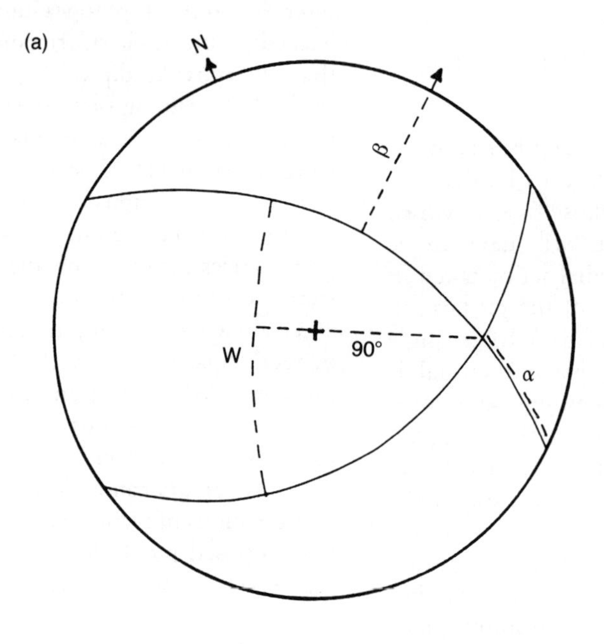

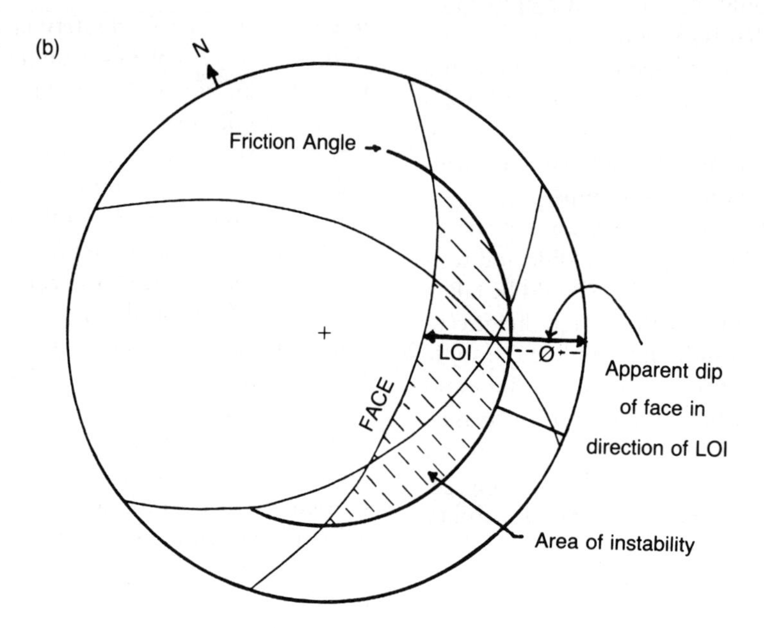

Fig. 5.33 (a) Stereonet showing the information needed to determine the wedge safety factors by Calder's method (1970). (b) LOI of a wedge relative to a pit face and angle of stability, ϕ.

5.33(b)) so that the LOI no longer daylights. In this figure, this will happen when the arc representing this new face passes outside the LOI point, i.e. closer to the net perimeter.

Block flow

Where hard rocks are under high stress, especially in the toes of benches, failure may take place. The presence of closely spaced parallel discontinuities dipping steeply into the rock face may exacerbate conditions. It is essential that such faces are not undercut during mining and that the broken rock pile produced by rupture and collapse is not removed.

Collection of geotechnical data

At this early stage in the design of an open-pit operation, the geotechnical data available to the geologist are often restricted to a few surface outcrops or exploration trenches and to exploration drill cores. In some cases, an exploration adit or old abandoned underground workings may be accessible. It is thus necessary that all these sources be studied in detail to determine:

(1) the location of faults;
(2) the dip and dip direction of all discontinuities in the rock mass (joints, open bedding, cleavage, schistosity, etc.);
(3) the lateral persistence (length) of these discontinuities;
(4) the surface shape of these discontinuities (planar, curved, irregular, stepped, etc.);
(5) the surface texture;
(6) the frequency of occurrence;
(7) the nature of any infillings which may reduce the shear strength of the planes;
(8) the relative strengths (uniaxial and triaxial compressive, etc.) of the rocks likely to be exposed in the proposed pit;
(9) depths of bedrock, weathering and hydrothermal alteration.

From this information, the geologist will then be able to compile bench by bench geological and geotechnical plans and also stereonets of discontinuity poles from which the different sets can be distinguished and their geographical location/development in the pit area determined. This information will thus give the mining engineers some indication as to the problems they might face and allow the production of pit designs in which the configuration of faces minimizes the probability of failure/collapse.

Particular reliance will have to be placed on drill-hole cores for subsurface information and it may be necessary at this stage to complete a programme of rock mechanics drilling, especially if the orebody is tabular and the drill-holes were located dominantly in the hangingwall as shown in Figure 5.34. Insufficient penetration of the footwall may have been achieved as a result. These additional holes could be drilled at a larger diameter ($>$ *NQ*) than used for exploration to produce larger diameter cores for geotechnical and rock mechanics studies and to allow the insertion of probes, piezometers or even pumps (for pumping tests and the determination of aquifer characteristics). These holes should be orientated in such a way as to intersect the major discontinuities at high angles ($> 30°$) to give a more statistically reliable number of intersections. Bias must be avoided and thus, if a constant drill-hole inclination is suspected of giving biased results, holes inclined in opposing directions should be drilled.

The information that should be recorded from drill cores can be summarized as follows:

(1) Lithological units and a qualitative assessment of rock strengths.
(2) Qualitative permeability/porosity estimates and measurements.
(3) Percentage core recovery by drill run and by lithological unit.
(4) Intersection angles of all discontinuities.
(5) Core-orientation data.
(6) Clockwise angular displacements of discontinuity low-points from the low-point of a reference plane in the core (e.g. cleavage, Annels and Hellewell, (1988)). This allows the 3D orientation of discontinuities to be determined by stereographic or mathematical methods.

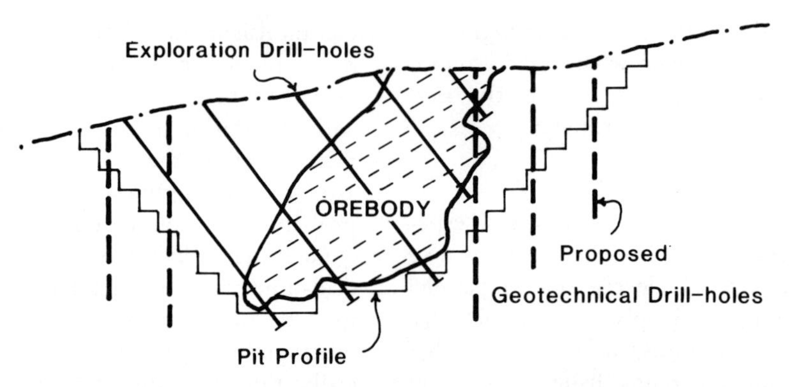

Fig. 5.34 Drilling to complete a geotechnical analysis of a potential open-pit site.

(7) Frequency of occurrence of each discontinuity set (expressed on a per metre basis).

(8) Integrated fracture density, d from:

$$d = L \times \Sigma_{i=1}^{n} 1/\sin \theta \qquad (10)$$

This removes the bias introduced by variable intersection angles between the different discontinuity sets. Over a given core length (L), the intersection angles θ_i of each discontinuity present are measured and the inverse of sin θ_i calculated and accumulated. The summation, when multiplied by the core length, gives the required integrated fracture density for the chosen interval.

(9) Rock quality designations (RQDs) over fixed length intervals or lithological units and calculated on the basis of:

$$RQD = 100(\text{total length recovered} - \text{total core in lengths} < 2 \times \text{core diameter}) / \text{length of interval}$$

Much of the numerical information listed above can be plotted on drill sections as histograms, or plotted as poles or stereonets.

Evaluation of geotechnical data

On the basis of the bench by bench geotechnical plans and the drill sections, the proposed pit area can be subdivided into structural domains in each of which the discontinuity pattern is essentially homogeneous. Design sectors can now be defined on the basis of these structural domains but also on the orientation of the pit walls. Each sector has a pit wall as one of its bounding limits (Figure 5.35). Design sectors can also be defined by vertical limits, if the upper bench plans show marked differences from those at the lower levels of the pit.

The poles of all discontinuities in each domain are plotted on a stereonet and the mean orientation of each set is determined using the centre of gravity of a cluster (perhaps using a counting/contouring net such as a Kalsbeek net) or by calculating the average direction of the poles in each cluster and the 50 percentile value of the dip angles from a cumulative frequency plot. This information is then used to test the pit wall in each design sector for potential simple and wedge failure using stereonets and safety factor calculations as described earlier. In the case of wedge failure, all possible combinations of pairs of discontinuities are compared with the face attitude and relevant stability angle. Excellent reviews of the procedures involved appear in Priest (1985) and in a Department of the Environment report published by Her Majesty's Stationary Office (HMSO, 1988). This latter publication also provides a useful review of the impact of hydrogeological and geotechnical factors in quarry and open-pit design and it is recommended that the reader consults this work for further information.

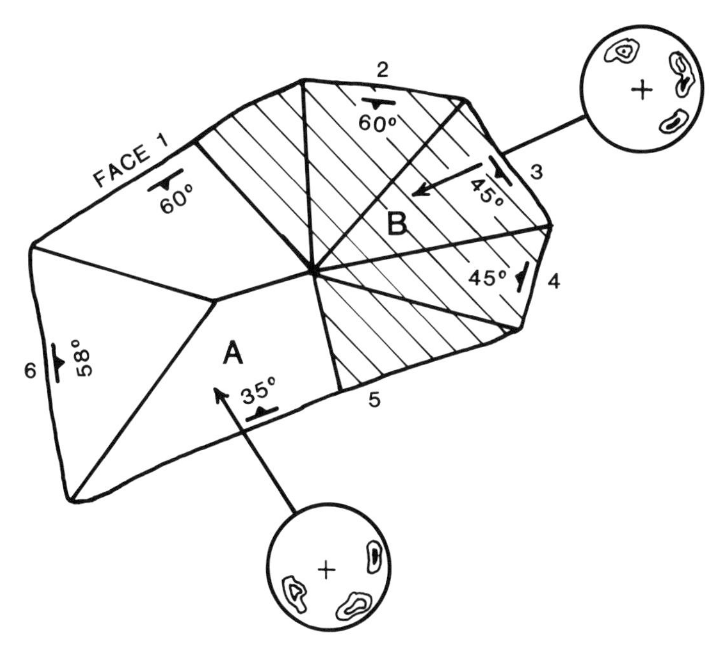

Fig. 5.35 Design sectors and structural domains (A and B) with associated pit walls and batters. Stereonets show poles of discontinuities.

The analysis of the individual design sectors will allow modifications to be made to the pit design. Not all will necessarily involve a deterioration in the economic potential of the pit as recommendations could now be made to increase the batters on certain walls to values well in excess of the 45° used in the preliminary design. These increases may well offset the effects of decreases in slope needed to stabilize certain other faces. Faces as steep as 65° have been used successfully in the past where both rock strength and geotechnical characteristics were favourable.

5.3.6 Mining and processing costs

In order to test a pit design for economic viability it is necessary to estimate the likely cost of mining the ore and waste and also of processing the ore. This exercise is dependent on the pit and mill production capacities envisaged and is, by its very nature, a 'guesstimate'. A careful analysis of similar operations in similar geological and economic environments will help to produce a reliable estimate. The various cost elements which will contribute to the mining cost will include:

(1) mining of overburden and rock (costs of dozing, ripping and blasting, as necessary);
(2) haulage of waste (unconsolidated overburden and competent waste) and ore to the respective stockpiles/tips;
(3) drainage, including pumping from the sump, the drilling of in-pit and peripheral boreholes, and the driving of drainage adits/drives;
(4) rehabilitation and environmental control;
(5) exploration drilling, assaying and geological overheads;
(6) a portion of the administrative overheads.

Elements which contribute to the cost of processing include:

(1) ore handling (e.g. stockpile blending);
(2) milling;
(3) concentrating;

(4) smelting/refining;
(5) smelter penalties – deleterious elements;
(6) environmental controls;
(7) tailings and slag disposal;
(8) administrative overheads.

In addition, the operation will face additional charges related to the marketing and shipment of concentrates or metals, interest charges, royalties, taxes and dividends (Chapter 6).

In order to gain an insight into the anticipated costs of mining rock in relation to production capacity, information from the *Canadian Mining Journal* (1988) review was again used. Figure 5.36 shows a plot of pit production capacity, expressed as tonnes per day (TPD), against mining cost in Canadian dollars per tonne ($PT). A regression line fitted to the data has the formula:

$$TPD = 45.3 \times 10^3 - 12.26 \times 10^3 \times \$PT \quad (11)$$

Hence, for a 20 000 TPD operation, the anticipated costs would be (45.3–20) × 10^3/(12.26 × 10^3) = 2.06 C$ or 1.73 US$ (at a current exchange rate of 1.19).

The same exercise was carried out for processing and milling costs for precious- and base-metal operations and the results are shown in Figures 5.37 and 5.38. The relevant regression equation for gold-silver operations, with a capacity of less than 3000 TPD, approximates to a straight line whose formula is:

$$TPD = 2340.1 - 68.96 \times \$PT \quad (12)$$

Hence, for a 1500 TPD plant, the cost would be (2340.1–1500)/68.96 = 12.18 C$ or 10.24 US$ (at current rates). The basemetal operations listed, however, showed a marked drop in operating costs as the tonnage increased. Up to 3000 TPD, these costs were not far removed from those for gold operations but after this tonnage, the rate of decline decreased sharply. Had more information from very large tonnage gold operations been available, it is possible that

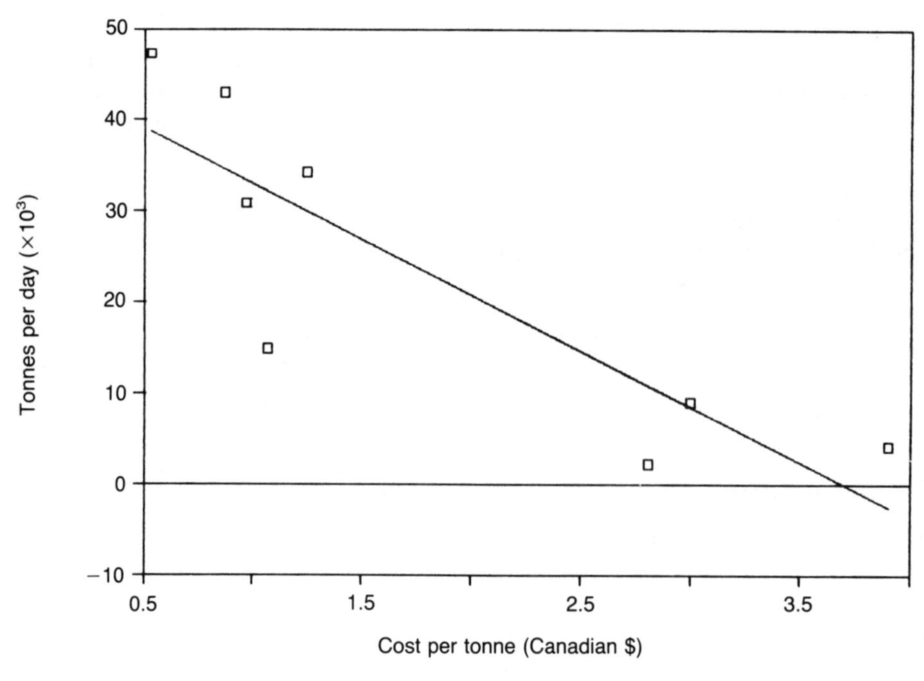

Fig. 5.36 Regression plot, using Lotus's PGraph option, of mining costs versus daily pit production rate (largely based on figures provided by the *Canadian Mining Journal*, 1988.) □, Actual value; ——, regression line.

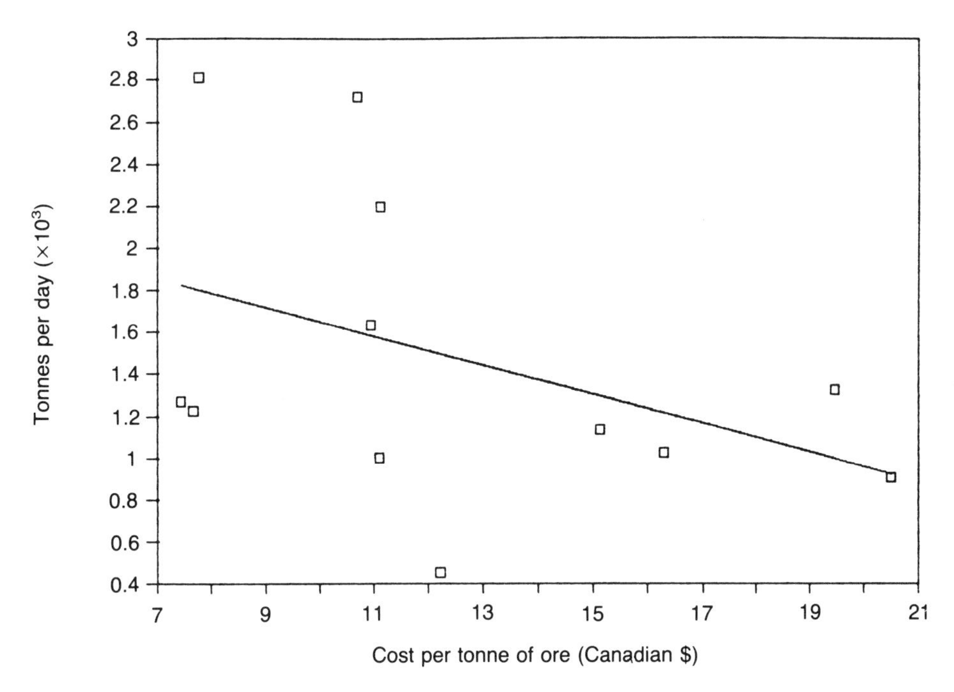

Fig. 5.37 Processing costs for gold-silver operations in Canada versus mill capacity. □, Actual values; ——, regression line.

they would have seen a similar trend. A plot of C$ per tonne against natural log of tonnes per day for the basemetal operations (Figure 5.38), also produced a linear regression with the formula:

$$\ln TPD = 9.826 - 0.229 \times \$PT \qquad (13)$$

Hence, a 1500 TPD plant would have an estimated processing cost of 10.97 C$ or 9.22 US$ (at current rates – July 1989).

It should be stressed that equations (11) to (13) only give a rough approximation as to the real costs and are presented merely as a guide for initial costings of potential operations. They are based on 1987 figures and hence inflation factors should be applied as relevant.

5.3.7 Valuation of ore blocks

Once grades have been assigned to ore-blocks in a 2D or 3D matrix, as described in section 5.2.3, the net revenue each will generate, once the costs of mining, milling, etc., have been deducted, must be calculated in order to proceed to the next stage in the evaluation exercise, i.e. the economic optimization (section 5.4). The block is given a dollar value on the assumption that it has already been exposed for mining so that stripping ratio and breakeven cut-off grades are irrelevant.

The information that is required to value a block can be summarized as follows:

(1) tonnage;
(2) grade;
(3) anticipated metallurgical recovery;
(4) content of penalty/deleterious elements;
(5) content of valuable by-products;
(6) current mining cost (+ overheads);
(7) current processing costs (milling, concentration, smelting, refining plus related overheads);
(8) anticipated daily pit and mill capacities.

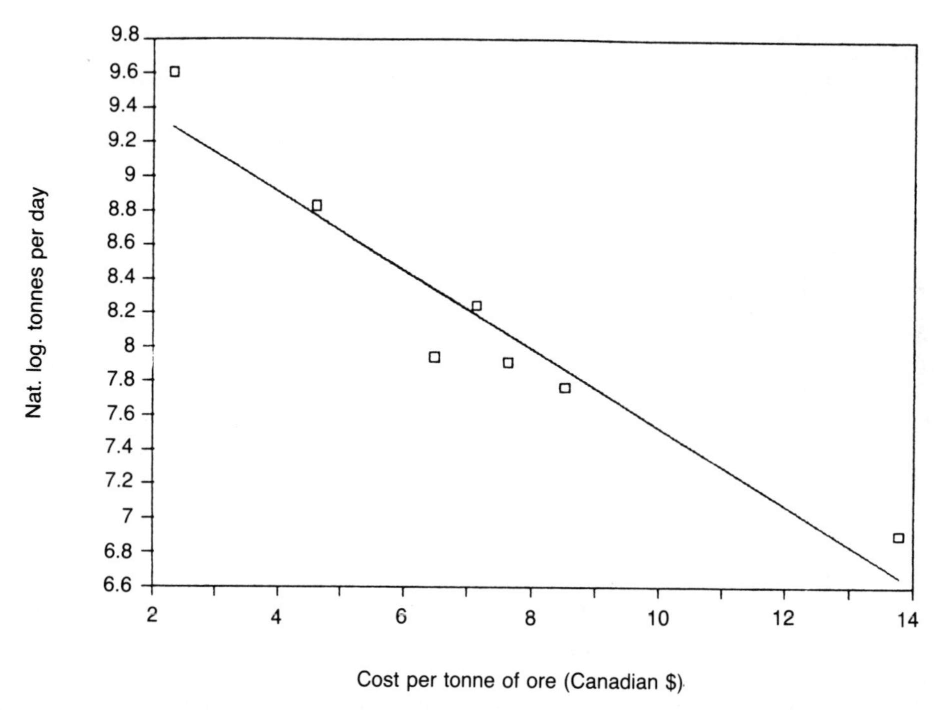

Fig. 5.38 Processing costs for base-metal operations in Canada and the USA versus the natural log of mill capacity. □, Actual value; ____, regression line.

If we assume that we are dealing with a gold operation whose pit capacity is 2000 TPD and whose stripping ratio is 1.1 : 1, then the mill throughput will have to be 950 TPD. Applying equations (11) and (12) in section 5.3.6, we have estimated mining costs per tonne of 3.53 C$ and a processing cost of 20.15 C$.

Metallurgical recoveries for copper, lead and zinc from polymetallic sulphide deposits range from 55% to 95%, 60% to 90% and 45% to 97% respectively. Recoveries of copper and molybdenum from porphyry deposits range from 80% to 90% and 40% to 88% respectively, whereas for gold, the typical range is from 85% to 97%. Thus, for our deposit, we could select a value of 93% as being realistic.

If we now consider a 10 × 25 × 25 m ore-block, representing 17 187.5 tonnes (tonnage factor = 2.75), with a gold grade of 7.3 g/t, then the contained gold is 125 468.75 g and the recovered gold is 116 685.94 g (or 3751.48 troy oz). At a gold price of 371 US$ per ounce, representing 441.49 C$, the total value would be 1 656 240.9 C$. The cost of mining and processing this block would, however, be 60 671.88 C$ and 346 328.12 C$ respectively, totalling 407 700 C$. Thus, the net revenue from this block would be +1 248 540.9 C$. This calculation has been made assuming that there are no penalty elements (e.g. As) or valuable byproducts (e.g. Ag) present. Note that this revenue will also have to cover the costs of the overburden stripping which for this pit averages 1.1 t of waste per tonne of ore.

Blocks, such as that described above, can be classified as ore if:

(1) their value is positive and thus breakeven blocks are excluded;

(2) if their value exceeds the costs by a given percentage, representing a required profit margin;

(3) if the loss incurred by treating them as ore is less than that which would be incurred if they were treated as waste.

5.3.8 Preliminary economic assessment of pit designs

In section 5.2.2 we considered a hypothetical gold deposit and produced an initial pit design suitable for its exploitation. We are now in a position to assess this design to see whether to continue its evaluation by additional drilling and whether to proceed to a more detailed design/feasibility stage.

Pit A

In Table 5.1 we saw that the total volume of ore and waste (excluding overburden) was 17 208 550 m^3. As the volume of the ore released to this pit was 593 687.5 m^3 (Table 5.3), the volume of waste rock is 16 614 862.5 m^3. We can thus calculate the stripping ratio as follows:

Material	Tonnage factor	Tonnes
Ore	3.8	2 256 012.5
Overburden	2.1	1 584 975.0
Waste rock	2.8	46 521 615.0
Waste	—	48 106 590.0

Hence the stripping ratio = 21.3 : 1(!)

Even at this early stage, therefore, the alarm bells are ringing. No attempt will be made to cost this operation for it is clearly a non-starter.

Pit B

Table 5.1 shows that the volume of ore and waste (excluding overburden) was 5 345 085 m^3. As the volume of the ore released to this pit was 501 800 m^3 (Table 5.3), the volume of waste rock is 4 843 285 m^3. We can thus calculate the stripping ratio as follows:

Material	Tonnage factor	Tonnes
Ore	3.8	1 906 840.0
Overburden	2.1	1 369 987.5
Waste rock	2.8	13 561 198.0
Waste	—	14 931 185.5

Hence stripping ratio = 7.83 : 1.

This is high but it is worth determining whether this pit could stand alone economically. The calculations and estimates used are as follows:

Total tonnage of rock removed = 15 468 038
Total tonnage of overburden moved = 1 369 987.5
Pit capacity = 10 000 TPD
Pit life = 4.8 years (350 days operational per year)
Required plant capacity = 1132.5 TPD
Estimated mining costs (equation (11), section 5.3.6) = 2.88 C$
Estimated processing cost (equation (12), section 5.3.6) = 17.51 C$
Estimated cost of overburden stripping = 1.00 C$
Mill recovery = 93%
Contained gold = 120 973.5 oz (Table 5.3)
Ore grade = 1.97 g/t

Breakeven stripping ratio

Recoverable value/t ore = 0.93 × 1.97 × 1.19 × 371/31.104
= **26 C$**

Production cost/t ore = 17.51 + 2.88
= **20.39 C$**

Stripping cost/t waste = (1 369 987.5 × 1.00 + 13 561 198 × 2.88)/14 931 185.5
= **2.71 C$**

Breakeven stripping ratio = (26.0 − 20.39)/2.71
= **2.07 : 1**

This is much lower than the estimated value for the pit design.

Breakeven COG

Applying equation (6) with profit = 0,

$$\text{BCOG} = \{[(7.83 \times 2.88) + 20.39] \times 31.104\}/(0.93 \times 371 \times 1.19)$$
$$= \mathbf{3.25\ g/t}$$

The actual grade would have to exceed this value if an overall profit is to be made at current prices and costs. Once again, Pit B fails in this respect.

Overall costs

Overburden	1 369 987.5
Mining rock	44 547 949.4
Processing	33 388 768.4
Total	79 306 705.3

Revenue

$120\,973.5 \times 371 \times 1.19 \times 0.93$
$= 49\,669\,989.2$ C$.

Profit

$49\,669\,989.2 - 79\,306\,705.3$
$= -29\,636\,716.1$ C$.

It is thus clear that the pit design, as it stands, has no chance of being profitable at current metal prices, especially as no allowance has been made for the dilution of ore expected for a flat-lying tabular deposit of this type and thus the increased milling costs. A gold price well in excess of 590 US$ would be required to renew interest in Pit B.

The problems facing this pit design could have been predicted at an early stage by the fact that that the maximum depth estimate for this pit, using BSR and the method described in section 5.3.2, would only have been 22.1 m. An alternative approach to the design of this pit would have been to determine the reserves bench by bench using cross-sections and accumulating from the top down. At the same time, the volumes of waste for each bench would be accumulated. The pit perimeter would rapidly expand outwards as incremental push-backs of the pit walls would be necessary as the pit bottom became deeper. The optimum pit bottom on each section would be that which still produces a profit. More sophisticated methods of pit optimization are dealt with in section 5.4.

It is also a pointless exercise to combine Pits A and B as the overlap is minimal, as explained earlier, and Pit A will incur an even greater loss than Pit B.

This exercise has highlighted the problem of trying to mine a flat tabular deposit by open-pit methods in an area where geographical constraints have been placed on the pit perimeter. Some better grade and thicker mineralization close to the fault (Figures 5.9, 5.10) cannot be exploited as a result. Further investment in drilling on this site is clearly not warranted unless the highway can be diverted, an expensive exercise, or a considerable rise in the gold price occurs.

5.4 ECONOMIC OPTIMIZATION OF PIT DESIGNS

Optimization algorithms can be subdivided into two main groups, (a) **rigorous**, for which mathematical proof of optimality is possible and (b) **heuristic**, which experience has shown works satisfactorily but which lacks rigorous mathematical proof.

5.4.1 Floating (moving) cone methods

These belong to the heuristic category and involve the economic analysis of ore- and waste-blocks falling within an inverted cone which is systematically moved through a block matrix with its apex occupying the centre of each block in turn. The methods are based on the premise that the net revenue from the ore contained in the cone offsets the cost of stripping waste to expose this ore. Individually, cones may be apparently unprofitable but when two or more overlap, a considerable proportion of the waste is shared and may result in a change in the economic status of the cones.

The block matrix is produced in the usual way by kriging or inverse distance weighting methods and then modified to allow for the surface topography. This model can now be examined with one of two variants of the floating cone method. The first involves the direct use of the block grades and an estimate of the operating cut-off grade, perhaps determined as outlined in sections 5.3.2 and 5.3.3. The slope angle of the cone is set to the required pit slope (e.g. 45°) and then the program starts at the top of the block matrix and locates the first block above *COG*.

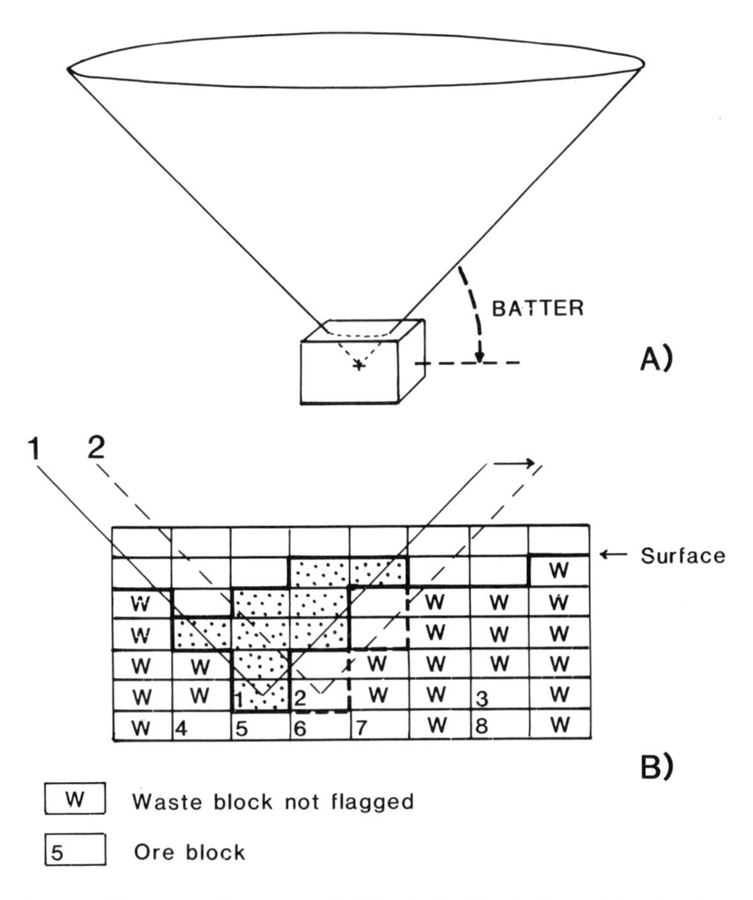

Fig. 5.39 (a) Inverted cone with apex at the centre of a block. (b) Blocks flagged for the first and second positions of the floating cone.

An inverted search cone is then positioned at the block centre (as in Figure 5.39(a)) and the number (NO) and mean grade (G_{ore}) of all ore-blocks, whose centres fall within the cone, are calculated together with the number of waste blocks (NW). The financial viability of this cone is then estimated from:

$$\text{Profit} = [\text{price} \times MR \times G_{ore} \times NO - (M + P) \times NO - (M \times NW)] \times BV \times TF$$

where 'price' is the current metal price, MR is the metallurgical recovery, M is the cost of mining and transporting each tonne of ore and waste, P is the cost of processing each tonne of ore, BV is the block volume and TF is the tonnage factor. Note, however, that if G_{ore} is expressed as a percentage, it must be divided by 100. If the profit value is positive, then all blocks in the cone are tagged and effectively removed from the block matrix so that a new surface profile would result as in Figure 5.39(b)). It must be remembered, however, that this is a 2D section and that additional blocks would have been removed at right angles to the plane of this diagram. The process thus simulates the mining of an open-pit. If the value is negative, then the matrix remains unchanged. The cone apex is now moved to the second block above COG in the bench, e.g. 2 in Figure 5.39(b), and the process repeated. In this example, only a few additional blocks would be tagged if the profit is positive for the new cone position. Had the first cone position indicated a negative profit then, after a positive value for the second block, the program would have returned to the first block to assess whether the tagging of waste blocks in this second cone had changed the status of the first. The technique is thus iterative and only terminates when all blocks have been covered in all benches and no further enlargement of the pit is possible, either laterally or downwards.

Once this process is completed, the tonnages of all tagged blocks are accumulated bench by bench, and on the basis of ore and waste, so that cumulative bench stripping ratios can be computed from the top bench down to the lowest. The last value would then represent the overall stripping ratio for the pit. Grades associated with each of the bench tonnages can also be determined. Re-runs of the program, for a range of different COGs, then allows a grade–tonnage curve for the different pit designs to be produced. Also, grade–tonnage curves can be produced for each pit which show the percentage of the overall tonnage above each grade class limit. This will produce a set of nested curves, one for each pit design. Such graphs allow the selection of the most suitable option for the proposed operation.

Programs such as PITPACK (Geostat Systems International Inc.) apply the second variant of the floating cone method. This involves the assignment of monetary values to each block, i.e. net revenue values (section 5.3.7) calculated using appropriate mining, milling, processing and transport costs. The result is an economic block inventory as opposed to a block grade inventory. Blocks with negative net revenue values are reassigned a negative value based on mining costs alone, as these blocks will not be sent to the mill, etc. Other than this, the method works exactly as described earlier. The slopes of the cone can be varied according to direction so that, for example, a 60° batter could be applied to a hangingwall face and a 45° batter to the opposite face. The negative and positive values of untagged blocks falling in each cone are accumulated to see whether they could be mined at a profit. The same iterative technique is used to maximize the size of the optimal pit.

The pit design produced can now be modified taking into account mining/technical constraints.

5.4.2 Lerchs and Grossmann optimization

Lerchs and Grossmann first proposed their unique method of open-pit design optimization in 1965. The objective of this method was to '. . . design the contour of a pit so as to maximize the difference between the total mine value of ore extracted and the total extraction cost of ore and waste'. They applied a programming technique which would provide a simpler, faster determination of the optimal mining configuration of blocks in cross-section. This 2D dynamic programming algorithm is explained in mathematical terms in their paper and in a later paper by Johnson and Mickle (1971) to which the reader is referred. This technique was further developed into a 3D algorithm which would take into account block values in longitudinal as well as transverse sections. It is the intention here to attempt to explain the methods in non-mathematical language with the aid of a series of diagrams.

Before attempting an optimization, it is essential to thoroughly understand the grade distribution and geometry of an orebody. This can be achieved by producing a block model of the deposit via such programs as DATAMINE. On cross-sections of this model, a manual interpretation and pit design can be accomplished which will give a general indication of the geological and mineable reserves and the ultimate stripping ratio. From this, realistic estimates of mining and processing costs can be made plus breakeven and operational cut-off grades.

2D Lerchs and Grossmann

This method provides a first approximation to an optimal pit as it only considers data on one section at a time, ignoring data on adjacent sections. Pit profiles on adjacent sections are thus produced independently and often need considerable adjustment to produce a coherent pit design by which time optimality has been lost. The method is also relatively easy to program and allows the production of a preliminary pit design. The following example has been produced using an 'in-house' program so that the stages in the process can be demonstrated.

Figure 5.40 is a graphics printout of a section of block grades (based on grade ranges) from a gold deposit in Canada which have been computed by inverse square distance weighting methods. Each is $25 \times 10 \times 10$ m in size, representing 7000 tonnes at an SG of 2.8 tonnes/m^3.

The first step in the design process is the production of a series of block sections on which grade values are plotted as in the example shown in Figure 5.41 (values rounded to one place of decimals). The second step is to calculate the net revenue for each block as in Figure 5.42, where minus values indicate that a net loss would be incurred on the mining and processing of the block concerned. Where this is the case, this value is replaced by the mining cost alone (-14×10^3\$). Assumptions made are that the mining costs are 2.0\$/t, processing costs are 5.5\$/t, the metallurgical recovery is 93% and the gold price is 371\$/oz. This calculation ensures that every block is classified as either ore or waste and will produce a design for a pit in which selective

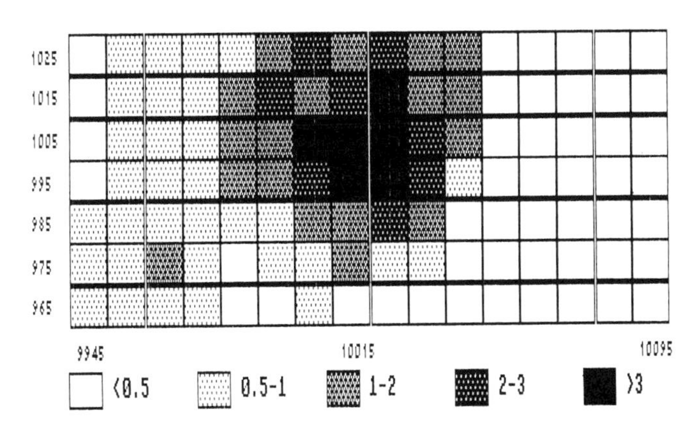

Fig. 5.40 Metallurgical classification of sectional block grades (g/t) for a British Columbian gold deposit. (Blocks $10 \times 10 \times 25$ m in size.)

	9945							10015								10095
1025	0	0.6	0.7	0.7	0.8	1.2	2.2	1.8	2.5	1.8	1.4	0	0	0	0	0
1015	0.5	0.6	0.7	0.8	1.2	2.6	1.7	2	3.3	2	1.6	0.2	0	0	0	0
1005	0.3	0.6	0.7	0.7	1.3	1.8	3.1	7.9	4.6	2.2	1.6	0.2	0	0	0	0
995	0	0.8	0.8	0.9	1.4	1.5	2	5.9	8.4	2.7	0.8	0.2	0.1	0.1	0	0.1
985	0.6	0.8	0.8	0.6	0.7	0.7	1.6	1.2	2.9	1.4	0.4	0.2	0.1	0.1	0.1	0.1
975	0.6	0.7	1.4	0.7	0.1	0.7	1	1	0.9	0.6	0.5	0.1	0.1	0.1	0.1	0.1
965	0.6	0.9	0.6	0.5	0.4	0.3	0.5	0.3	0.3	0.4	0.1	0.1	0.1	0.1	0.1	0.1

Fig. 5.41 Sectional block grades rounded to one place of decimals.

1025	-14	-14	2	3	11	38	121	88	141	88	56	-14	-14	-14	-14	-14
1015	-14	-14	-14	8	39	149	82	105	203	101	75	-14	-14	-14	-14	-14
1005	-14	-14	-14	-14	49	89	192	559	301	121	69	-14	-14	-14	-14	-14
995	-14	7	7	17	54	63	105	407	603	158	10	-14	-14	-14	-14	-14
985	-14	9	6	-14	2	-14	70	44	171	59	-14	-14	-14	-14	-14	-14
975	-14	2	55	4	-14	1	22	28	18	-14	-14	-14	-14	-14	-14	-14
965	-14	14	-14	-14	-14	-14	-14	-14	-14	-14	-14	-14	-14	-14	-14	-14

9945 10015 10095

Fig. 5.42 Net revenue values for blocks.

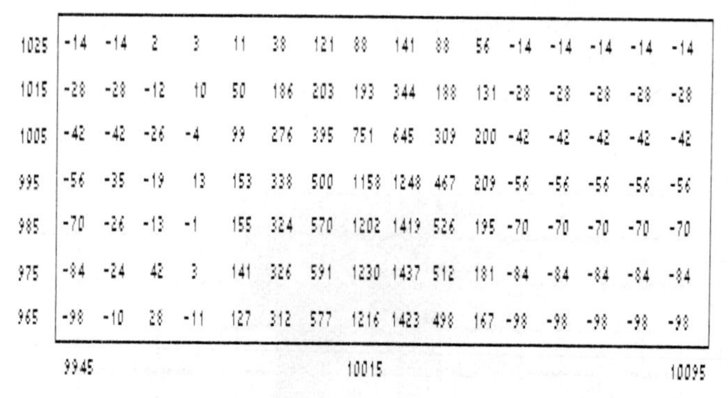

1025	-14	-14	2	3	11	38	121	88	141	88	56	-14	-14	-14	-14	-14
1015	-28	-28	-12	10	50	186	203	193	344	188	131	-28	-28	-28	-28	-28
1005	-42	-42	-26	-4	99	276	395	751	645	309	200	-42	-42	-42	-42	-42
995	-56	-35	-19	13	153	338	500	1158	1248	467	209	-56	-56	-56	-56	-56
985	-70	-26	-13	-1	155	324	570	1202	1419	526	195	-70	-70	-70	-70	-70
975	-84	-24	42	3	141	326	591	1230	1437	512	181	-84	-84	-84	-84	-84
965	-98	-10	28	-11	127	312	577	1216	1423	498	167	-98	-98	-98	-98	-98

9945 10015 10095

Fig. 5.43 Downward accumulation of revenue values for each column (M_{ij}).

mining could be applied. It is equivalent to the application of a strict COG with all blocks below this value being sent direct to the waste tip.

Having tabulated the block values, these are then accumulated for each column from the top down, as in Figure 5.43. These are Lerchs and Grossmann's M_{ij} values (see Johnston and Mickle, 1971) where '*i*' is the row number and '*j*' is the column number. Next, commencing with the first column of blocks (starting at the top), the value of P_{ij} is calculated for each block where,

$$P_{ij} = M_{ij} + \max\,(P_{i+r,\ j-1}) \qquad \text{(a)}$$

and where r is set to -1, 0 and $+1$. The second term in this equation determines which is the largest value of P_{ij} in the three closest blocks in the column to the left of the block being evaluated (defined by the current value of '*i*' and '*j*'). Once this has been determined, it is added to the M_{ij} value of this block. A slight modification to this has to be made so that the P_{ij} values for column 1 can be calculated. It is assumed that all block grades to the left of this column are zero and thus the net profit of each is -14 ($\times\ 10^3$). The P_{ij} value of each block in column 1 is thus the total net profit of all those blocks that have to be removed to expose this block, excluding the immediately overlying blocks whose value is already accounted for in the block M_{ij} value. Referring to Figure 5.43, block 1,1 does not require any additional mining to expose it, therefore its P_{ij} value is -14 (Figure 5.44). In the case of the second block, one block of waste has to be removed to its left (45° slope) to expose it, therefore the P_{ij} value is $-28 + -14 = -42$ (as in Figure 5.45). For the third block, $P_{ij} = -42 + -(3 \times 14) = -84$. This process is repeated for the whole of the column. For all the other columns,

1025	-14	-14	2	4	23	100	419	690	1190	2145	3079	3916	4313	4358	4344	4330
1015	-42	-42	-26	12	62	298	602	1049	2057	3023	3930	4327	4372	4344	4330	4316
1005	-84	-84	-68	-30	112	399	857	1713	2835	3799	4355	4400	4358	4330	4302	4288
995	-140	-119	-103	-55	123	462	962 (4,7)	2190	3490	4156	4442	4386	4344	4302	4274	4246
985	-210	-166	-132	-104	100	448	1031	2241	3688	4233	4428	4372	4316	4274	4232	4204
975	-294	-234	-124	-121	36	425	1039	2269	3707	4219	4414	4344	4288	4232	4190	4148
965	-392	-304	-206	-135	5	348	1003	2255	3693	4205	4386	4316	4246	4190	4134	4092

9945 10015 10095

Fig. 5.44 Block profit values, P_{ij}.

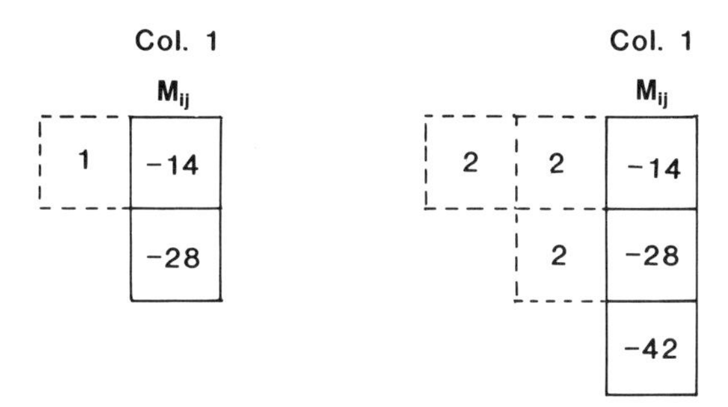

Fig. 5.45 Calculation of P_{ij} values for column 1 in the sectional block model.

the P_{ij} values are calculated using equation (a). For block 4,7 (i,j) in Figure 5.44, the values to the left of the block are as shown in Figure 5.46. The largest is 462 hence this value is added to 500 (M_{ij} – Figure 5.43) to obtain the block P_{ij} value of 962 ($\times 10^3$). Similarly, the value of block 3,8 is 1713 ($\times 10^3$). In each case, an arrow is drawn from the block under evaluation to the block with the highest profit level. Figure 5.47 shows all these tie-lines combined on to one diagram. An area is outlined which is centred on block 4,7 and which shows a line entering from the top right and another leaving to the left as was determined in Figure 5.46. By following these tie-lines, a series of optimal pits are defined as each represents the line of maximum profit for a pit to the left of this line. Each of these nested pits represents a design down to a successively deeper pit bottom. The pit which maximizes the profit margin is that which has a bounding contour extending downwards from the highest value of P_{ij} on the top line, i.e. 4358 in block 1,14. This value represents the profit in mining all blocks to the left of the bounding contour, down to the pit bottom, when all the ore-grade blocks are selectively

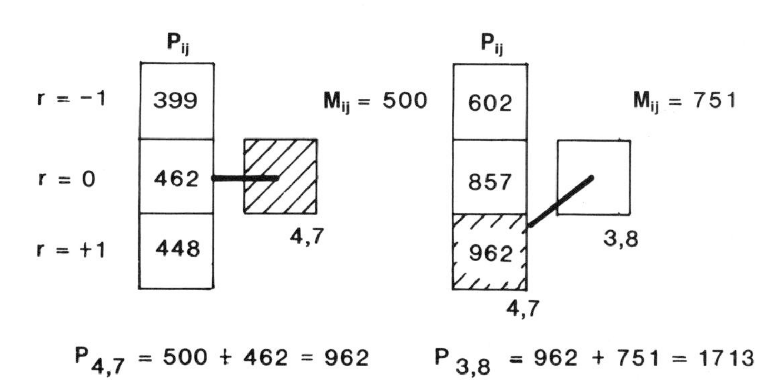

Fig. 5.46 Calculation of P_{ij} values for blocks 4,7 and 3,8.

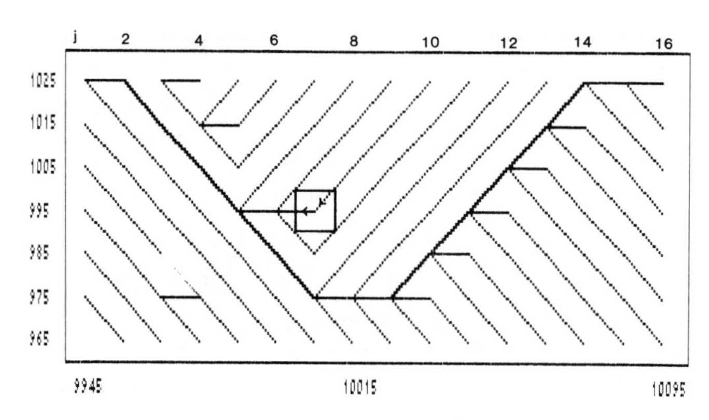

Fig. 5.47 Possible pit outlines with the optimal pit highlighted.

processed. Note, however, that the ore-blocks have been selected using what is effectively a break-even cut-off as no profit margin was built into the calculation. Had this been done, additional blocks would probably have been classified as waste.

3D Lerchs and Grossmann

Alford and Whittle (1986) provide a readily understandable description of the method and thus the following account is heavily reliant on this work.

The aim of this optimization procedure is to define that combination of ore- and waste-blocks which gives the highest value for the metal recovered for a specified pit slope. It thus produces the true optimum pit as no other combination will give a higher value. The dimensions chosen for the individual blocks in the model are usually based on the need to delineate the orebody as precisely as possible; on bench heights, SMUs and mining sequence chosen; on the confidence of block grade estimates (controlled largely by drill-hole spacing) and the capacity and cost of the available computer facility. Many typical block models would thus contain from 10^5 to 10^7 blocks. Such data bases would involve enormous run times for optimizations programs and hence it is common practice to group these smaller blocks into larger units and then to add the net revenue values ($) of the constituent blocks to produce a new block value. In practice, it has been found that this has little effect on the resultant pit design and it allows the use of PCs rather than mainframe computers. Generally 5×10^4 to 10^5 blocks are adequate for the exercise.

For each block in the model, the program (e.g. 3D Whittle Programming) determines which other blocks (in 3D) must be removed to uncover it. This requires the user to define pit slopes on the basis of non-overlapping subregions of the model. Figure 5.48(a) shows a cross-section of an inverted cone whose slope angles have been set to those required for the pit slope. All blocks which must be mined, on one section through this cone to expose the block at the apex, are highlighted. These are stored by listing pairs of block ID numbers, known as 'structure arcs', so that the first number represents the block to be mined and the second, that which must be mined to uncover it. Figures 5.48(b) and (c) show how structure arcs are used to define the blocks which must be mined. The value of each block, once it has been uncovered, can then be determined by assigning waste-blocks a negative value based on the cost of blasting, digging and haulage, and the ore-blocks a positive value based on the value of the contained metal (corrected for metallurgical recovery) minus the cost of mining and processing.

The Lerchs and Grossmann method progressively generates lists of related blocks, like the

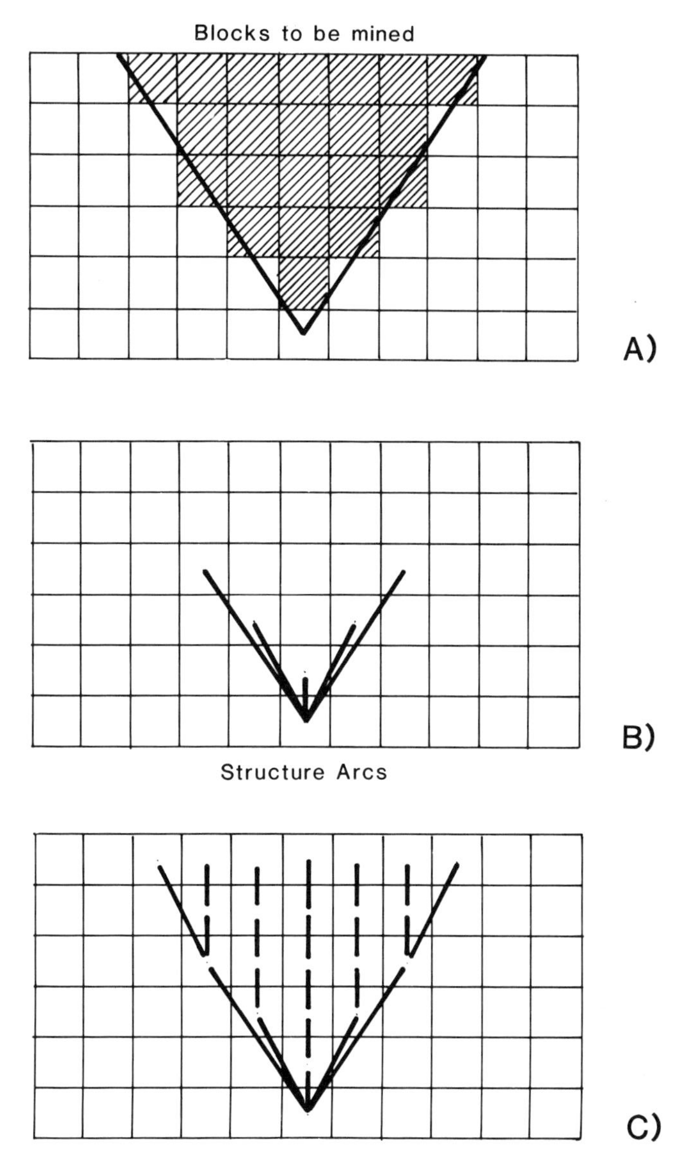

Fig. 5.48 (a)–(c) 2D representation of the use of structure arcs to define blocks within an inverted cone.

branches of a tree, in which branches are flagged as 'strong', if the total of their block values is positive, or 'weak', if they contain waste-blocks. The program then searches for structure arcs where some part of a strong branch underlies a weak branch. If found, the two branches are restructured by combining them into one (which may be either strong or weak) or breaking a portion of one branch and adding it to the other. The procedure is repeated until no structure arc goes from a strong branch to a weak one. All strong branches are then combined to form the ultimate pit design. A listing of the ore- and waste-blocks within the optimal pit limits is then

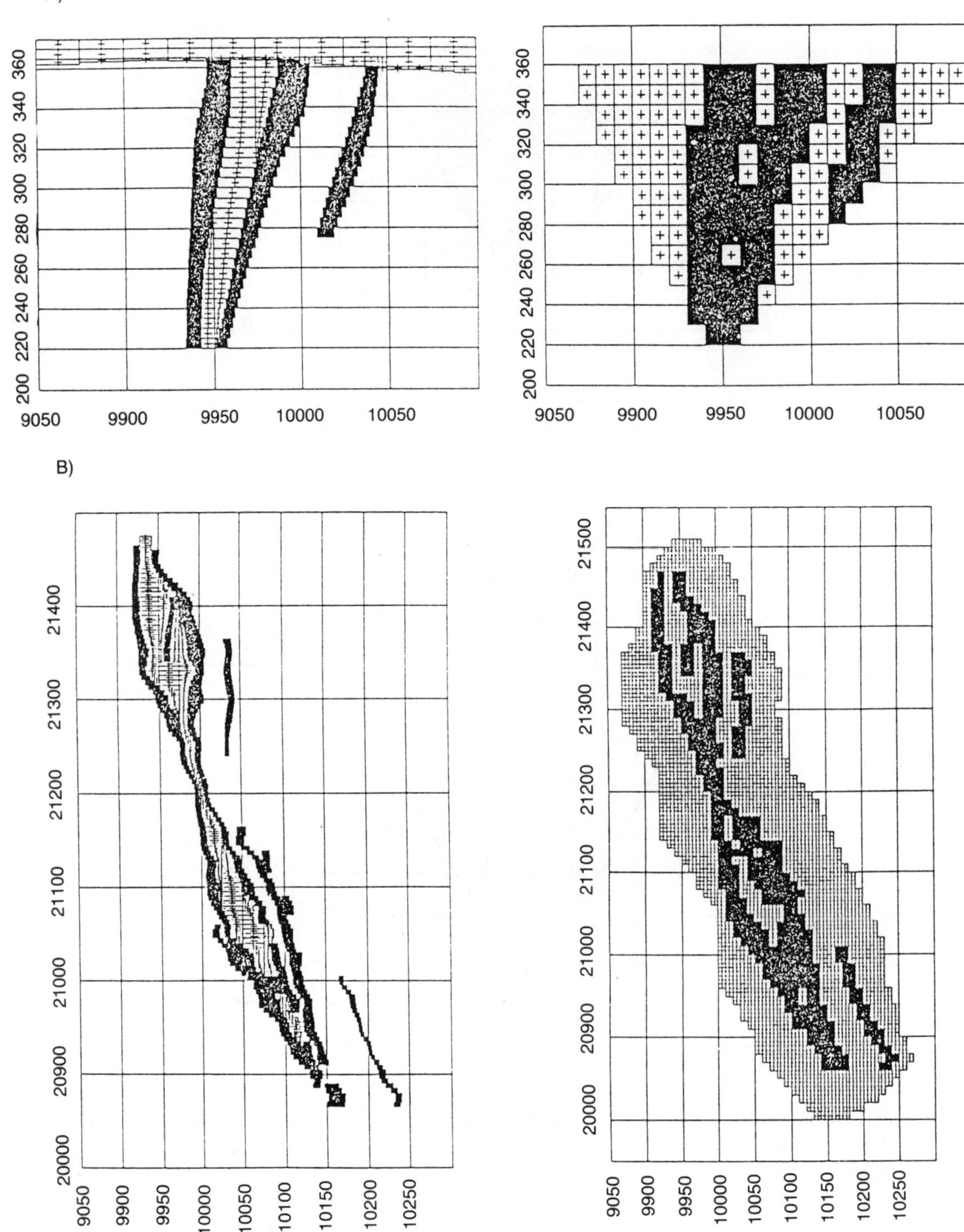

Fig. 5.49 (a) Geological model section (with topographic model) and the equivalent optimized pit profile for the Hampton Boulder Gold Deposit (from Alford and Whittle, 1986). (b) Geological model plan and optimized pit limits.

produced. The program has the facility to produce a series of pits each down to a different pit bottom and thus it simulates mining with incremental push-backs of the pit limit. Cross-sections in three orthogonal planes can now be produced showing, for each block, the phase number in which it will be mined.

As was mentioned earlier, the economic factors used were initially estimated, perhaps through a manual pit design. Many are, however, dependent on the design and thus, as the optimization produces a different design, the parameters will be different. An iterative approach is thus used in which the results of the initial optimal pit design are used to provide a new set of parameters for a second run and so on, until a satisfactory result is achieved. A final pit design will then be produced in which the pit perimeter is smoothed out and to which haul/access roads are added using interactive graphics/digitization. Alford and Whittle record an example of the technique as applied at the Hampton Boulder gold deposit in Australia from which Figure 5.49 is extracted. Here, three ore-zones are contained in steep-dipping, sub-parallel, porphyry units. Figure 5.30 represents another example of an optimized pit design, although this was produced by DATAMINE's Lerchs–Grossmann facility.

The 3D Lerchs and Grossmann method has some drawbacks in that it allows only the consideration of a single attribute (the economic value of the block) and that it precludes the separate consideration of different metals in a polymetallic orebody or the use of ore blending. Constant economic parameters are used for all blocks in the model whereas mining costs and metal prices may vary with time as the pit deepens. A detailed knowledge of the order in which blocks are to be mined would be necessary to allow for this. Finally, all blocks that can be economically mined are included in the optimal model and high grading is precluded.

The problem with the optimal model is that, should any of the economic factors change, then the value of the pit changes and it is no longer optimal. Whittle (1988) has described a method (4D Whittle Programming) by which the complex relationship between these factors can be resolved down to a level which is within the capacity of present computer technology/capacity. The various factors involved were discussed earlier in section 5.3.7, but they are listed again so that the basis of this technique can be examined in more detail.

Tonnes of rock in each block .... (1)
Tonnes of ore in each block .... (2)
Units of metal in each block .... (3)
Proportion of metal that can be recovered ... (4)
Price for a unit of metal produced (minus delivery costs) .... (5)
Cost of mining a tonne of rock (including overheads) .... (6)
Cost of processing a tonne of ore (including overheads) .... (7)

The value of a block is thus:

$$\{(3) \times (4) \times (5) - (2) \times (7)\} - (1) \times (6). \quad \text{(b)}$$

If this value is negative, the portion in brackets is set to zero and the new value assigned to the block (waste). Most likely ranges can be set for each of the factors (5), (6) and (7) so that a range of pit designs can be produced. A mining strategy can thus be devised which minimizes the risk due to changes in economic conditions. Factors (6) and (7) can be estimated over the life of the operation on the basis of projected inflation rates, but the price (5) is less easy to predict. The manipulation of seven different factors is a massive task but Whittle explains how, by dividing formula (b) above by the cost of mining a tonne of waste, the impact of inflation can be effectively removed. The new equation is thus:

$$\text{Value} = \{(3) \times (4) \times (5)/(6) - (2) \times (7)/(6)\} - (1) \quad \text{(c)}$$

In this, (7)/(6) is dimensionless and any inflation factors built in cancel out. This ratio is now referred to as (8). If the ratio (5)/(6) is inverted, it becomes equivalent to grade, as the dollar units

Table 5.4 Typical output from 4D Whittle programming showing cash-flows for the worst case mining schedule

Analysis of Four-Dimensional Pit Optimization Results file

Results file: WPRES.DAT
Price of metal ($/unit): 18.00
Cost of mining ($/t): 1.50
Maximum mining per period (t): 2000 000
Discount rate (% per period): 10.00
Pit number 13 (D)

Processing method	Ore type	Recovery ratio	Processing cost	Cut-off	Processing method	Maximum tonnes
Mill	Ore	0.950	14.03	1.71	Mill	350 000
Heap	Ore	0.650	4.80	0.41	Heap	1 000 000

Worst case schedule, with each bench completed before the next is started:

Period		Method	Tonnes	Metal	Grade	Cash-flow	Discounted cash-flow
1	Rock		2 000 000			−3 000 000	−3 000 000
		Mill	199 782	489 767	2.45	5 573 076	5 573 076
		Heap	677 474	545 277	0.80	3 127 866	3 127 866
						5 700 942	5 700 942
2	Rock		1 927 916			−2 891 874	−2 602 687
		Mill	350 000	991 251	2.83	12 041 650	10 837 485
		Heap	856 021	711 943	0.83	4 220 832	3 798 749
						13 370 608	12 033 547
3	Rock		1 493 085			−2 239 627	−1 814 098
		Mill	350 000	1 107 500	3.16	14 029 497	11 363 892
		Heap	656 597	538 579	0.82	3 149 711	2 551 266
						14 939 580	1 210 060
4	Rock		1 222 022			−1 833 033	−1 336 281
		Mill	350 000	1 182 691	3.38	15 315 269	11 164 831
		Heap	625 095	486 978	0.78	2 697 183	1 966 247
						16 179 419	11 794 797
5	Rock		1 232 687			−1 849 031	−1 213 149
		Mill	251 520	718 518	2.86	8 759 094	5 746 842
		Heap	743 935	609 797	0.82	3 563 743	2 338 172
						10 473 807	6 871 865
Totals:	Rock		7 875 710			−11 813 565	−9 966 215
		Mill	1 501 302	4 489 728	2.99	55 718 586	44 686 127
		Heap	3 559 123	2 892 575	0.81	16 759 336	13 782 299
						60 664 357	48 502 211

Table 5.5 Typical output from 4D Whittle programming showing cash-flows for the best case mining schedule

Analysis of Four-Dimensional Pit Optimization Results file

Results file: WPRES.DAT
Price of metal ($/unit): 18.00
Cost of mining ($/t): 1.50
Maximum mining per period (t): 2000 000
Discount rate (% per period): 10.00
Pit number 13 (D)

Processing method	Ore type	Recovery ratio	Processing cost	Cut-off	Processing method	Maximum tonnes
Mill	Ore	0.950	14.03	1.71	Mill	350 000
Heap	Ore	0.650	4.80	0.41	Heap	1 000 000

Best case schedule, with inner pits always mined out first:

Period		Method	Tonnes	Metal	Grade	Cash-flow	Discounted cash-flow
1	Rock		2 000 000			−3 000 000	−3 000 000
		Mill	260 050	664 197	2.55	7 710 563	7 710 563
		Heap	752 793	622 066	0.83	3 664 769	3 664 769
						8 375 322	8 375 332
2	Rock		1 379 960			−2 069 940	−1 862 946
		Mill	350 000	1 043 259	2.98	12 930 976	11 637 879
		Heap	660 811	566 317	0.84	3 337 020	3 003 318
						14 198 056	12 778 250
3	Rock		831 017			−1 246 525	−1 009 685
		Mill	350 000	1 178 349	3.37	15 241 011	12 345 219
		Heap	349 230	301 761	0.86	1 854 295	1 501 979
						15 848 781	12 837 512
4	Rock		951 978			−1 427 967	−1 040 988
		Mill	350 000	1 172 342	3.35	15 138 293	11 035 815
		Heap	388 030	341 342	0.88	2 131 151	1 553 609
						15 841 477	11 548 437
5	Rock		1 934 015			−2 901 023	−1 903 361
		Mill	177 837	402 831	2.27	4 394 254	2 883 070
		Heap	1 000 000	787 001	0.79	4 407 913	2 892 032
						5 901 145	3 871 741
6	Rock		778 740			−1 168 111	−689 758
		Mill	13 415	28 751	2.14	303 489	179 207
		Heap	408 259	284 088	0.70	1 364 188	805 540
						499 567	294 989
Totals:	Rock		7 875 710			−11 813 565	−9 506 738
		Mill	1 501 302	4 489 728	2.99	55 718 586	45 791 753
		Heap	3 559 123	2 892 575	0.81	16 759 336	13 421 246
						60 664 357	49 706 262

cancel leaving a value (9) equal to units of product per number of tonnes. Formula (c) can thus be rewritten as:

$$\text{Value} = \{(3) \times (4)/(9) - (2) \times (8)\} - (1). \text{(d)}$$

Thus the problem is significantly simplified as only a possible range of values for factor (9) need be considered.

If (9) is decreased, the value of the ore-blocks increases (the waste-blocks remain unaffected) and the new design so produced totally encloses the original design. Repeating this exercise will produce a set of up to 40 nested designs. Computing time is reduced by using the highest likely value of (9) which will generate the largest pit, and then excluding all blocks outside it. Graphs of optimal pit volumes, ore and waste tonnages and average grades can then be prepared for each value of factor (9). If a realistic cost of mining waste can be assessed, then these reserve values can be plotted against metal price (5) (for (5) = (6)/(9)). Examples are given in Whittle (1988). The designs can now be examined on the basis of a range of economic scenarios. Also, the sensitivity of factor (8), i.e. (7)/(6), can be tested by rerunning the program with (8) − 20% and (8) + 20%. The impact of mining sequence, for a given set of economic criteria and production rates, can also be tested. Table 5.4 shows the discounted (10%) and undiscounted cash-flows for a worse-case mining schedule which requires each bench to be completed to the ultimate bench contour before commencing work on the underlying bench. Table 5.5 is a repeat run for the best-case schedule in which inner pits are mined out first (each defined by a progressively lower pit base), i.e. the incremental push-back method. A small improvement in profitability has resulted.

REFERENCES

Alford, C. G. and Whittle, J. (1986) Application of the Lerchs–Grossmann Pit Optimization to the Design of Open-pit Mines, in *AusIMM/IE Aust Newman Combined Group Large Open Pit Mining Conference, October 1986*, pp. 201–7.

Annels, A. E. and Hellewell, E. G. (1988) The orientation of bedding, veins and joints in core; a new method and case history. *Int. J. Min. Geol. Eng.*, **5**, p. 307–20.

Calder, P. N. (1970) Slope stability in joined rock. The *CIMM Bull.*, **LXXIII**, p. 586–90.

Canadian Mining Journal (1988) Reference manual and buyers' guide, pp. 47–56.

HMSO (1988) *Technical Review of the Stability and Hydrogeology of Mineral Workings* (Report for the Department of the Environment), HMSO, London.

Henley, S. and Wheeler, A. J. (1988) The use of solid modelling in underground mine design, in *Computer Applications in the Minerals Industry, First Canadian Conference* (eds K. Fytas, J.-L. Collins and R. K. Singhal), Balkema, Rotterdam, pp. 359–364.

Johnson, T. B. and Mickle, D. G. (1971) Optimum design of an open-pit – an application in uranium *CIM*, **12**, (Spec. Vol.: Decision Making in the Mineral Industry), 331–338.

Lane, K. F. (1988) *The Economic Definition of Ore: Cut-off Grades in Theory and Practice*, Mining Journal Books, London, 149 pp.

Leigh, R. W. and Blake, R. L. (1971) A iterative approach to the optimal design of open pits. *CIM*, **12**, (Spec. Vol.: Decision Making in the Mineral Industry), 254–260.

Lerchs, H. and Grossman, I. F. (1965) Optimum design of open-pit mines. *Trans. CIM Bull.*, **58**(633), 47–54.

Priest, S. D. (1985) *Hemispherical Projection Methods in Rock Mechanics*, George Allen and Unwin, London, 124 pp.

Roberts, D. I., Hoek, E. and Fish, B. G. (1972) The concept of the mammoth quarry. *Quarry Managers' J.*, **56**, 229–38.

Taylor, J. B. (1971) Incorporation of access roads into computer-generated open-pits. *CIM*, **12**, (Spec. Vol.: Decision Making in the Mineral Industry), 339–43.

Walton, G. and Atkinson, T. (1978) Some geotechnical considerations in the planning of surface coal mines. *Trans. A* IMM, **87**, 147–71.

Whittle, J. (1988) Beyond optimization in open-pit design, in *Computer Applications in the Minerals Industry, First Canadian Conference* (eds K. Fytas, J.-L. Collins and K. Singhal), Balkema, Rotterdam, pp. 331–37.

BIBLIOGRAPHY

Barnes, R. J. and Johnson, T. B. (1982) Bounding techniques for the ultimate pit limit problem, in *Proceedings of the 17th APCOM Symposium* (eds T. B. Johnson and R. J. Barnes), SME of AIME, New York. pp. 263–73.

Carlson, T. R., Erickson, J. D., O'Brian, D. T. and Pana, M. T. (1966) Computer techniques in mine planning. *Mining Eng.*, May, 53–56.

Crawford, J. T. (1979) Open-pit limit analysis – some observations on its use, in *Proceedings of the 16th APCOM Symposium* (ed. T. J. O'Neil), SME of AIME, New York, pp. 625–634.

Francois-Bongarcon, D. and Marechal, A. (1977) A new method for open-pit design: parametrization of the final pit contour, in *Proceedings of the 14th APCOM Symposium* (ed. R. V. Ramani), SME of AIME, New York, pp. 573–583.

Grosz, R. W. (1969) The changing economics of surface mining: a case history, in *A Decade of Digital Computing in the Minerals Industry* (ed. A. Weiss), SME of AIME, New York, pp. 401–440.

Kim, Y. C. (1978) Ultimate pit limit design methodologies using computer models – the state of the art. *Min Eng.*, **30**(10), 1454–9.

Koenigsberg, E. (1982) The optimum contours of an open-pit mine: an application of dynamic programming, in *Proceedings of the 17th APCOM Symposium* (eds T. B. Johnson and R. J. Barnes), SME of AIME, New York, pp. 274–287.

Lemieux, M. (1979) Moving cone optimizing algorithm, in *Computer Methods for the 80's in the Mineral Industry* (ed. A. Weiss), Port City Press, Baltimore, pp. 329–345.

Lemieux, M. (1977) A different method of modelling a mineral deposit for a three-dimensional open-pit computer design application, in *Proceedings of the 14th APCOM Symposium 1976)* (ed. R. V. Ramani), SME of AIME, New York, pp. 557–572.

Lipkewitch, M. P. and Borgman, L. (1969) Two-and three-dimensional pit design optimization techniques, in *A Decade of Digital Computing in the Minerals Industry* (ed. A. Weiss), SME of AIME, New York, pp. 505–523.

Marino, J. M. and Slama, J. P. (1972) Ore reserve evaluation and open-pit planning, in *Proceedings of the 10th APCOM Symposium* (eds M. G. D. Salamon and F. H. Lancaster), South African Institute of Mining and Metallurgy, Johannesburg, pp. 139–144.

Phillips, D. A. (1972) Optimum design of an open-pit, in *Proceedings of the 10th APCOM Symposium* (eds M. G. D. Salamon and F. H. Lancaster), South African Institute of Mining and Metallurgy, Johannesburg, pp. 145–147.

Robinson, R. H. and Prenn, N. B. (1972) An open-pit design model, in *Proceedings of the 10th APCOM Symposium* (eds M. G. D. Salamon and F. H. Lancaster), South African Institute of Mining and Metallurgy, Johannesburg, pp. 155–63.

Wellmer, F. W. (1989) *Economic Evaluations in Exploration*, Springer, Berlin, 180 pp.

6

Financing and Financial Evaluation of Mining Projects

E.G. Hellewell

6.1 INTRODUCTION

In the previous chapters a range of techniques to assess the essential characteristics of mineral deposits has been considered. In particular, the quantity, quality and distribution of the potentially valuable components of a deposit are fundamental elements without which a decision to exploit cannot be soundly based. However, the geological factors alone are insufficient to determine whether a mining project should be established.

In all but a few exceptional cases, an adequate financial return from a mining project is the essential criterion which must be fulfilled before an affirmative decision to exploit is taken. These exceptional cases can arise where the working of a deposit is ancillary to a major civil engineering operation, for example in a land reclamation scheme or where an orebody is exposed in the course of construction work.

The vast majority of mineral exploitation projects are therefore undertaken for financial gain and the geological characteristics of the deposit are but one factor of many which collectively determine a project's profitability. Furthermore, it is not simply the total profit generated by exploitation which will govern the decision. In order for the project to produce profits, capital investment is necessary to fund the initial development of the orebody and establish the mining operation. The decision should therefore be based upon some measure of a project's profitability in relation to the capital investment and having due regard to the financial risk.

Although an adequate financial return is the first essential criterion, it should not be assumed that if this is fulfilled a decision to commence mining will automatically follow. Legal, social and political considerations are all relevant and in some instances a problem in one of these areas may prove intractable. In the UK, the development of land (including mining operations) is legally constrained principally by the Town and Country Planning Act (1971). Under this legislation, responsibility for granting planning permission, without which a mining operation is illegal, is vested in the County Councils which act as the Mineral Planning Authorities.

Financial evaluation must be regarded as a dynamic operation and not simply a 'once and for all' event. When examining a particular deposit, alternative strategies can be tested, including alternative mining methods, different production rates, various processing techniques, etc. Clearly, some combinations will produce superior financial results and the evaluation is used to select the optimum. For projects which have previously been rejected on economic grounds, periodic reappraisals can be made in the light of changed economic circumstances or technical innovation.

6.2 FINANCIAL ASPECTS UNIQUE TO MINING PROJECTS

Although mining is part of the wider industrial and commercial activity undertaken by a developed society, there are certain financial aspects unique to mining ventures which are not shared by other business enterprises. An understanding of these special features is essential if a satisfactory mining project evaluation is to be achieved.

The most significant feature which sets mining projects apart from every other commercial enterprise lies in the nature of its principal and indispensable asset, the mineral deposit. By the very act of mining, the deposit is depleted and when exhausted cannot be replaced. The term exhausted in this context does not necessarily imply that all the mineral has been extracted, but rather that continued exploitation is judged to be either uneconomic, or technically impracticable, at that particular point in time. There are many instances, of course, where a deposit has supported a series of separate mining projects stretching back in time perhaps hundreds of years. The successive mining phases may have been stimulated by changed economic circumstances or through improved technology. However, it is an inescapable fact that, in contrast to other business enterprises which are considered to have an unlimited life span, mining projects are finite life operations. In consequence, the profit stream generated by a mining project is of finite length. Furthermore, that portion of the capital used to purchase the deposit and the necessary working rights is lost when mining ceases.

A large proportion of the capital investment in a mining project is tied up in works necessary for the development of the mineral deposit. In the case of an underground mine, access to the deposit has to be gained by shafts, drifts or spiral ramps and the deposit is then opened up for working by a network of underground development roadways. In the event of the mine failing, the capital invested in these works is entirely lost. Furthermore, only a small fraction of the capital invested in the buildings, plant and machinery may be recovered due to the specialist and often purpose-designed nature of these fixed assets. In instances where the mine is situated in remote areas, the recovery of plant and machinery may be uneconomic and total loss is inevitable.

In all but the simplest surface mining schemes, there is a long lead-time before revenues are generated, this being especially true for deep mining operations. Few large deep underground mines can be developed in less than 5 years and some have taken considerably longer. During this development period capital is tied up in a non-revenue-generating operation and, in conseqence, an alternative investment opportunity is thus denied.

In addition to the initial period of capital investment when the mine is being developed, additional large capital expenditures may be required at intervals during the life of the mine. These may be to maintain productive capacity, to maintain profitability or for technical reasons, such as the provision of a safe working environment underground when, for example, a new ventilation shaft has to be sunk.

Mining operations are generally inflexible and are thus unable to respond rapidly to changed circumstances. For example, a mine may have its production constrained by limited underground haulage and winding capacity and be unable to take advantage of a sudden very profitable short-term increase in demand. On the other hand, for a mine to carry considerable spare production capacity with the intention of meeting what may be only infrequent demand surges is a clear under-use of assets and underemployment of capital.

In addition to the normally accepted commercial risk, there is always an additional risk element due to unknown or unforeseen adverse geological and mining circumstances. For example, in deep mining there is always some uncertainty with regard to the behaviour of rock masses when ore extraction commences, which if unfavourable may have a profound effect on the profitability of continued working.

By its very nature, a mining venture can only be sited where an ore-deposit occurs. There is, therefore, no possibility of taking advantage of a particular geographical location which could be favourable from climatic, transport, labour, political, investment incentive and many other factors, all of which can have a significant bearing on profitability.

Environmental issues are now assuming greater importance in modern society and all industrial, commercial and mining operations have, to a greater or lesser extent, adverse environmental consequences. A larger range of potential environmental problems are however inherent in the mineral extractive industry, which includes both deep mining and open-pit operations. Most operators now adopt a positive attitude to environmental matters and recognize their responsibility to society. This has resulted not only in larger recurrent expenditures to meet increasingly stringent and often legally enforceable environmental standards during the project's life, but also sizeable expenditures on site restoration and land after-care following cessation of working.

6.3 CAPITALIZATION OF MINING PROJECTS

6.3.1 The need for capital

A mining project is no different from any other business enterprise in its requirement for long-term funds, referred to as capital, the capital being needed for two distinct purposes. The majority of the funds purchase the fixed assets of the business without which the enterprise is incapable of generating revenue. The fixed assets therefore are the wealth generators and the type of assets will depend upon the nature of the business. In the case of a mining operation, the fixed assets will typically consist of the mineral deposit and its accesses, land, buildings, plant and machinery, vehicles, etc. In addition to the fixed assets a smaller proportion of the long-term funds supply the working capital of the business, without which a business is unable to meet its immediate debts. Working capital is strictly the difference between the current assets (mainly stock, debtors and cash) and the current liabilities (mainly creditors). Consideration of the nature of working capital will indicate that it does not generate wealth. Good management will therefore control working capital carefully, always ensuring there is sufficient cash to meet immediate debts, thus avoiding the danger of creditors forcing the business into liquidation, yet at the same time limiting the volume of this unproductive capital. The capital needs of the business can be met by issuing shares (share capital) and additionally in some instances by obtaining long-term loans (loan capital). In certain circumstances, government grants may be available, for example where it is considered desirable to stimulate economic activity in a particular region. Although initially such grants fall outside the strict definitions of either share capital or loan capital they eventually become part of the shareholders' interest.

6.3.2 Share capital

The permanent capital of a business is provided by the shareholders who, through their subscription for shares at the time of formation or through subsequent share purchase, are the owners of the business. This share capital is often referred to as 'risk capital' since, in the event of business failure, the owners do not receive any payment from the liquidation of assets unless all other creditors' claims have been met. The shareholders may therefore lose their total investment and in circumstances where a business is not operating as a company claiming limited liability status the shareholders may have to use their personal assets to satisfy creditors. The term 'permanent capital' has arisen because, in the case of a limited liability company, legal restrictions are placed on the repayment of the capital to the shareholders during the life-span of the business except in special circumstances or upon liquidation. The shareholders receive dividends on their investment but the payment of dividends is fundamentally dependent on the business making profits. In the case of limited liability companies, there are legal restrictions on dividend distributions which relate to making good previous losses. There are three main types of shares, preference shares, ordinary shares and deferred shares, each representing a different level of risk. In the event of liquidation, preference shareholders rank for payment before other shareholders and therefore preference shares represent the least risky form of business ownership.

Preference shares

These carry a fixed rate of dividend which is usually expressed as a percentage of the nominal value of the share. The preference shareholders, although they rank for payment of dividend before the ordinary shareholders, have no legal redress should the directors recommend that no dividend distribution be made. In that event, the preference shareholders who normally have no voting rights are entitled to vote at the general meeting of shareholders. However, if the preference dividend is passed, then no dividend can be declared on any other type of share.

Preference shares may also carry one or more of the following attributes:

Cumulative – previous passed dividends can be recouped in profitable years and any arrears of preference must be paid before any other distributions are made.
Participating – an additional variable dividend, usually related to the size of the ordinary dividend is paid along with a fixed dividend.
Convertible – preference shareholders have the option of converting their holding into ordinary shares.
Redeemable – the capital invested is normally repayable at the nominal value in a specified year or between two future dates.

Ordinary shares

The majority of the share capital is usually made up of ordinary shares. Ordinary shareholders are, with few exceptions, entitled to vote at general meetings and therefore exercise, in theory at least, their control of the business. Ordinary shareholders are entitled to all the profits of the business after preference dividends and tax (in the case of companies) have been deducted. However, the whole of the profit attributable to the ordinary shareholders is not necessarily distributed as dividends but can be retained to form the revenue reserves.

When a business does not require the whole of the proceeds of an ordinary share issue, for example when a long exploration programme is envisaged, partly paid shares may be issued. This means that the investor does not pay the full nominal value immediately, but is placed under an obligation to subscribe the uncalled portion at the request of the business. Sometimes ordinary shares are issued at a premium over their nominal value. In the case of a limited liability company the premium element must be shown separately on the balance sheet and repayment of the premium to shareholders is forbidden except on liquidation.

Deferred shares

These are somewhat uncommon and represent the highest risk to the holders. No dividend is payable unless the dividend to ordinary shareholders has reached a predetermined level or in some instances there is a deferment period after issue when no dividend is paid.

6.3.3 Loan capital

Loan capital is often referred to as debt capital, reflecting the legal requirement that almost all borrowings have to be repaid by a fixed date. Irredeemable loans do exist but these are rare. The general requirement to repay loan capital contrasts sharply with the share capital, which under normal circumstances can only be returned to the shareholders upon liquidation of the business. The loan capital holders are entitled to receive interest on the loans and this must be met irrespective of the profitability of the business. Strictly, it is liquidity and not profitability which determines whether creditors can be met, but in the longer term an unprofitable business will inevitably have liquidity problems. The rigid interest payment regime contrasts sharply with dividends on shares which can be deferred or omitted altogether.

If the business is unable to repay the loan at the appointed time or to provide the periodic interest payments, then the loan capital holders may bring legal proceedings against the business for recovery of debt. In the interim, the shareholders may lose control of the business. Ultimately the business may be forced to cease trading and its assets liquidated to meet these claims. Loan capital therefore carries a lower risk than share capital, but from the point of view of the business, loan capital increases the risk of failure. In consequence, inclusion of loan capital within the capital structure, known as 'financial gearing' or 'leverage', increases the risk to shareholders. A useful definition of gearing is the ratio of the loan capital to the total capital, the figure often being expressed as a percentage. The higher the gearing the greater the risk to which shareholders are exposed, this being particularly the case in times when adverse trading conditions prevail. During periods of prosperity, however, a high proportion of loan capital can increase dramatically the returns to ordinary shareholders. The effect of gearing on the return to ordinary shareholders of a company under different trading conditions is presented in Table 6.1.

The table shows that in a good year a company earning a 30% return on its capital invested can, with 50% gearing, provide a return of 32.5 pence per share to its ordinary shareholders after paying interest at 10% on its borrowings. This compares with a return of 19.5 pence per share if the company is capitalized entirely by ordinary shares, an increase to the shareholders of some 67%. In a bad year, when the company is only achieving a rate of return of 5% on capital invested, the entire operating profit is completely absorbed by the interest payments leaving nothing for the shareholders. Borrowing therefore increases the volatility of profits and the risk of borrowing is further increased when variable interest rates are involved. The directors face a difficult decision in determining the optimum gearing. There is the incentive to maximize debt for the good trading years, yet ensuring that the shareholders do not lose control during adverse trading conditions.

There are broadly two categories of loan cap-

Table 6.1 Effect of gearing on returns to ordinary shareholders

Financial gearing:	0%		50%	
Share capital (£1 ordinary):	£2000 000		£1000 000	
Loan capital (10% interest):	nil		£1000 000	
	Good year	Bad year	Good year	Bad year
Rate of return	30%	5%	30%	5%
Operating profit (£000)	600	100	600	100
Interest charges (£000)	nil	nil	100	100
Pre-tax profit (£000)	600	100	500	nil
Tax say 35% (£000)	210	35	175	nil
After-tax profit (£000)	39	65	325	nil
Earnings per share (pence)	19.5	3.25	32.5	nil

ital, that which is obtained directly from the general public and that which is obtained from banks and other financial institutions. However, in principle there is no difference between a loan raised from the general public and a loan from a bank or other financial institution. In the latter case the bank is effectively acting as an independent intermediary between the investing public and the borrower, with negotiations being undertaken in private.

If the loan is secured against the assets of the business, the loan is termed a debenture. Debentures may be secured by fixing or floating charges. In the former case, the debenture deed places a charge on specific assets, the most common being land and buildings and the debenture is then known as a mortgage debenture. The debenture holder has a legal interest in the specified assets which cannot be sold by the borrower unless the debenture holder relinquishes the charge. This is unlikely to occur unless an equally sound security is offered to the lender. The loan may be secured against unspecified assets in which case there is a floating or general charge on the assets. With a floating debenture, the holder has no legal interest in the assets unless the borrower fails to meet conditions set out in the debenture deed, for example if a default occurs in respect of interest payments, liquidation of the business, etc.

Loans which are not secured against assets of the business are referred to as unsecured loan stock. In the event of liquidation, the holders of this stock rank equally with other unsecured creditors. Debentures and unsecured loan stock are sometimes issued as convertibles and from a company's viewpoint this can be advantageous when compared to convertible preference shares since the interest charge on the former is deducted before Corporation Tax is computed.

6.4 FINANCIAL MODEL OF A MINING PROJECT

The capitalization of a project is clearly an important aspect, for without properly structured adequate funding the risk of failure is increased. However, capitalization is not of continuous concern during a project's life-span but is a matter which requires careful attention at the outset and subsequently is only reviewed at infrequent intervals. In contrast, the movement of funds or cash-flows resulting from the operation is a continuous and dynamic process. A sound understanding of these movements and their interrelationships is essential for a successful financial appraisal. The widespread availability of computer-based financial evaluation packages, although removing the drudgery of the arithmetic process, still demands a clear appreciation of the movement of funds and a knowledge of accounting procedures and terms if spurious results and misleading conclusions are to be avoided.

Figure 6.1 presents a financial model of a typical project showing the main movements of funds as the operation is progressing.

The capital and grants (if any) supplied to the project form a cash pool from which expenditure can be made. Expenditures are classified into two categories, these being capital expenditure and revenue expenditure. A clear distinction between the two has to be made since their respective treatments for the determination of profit (and also for tax purposes in most countries) is entirely different. In contrast to capital expenditure, which establishes an asset having a life exceeding that of the accounting period (trading period) in which the expenditure is made, revenue expenditure is entirely consumed during the accounting period either directly or indirectly in producing the revenue of the project. Expenditure for certain functions need not be exclusively either revenue or capital expenditure. Wages paid to stopers engaged in mineral production underground, which will ultimately provide revenue, are classified as revenue expenditure but wages of shaft-sinkers, who are establishing an asset (the shaft) would be considered as capital expenditure. The test, therefore, is the purpose for which the expenditure is being made.

Major categories of capital expenditure in a mining project would be the purchase and establishment of fixed assets, viz. land, the mineral

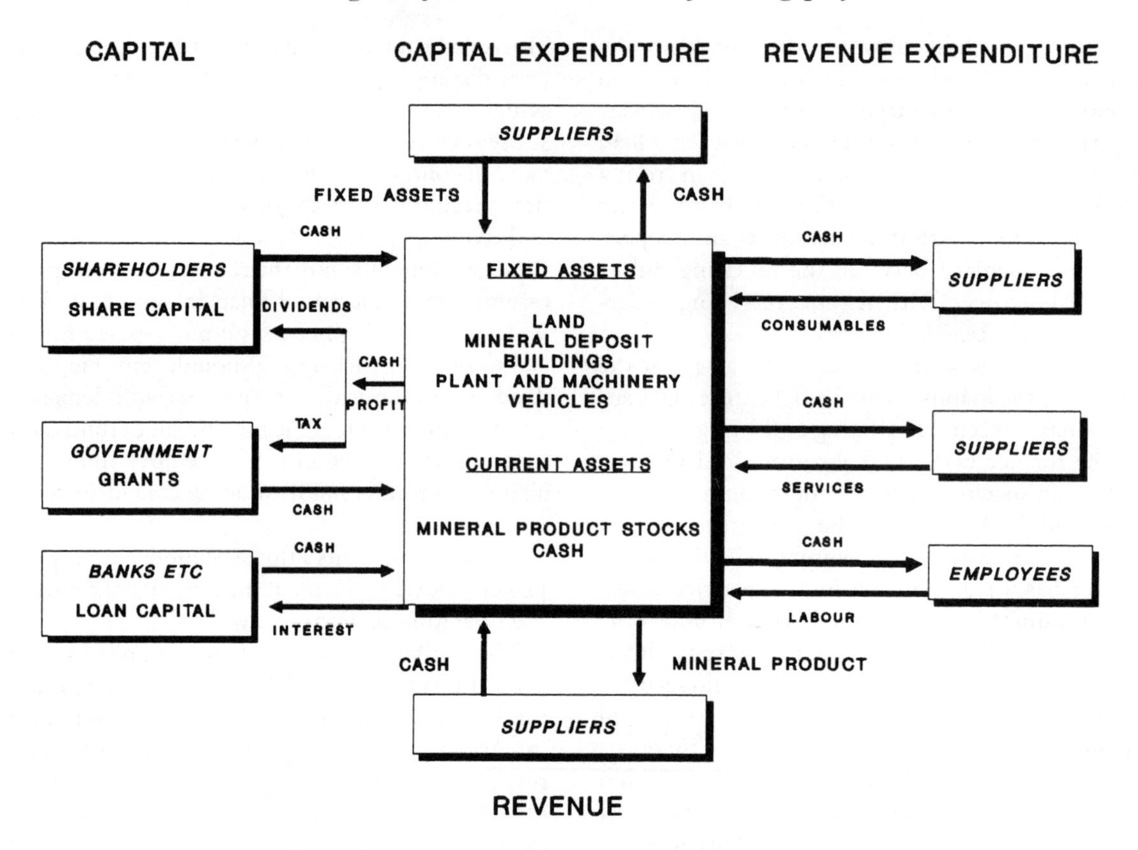

Fig. 6.1 A financial model of a typical mining project.

deposit and the necessary working rights, winning access to the deposit, buildings, plant, machinery and the provision of the essential infrastructure. The nature of capital expenditure is such that the cash-flows are usually large but generally of infrequent occurrence. From the point of view of the project, these fund movements can be considered negative as they are moving out of the project.

In contrast, revenue expenditures which are sometimes referred to as operating costs are usually smaller but occur regularly. Wages and salaries of employees, other than those engaged in establishing assets, is probably the most regularly occurring and largest category of revenue expenditure. Another readily identifiable category of revenue expenditure arises from the purchase from suppliers of consumables such as lubricating oils, fuels, power, reagents, explosives and the host of small items needed to run the operation. An item of revenue expenditure in this category, which is peculiar to mining projects, is the mineral lease payment to a lessor in circumstances when the freehold of the mineral and its associated working rights have not been purchased outright for a capital sum. A further revenue expenditure category would be for services such as insurance, rates, contract drilling, test work, etc. Since revenue expenditure involves the movement of funds out of the project, these are considered negative.

When the capital expenditures necessary to establish the project have been made and the project has become operational through revenue expenditures, mineral products will become available for sale. The funds generated by sales from the stocks of the mineral products are termed revenue, sales or turnover. From the point of view of the project, the revenue represents a movement of funds into the project and is therefore considered positive. In order for the project to operate effectively, not all the initial capital will be spent on fixed assets, a smaller proportion will remain within the project to finance the stocks of mineral product, to act as an immediate cash reserve and to fund any difference between the amount owed to the project by trade debtors which is not matched by the amount which the project owes to its trade creditors. This unproductive capital is referred to as working capital or net current assets.

The magnitude of the revenue, in relation to the revenue expenditure and interest charges on any loan capital, is the critical determinant of a project's financial health. For the project to remain viable in the shorter term, the revenue (R) must exceed the sum of the revenue expenditures (R_e) and the loan capital interest payments (I) (if any), each being taken on an annual basis. This statement can be expressed mathematically as:

$$R \geqslant R_e + I$$

This does not mean, however, that a project which cannot meet the above criterion in a particular trading year will automatically fail. Quite the contrary, there are many businesses which experience temporary cash-flow problems on the revenue side yet survive and continue to flourish. However, repeated failure to generate sufficient revenue to cover revenue expenditure and interest charges will inevitably lead to a winding up of the business, either voluntary or compulsory.

Despite the fact that a project may satisfy the above expression, it does not follow that the project will be allowed to continue in the longer term. The purpose of a business enterprise in a competitive, largely unregulated economy, is to provide profits for the owners (shareholders). If the profits are insufficient to satisfy the demands of the shareholders, they, as owners, have the ultimate right to close down the operation. There are several measures of profit which are relevant to project appraisal. The operating (trading) profit (P_o) is broadly the difference between the revenue (R) and the sum of the revenue expenditures (R_e) and depreciation (D), these being taken on an annual basis. This can be written as:

$$P_o = R - (R_e + D)$$

It is particularly important to note that, whereas the total annual revenue expenditure is used in the computation of profit, capital expenditures are completely excluded. It is only the depreciation charge for the year which can be properly set against the revenue. Depreciation at its simplest can be thought of as the loss of value suffered by the assets in the course of running the project during the year. A further important point is that depreciation does not involve a movement of funds and is best considered as a non-cash transaction.

When the annual loan interest charge I is deducted from the operating profit the result is the net profit before tax attributable to the shareholders, usually referred to as the pre-tax profit (P_p). Mathematically this can be written in full as:

$$P_p = R - (R_e + D + I)$$

One further aspect which is of direct concern to the shareholders is a possible tax charge on the profits. Internationally there is no uniform tax regime and much depends on the fiscal policy adopted by the particular government in power at the time. Corporate taxation is an exceedingly difficult and complicated area and only the briefest outline can be given of that pertaining to the UK. Here, corporation tax was introduced in 1965 as a separate tax on companies. Adjustments are made to the net profits before tax in order to obtain the company's taxable income. The major adjustment is with regard to book depreciation which is replaced by capital allowances (C), these being effectively statutory depreciation rates on fixed assets. Broadly, taxable income (TI) can be written as:

$$TI = R - (R_e + C + I)$$

It will be noted that, whereas the whole of revenue expenditure made in the year is allowed for tax purposes, only that proportion of capital expenditure allowed by the capital allowances provisions can be set against revenue.

The corporation tax charge (T) is then taken as some function of the taxable income.

$$T = f(TI)$$

From the point of view of the project, the tax charge represents a movement of funds out of the project and is therefore considered negative.

The net profit after tax attributable to the shareholders, referred to often as the after-tax profit (P_a), as a first approximation, can then be obtained from the following:

$$P_a = R - (R_e + D + I + T)$$

It is necessary to qualify the after-tax profit equation above by saying that the tax chargeable in a given year is dependent upon the timing and amount of dividends paid to shareholders. However, assuming that the tax charge could be paid in the same year that the profits are earned, then the annual after-tax profit is the measure of the change in the shareholders' wealth as a result of their investment in, and ownership of, the project. The whole of the after-tax profit is entirely attributable to the shareholders and as such could be distributed as dividends if the cash position allowed. Usually a company will not distribute the whole of the profit and the undistributed portion, often called retained earnings, accumulate to form the revenue reserves which are available for reinvestment.

In practice, the various elements which contribute to the after-tax profit, and the after-tax profit itself, are to be found in the financial statement known as the profit and loss account. This is drawn up at the end of each trading period which is usually of 1 year's duration. A further financial statement relevant to project evaluation is the balance sheet. Although this statement is drawn up on the same date as the profit and loss account, the balance sheet does not refer to financial events which have occurred during the period, but it simply lists the assets and liabilities of the project at the specific date. Of particular importance is the capital employed in the project and this is the sum of the fixed assets and the working capital shown on the balance sheet at that date. It will be realized that the capital employed will vary during the project's lifespan, mainly as a result of depreciation reducing the book value of fixed assets.

6.5 FINANCIAL EVALUATION TECHNIQUES

Prior to discussing the various evaluation techniques, some consideration needs to be given to the purpose and limitations of this exercise. The fundamental aim of an evaluation is to determine whether or not a project is worthwhile in financial terms. Several techniques are in common use, each employing a different basis upon which the project is examined and compared with a predetermined standard. Those projects reaching or exceeding the standard are then considered financially worthwhile. In the case where there are several alternative projects for consideration, financial evaluation can be used to rank these in order of preference. It is however necessary to reiterate the caution raised in the introduction to this chapter. The fact that a project is demonstrated to be financially worthwhile is of itself an insufficient basis for the project to go ahead. Financial attractiveness is but one element of many relevant to the decision process and financial evaluation must be regarded as a tool to assist in the proper consideration of all pertinent factors. Clearly though, a project which fails to meet the financial criterion will, in all but the most exceptional cases, be rejected irrespective of its merits in other directions.

In the remainder of this chapter, it is assumed that, after due consideration of the relevant technical factors, an exploitation design has been formulated by a project team and a financial evaluation is now required. Irrespective of the financial evaluation technique to be employed, estimates of the various financial elements of the model previously discussed are first made. In practice, an integrated approach is adopted whereby specialists having detailed knowledge of the financial implications of the various aspects of the project each make a contribution. For example, the mineral economist will provide estimates of the likely pattern of demand for the product(s) and the likely selling prices, therefore allowing a revenue projection to be prepared. Other specialists will provide estimates of capital expenditures and their timing, revenue expenditures appropriate to the chosen mineral exploitation design, taxation, etc. An outline of the necessary financial data required for each evaluation technique is considered when the basis of each method is discussed in the following sections.

There are four financial evaluation techniques in common use, these being:

1. return on capital employed;
2. payback period;
3. discounted cash-flow net present value;
4. discounted cash-flow internal rate of return.

It is true to say that, despite disadvantages inherent in the first two methods, these traditional approaches still dominate, although the trend is towards one or other of the discounted cash-flow variants. This move has been facilitated by the ready availability of a large number of computer-based packages which remove much of the tedium of repetitive calculation associated with these methods. Furthermore, decision-makers in industry generally have an increasing awareness of the underlying principles of discounted cash-flow, and although misconceptions still arise, they are now more willing to accept the techniques which have been developed.

In order to compare the four evaluation techniques and to highlight the differences between them, the same project, that of a hypothetical short-life mine named the Eureka Mine Project, will be examined using each method in turn. Detail costings of the project are outside the scope of this chapter, the aim being to illustrate the principles underlying each technique.

6.5.1 Return on capital employed

This technique is based upon the concept inherent in the well-known financial analysis ratio 'return on capital employed' (ROCE), which is the customary measure of business profitability. The method is sometimes called the accounting rate of return. The ROCE, however, can be defined in a variety of ways and there is unfortunately no one definition which has gained universal acceptance. One common definition is:

ROCE = (average after-tax profit/average capital invested) × 100%

The average after-tax profit is obtained from projected profit and loss accounts whereas the average capital employed is extracted from projected balance sheets, each of these financial statements being prepared on an annual basis for the duration of the project.

This evaluation technique can be used in two ways. First, it can assist in evaluating a single project where acceptance is dependent upon the ROCE exceeding the minimum rate of return demanded by shareholders. Second, in the case where several projects have to be ranked in order of preference, the greater the ROCE the more desirable the project.

The projected profit and loss accounts for the 4 years of operation of the Eureka Mine Project are given in Table 6.2, the main components being self-explanatory. The exceptional item arising in year 4 is the recovery of the working capital at the end of the project. The extracts from the projected balance sheets given in Table 6.3 show the fixed asset values at the start and end of each year together with the average working capital. The sum of the fixed assets and working capital is equal to the capital employed which can be determined on an annual basis and an average over the life of the project then found. The ROCE is then simply computed from the average annual profit and average capital employed over the project's life-span as below:

$$\begin{aligned} \text{ROCE} &= 152\,500 \times 100\% \,/\, 587\,500 \\ &= 26\% \end{aligned}$$

Table 6.2 Eureka Mine Project – projected profit and loss accounts

	Year			
	1	2	3	4
Revenue	*840*	*950*	*1040*	*870*
less				
Revenue expenditure	450	480	500	460
Depreciation	250	250	250	250
Operating profit	140	220	290	160
less				
Interest charges	20	20	20	20
Exceptional item	—	—	—	(100)
Pre-tax profit	120	200	270	240
less				
Tax	40	50	70	60
After-tax profit	80	150	200	180

Figures given are £000.
Average annual after-tax profit = £152 500.

Table 6.3 Eureka Mine Project – extracts from projected balance sheet

	Year			
	1	2	3	4
	Capital employed £000			
Fixed assets				
Start of year	1000	750	500	250
less depreciation	250	250	250	250
End of year	750	500	250	nil
Average fixed assets	875	625	375	125
Average working capital	100	100	100	50
Capital employed	975	725	475	175

Average capital employed = £587 000.

Although the technique is well founded upon accepted accounting practice and therefore readily understood by most managements, it suffers from several disadvantages. Unfortunately, there are alternative bases for ROCE in addition to the one used for the Eureka Mine Project and there is therefore the possibility of manipulation to produce the desired result. When presenting a case based on ROCE, the basis of calculation should always be stated. Clearly, considerable caution should be exercised when interpreting a ROCE in the absence of such information. Irrespective of the definition adopted, the method suffers from the disadvantage that the timing of the profits and duration of the profit stream are ignored.

6.5.2 Payback period

As its title suggests, this technique simply determines the time taken for the initial capital invested, a negative cash-flow, to be recovered by the stream of annual positive cash-flows, these being the annual surplus of revenues over revenue expenditures after deduction of tax. It will be noted that this method utilizes cash flows and not profits, therefore there is no necessity to estimate depreciation rates.

The method can be used in two ways. First, it can assist in evaluating a single project where acceptance is dependent upon the payback period being shorter than a predetermined standard. Second, it can be used to rank projects in order, the shorter the payback period the more desirable the project.

The projected cash-flows for the Eureka Mine Project are summarized in Table 6.4. Cash-flows into the project, by convention, are considered positive whilst conversely, cash-flows out of the project are treated as negative. Although revenues are being generated and revenue expenditures made on a continuous basis throughout each year, it is assumed, for the sake of simplicity, that these totals arise at the end of each year, that is at times 1, 2, . . . years. Similarly, it is assumed that the payment of tax occurs at each year end. On the other hand, the capital expenditures are assumed to be incurred at the very start of the project, that is at time 0. It should be noted that the interest charge on the loan capital is excluded despite this being a cash-flow. If this item were to be included, double counting would occur since it is a payback

Table 6.4 Eureka Mine Project – projected cash-flows

	Time (yr)				
	0	1	2	3	4
Cash in (+ve)					
Revenue (turnover)	—	840	950	1040	870
Recovery of capital	—	—	—	—	—
Total cash in	—	840	950	1040	870
Cash out (−ve)					
Fixed assets	1000	—	—	—	—
Working capital	100	—	—	—	—
Revenue expenditure	—	450	480	500	460
Tax	—	40	50	70	60
Total cash out	1100	490	530	570	520
Net cash flow	(1100)	350	420	470	450

Figures given are £000.

of part of the capital. The net cash-flows, being the difference between the total cash in and the total cash out, are easy to determine. Table 6.5 takes these net cash-flows and presents the cumulative cash-flows from which it can be determined that the payback period in this example is just under 3 years.

The method suffers from the major disadvantage that it does not consider either the project's profitability, the total return generated or the timing of the cash-flows. The method is also ambiguous in circumstances where the cash-flows are reversing over a period of years, for in these cases the initial capital investment, and conequently the payback period, are hard to define.

6.5.3 Discounted cash-flow

In contrast to the previous methods, which ignore the timing of profits and cash-flows, the discounted cash flow (DCF) techniques recognize the time value of money. In an unregulated economy there will be a group of individuals (lenders) who are prepared to hire out money to other individuals (borrowers) provided that both parties can reach agreement on the terms of the transaction. One of the important considerations

Table 6.5 Eureka Mine Project – cumulative cash-flows and payback period

	Time (yr)				
	0	1	2	3	4
Net cash-flow	(1100)	350	420	470	450
Cumulative cash-flow	(1100)	(750)	(330)	140	590
	Payback period = 3 years				

Figures given are £000.

is the reward the lenders receive for foregoing an alternative use of their money which, if the transactions were made directly between the parties, will be equal to the charge which the borrowers are prepared to pay for the temporary use of that money. The reward or charge is known as the interest rate and it is a proportion of the sum involved per unit time (usually 1 year). The interest rate can be expressed as a percentage per annum or alternatively as a decimal per annum, the difference simply being a factor of 100. If a sum of £1 is lent for n years at an interest rate r (expressed as a decimal per annum), then the amount at the end of this period using the familiar compound interest principle will be £$(1 + r)^n$. This is equivalent to saying that £1 now (at present) is worth £$(1 + r)^n$ in n years time. Conversely, a sum of £1 in n years time is worth £$(1 + r)^{-n}$ now, or a sum of £1 has a present value of £$(1 + r)^{-n}$. The term $(1 + r)^{-n}$ is known as a discount factor and all future sums can be expressed in present value terms, i.e. discounted by multiplying them by appropriate discount factors. The interest rate r is known as the discount rate.

Discounted cash-flow net present value (NPV)

The net present value (NPV) of a project can be found as the sum of its discounted cash-flows. If A_i are the project cash flows arising at times 0 to n years and r is the discount rate then,

$$NPV = \Sigma_{i=0}^{n} A_i(1 + r)^{-i}$$

In the case where a single project is being considered if the NPV is zero or positive then the project can be accepted. When projects have to be ranked in order of preference, the larger the NPV the more desirable the project.

Using the net cash-flows for the Eureka Mine Project already derived in Table 6.4, these are discounted assuming a discount rate of 15% p.a. (0.15 p.a. as a decimal) and the results presented in Table 6.6. The NPV of the project is easily found by summation. It will be noted that the interest charge on the loan capital has again been excluded as this represents a return on the loan portion of the capital.

The Eureka Mine Project gives a NPV of £88 397 and is therefore acceptable according to the NPV rule which requires a project to have a zero or positive NPV. This represents the immediate increase in wealth which would result from a capital investment of £1100 000 in the project, the capital being raised at 15% p.a. This can be illustrated in Table 6.7 which presents a hypothetical lenders account showing an initial borrowing of £1188 397. The annual positive cash-flows generated by the project are sufficient to service the debt at 15% p.a. and repay the capital borrowed. The difference between the borrowed sum of £1188 397 and the sum of £1100 000 invested in the project, i.e. £88 397, is the immediate increase in wealth due to the project. An alternative interpretation is that the sum of the NPV (£88 397) and the initial capital investment (1100 000), i.e. £1188 397, is the highest price which could be paid for the project without

Table 6.6 Eureka Mine Project – discounted cash-flows and NPV @ 15% discount rate

	Time (yr)					
	0	1	2	3	4	NPV
Net cash-flow	(1100)	350	420	470	450	
Discount factor $(1 + 0.15)^{-i}$	1	8700	7561	6575	5718	
Present value	(1100)	304	318	309	257	88

Figures given are £000.

Table 6.7 Eureka Mine Project – lenders account at an interest rate of 15%

Time (yr)	Opening balance*	Interest @ 15%*	Total debt*	Repayment from projected cash-flow*	Closing balance*	Time (yr)
0	1188	178	1366	350	1016	1
1	1016	152	1169	420	749	2
2	749	112	861	470	391	3
3	391	59	450	450	nil	4

*Figures given are £000.

suffering financial loss assuming that finance is costing 15% p.a.

Discounted cash-flow internal rate of return (IRR)

The discounted cash-flow internal rate of return (IRR), which is sometimes called the DCF yield, is defined as that interest rate r which, when used to discount the cash-flows of a project, produces a net present value of zero. It can be described mathematically as the solution of the following equation:

$$0 = \Sigma_{i=0}^{n} A_i(1 + r)^{-i}$$

Since this is a polynomial, the solution can be obtained by an iterative technique or by graphical means. The IRR for the project can then be compared with a predetermined minimum rate of return and accepted if the IRR is equal to, or is in excess of, this minimum.

Taking the cash-flows for the Eureka Mine Project which have been used previously, these are discounted at rates of zero, 5, 10, 15 and 20% and the corresponding NPVs presented in Table 6.8. It can be seen from the table that the project has an IRR greater than 15%, the corresponding NPV being +£88 000, but less than 20% corresponding to a negative NPV of −£27,000. Plotting the NPV against the discount rate r in Figure 6.2 results in the curve intersecting the discount rate axis at a value of 18.7%, this value being the IRR for the project.

Particular note should be taken of the fact that the IRR is quite different from the ROCE which was discussed previously. Furthermore, the IRR is not the rate of return on the initial capital

Table 6.8 Eureka Mine Project – cash-flows discounted at various rates

	Time (yr) 0	1	2	3	4	NPV
Net cash-flow	350	420	470	450		
PV (discounted @ 0%)	(1100)	350	420	470	450	590
PV (discounted @ 5%)	(1100)	330	381	406	370	390
PV (discounted @ 10%)	(1100)	318	347	353	307	225
PV (discounted @ 15%)	(1100)	304	318	309	257	88
PV (discounted @ 20%)	(1100)	292	292	272	217	(27)

Figures given are £000.

Table 6.9 Eureka Mine Project – meaning of internal rate of return

Time (yr)	Capital invested	Return on capital @ IRR of 18.7% p.a.	Cash-flow	Capital recovered	Capital invested	Time (yr)
0	1100	206	350	140	956	1
1	956	179	420	241	714	2
2	714	134	470	336	378	3
3	378	71	450	379	(1)*	4

Figures given are £000.
*Due to rounding.

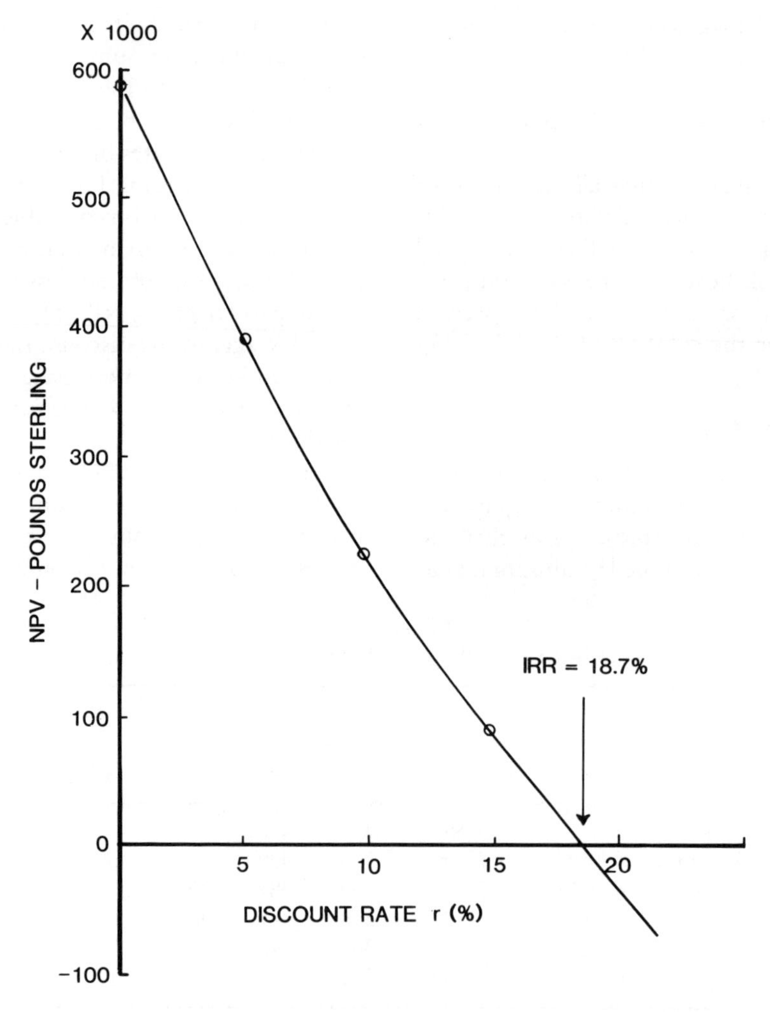

Fig. 6.2 Plot of *NPV* versus discount rate for the Eureka Mining Project.

investment as often supposed, but is the return on the unrecouped part of the capital. One portion of each positive cash-flow provides a partial return of the capital whilst the remainder is the return on the unrecovered capital. Table 6.9 illustrates this essential concept.

Whilst the IRR technique can certainly determine whether a particular project is financially worthwhile, it cannot be used to rank projects in order of preference. A contrary and incorrect view is, however, widely held. Briefly, the reason for the inability to rank projects lies in the implicit assumption in the IRR method that, as each portion of the capital is recovered, it can be reinvested for the remaining life-span of the project at an interest rate equal to the IRR of the project. A further illustration is provided in Figure 6.3 which shows NPV v. discount rate for two projects A and B. The projects rank equally at a discount rate of 11%. Below this figure Project A is preferred whilst at a higher figure Project B becomes preferable, the respective internal rates of return of 14.5% and 18% being irrelevant. A further disadvantage of the IRR method arises if the polynomial has multiple roots giving rise to multiple internal rates of return. This situation can arise where a relatively large cash outflow occurs near the end of the project life-span.

In the preceding discussion of the two variants of discounted cash-flow, several references have been made to the discount rate without considering its meaning in detail or how it is to be

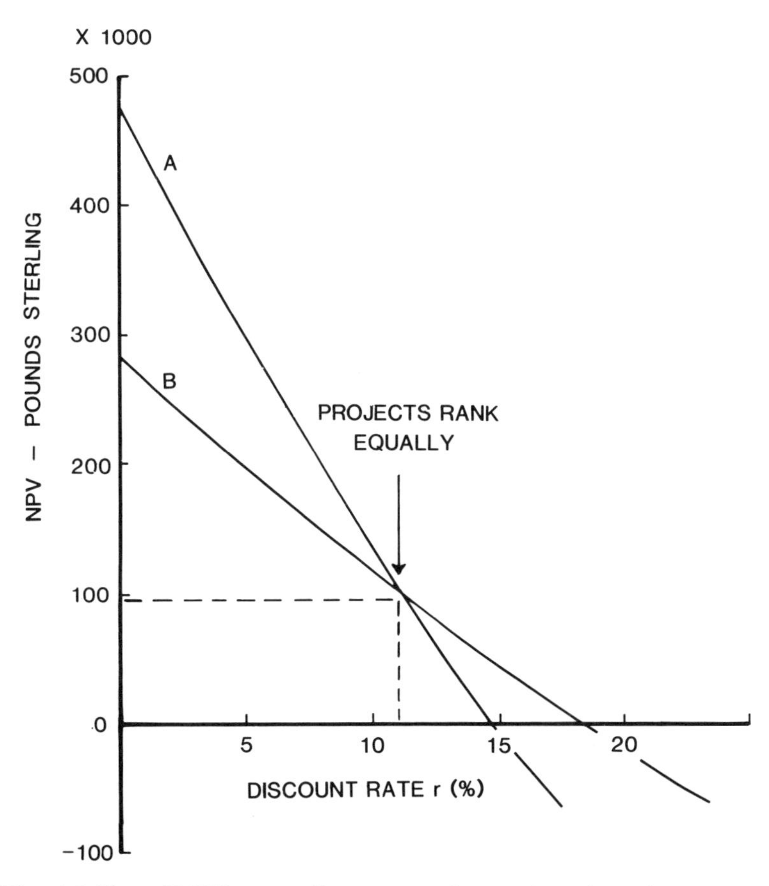

Fig. 6.3 Plot of *NPV* versus discount rate for two hypothetical projects A and B.

determined. In the NPV variant the rate is usually referred to as the cost of capital whereas in the IRR method it is often referred to as the hurdle rate. The rate, however, is the same in both cases and is most usefully and correctly called the cost of capital. The cost of capital to a business can be defined as that interest rate which when used to discount all the cash-flows of the business neither increases nor decreases the wealth of the business. In the NPV method, if the cash-flows of a project are discounted at the cost of capital and the NPV is found to be positive, then clearly the project is financially worthwhile because it is generating a return in excess of the cost of capital to the business, the magnitude of the NPV indicating the present value of the wealth increase. In the alternative IRR method if the computed internal rate, which is the earning power of the project, is greater than the cost of capital then the project is worthwhile, but in this instance no indication of the magnitude of the wealth increase is obtained. A business can obtain its capital from a range of sources, these being discussed previously, each source having an associated cost. The various classes of share capital will each have a different cost to the business related to the expected returns to shareholders, whilst different forms of loan will attract different interest rates. Even retained earnings, which appear to be a 'free' source of capital, have a cost, this being effectively the return foregone in an alternative investment. In principle, therefore, a weighted average cost must be obtained which is related to the proportion and cost of each constituent source. Furthermore, in the case of loans, since interest charges on this capital are deductible in the profit and loss account before the tax assessment is made, the cost of capital must reflect this fact. There are many theoretical and practical difficulties in determining the cost of capital and a detailed consideration is beyond the scope of this chapter. Reference should be made to textbooks on finance and investment appraisal for further information on this subject.

An additional problem can also arise with any appraisal method which involves the time value of money. The discussion so far has assumed that the purchasing power of the currency will remain constant during the life-span of the project. If this is not the case then the arguments developed must be modified to allow for this further factor. Usually the general price changes in the economy are increases and are referred to as price inflation. Perhaps the simplest way of dealing with this aspect is to estimate future cash-flows assuming that inflation will occur as a specified rate and use the market rate of interest as determined by the weighted cost of capital. The market rate of interest will have the market's perception of future inflation already included in the figure and no further adjustment is required to this interest rate. Clearly problems arise when different components of the cash-flow are subject to different and or variable inflation rates and a full discussion of this is beyond the scope of this chapter.

BIBLIOGRAPHY

Barnes, M. P. (1980) *Computer Assisted Mineral Appraisal and Feasibility*, American Institute of Mining, Metallurgical and Petroleum Engineers Inc., New York, ch. 4, 167 pp.

Gentry, D. and O'Neil, T. J. (1984) *Mine Investment Analysis*, Society of Mining Engineers, American IMM and Petroleum Engineers Inc., New York, 502 pp.

Lumby, S. (1984) *Investment Appraisal*, Van Nostrand Reinhold (UK), Wokingham, 323 pp.

Merrett, A. J. and Sykes, A. (1973) *The Finance and Analysis of Capital Projects*, Longman, London, 573 pp.

Reid, W. and Myddelton, D. R. (1988) *The Meaning of Company Accounts*, Gower, Aldershot, 351 pp.

Samuels, J. M. and Wilkes, F. M. (1981) *Management of Company Finance*, Thomas Nelson, Walton on Thames, Surrey, 497 pp.

7

Grade Control

7.1 INTRODUCTION

In some mining operations, one of the most important roles played by the geologist is the supervision of grade control. In an open-pit operation, this involves the sampling of blast-hole cuttings produced by down-the-hole hammer drills and the classification of bench reserves into ore, low grade or waste material, or into various metallurgical types. In some cases, grade control may also involve the sampling of truck or shovel loads to ensure that rock is assigned to the correct stockpile or waste dump. In underground operations, grade control may involve the mapping and sampling of stope faces, sampling of tram car loads or draw-point muckpiles, broken rock at a recently blasted face, jackhammer cuttings or diamond drill cores. It is the geologist's job to ensure that mining is closely following the mineralized zone and that overbreak during stoping is kept to a minimum. The main purpose of grade control is thus to ensure that material being fed to the mill is of economic grade and that large fluctuations in grade are minimized by blending ores from different benches, or parts thereof, or from different stopes. It is essential that mill feed be kept as close as possible to that called for in the original design specification of the mill and concentrator. Regular reconciliations will be required between the estimated stope grades, the grades indicated from stope/truck sampling and those reported by the mill. It is essential that this is undertaken so that modifications can be made to sampling practice or to the methods or parameters used to calculate grade, tonnage or contained metal.

The following discussion of grade control techniques will be subdivided on the basis of open-pit and underground operations and will be largely undertaken via a series of case histories.

7.2 OPEN-PIT OPERATIONS

7.2.1 Sampling of blast-hole cuttings

Open-pit benches are usually in the range 6–15 m thick and are usually drilled by blast-hole rigs along lines parallel to the free face and at spacings of between 3 m and 10 m, depending on the degree of weathering and rock type (Figures 7.1, 7.2). Square, rectangular or offset (parallelogram) grids of holes thus result. Holes are usually 17 cm (6.75 in) to 31 cm (12.25 in) in diameter and are drilled through the bench for a short distance into

(a)

(b)

Fig. 7.1 (a) Blast-hole rig at RTM's Cerro Colarado Cu-Au Pit in Spain. (b) Blast-hole drilling in a bench at the Ingerbelle Pit, British Columbia.

Fig. 7.2 Blast-hole drilling in a bench in the Zortman/Landusky operation, Montana. (Photograph courtesy of Pegasus Gold Corporation.)

the underlying bench. This 'sub-grade' drilling, as it is called, is usually in the range of 0.5–1.5 m and is designed to ensure that no hard toes are left to impede the operations of the shovel/bucket at the face. Because of the potential contamination of the pile of cuttings (Figures 7.3, 7.4) by this sub-grade mineralization, it is necessary to either remove the sampling device once the base of the bench is reached or scrape off the top layer of the pile before taking the sample.

Many different methods have been devised for sampling the cuttings. These range from the use of trowels (Figure 7.5), or shovels, to remove equal quantities of material from different sides of the pile to the use of wedge-shaped sample cutters (Figure 7.6(a)) or pipe cutters (Figure 7.6(d)). At Montana Resources' Continental Pit, near Butte in Montana, 10–12 lb of cuttings are collected in a box-shaped cutter made of sheet steel approximately 13 in high × 18 in long × 2 in wide, which is placed radially from the collar of the hole to be drilled (Figure 7.6(b)).

A variation of this technique is used at the Golden Sunlight Mine (Placer Dome US, Inc.), also in Montana, where the cutter is designed to match the profile of a typical pile of cuttings and which collects 18–20 lb of sample for analysis (Figures 7.6(c), 7.7). Prior to this, a pipe cutter had been experimented with but this had been rejected in favour of the profile cutter. The former consisted of a 3–4 in diameter and 2 ft 6 in long metal pipe with an aperture cut out of its side as in Figure 7.6(d). The pipe is pushed into the pile and then tilted backwards to retain the sample (Figure 7.8).

At Brenda Cu-Mo Porphyry, at Peachland in British Columbia, tests have been made to determine the best sampling technique using wedge-shaped cutters, shovel cuts and pipe cuts. The latter method was finally adopted as it is quicker,

Fig. 7.3 Blast-hole cuttings at the Montana Tunnels operation, Montana, USA.

although perhaps slightly less accurate. The mine samplers collect approximately 5 lb of cuttings from 8 to 12 pipe cuts around the pile, once the subgrade has been removed.

At the Nickel Plate Mine, Hedley, British Columbia (Mascot Gold Mines Ltd.) an 'auto-sampler' has been designed which is based on an original design by Cominco. The sampler (Figure 7.6(c)) consists of a 5 in diameter PVC pipe 2 ft long which is suspended at a fixed angle beneath the drill-deck and which catches 2–5 lb of sample per 10 ft length of hole drilled. The cuttings enter the tube via a 10 in long aperture. It has been found that the length of pipe, the size of the aperture and the angle of inclination are crucial and are specific to each deposit. Numerous modifications were made until the results of sample assays gave good agreement with those produced by taking standard cuts from the piles. It has been found that the assays obtained by this method give good block-grades which are in close agreement with the recovered grade of the block after mining. The major advantage of the method is that mine samplers are no longer needed, for the driller collects his own samples through a hinged plate in the drill-deck. The samples are bagged immediately and sent to the assay laboratory. Considerable improvement in the 'turn-around' time is thus gained which is critical for an open-pit operation.

Although care should be taken to obtain a representative sample from the pile of cuttings, it must be remembered that, for a standard 7.5 × 7.5 m pattern of drilling in a 12 m bench, using 25 cm diameter holes, the amount of rock hanging on each hole is approximately 1850 t. A 2 kg sample from a pile of cuttings is then taken which is assumed to be representative of 1.6 t of drilled rock. How representative this sample is depends partly on the quality of sampling and partly on the uniformity, or otherwise, of the mineralization in the block.

A comparison of routine blast-hole cutting grades with those produced by repeat sampling was undertaken at Brenda (Johnston and Blackwell, 1986) to investigate the quality of sampling and to search for evidence of bias. A plot of routine copper grade against check sample grade gave a regression line with a slope of 45° which was encouraging but the spread of points increased with grade (Figure 7.9). It was suspected that this could be due to bias produced by segregation of particle sizes during air flushing of the cuttings. The piles were thus sectioned vertically and horizontally and grades determined for each particle size distribution. The finer fractions

Fig. 7.4 Close-up of blast hole rig at Montana Tunnels showing cuttings.

clearly showed higher grades than the coarser fractions, but no systematic particle size distribution could be recognized in the pile. It was thus concluded that the grade variability was due to natural causes, i.e. the variability in grade of the *in situ* rock.

7.2.2 Definition of grade control limits and reserves

The methods by which grade control lines are drawn on bench plans and then flagged in the pit, and the techniques used to compute the reserves in each category, vary from pit to pit depending on local conditions. This subject will thus be covered by presenting a series of case histories.

Case history I – Golden Sunlight Mine, Montana

This operation possesses one of the largest gold reserves in northwestern USA (53 million tons @ 0.05 oz/ton) with an overall stripping ratio of 5.5 : 1. Sixty per cent of the ore occurs within a pipe-shaped mass of breccia and the remainder in the adjacent country rocks (latite porphyry and Precambrian sedimentary rocks). Ninety-five per cent of the gold is free (5–200 μm) and is associated with pyrite. Some minor copper is also present which is probably derived from the sedimentary rocks.

Blast-hole drilling varies with rock type, and is approximately on a 15 × 15 ft pattern. Six and three-quarter inch

Fig. 7.5 Trowel sampling of blast hole cuttings, Montana Tunnels.

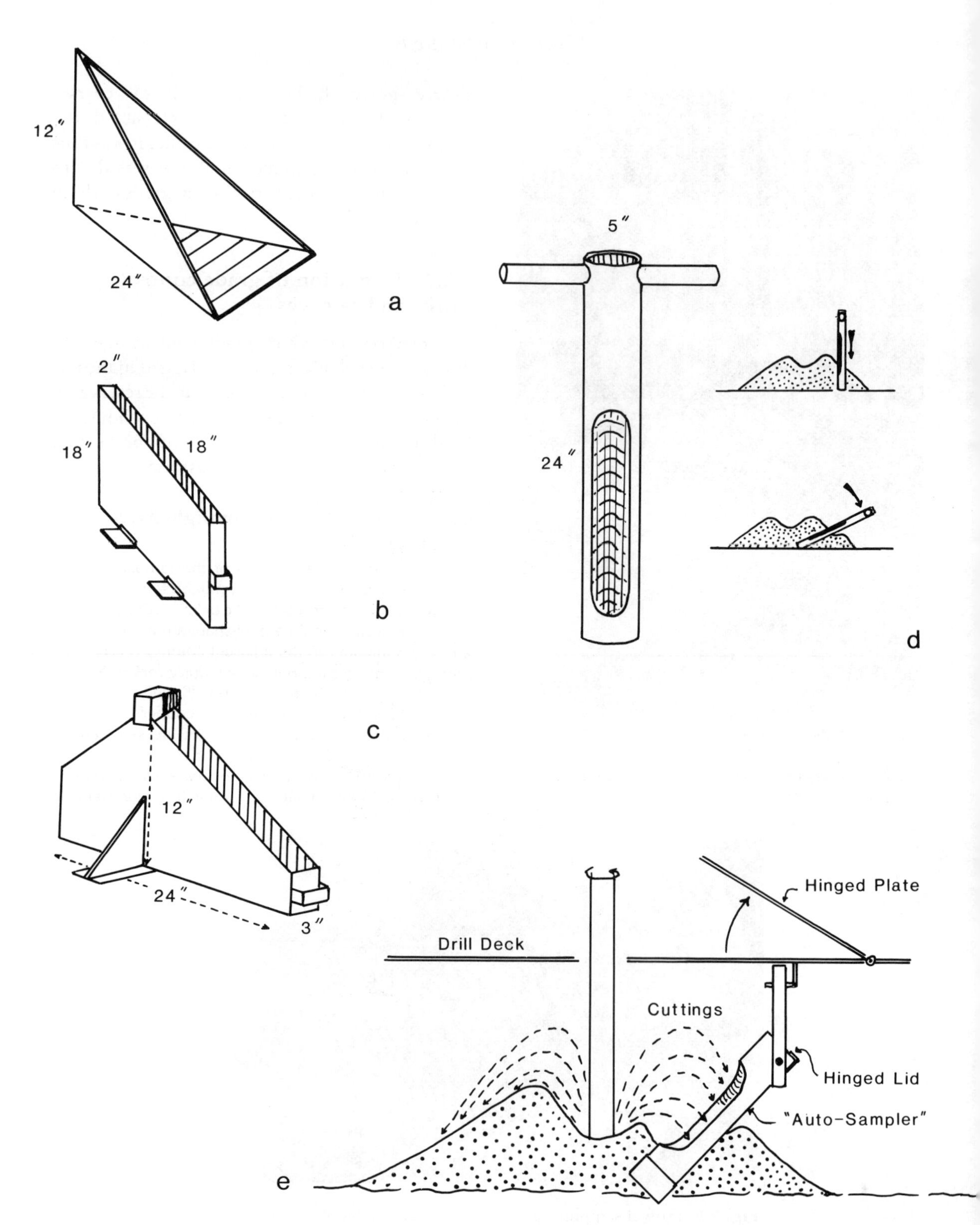

Fig. 7.6 Sampling devices for blast-hole cuttings. (a) Wedge cutter. (b) Box cutter. (c) Profile cutter. (d) Pipe cutter. (e) Auto-sampler beneath drill-deck.

Fig. 7.7 A profile and pipe cutter as used at Golden Sunlight Mine, Montana.

Fig. 7.8 A pipe cutter in use at Golden Sunlight Mine, Montana.

holes are drilled to depths of 30 ft to allow 5 ft of subgrade drilling beneath 25 ft benches. The samples from each hole are assayed by AAS for Au and Cu but every fifth hole is fire assayed. These fire assays are then ratioed with the respective routine AAS values to produce a factor which is then applied to those samples which have not been fire assayed. Generally, it is found that an up-grade of as much as 30% may occur. The copper content of the mineralization is critical for once this exceeds 0.02%, it changes the grade control classification. Above this level, the consumption of cyanide in the plant becomes excessive. The cut-off grade for gold ore is variable depending on price and copper content.

The current (1989) classification of mineralization is detailed in Table 7.1.

The high copper ore is either stockpiled, or blended with low copper ore, to produce a suitable mill feed.

The values used to assign blocks of ground to one or the other of these categories are based on point kriging on an 8 × 8 ft grid over a specified area of a bench. An assay data file is updated by the input of blast-hole coordinates, via a digitizing tablet, together with grade data. The limits of the area of the blast are also defined using the digitizer. The kriging program cuts any high grade gold values to 0.5 oz/ton and then uses standard relative semi-variogram parameters to krige the defined

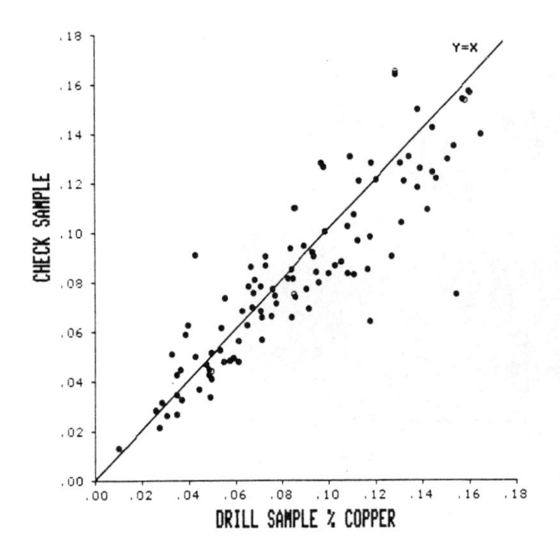

Fig. 7.9 Comparison of check sample copper grades with routine blast-hole sample grades, Brenda, British Columbia. (From Johnston and Blackwell, 1986.)

area. The current relative semi-variogram for the mine conforms to a nested spherical scheme model whose parameters are Co = 0.5, $C_1 = 0.15$, $C_2 = 0.1$, $a_1 = 50$ ft and $a_2 = 160$ ft. Figure 7.10 shows a typical example of kriged gold grades in a bench, in this case all grades run less than 0.02% Cu. This diagram shows the position of the free face of the bench at the time and the back wall of the pit. The geologist positions contours between values of 0.019 and 0.020 and between 0.014 and 0.015, to define areas of ore, high grade waste and waste. Having done this, grade control lines are drawn in as straight lines smoothing out the minor fluctuations in the original contours (Figure 7.10). In defining ore, the smallest mining unit (SMU), a 25 ft square block, and geological features are taken into account. The average of all kriged grades in each defined area can then be calculated, along with digitized areas, so that tonnages and grades for each grade control type in the bench, or current blast area, can be specified.

The corners of each of the grade blocks can then be marked in the pit using stakes to which coloured triangles are stapled, each pointing in the direction of the relevant ore type or waste.

Case history II – Montana Tunnels, Jefferson City, Montana (Pegasus Gold Inc.)

This open-pit is exploiting gold, silver and base-metal mineralization in a diatreme consisting of quartz-latite breccias. The sulphides occur as disseminations and fracture fillings and veins within these breccias. Reserves remaining at the end of 1987 were quoted as 50.3 million tons @ 0.025 oz/t Au, 0.49 oz/t Ag, 0.25% Pb and 0.65% Zn.

However, early in 1989, the reserves were recalculated in response to a change in the pit design necessitated by the results of additional infill drilling and an analysis of slope stability. This exercise decreased the reserves to 37.6 million tonnes @ 0.7 g/t (0.02 oz/ton) Au but with little change in the base-metal grades. A deterioration in stripping ratio also resulted from 2.9 : 1 to 3.3 : 1.

The blast-hole pattern for 30 ft benches is 15 × 15 ft in the diatreme and 12 × 12 ft in the quartz-latite. Three scoops of cuttings (Figure 7.5) are taken from each pile and combined to form a blast-hole composite. Three separate AAS assays for gold and base-metals are then made on each of these composite samples and, if the gold results are erratic, a further three analyses are made or a single fire assay. Every fifth hole is, however, routinely fire-assayed for Au and Ag. Erratic fluctuations in grade reflect the presence of free gold particles producing a high nugget effect. Individual average metal grades for each hole are then converted to gold equivalent values by the application of factors which are based on mill recoveries and smelter value of metal produced. No allowance is made for copper, which is not recovered. Typical factors for March 1988 (now obsolete) are incor-

Table 7.1 1989 Classification of mineralization

Ore type	Grade	Flag
High copper ore	⩾0.02% Cu, ⩾0.027 oz/t Au	Green
High copper, low grade sub-ore	⩾0.02% Cu, 0.020–0.026 oz/t Au	Yellow Ore
	<0.02% Cu, >0.02 oz/t Au	Red
High grade waste	0.015–0.019 oz/t Au	White
Waste	<0.015 oz/t Au	Blue

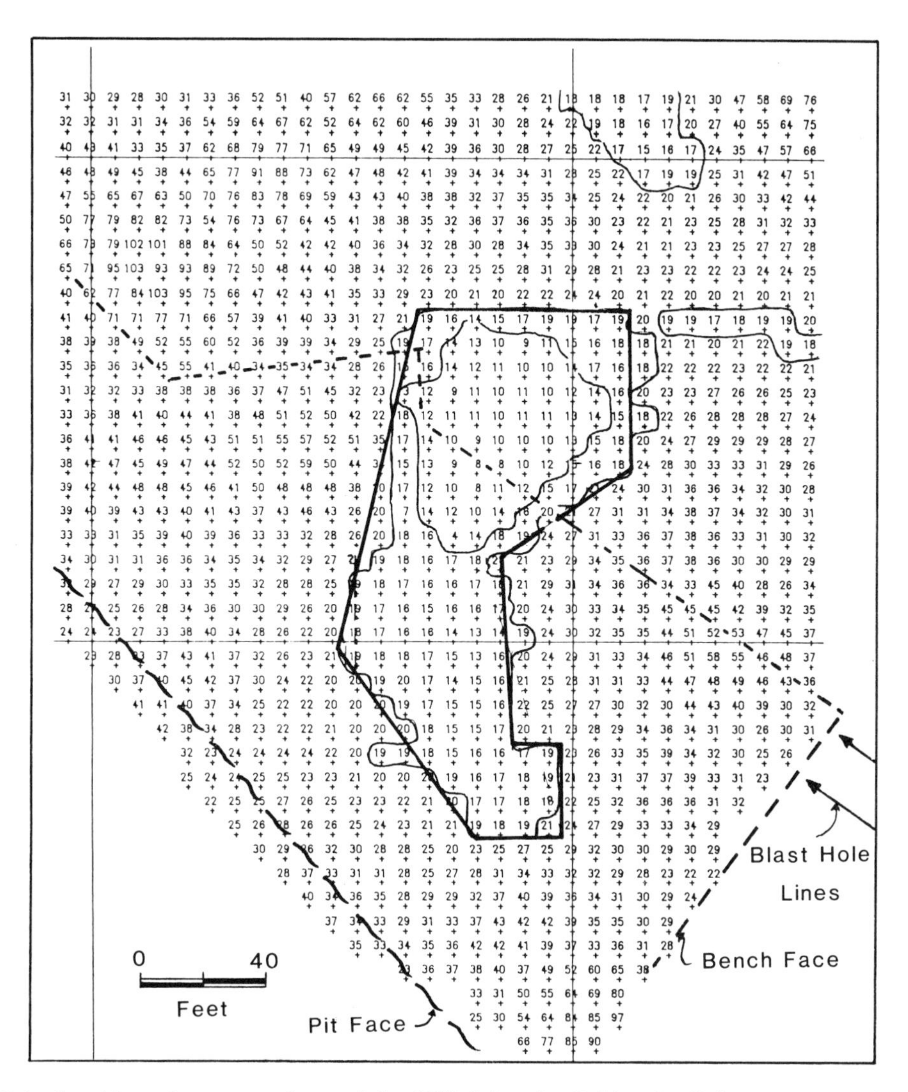

Fig. 7.10 Kriged gold grades on portions of the 5550 ft bench, Golden Sunlight, Montana, showing waste (<0.015 oz/t), high grade, low copper waste (0.015–0.020 oz/t) and ore (>0.02 oz/t) together with grade control lines.

porated in the following equation:

Total gold equivalent value =
Au + Ag/91 + Pb/400 + Zn/75

A typical blast-hole grade would be 0.036 oz/ton Au, 0.630 oz/ton Ag, 0.29% Pb, 1.29% Zn which would produce a gold equivalent of:

0.036 + (0.63/91) + (0.29/400) + (1.29/75) =
0.061 oz/t

If the gold assay value exceeds 0.2 oz/t, then it is cut to this level.

The ends of lines of blast-holes are surveyed in the pit and then plotted on the relevant bench plan, intermediate holes being assumed to be uniformly spaced. All holes are then digitized to produce an assay file of grades and coordinates. A program called 'Poly Cad', produced by Geostats Systems of Lakewood, Colorado, is then used to create polygons around each hole within specified areas of the current bench. At the edges of the data

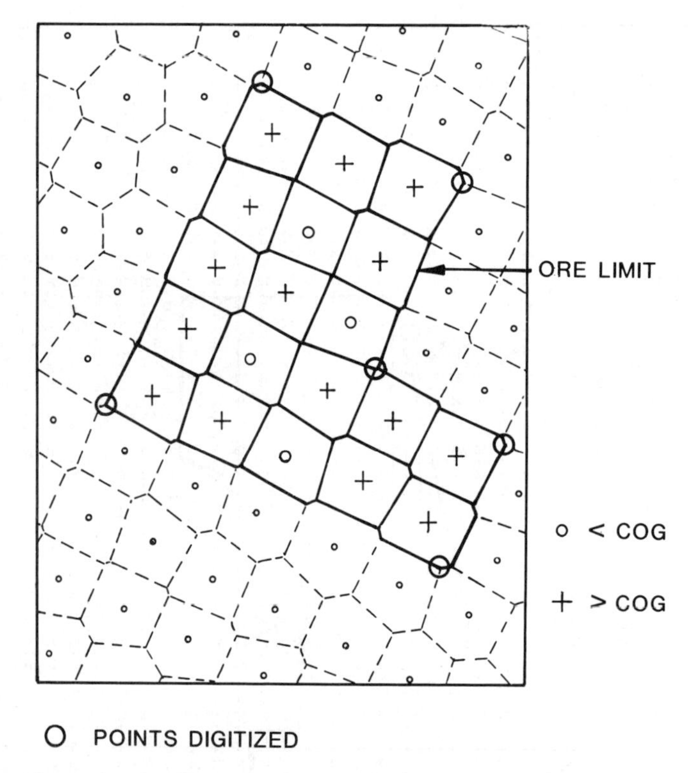

Fig. 7.11 Blast-holes and associated polygons at Montana Tunnels Gold Mine, Montana, showing those above COG ('+') and the limits of ore as defined by the geologist.

fields the polygons are extrapolated for a distance of 15 ft. The program compares each gold equivalent grade with the current COG (0.6 g/t Au eq.) and marks all holes exceeding this value with a cross, lower-grade holes remain as a dot (Figure 7.11). The grade values of any particular hole appearing on the monitor can be examined by moving the screen cursor to its position. By scrutiny of the distribution of positive and negative polygons, the limits of ore- and waste-blocks can be defined using the CAD software and the screen cursor. The program then kriges the grades in these blocks and produces overall grades and tonnages. Experience has shown that the mill feed grade is higher than that calculated from the blast-holes within equivalent ore-blocks. This is perhaps partly explained by the presence of subvertical veinlets which have been missed by the vertical blast-holes.

The ore-blocks are flagged in the pit and are then mined by use of a bulldozer which pushes broken ore to a loader. Unfortunately this produces considerable smearing of ore and waste which could be partly avoided by use of a shovel.

Case history III – Brenda, Peachland, British Columbia

This open-pit is exploiting a copper-molybdenum porphyry in the biotite-hornblende-quartz diorite marginal facies of a granodiorite stock. Mineralization is concentrated in the most heavily fractured areas and, indeed, most of the sulphides occur as infillings of a series of subvertical fractures. The pit, which commenced operations in 1970, is exploiting reserves of 108.7 million tons grading 0.165% copper and 0.04% molybdenum which will be extracted at a stripping ratio of 0.5 : 1.

Twelve and a quarter inch blast-holes are drilled in the pit at intervals of 24 ft parallel to the face and 30 ft perpendicular to it. The cuttings are analysed for copper and molybdenum and a copper equivalent value is calculated for each hole. Hole locations on the bench plans are coloured on the basis of whether the relevant grade meets the requirement for Mill Feed (red), Low Grade Stockpile Ore (green) and Waste (blue). Figure 7.12 shows a portion of the 5060 ft bench in 1977 when these categories were defined on the basis of ≥0.4%, 0.3–

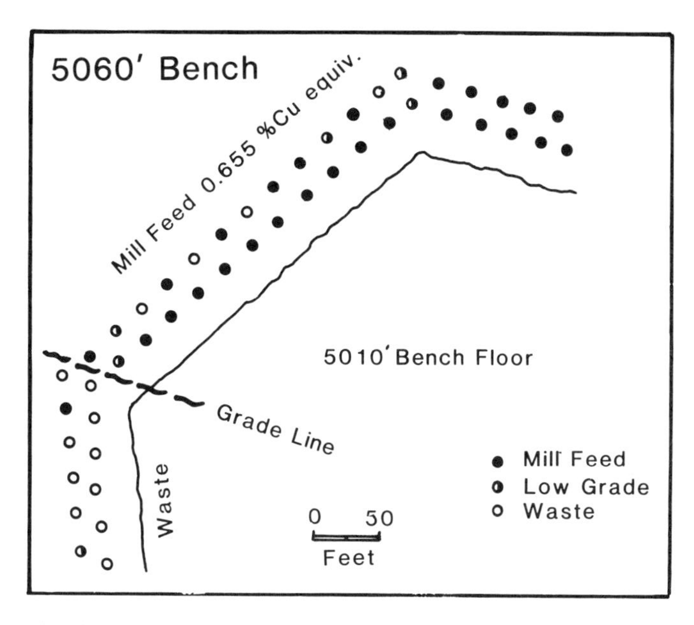

Fig. 7.12 Blast-hole grades classified as ore (≥ 0.4% Cu equiv.), low grade (0.3–0.39% Cu equiv.) or waste (< 0.3% Cu equiv.), Brenda 5060 ft Bench.

0.399% and <0.3% respectively. The grade control line has been drawn on this plan using groups of holes to guide its location rather than single holes and, if possible, they are drawn at right angles to the free face. Insufficient holes in the Low Grade category are present to warrant the definition of this ore type on the bench and thus only waste and mill-feed are outlined. The grades of the material blocked out in this fashion are calculated by averaging hole values and the mill is then notified of the expected grade of ore to be trucked in the immediate future. After blasting, the grade lines are marked on the broken rock with stakes connected with string and coloured ribbons. These also hang vertically down the face to help the shovel operators discriminate between grade control types.

Case history IV – Ingerbelle, Penticton, British Columbia

This pit, operated by Similkameen Division of Newmont Mines Ltd, is now exhausted but was working a low grade prophyry copper deposit in highly altered Triassic andesitic volcanics (Nicola Group) and intrusives close to the SW flank of the Lost Horse Intrusion, a porphyritic monzonite-syenite stock. Mineralization was highly erratic and required careful grade control procedures. Although the sulphides (chalcopyrite and pyrite) are dominantly disseminated, locally they may be concentrated in randomly orientated fractures.

Although a total mineral inventory of 73 million tons existed, Raymond (1979) states that only 30 million tons of this was of ore grade at a 0.2% copper cut-off. The overall ore grade was estimated as being 0.53%. The pit was worked using 40 ft benches and the final pit had an overall stripping ratio of 2.6 : 1. Mining took place at approximately 8 million tons per annum using 9.8 in blast-holes on approximately 22 ft centres, two rows at a time (Figure 7.1(b)).

Ingerbelle has been chosen as a case history for it demonstrates an interesting evolution of grade control techniques (Raymond, 1979). Raymond's paper is recommended to the reader as a classic example of the use of geostatistical techniques in grade control. What follows is a synthesis of this paper and information gained by the author on a visit to the mine in 1977.

In 1971, at the commencement of mining, all blast-hole grades in the pit were colour coded relative to the cut-off grade and then polygons were drawn around each ore grade hole as shown in Figure 7.13. Because of the rapid fluctuation of grade between ore and waste, the polygons defined areas in the bench whose sizes were often less than, or close to, the SMU for a 10 yd^3 shovel. Grades of ore-zones were then calculated by averaging all blast-hole grades contained within their boundaries. Attempts to flag and mine these small, rather irregular, areas were largely unsuccessful as poor correlation was obtained between estimated and recovered grades. The latter were 20% lower than anticipated.

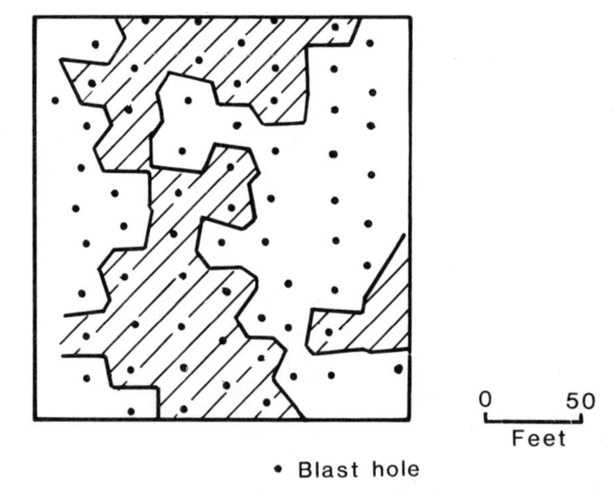

Fig. 7.13 Polygons fitted to individual blast hole grades outlining areas of the Ingerbelle bench exceeding 0.20% (the cut-off grade). After Raymond 1979.

In 1973 grab sampling from the production shovel was instituted so that a composite sample was produced which was taken to be representative of five or six truck loads. Each sample was crushed and pulped using a mobile mill in the pit (Figure 7.14) and analysed by XRF techniques in an adjacent portable unit, giving a 15 minute turn-around. This allowed control of the final destination of the trucks. However, the comparisons between the recovered and estimated grades showed that the mill feed grade was still being overestimated by 15%. Significantly, high grade areas in the pit were losing tonnage to the waste tip while in low grade areas, waste material was being classified as mill feed. Thus, in 1975, grab sampling was restricted to areas of marginal grade, whereas low and high grade areas were estimated manually by grouping and averaging blast-hole assays. This resulted only in a marginal improvement and thus in 1976 a manual random kriging method was introduced using 50 ft × 50 ft × 40 ft ore-blocks.

Figure 7.15(a) shows such a 50 ft × 50 ft grid superimposed over the blast-hole grades so that, on average, four to five holes lie in each block. To reduce the effect of the variability of the blast-hole grades, block averages were calculated as in Figure 7.15(b). Directional semi-variograms of block grade were then computed for groups of blocks constituting 500 ft × 500 ft areas. These indicated that an almost isotropic model existed for Ingerbelle. The decision was made to use random kriging to re-evaluate each block grade, using only this block and the surrounding eight blocks (e.g. block D3 in Figure 7.15(b)). As only limited distances were involved, it was also decided to apply a Linear model to the semi-variogram shown in Figure 7.16 as at low lags it is almost linear. Figure 7.17 presents the weighting factors (a_1, a_2 and a_3) applied to each of the nine blocks, which are a function of the ratio of the intercept *Co* of the semi-variogram to its slope. Figure 7.17 also shows the weighting factors for a block surrounded by peripheral blocks on three sides and for another surrounded on two sides only. In the example in Figure 7.16, the intercept is 0.025 (% Cu)^2 while the slope is 0.016 per 100 ft, giving a ratio of 1.56/100 ft. However, a standard ratio of 0.7/100 ft was adapted for the pit as a whole on the basis of the results of a more detailed study of many 500 ft × 500 ft blocks.

Fig. 7.14 A mobile mill unit in the Ingerbelle Pit used to speed up the preparation of grade control samples.

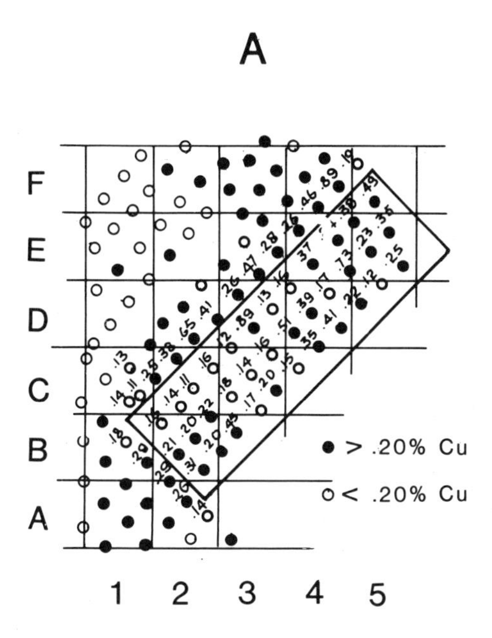

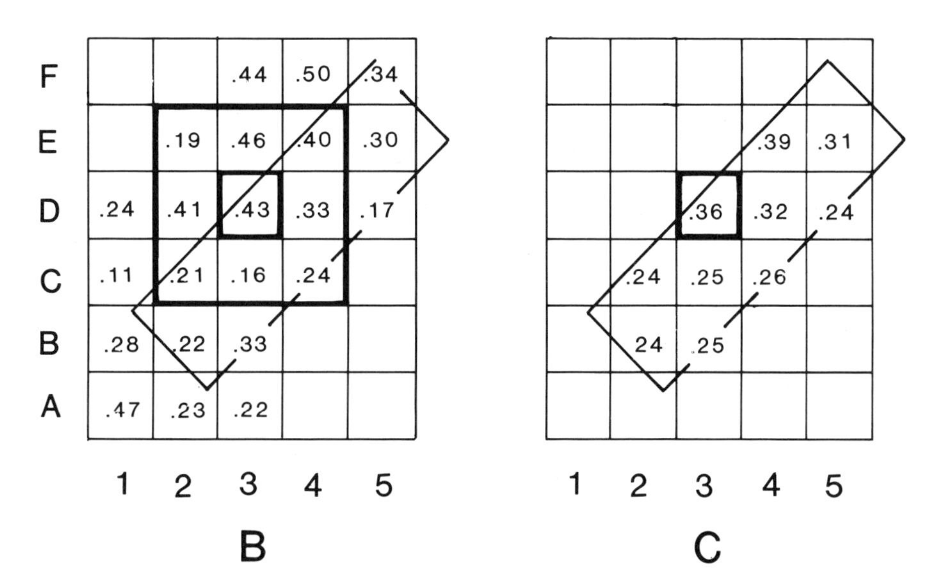

Fig. 7.15 (a) Blast-hole grades (% Cu). (b) Block average grades. (c) Kriged grades, Ingerbelle Pit, British Columbia.

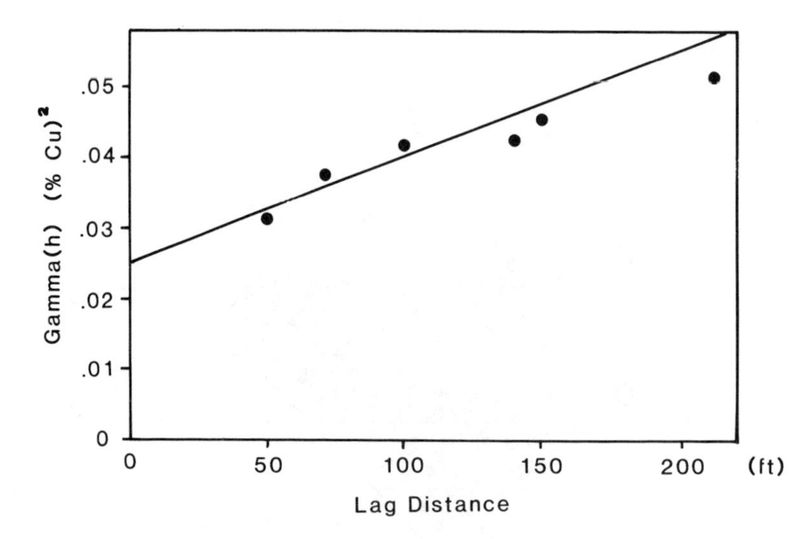

Fig. 7.16 Linear semi-variogram used for random kriging of 50 × 50 ft blocks.

The method of weighting the block grades can be illustrated by taking the group of nine blocks outlined in Figure 7.15(b) with a central grade of 0.43. For this central block, model A is used on Figure 7.17. The kriged grade is thus:

$$G_{0.43} = (0.43 \times 0.42) + [(0.46 + 0.33 + 0.16 + 0.41)/4] \times 0.38 + [(0.19 + 0.40 + 0.21 + 0.21)/4] \times 0.20 = 0.36\%$$

where 0.42 is the weighting factor (a_1) for the central block, 0.38 the factor (a_2) for the top and side blocks shared four ways and 0.20 the factor (a_3) for the corner blocks also shared four ways. Note that $a_1 + a_2 + a_3 = 1$. If the group of nine blocks had been surrounded by barren material to the N and W, then the calculation for block E2 (0.19%), using model C, Figure 7.17, would have been:

$$G_{0.19} = (0.19 \times 0.575) + [(0.46 + 0.41)/2] \times 0.355 + (0.43 \times 0.07) = 0.29\%$$

and the calculation for block E3 (0.46%), using model B, Fig. 7.17, would have been:

$$G_{0.46} = (0.46 \times 0.48) + [(0.19 + 0.40)/2] \times 0.31 + [(0.41 + 0.43 + 0.33)/3] \times 0.2 = 0.39\%$$

Figure 7.15(c) shows the results of random kriging for the current blast area (outlined). Each can now be weighted by the portion of each block within this area to gain an overall grade.

Those blocks with kriged grades lying between 0.15% and 0.29% would have been grab sampled and for these, a grade equal to the sum of 50% of the grab and 50% of the kriged grade would have been assigned. On this basis, these marginal blocks would have been sent to the waste tip or to the mill. Blocks with kriged grades of less than 0.15 would be automatically sent to waste. This method resulted in an increase of ore tonnage sent to the mill without further grade losses.

Late in 1976, the system was computerized to speed up the updating of the bench reserves as new information became available. This time isotropic spherical scheme models were fitted to the semi-variogram and kriged grades calculated for 12.5 ft centres. A typical semi-variogram from Raymond (1979) is shown in Figure 7.18. This highlights the very high nugget effect of this deposit ($\varepsilon = 0.92$) and a relatively small range ($a = 160$ ft), both of which would have contributed to a poor precision for the estimates. The block grades were then contoured manually to define the ore limits in the bench, prior to flagging in the pit. In 1977, grab sampling of marginal areas was abandoned as it was found that it provided little advantage over the use of kriged block grades alone. Raymond reports that over an 11 month period (1977–78) milled tonnages and grades were within 2% of the kriged estimates, a remarkably good achievement. To reduce computing costs for daily ore control, kriging was eventually replaced by inverse distance weighting using $1/(d^3 + k)$ as a weighting factor, where d is the distance between each sample and the point being

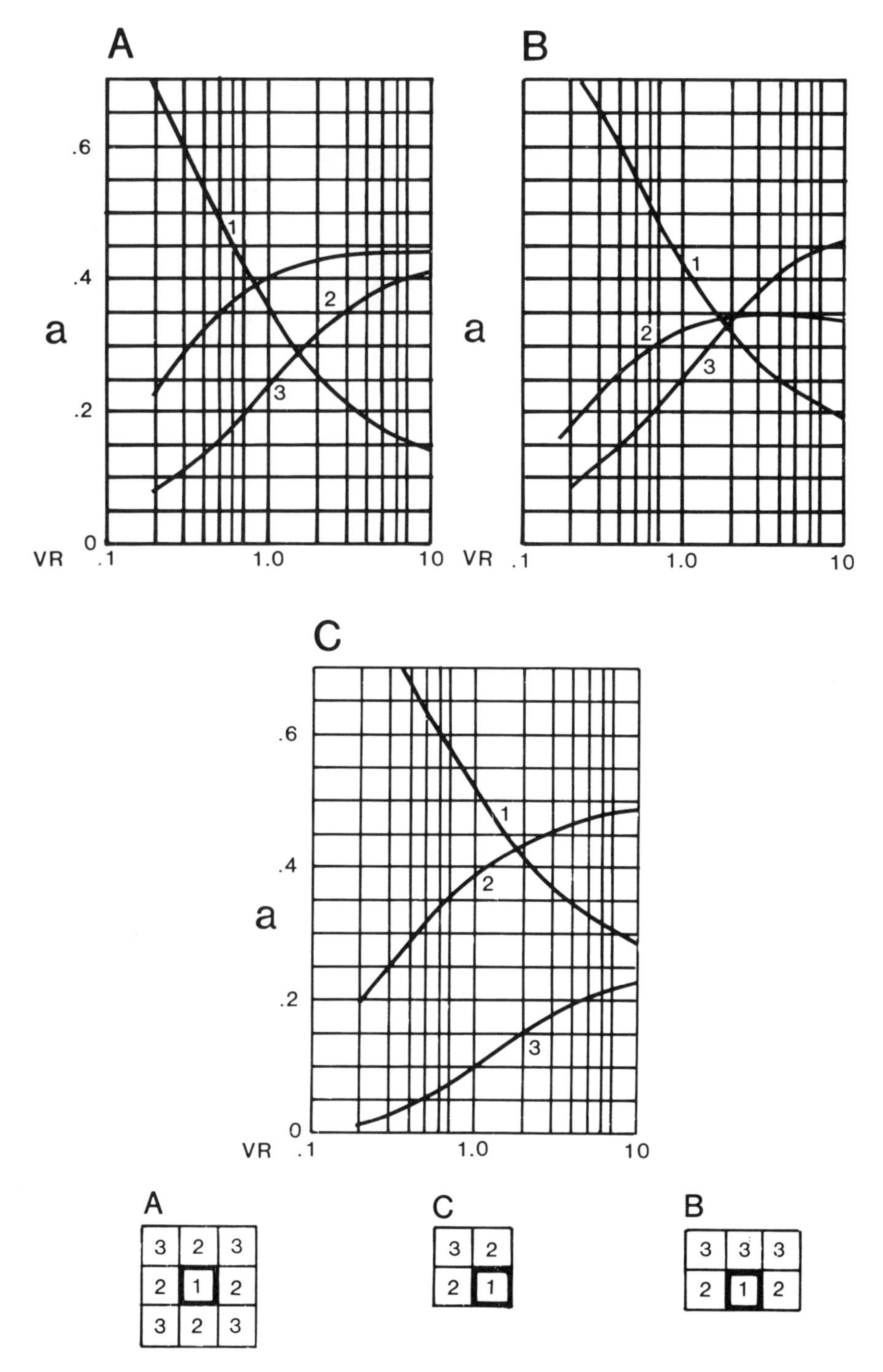

Fig. 7.17 Weighting factors (*a*) for random kriging of ore blocks, Ingerbelle. *VR* = variogram ratio (intercept/(slope/100)). A, Central block. B, Side block. C, Corner block.

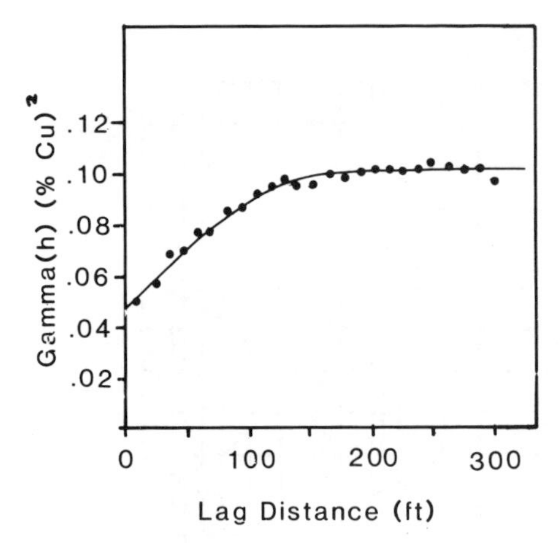

Fig. 7.18 Spherical scheme semi-variogram model for blast-hole grades. After Raymond 1979.

evaluated and k is a constant determined by comparing the results of kriging and distance weighting for the same areas of the orebody. Presumably k was calculated for a large number of points by putting:

$$\Sigma[G_i \times (1/d_i^3 + k)]/\Sigma(1/d_i^3 + k) = G_{kriged}$$

in each case. 'k' was found to be remarkably constant reflecting the uniformity of sample spacing and the constancy of the semi-variogram throughout the deposit.

Case history V – Tynagh Mine, Eire (Irish Base Metals Ltd)

The open-pit at Tynagh is also exhausted, but was mainly working secondary ore (Zone I) consisting of sulphides and 'oxides' of lead, zinc, silver and copper hosted by Carboniferous limestones. The highly variable nature of the ore from the metallurgical point of view, and the intimate association with waste, required that a careful grade control procedure had to be established to ensure that trucks reached the correct destination and that blending produced the optimum mill feed (Fitzgerald and Oram, 1969). In order to meet the demands of blending, and to allow selective mining of ore-blocks, a SMU of 25 × 25 ft in a 20 ft bench was selected. This meant that several faces could be worked at the same time (Figure 7.19).

The ore in the pit was classified by metallurgical treatment types which are summarized below:

1. Sulphide – <15% of metal as oxide (Pb + Zn concentrate produced).
2. Sulphide plus copper – <15% of metal as oxide, >0.6% Cu.

Fig. 7.19 Zone I pit, Tynagh Mine, Eire, showing selective extraction of ore-blocks in a bench. (From Fitzgerald and Oram (1969), reproduced by courtesy of the Institution of Mining and Metallurgy.)

3. Lead oxide – >15% of Pb as oxide (Pb sulphide and oxide concentrate produced):
   (a) high grade >8% Pb;
   (b) low grade <8% Pb.
4. Lead oxide plus copper – >15% of Pb as oxide, >0.6% Cu.
5. Mixed oxide – oxide ores rich in Zn not falling into any other oxide category.
6. Zinc oxide – only zinc present, >50% of zinc metal as oxide.
7. Silver ore – usually a by-product of above but if Pb < 2.7%, the Ag content will govern whether the mineralization is ore or waste.

If Ag > 5.8 oz/ton and Pb very low – silver ore
If Ag > 9 oz/ton and Pb very low – high grade silver ore.

The classification outlined above was complicated further by the fact that the ore was residual and was largely in the form of a 'black mud'. As a result, the percentage of slimes was important; a high percentage reduced metal recovery.

Benches were drilled using air flush diamond drilling on 25 ft centres and the core was recovered from the inner tube by a hand-operated extruder. The core recovered was then measured and the wet and dry weights recorded from which the density and moisture content were calculated. Individual 25 ft × 25 ft blocks, with drill-holes at each corner, were then evaluated by weighting the various grades of the four corner holes by ore length over the bench thickness and tonnage factor (density). The latter usually varied between 11.5 and 17.0 ft^3/ton. Tonnages of ore and waste were also produced on the basis of their average lengths in each of the corner holes. The calculated grades allowed each block to be assigned to one of the treatment types listed earlier. The end product was thus a series of bench maps for each of the treatment types. Each block on these maps contains details of its contained reserves, as in Figure 7.20.

Groups of ore-blocks were then combined in different ways until an overall weighted grade was attained which was close to that required. These groups would then be flagged in the pit so that they could be worked and trucked simultaneously to the relevant stockpile on an asphalt pad. Further blending could also be achieved by selective trucking to the mill from the various stockpiles.

Case history VI – Western Australian Gold Mines (Kalgoorlie)

Grade control in Western Australian open-pit gold operations is somewhat different to that described for the North American examples in that the open-pits are generally working small (<5 m t and sometimes <1 m t) deposits at shallow depths (e.g. <100 m) in areas where weathering is deep 40–90 m. Grades are usually in the 1–5 g/t range but may be highly erratic, with ore-zones intimately intermingled with waste zones. Highly selective mining is thus necessary on the basis of careful grade control. Because a large proportion of the ore is soft and rippable and directly mineable by front-end loaders or hydraulic excavators, blast-hole sampling is rarely applied. In order to facilitate selective mining, each bench is subdivided into flitches which may be only 1.25 m thick in the upper levels, and 2–5 m (frequently 2.5 m) in the lower levels. These thicknesses are controlled partly by the limited reach of the hydraulic excavators, and partly by the lack of vertical continuity of the gold grades. Areas that have been defined as ore are selectively mined and then the remainder of the flitch stripped as waste.

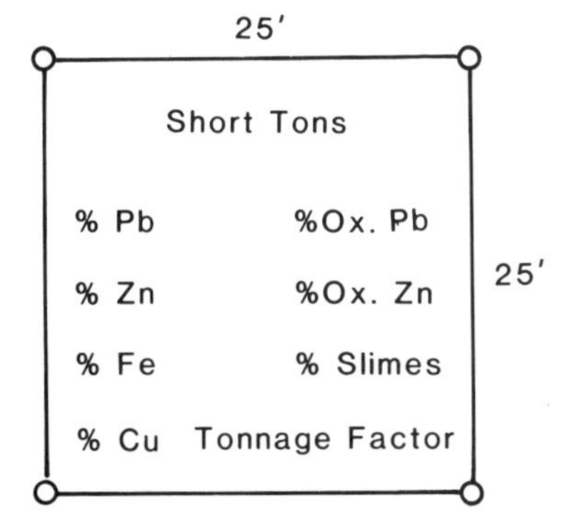

Fig. 7.20 Ore-block in Tynagh Open-pit, Eire, showing information from the four corner-holes.

Grade control practices are thus largely aimed at accurately defining lithological and grade zone contacts, the average grades of each of the ore categories defined in the pit bench, the grade of material in stockpiles and finally, the maintenance of stockpile reserves by the application of flexible selective mining in the pit. What follows represents a composite picture of the methods used in Western Australia and does not necessarily apply to each and every pit.

The initial drilling of these deposits to define the main ore-zones and to provide the data for the initial feasibility studies, may be on centres as wide as 40 × 40 m (e.g. Paddington I) but at later stages this is closed down to 40 × 20 m and then to 20 × 10 m to define zones of internal waste and more accurately delineate the boundaries of the ore-zones. This exploration drilling is then augmented by grade control drilling done during the operational phase. At this time, either reverse circulation (RC) or airtrac drillng rigs are used to reduce the drill spacing down to 10 × 10 m (e.g. Phar Lap) or 10 × 5 m (e.g. Haveluck). The hole depths are usually limited to three to four bench thicknesses and samples are taken every 1–2 m so that information is available for grade control on the current bench and for planning of mining on the underlying benches. The results of the grade control drilling are then combined with the preproduction drilling data to revise the kriged block model for the pit (produced by such programs as SURPAC and DATAMINE). RC drilling is thought to give more reliable grades than airtrac drilling, but both are found to give a poor estimate of the mill feed grade (based on mill recoveries) even when the drilling grid is reduced to 5 × 5 m.

In most operations, the above drilling is backed up by the use of dozer rip channel samples or samples produced by a continuous trencher (e.g. Ditch Witch – section 2.11). These samples are usually taken at 1–2 m intervals along lines perpendicular to the ore-zones and 5 m apart. The Ditch Witch trenches are often placed between the drill lines and also between individual drill-holes (Figure 7.21) to accurately define lithological or grade zone contacts. Where the ore becomes harder,

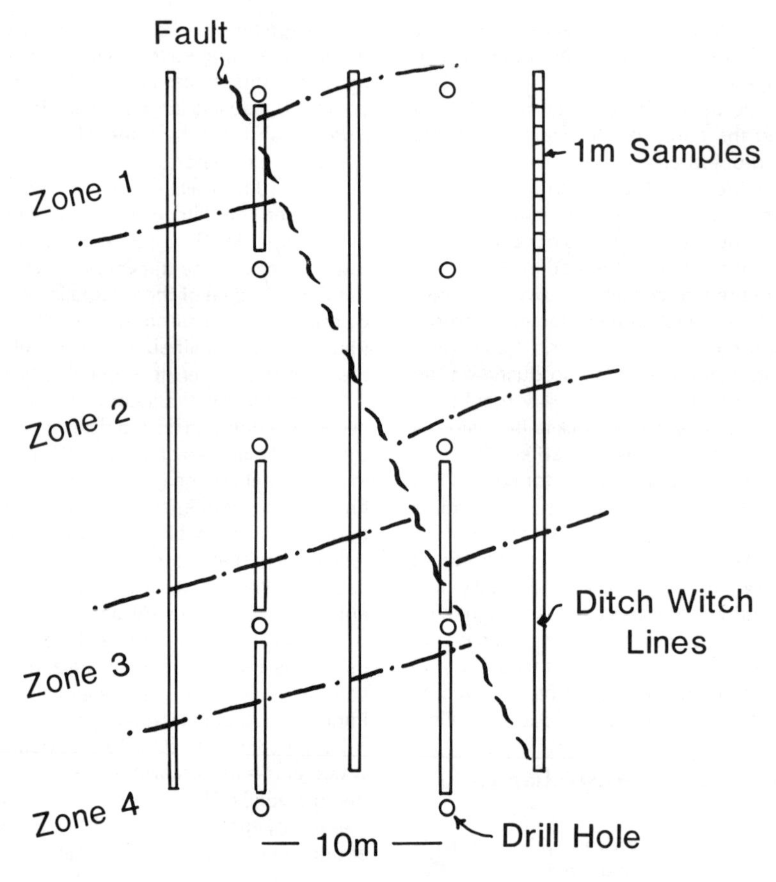

Fig. 7.21 Ditch Witch trench lines used to augment drilling in a bench.

especially in the lower benches, it may be necessary to revert to airtrac drilling once again. On the basis of this sampling, the geologist is able to produce a geological plan of the bench, together with a grade control plan, on which various grade categories can be colour coded. As most mines work to a COG of 1 g/t, these categories may be defined as follows:

<0.5 g/t	Waste
0.5–0.99 g/t	Low grade stockpile
>1.0 g/t	Ore

The ore category can itself be subdivided, as at Ora Banda, into 1–2.49, 2.5–4.99 and >5 g/t ore, or as A, B, C, D ore as at Reedy's Gold Mine (1–2, 2–3, 3–4 and >4 g/t) respectively, or as Class I (>2 g/t) or Class II ore (1–2 g/t) as at Phar Lap Gold Mine. This categorization allows ore to be stockpiled separately and then blended to produce a constant grade feed (e.g. 4 g/t). Although this grading process was originally done manually, it is now increasingly done using commercial or in-house computer software.

Ditch Witch sampling is generally accepted as giving a more reliable grade, for it allows more accurate delimitation of narrow waste and ore-zones that may be intermingled. The samples are less affected by the erratic gold grades and the high nugget effect typical of Western Australian gold deposits. The technique (illustrated in Figures 7.22–7.25) is particularly suitable for tabular steep-dipping ore-zones whose bench intercepts would be poorly defined by vertical RC or airtrac drilling.

At Gimlet South, Ditch Witch sample grades are grouped into areas so that the majority of samples in each area lie in one grade category. Where doubt exists as to the true grade of an area, instructions will be given that all trucks servicing this area should take their loads to a 'truck-dump'. Here each load will be dumped on a concrete pad as a separate pile and each will be sampled independently by taking a 2 kg composite from all sides

Fig. 7.22 Ditch Witch 7510 continuous trencher at Ora Banda Open-pit in Western Australia. This model is fitted with a saw and chutes for spoil discharge. (Photograph courtesy of R.G. Bird.)

Fig. 7.23 Ditch Witch R100 with a digging chain at a Western Australian gold operation. Colour changes in spoil heaps reflect lithological changes in the pit floor. (Photograph courtesy of R.G. Bird.)

Fig. 7.24 Ditch Witch 7510 using a chain to cut a sampling trench at Marvel Lock Gold Mine, Western Australia. Augers are being used to distribute the spoil. (Photograph courtesy of The Charles Machine Works, Inc., Oklahoma, USA.)

Fig. 7.25 Close-up of a sample trench at Ora Banda. Ridges in the trench bottom reflect hard quartz veins. (Photograph courtesy of J. Davis.)

of the pile. The grade of this sample then determines whether the pile will be assigned to the mill or to the waste dump (indicated by flagging with coloured tape).

7.3 UNDERGROUND OPERATIONS

Grade control in underground mines can be accomplished in a variety of ways depending on the nature of the ore, its attitude and on the mining method being employed. Sampling could involve diamond drilling, blast-hole drilling, channel or chip sampling, drawpoint or muckpile sampling, or grab sampling from tram cars or conveyor belts. All of these methods have been discussed earlier in Chapter 2. Some of the ways in which these methods have been used for grade control will be discussed via a series of case histories.

7.3.1 Grade control case histories

Case history I – Kerr Addison Gold Mine, Ontario, Canada

This 1000 ton per day operation is mining gold ore associated with a major regional fault, the Larder Lake

Break, and in particular with green carbonate-fuchsite zones in altered basic volcanics ('green-carbonate ore') and with pyritized dacitic to rhyolitic lavas ('flow ore'). In the former ore type, the grade of gold is proportional to the percentage of quartz veins present. These veins generally occur as flat-lying stringer veins across the dip of the ore zone (72°N), or as south-dipping veins with the same strike as the ore-zone, or as steep easterly dipping veins at right angles to the strike. In the case of the 'flow ore', the gold tenor is dependent on the abundance of pyrite and on its fineness (the coarser the pyrite, the lower the gold grade).

This well-established mine has produced to date 37.5 million tons of ore yielding 10.2 million ounces of gold, but its production is now waning as its operational life comes to an end. Production in 1985 was 320 000 tons at an average recovered grade of 0.127 oz/ton. At this time, the reserves stood at only 803 000 tons @ 0.121 oz/ton. Stoping is by 'cut and fill' methods using 8 ft lifts and a minimum stoping width of 4 ft, and by blast-hole open stoping with a 4 ft burden.

Areas of the orebody to be mined by cut and fill methods are evaluated using plan sections at 30 ft vertical intervals. During mining of each stope, the geologist produces a set of plans for each 8 ft lift showing the interpolated location of the assay cut-offs. The miners then use these plans to control mining. No manpower is available now to mark up these cut-offs by visual inspection, as was the case in earlier days. Grade control is undertaken by the use of horizontal chip sampling of each breast in the current lift prior to blasting (Figure 7.26). This is augmented by two +10° jackleg holes drilled to depths of 9 ft, one into the hangingwall and one into the footwall of the stope after each 8 ft round is blasted and mucked. These holes are sludge sampled over 3 ft intervals. If significant ore-grade material appears in a wall of a stope, then this wall is also chip sampled as shown in Figure 7.26 prior to slyping (slashing) of this wall.

The spot heights are surveyed on the back of the stope at the end of each month and their elevations above the underlying mine level are plotted on the lift plan. These elevations are averaged and subtracted from the average for the underlying portion of the underlying lift. This gives the average height of the current lift. The plan outlines of the stope are also surveyed and the area determined by planimetry. From this, the tons broken during the month can be calculated whereas the grade of this ore is determined by the length-weighted average of all chip samples in this portion of the stope ('stope grade'). All cars from the stope are also grab-sampled and from this information the tonnage and grade are estimated, together with the number of contained ounces of gold ('trammed ounces'). At the mill, the tons milled during the month are also recorded together with gold recovered hence:

Mill feed grade = ounces recovered/tons milled

The ratio ounces recovered : ounces trammed is then calculated. Generally this lies between 1.1 and 1.25, indicating that the calculated grades from the grab sampling are lower than the mill feed grades. The 'Stope Grades' from the chip sampling are thus adjusted so that, for the tonnage mined, the number of contained ounces agrees with the mill recovery as in the worked sample below:

Planimetered tons	= 115 200	
Chip sample weighted grade	= 1.35 oz/ton	
Tons trammed	= 110 990	oz trammed
Grade sample grades	= 1.30 oz/ton	= 144 287
Tons milled	= 111 775	
Mill recovery	= 154 250 oz	
Mill feed grade	= 154 250/111 775 = 1.38 oz/t	

Ounces recovered/ounces trammed = 154 250/144 287 = 1.07

Revised monthly stope tonnage = 115 200 @ a new grade of 1.34 oz/t (equivalent to 154 250 oz of contained gold as recovered in the mill)

In reality, several stopes would be in operation over the month and a mine-weighted value is thus produced for the grade together with total tons broken. A running total and grade is also determined which, at the end of the year, should give a gold content comparable to that recovered by the mill.

Case history II – Teck-Corona Gold Mine, Hemlo, Canada

This mine is exploiting part of an E–W 2 km long, strata-bound gold deposit hosted by amphibolite facies meta-sediments of the Abitibi-Wawa greenstone belt (2600 Ma). The whole deposit is estimated to contain 75 million tonnes of ore grading 7.2 g/t Au and, to date, it has a proven dip-length of 1.5 km with dips ranging from 50° to 65°N. As well as Teck-Corona, which has reserves of 7.6 million tonnes at 12 g/t, there are two other operations on the same deposit, Page Williams and Golden Giant (Noranda). The ore consists of pyrite and molybdenite, as well as native gold, and locally there are high concentrations of barytes. Significant amounts of Hg, Ag, V, Sb, As and Zn are also present. The ore-zone is characterized by the appearance of muscovite, at the expense of kyanite, and the development of a grey hue due to molybdenite, which has a very close association with the gold.

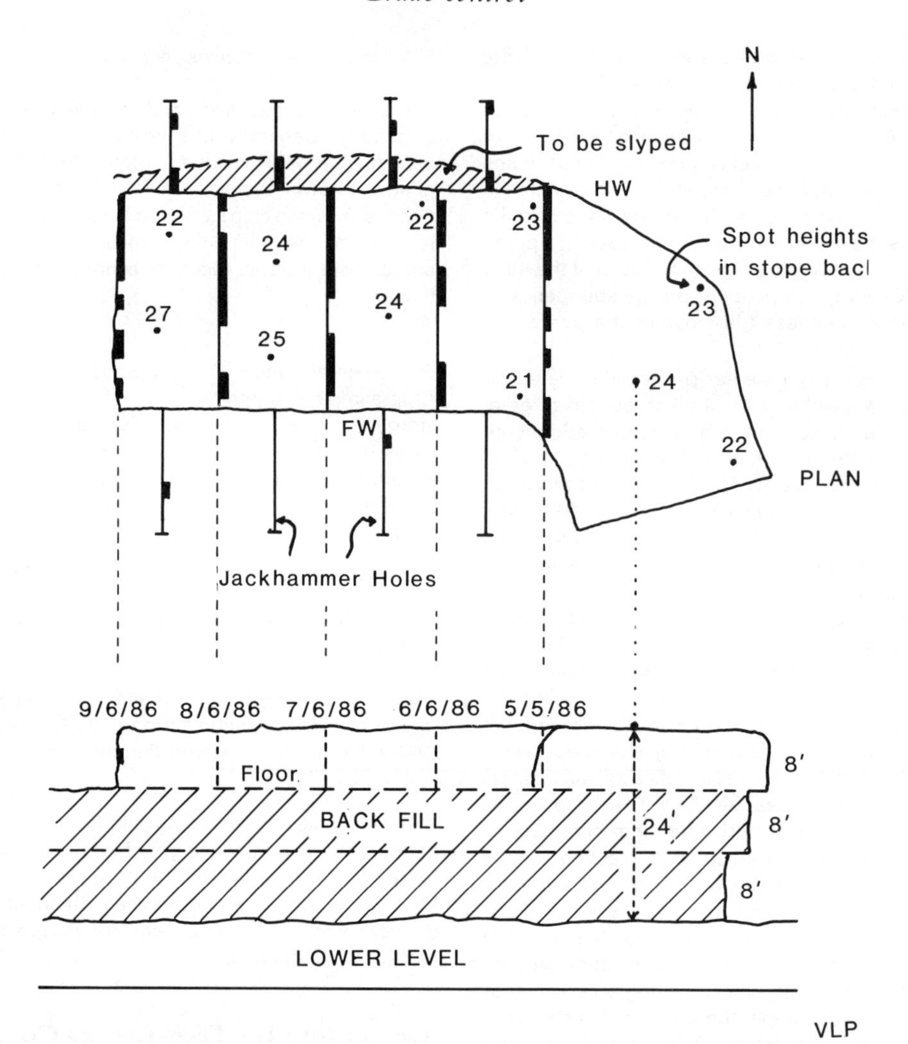

Fig. 7.26 Breast chip sampling in a cut and fill stope at Kerr Addison Gold Mine, Ontario, Canada.

Mining of the East Ore Zone (Figure 7.27) at Teck Corona is via cut and fill methods using 100 m level spacings, 3 m (10 ft) lifts in the stopes and a minimum stoping width of 2.44 m (8 ft) for an orebody which varies in thickness from 2.5 m to 5.0 m. Gold grades are remarkably uniform and predictable and this part of the deposit has more in common with stratiform base-metal mineralization than with typical Archean shear zone type gold deposits.

Grade control procedures involve the visual delimitation of cut-offs in the breasts of each lift, usually on the basis of the abundance of pyrite and the appearance of molybdenite. These are marked by the geologist using aerosol spray paint. Blast-hole cuttings are also sampled at 0.7 m intervals while chip sampling is undertaken at 3 m intervals in the first two lifts in a new stope. The latter is to check compatibility of the stope grades estimated by polygon methods (section 3.8 and section 8.4.4) with those for the mill feed. Thereafter, sampling is undertaken at 10 m intervals to the top of the block. Chip samples of the breast are taken at right angles to the assay cut-offs (Figure 7.28).

If a breast sample is not taken, then a back sample is used and the true thickness estimate made by tape. The stope width is also measured, plus the inclined length of the stope, to enable volumes of ore broken to be calculated.

A grade control balance is made each month using a

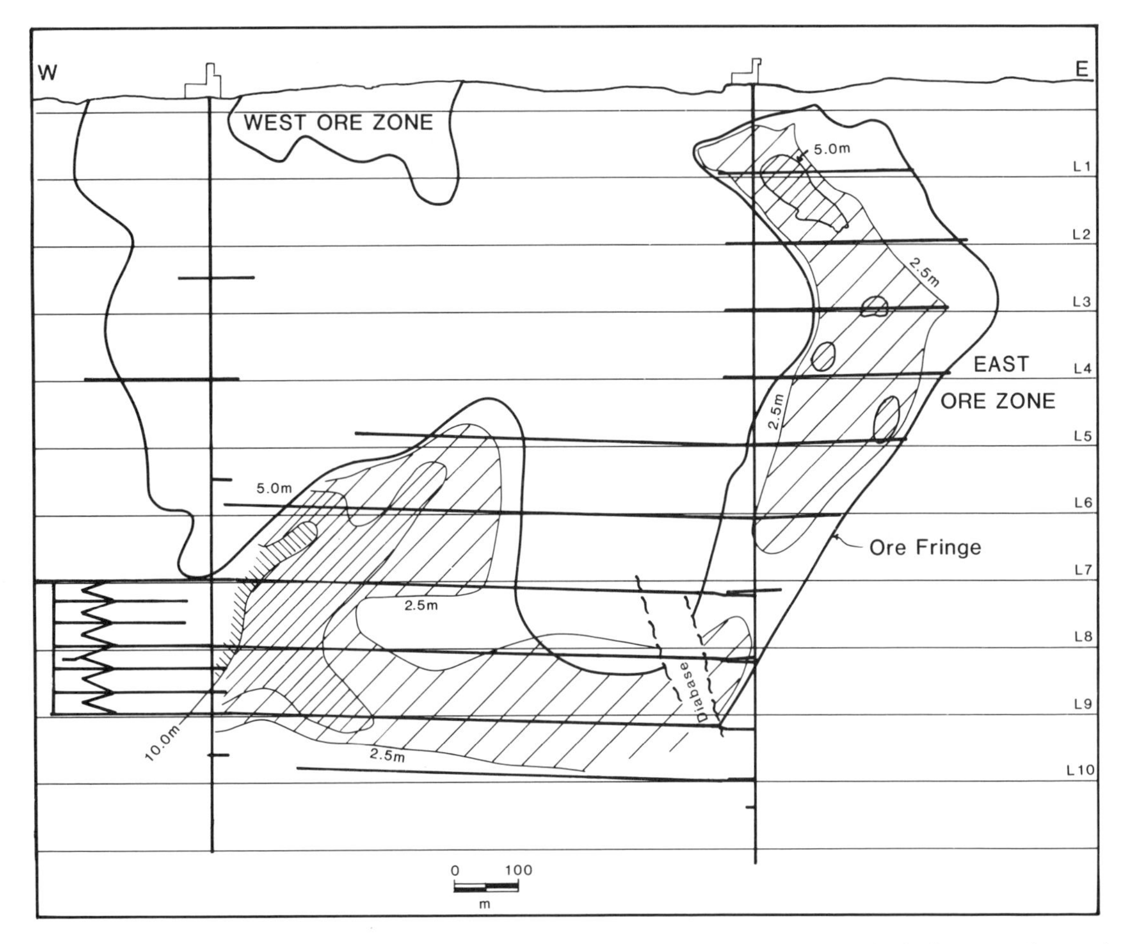

Fig. 7.27 Vertical longitudinal projection of Teck-Corona Mine, Hemlo, showing orebody fringes and isopachs.

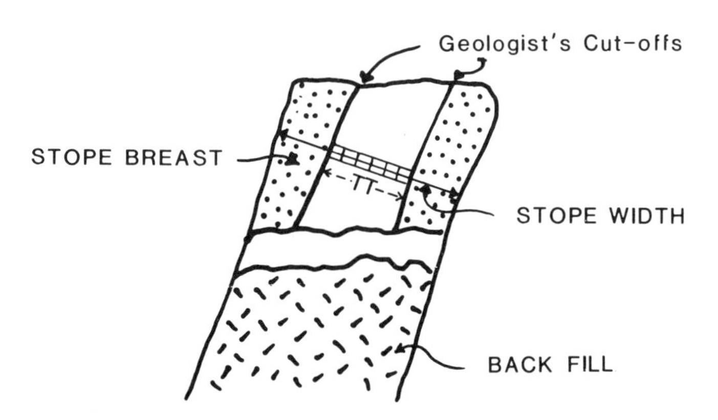

Fig. 7.28 Cross-section of 'cut and fill' lift at Teck-Corona showing breast sampling. TT = true thickness of ore zone.

document similar to that shown in Table 7.2. This requires some explanation, which is given below:

'Length in ore' . . . the total strike length of each lift mined during the month.

'True width of ore' . . . the average of chip sample ore lengths.

'Stope width' . . . the average of all stope widths measured at the chip sample locations.

'Dilution of ore' . . . the total waste thickness divided by the total ore thickness expressed as a percentage (usually 22–53%).

'Stope overbreak' . . . this is the dilution above the minimum stoping width (MSW) and is a measure of the efficiency of mining. It is thus the dilution over 2.44 m calculated as follows:

$$[(\text{Stope width} - 2.44) \times 100]/2.44$$

Where the ore zone is greater than the MSW then the calculation is:

$$[(\text{Stope width} - \text{ore width}) \times 100]/\text{ore width}$$

'Tonnage' . . . ore broken in each lift in each stope which is combined to produce 'Total broken'.

'Grade' . . . the diluted grade over the stope width. Note that in 1986 the mine had made a tentative step towards metrication using oz/metric ton (tonne)!

Table 7.2 Typical grade balance sheet, TeckCorona

ORE BROKEN

	Length in ore (m)	True width of ore (m)	Stope width (m)	Dilution of ore (%)	Stope overbreak (%)	Tonnage (tonnes)	Grade (oz/tonne)
Level 1 Stope							
Level 2 Stope							
Level 3 Stope		TOTAL BROKEN					

ORE TRAMMED

Level 1 Stope . . . . tonnes @ . . . oz/mt
Level 2 Stope . . . . tonnes @ . . . oz/mt
Level 3 Stope . . . . tonnes @ . . . oz/mt
Level 3 Sump Slimes . . . . tonnes @ . . . oz/mt

TOTAL TRAMMED . . . . tonnes @ . . . oz/mt

PRODUCTION BALANCE

COB reserve, (Date) . . . . tonnes @ . . . oz/mt
Ore hoisted . . . . tonnes @ . . . oz/mt
COB reserve, (Date) . . . . tonnes @ . . . oz/mt

ADJUSTED MILL FEED . . . . tonnes @ . . . oz/mt

BROKEN RESERVE

Level 1 Stope . . . . tonnes @ . . . oz/mt
Level 2 Stope . . . . tonnes @ . . . oz/mt
Level 3 Stope . . . . tonnes @ . . . oz/mt

BROKEN RESERVE . . . . tonnes @ . . . oz/mt

'Ore trammed' ... this is based on number of cars trammed and grab sample grades but also includes slimes collected from the sump at the lowest production levels (representing losses during stope drainage – drill cuttings).

'Production balance' ... this is based on the coarse-ore-bin (COB) reserve at the beginning of the month, the amount hoisted from this underground bin and the reserve at the end of the month. The last value is the first value plus the 'total trammed' minus the amount hoisted. Grades are weighted by tonnages to obtain the final COB grade.

'Ore hoisted' ... this is the calculated mill feed (e.g. 0.24 oz/mt) but the actual mill feed grade, based on weightometer readings and recovered gold values, is usually different (e.g. 0.229 oz/mt). In this example, there is a discrepancy of 4.8% so the calculated mill feed is converted to an 'adjusted mill feed' so that the gold content matches the mill value. Thus, if the original value was 19 250 t at 0.24 oz/mt then the new value would be 20 175 t at 0.229 oz/mt.

'Broken reserve' ... this is the total amount of broken rock in the stopes at the end of the month which is waiting to be scraped and trammed to the coarse ore bin.

Every 6 months, the total 'adjusted mill feed' for this period is compared with the value for the stoped area as determined from the polygon VLP. A close agreement is sought.

Case history III – J-M Reef, Stillwater, Montana (Chevron Resources)

The J-M Reef is a platinum/palladium-bearing reef containing pyrrhotite, pentlandite, chalcopyrite and secondary pyrite mineralization at the top of the Lower Banded Series of the Stillwater Complex in Montana (section 8.3). The reef package consists of plagioclase and plagioclase-olivine cumulates and olivine-bronzite pegmatoids, often heavily sheared and serpentinized. The ore grade material occurs in plunging lenses or pods in a tabular zone which dips at between 55° and 90° to 110° (average dip 70°), and which is mined by cut and fill methods. Individual lifts are developed along strike from central ore-pass and manway raises for distances of up to 150 ft or 46 m (Figure 8.7). The average horizontal stoping width is 7 ft (2.1 m) while the MSW = 4 ft (1.2 m). Stope planning is based on limits defined by the geologist on the basis of kriged block grades (section 8.3).

Grade control was initially undertaken by grab sampling 25 lb samples from 10 ton trucks. However, this proved to be very time consuming due to the time needed to dry these very wet samples. It was thus decided to resort to chip sampling of stope faces between each round. Once the previous blast has been mucked, the geologist marks the footwall and hangingwall contacts with aerosol spray paint on the basis of visual inspection of the distribution of sulphide. It has been found that, if visible sulphides are present, then the grade usually exceeds the established COG of 0.3 oz/ton combined Pt + Pd. The intervening zone (Figure 7.29) is then subdivided into intervals of less than 3 ft horizontal width which are based on lithological changes, the presence of faults/shears and on variations in the abundance of sulphides. Each is chip sampled at approximately chest height so that a sample bag is filled with material. The sampling is done from the footwall to the hangingwall to be consistent, so that the first sample bag/number in a sequence is always the footwall sample. A drawing is then made of the face, based on accurate measurements, on which the main elements of the geology are inserted. A series of codes are used to represent each rock type. Areas of visible sulphide are marked with red dots. The location of each sample is also shown. An assessment is made as to whether the current face is in ore, low grade or waste and it is then sprayed with 'O', 'LG' or 'W' accordingly.

The distance of the face to the nearest side of the raise (manway) is measured to determine the face advance since the previous face sampling. The width of the stope is also measured at chest height 3 ft back from the face (approximately at the centre of the round) to determine how close the miners have kept to the geologist's stope width. Face profiles are digitized to determine their areas and, with the calculated advance, the volume of rock broken can be calculated. The grade assigned to this volume is then the length-weighted grade of the chip

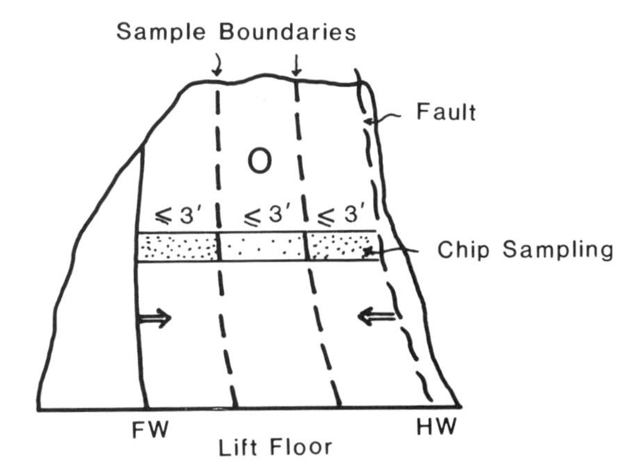

Fig. 7.29 Geologist's face section showing geology and location of grade control samples at Stillwater PGM Mine, Montana. Stippling indicates intensity of sulphides.

samples bounding it. On initial lifts, the face may be advanced for two to three rounds into waste, beyond definite ore, before the lift is stopped. Subsequent lifts are stopped on the basis of experience on previous lifts and on the disappearance of visible sulphides. Internal zones of waste as defined by stope drilling and kriging, and those which are greater than 25 ft (7.6 m) in length, are selectively mined and trammed. Any waste zone less than this is taken as acceptable dilution and is trammed with ore.

The geologist plots a plan showing the face positions, the ore widths and the main geological features. Once the stope lift is declared complete, this information is transferred to a dilution plan based on accurate stope surveys (Figure 7.30). The ore-zone is marked with red cross-hatching and the geologist's stope width marked in green. Authorized dilution (near manways, etc.) is marked with green cross-hatching. This map then allows the stope over-break to be calculated by digitizing the actual (surveyor's) stope limits and the geologist's limits. This allows a percentage overbreak to be calculated which is usually in the range of 20–30%, and the average mined width. Each new lift is then plotted on a VLP showing the kriging stope limits and all previous lifts. Each lift is annotated with tons of ore, low grade and waste and with percentage dilution, while a cumulative dilution, and tons in each category, are updated as each new lift is mined. Weighted grades are also calculated for each lift which eventually allow a direct comparison with stope estimates and with mill feed.

Case history IV – Jardine Gold Deposit, Gardiner, Montana

At the time of writing, the mine was in a preproduction phase and had quoted reserves of 1.0 million tons @ 0.3 oz Au/ton. The gold mineralization is associated with a stratiform iron formation enclosed within a package of meta-turbidites, now biotite and quartz-biotite schists with cummingtonite. Bedded pyrite and pyrrhotite mineralization is invaded by subconcordant quartz-arsenopyrite (± scheelite) veins, 2 cm to 2 m thick, and porphyroblasts of arsenopyrite. The gold is very fine grained and intimately associated with arsenopyrite. The mineralized zone is in refolded isoclinally folded strata and very detailed drilling and careful grade control is

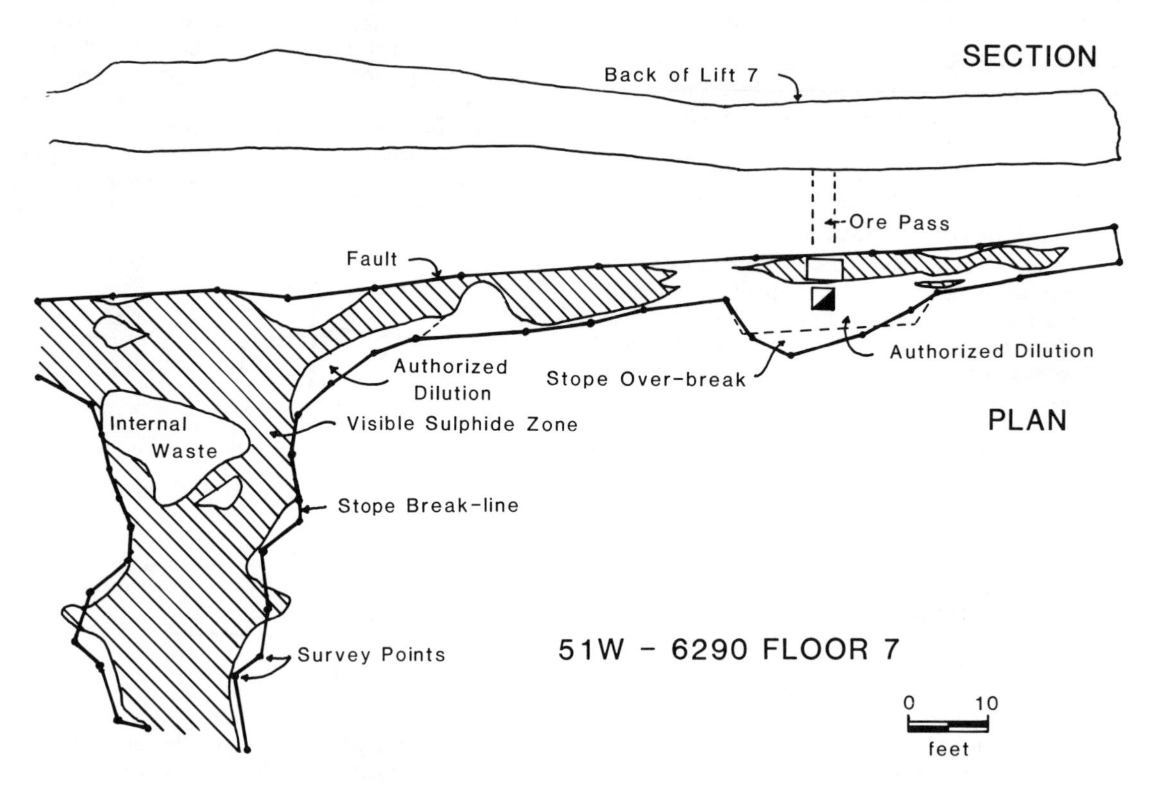

Fig. 7.30 Grade control in 'cut and fill' stopes at Stillwater Pt-Pd Mine, Montana.

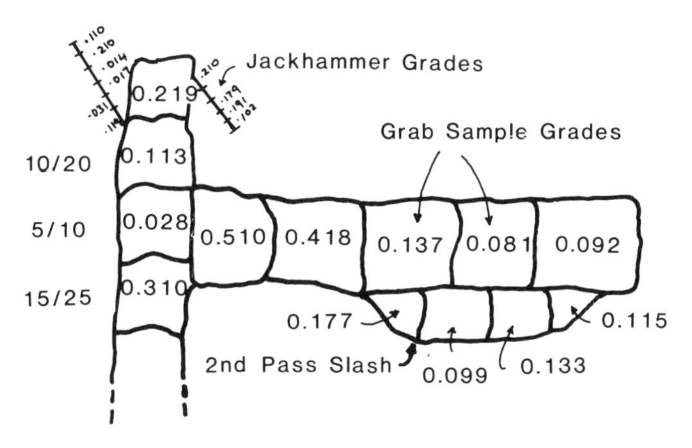

Fig. 7.31 Blast grade plan typical of those produced at Jardine, Montana, 5/10 = percentage pyrrhotite/ percentage arsenopyrite.

necessary to ensure profitable mining of the ore. It will be necessary to switch from mining method to mining method as the attitude of the ore changes. Mining will probably involve both 'cut and fill' and 'room and pillar' methods.

Each round at the advancing face is assigned a grade based on one shovelful of sample from each truck load of ore leaving the muckpile. This allows a blast grade plan to be drawn as in Figure 7.31. Jackhammer hole cuttings are also collected and analysed from up-holes drilled into the overlying rock to check for upward continuity of mineralization. After each blast, the geologist measures up the face and the advance, from which the tonnage mucked can be calculated. A scale diagram is made of the face on which the iron formation and associated veins are drawn, together with estimates of the percentage pyrrhotite and arsenopyrite for the whole face. The hangingwall and footwall grade lines are then sprayed on to the face together with the centre-line and drive profile for the next round. From the blast-hole plan, overall tonnages and tonnage weighted grades can then be calculated for a given period of mill feed or accumulation on the stockpile.

Case history V – Sullivan Pb-Zn Mine, Kimberley, British Columbia

This stratiform pyrrhotite-pyrite-sphalerite-galena deposit has a generally flat dip and is being mined by the Cascade mining method in which cone-shaped draw points are blasted beneath the ore to induce gravitational feed of the broken ore to the under-lying gathering drive. Each round blasted in the stope has a 7–10 ft burden and the tonnage can be determined by planimetry of blast-hole sections. The determination of stope tonnages and grade for Sullivan has been discussed elsewhere (section 3.6.3).

Grade control is by visual estimates of the grade of the muckpile at the draw points and an examination of the lithologies and mineralization of the rock fragments. This is aided by the production of draw-down charts which are regularly updated from car and slusher counts provided by production crews. This allows a comparison of tonnage drawn from each drawpoint and the tonnage blasted. Once the two figures begin to converge, careful examination of the drawpoints is made to detect a fall-off of grade as hangingwall overbreak begins to dilute the core. Grab sampling of cars allows a continual check against the anticipated grade calculated for the stope. Generally it is found that sampling grades agree with mill grades to within 10% or 1% Pb + Zn over a monthly period.

Car sampling and counts yield the tonnage and grade for the ore produced which, at the month's end, is compared to the tonnage and grade data supplied by the mill. This allows corrections to be made to the standard tonnage per car factors used and to the assays. Once this has been done, each ton of the ore and metal are credited back to individual blocks (or pillars, as a considerable portion of the ore is coming from pillar wrecking).

7.3.2 Grade control and grade cutting

Information obtained during grade control operations can be put to good use in that an analysis of predicted and mill grades over long periods of

Table 7.3 The impact of cutting levels on grade estimates

Period	Level of cut	R	Comment
1952–1960	0.5 oz/t	1.036	Cut slightly too low
1960	0.5 oz/t	1.144	High grade Fuchsite ore being mined, therefore cut much too low
1961–1966	1–2 oz/t*	1.052	Some improvement obtained
1972–1974	1–2 oz/t*	1.001	Good correlation
1974–1983	1–2 oz/t*	0.892	Lower grade ore being worked due to high Au price plus less Fuchsite ore

*Different grades are applied to different stopes working different ore types.

time can give an empirical determination of the optimum level for cutting grades. This should be done in conjunction with an examination of the nature of the ore being mined over these same periods and of fluctuations in the metal price and the impact this had on mining.

At Dome Gold Mine in Ontario, Canada, the cutting of grades is based on production experience over periods of time since 1952. Comparisons have been made between the level of cut and the ratio (R) of grades obtained from gold recovered plus gold lost to tailings to grades estimated from car sampling underground. Table 7.3 summarizes the results of this study.

It is evident that, over the period 1974 to 1983, the cut was too high for reasons explained under 'comments'. However, another factor playing a role was the increased blockiness of ore from blast-hole stopes. It was found that the fines collected during grab sampling are richer in gold than the larger blocks causing a positive bias. As a result, such stopes have now had their car grades cut to 0.33 oz/ton in order to increase R towards unity once again.

REFERENCES

Fitzgerald, D. H. B. and Oram, R. A. J. T. (1969) Grade control at Tynagh mine, Ireland, using data processing techniques, Paper 31, in *Ninth Commonwealth Mining and Metallurgical Congress*, IMM, London, 25 pp.

Johnston, T. G. and Blackwell, G. H. (1986) Short and long term open pit planning and grade control, in *Ore Reserve Estimation Methods, Models and Reality (Proceedings of the CIMM Symposium, Montreal)*, (eds M. David *et al.*), pp. 108–129.

Raymond, G. F. (1979) Ore estimation problems in an erratically mineralized orebody. *CIM Bull.* **72**, (806), 90–98.

BIBLIOGRAPHY AND REFERENCES

Alexander, P., Geldard, D., O'Beirne, W. and Payne, P. (1986) Grade Control at Phar Lap and Haveluck Mines Meekatharra: Techniques, Successes and Failures, in *The AusIMM Perth Branch, Selective Open Pit Gold Mining Seminar*, pp. 103–119.

Canadian Institution of Mining and Metallurgy (1968) *CIM*, **9** (Spec. Vol.: Ore Reserve Estimation and Grade Control), 321 pp.

Haugh, G. R., Finn, D. and Bryan, R. C. (1988) Geostatistical grade control at the Candelaria Mine. *Eng. Mining J.*, **189**, 52–57.

Kornze, L. D., Faddies, T. B., Goodwin, J. C. and Bryant, M. A. (1985) Geology and geostatistics applied to grade control at the Mercur Gold Mine, Mercur, Utah, in *Applied Mining Geology: Sampling and Grade Control*, (ed. B. A. Metz), Soc. Min. Eng. of Am. IMM and Pet. Eng., New York, ch. 4, pp. 45–55.

Sandercock, I. H. and Amos, Q. G. (1986) Grade Control Techniques at Paddington Gold Mine, in *The AusIMM Perth Branch, Selective Open Pit Gold Mining Seminar*, pp. 89–102.

8

Ore-evaluation Case Histories

8.1 INTRODUCTION

The aim of this chapter is to bring together a series of examples of the way different mining companies determine their ore-reserves. A wide range of mineral deposits are covered from industrial minerals to basic metals; from precious metals to sands and gravels. The techniques discussed also range from the more sophisticated geostatistical and computerized approach to the more straightforward manual computations using geometrical and/or statistical methods.

8.2 CASE HISTORY – WHITE PINE COPPER MINE, MICHIGAN, USA

8.2.1 Location

The White Pine Mine, operated by the Copper Range Company Ltd, is a stratiform copper deposit of syndiagenetic origin. It is located in the Keweenaw Native Copper District close to Lake Superior in Michigan, USA (Figure 8.1).

8.2.2 History

Although mining on site first began in 1865, it was not until metallurgical developments took place allowing the recovery of fine-grained chalcocite, and also the onset of the Korean War, that mining on a significant scale was able to begin in 1953. By 1975 production had reached a peak at 23 000 tons/day. Economic conditions forced the mine to close in 1982 after overall production had reached 138 million tons @ 1.14% copper. An employee buy-out in 1986 allowed production to continue with reserves standing at 184 million tons @ 1.1% Cu and 6.77 g Ag/ton. Figure 8.2 shows the extent of the underground workings at the time of writing which cover an area of 10.6 miles^2. This mine is only able to exploit such low grade material because of the relatively predictable nature of the mineralization, the lack of major tectonic disturbance and the ease of mining by 'room and pillar' methods. Throughout its life, however, very close geological control has been necessary to minimize dilution.

8.2.3 Geology

The mine lies in a succession of Middle and Upper Keweenawan rocks (1100–1225 Ma) which commence with the Portage Lake Volcanics and pass up via the Copper Harbor Conglomerates (200–2000 m thick) into the graphitic shales, siltstones and sandstones of the Nonesuch Shale (40–215 m thick). These are overlain by red Freda Sandstones exceeding 3660 m in thickness. Structurally, these rocks are dominated by the Porcupine Mountain Anticline in the N, the Iron River Syncline in the S and by several NNW striking high angle faults (e.g. White Pine Fault – Figures 8.1, 8.2).

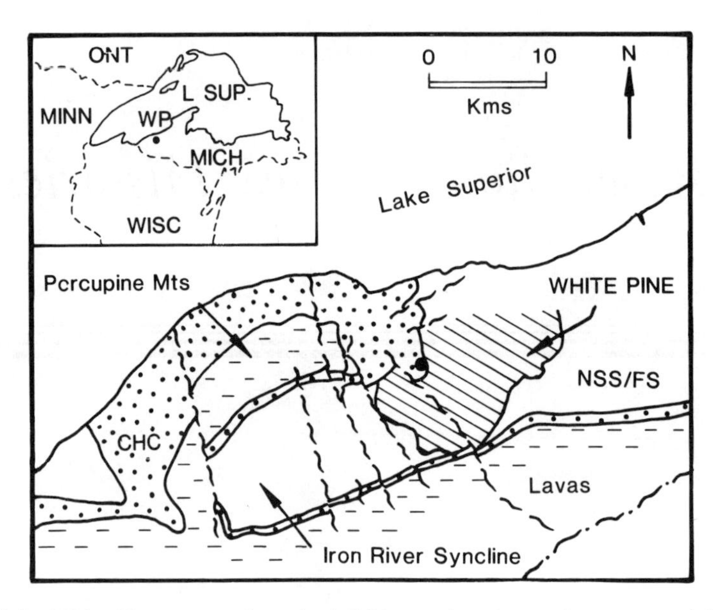

Fig. 8.1 Location of the White Pine copper deposit, Michigan, showing major structural features of the region. NSS/FS, Nonesuch Shale/Freda Sandstone; CHC, Copper Harbor Conglomerate.

Potentially economic mineralization, varying from 0 m to 20 m in thickness, lies in the basal portion of the Nonesuch Shale and also in the upper section of the Copper Harbor Conglomerate (Lower Sandstone) where mineralization can vary from 0 m to 6 m in thickness. The mineralized stratigraphic section has been subdivided into lithological units (Figure 8.3) with shales being assigned odd numbers and siltstones and sandstones even numbers. Basically, the succession consists of two main cycles, 2–3 m thick, separated by an erosionary surface. Each commences with a basal arkosic sandstone which is overlain by a sequence of laminated to massive siltstones and graphitic shales.

The mineralization is dominated by fine-grained chalcocite which may occur as disseminations, particularly in siltstone seams, and as nodules. Locally high concentrations of native copper also occur as disseminations in chloritic facies of the Copper Harbor Conglomerate, as arborescent growths in carbonate (± barytes) veins and as large plates or disseminated flakes along bedding planes. High concentrations appear to occur in the proximity of faults where fracturing has been intense and are thus thought to be the product of late stage hydrothermal activity.

Ore-grade mineralization tends to be concentrated in specific beds (e.g. 21, 23, 26, 41, 43 in Figure 8.3) separated by low grade to almost barren intervals. Within each bed, however, grades are remarkably uniform. The ore-reserve method described below was thus devised to take this feature into account. It provides another example of the use of stratigraphic slicing which is ideally suited to sediment-hosted stratiform base-metal deposits.

8.2.4 Ore reserves

At an early stage in the evaluation of the deposit, drilling was based on 1000 ft centres and holes terminated 20–30 ft into the Copper Harbor Conglomerate. Later drill spacings were increased to 3000–5000 ft with infill drilling undertaken on 1000 ft centres immediately prior to mining. Now drilling is on 3000 ft centres because of cost considerations and the proven predictability and continuity of grade and thickness. A total of 507 drill-holes provide the data base for the reserves. Although originally the sampling of cores was on the basis of a standard 1 ft length, it is now controlled by lithological changes while internal grade changes in each lithology are reflected by individual samples ranging between 1 in and 24 in in length. Only the copper assays are used to define mineable intervals and economic blocks of ground. This is partly due to the fact that, prior to 1972, only samples with copper grades in excess of 1% were sent for silver assay. At that time it was thought, erroneously, that there was a direct relationship between the concentration of the two metals. Although ore-zones are defined on the basis of a flexible 1% copper cut-off grade, lower grade

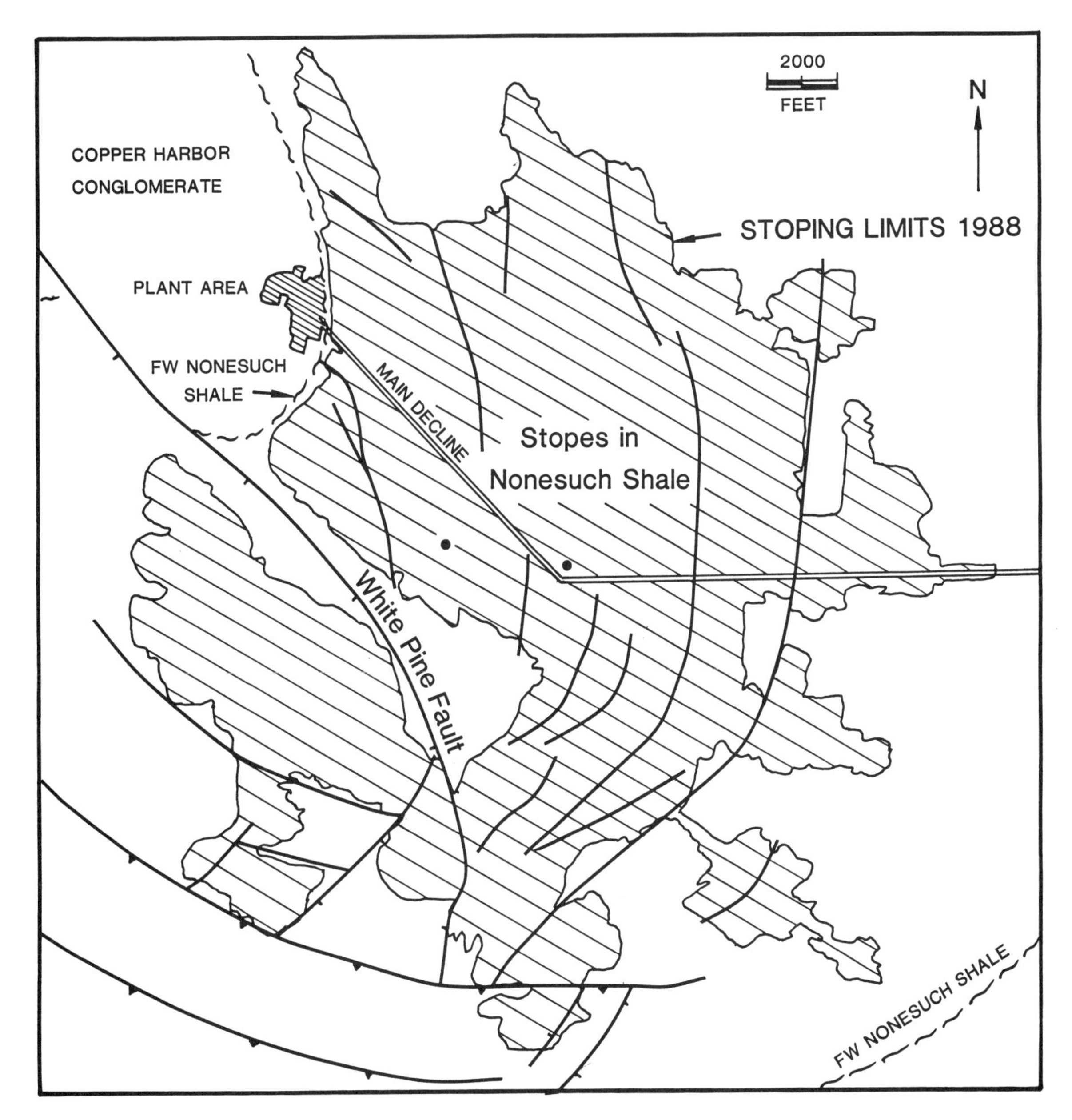

Fig. 8.2 The distribution of stoped areas within the Nonesuch Shale at White Pine. Also included are the main fault systems.

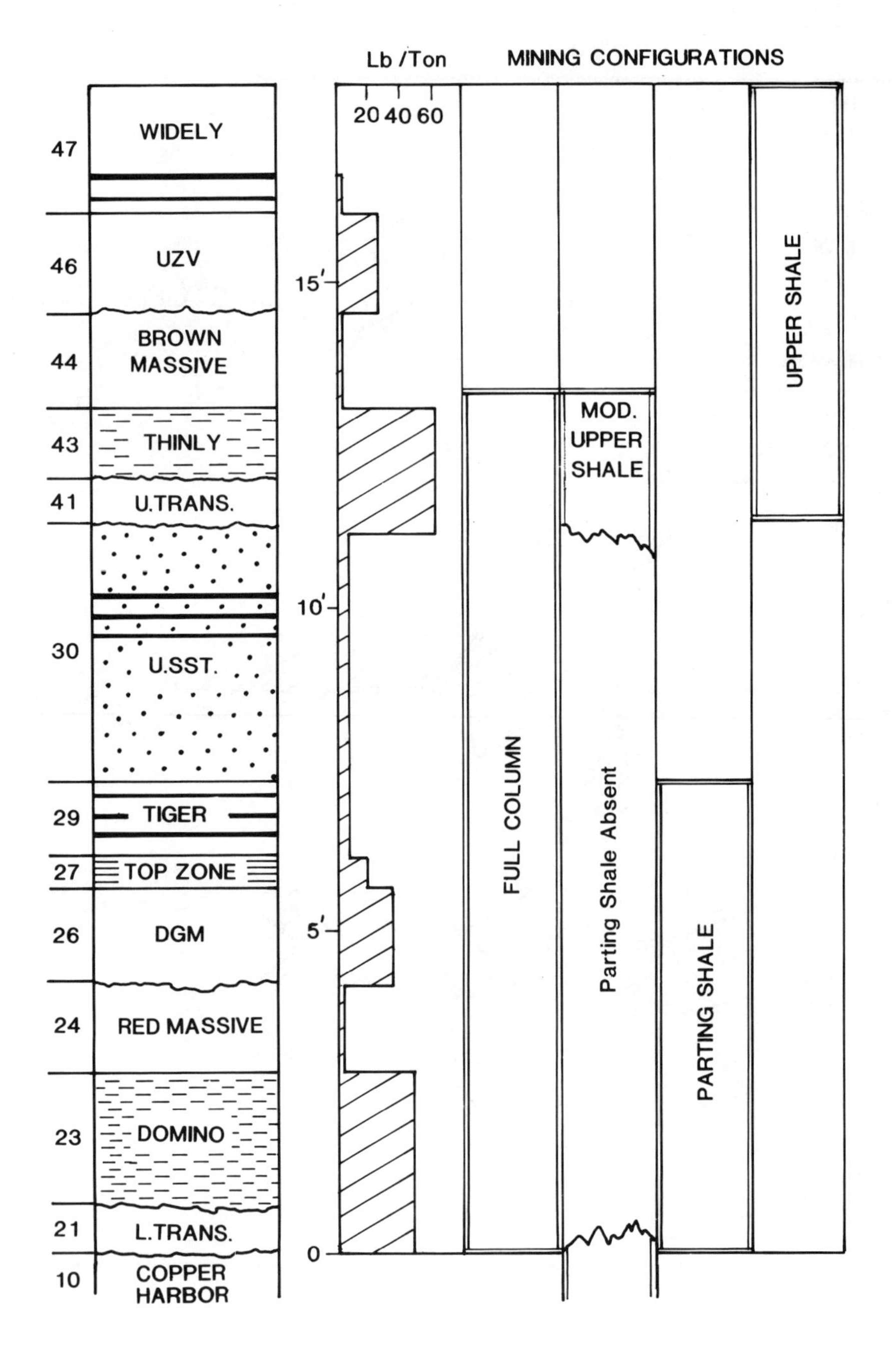

Fig. 8.3 Stratigraphic and mining units within the Nonesuch Shale at White Pine.

material ('incremental tonnage') is included by the geologist to maximize copper production. Other factors also taken into account are the availability of services and the location relative to the nearest conveyor belt used for moving broken rock.

Initially the reserves were calculated using uncut assay values and the polygon method of ore-reserves but this gave a poor estimate of the *in situ* reserves. Now, as indicated earlier, calculations are made on a bed by bed basis using inverse distance weighting methods. Individual high assays are thus smoothed out by the inclusion of data from adjacent drill-holes intersecting the same bed. The length-weighted average grade (lb/ton) of each intersection of each bed is calculated and the information stored on disk file. Inverse cube distance weighting methods are then used with a circular search area, 5000–10 000 ft in radius, which is sub-divided into quadrants. Each quadrant is searched for a user-specified number of holes lying closest to the centre of the search area and the ore-block under evaluation. The end product is an ore block map for each bed, based on a 100 × 100 ft grid. All information is presented by vertical projection on to a horizontal plan and hence vertical thicknesses values are presented with weighted grades.

The computer listing of the bed grades and thicknesses for each ore-block allows them to be combined in four different ways depending on the mining configuration to be used (Figure 8.3). There are four possibilities:

1. Parting Shale only; beds 21–29 inclusive, 2.6 m section.*
2. Upper Shale also mined at the same time as the Parting Shale but as a separate stope; beds 41–47, >2.3 m section.
3. Modified Upper Shale (where the Lower Parting Shale is absent) plus underlying Lower Sandstone; beds 41 and 43 + 'Dimple' at base of 44 + top of 10.
4. 'Full Column'; beds 21 ('Lower Transition') to base of 44; 4.6 m section.

The combination of beds into a mining configuration is done by stipulating the top and the bottom beds. The selection of the assay cut-offs is not made on the basis of grade but on the configuration which will give the maximum number of pounds of copper per column. This increases the number of tons milled but this is no problem for the mill which has excess capacity at the present time (14 000 tons mill feed per day in September 1988 versus 23 000 tons per day in 1975). Incremental tons grading 8–12 lb/ton (0.4–0.6% Cu) will pay for themselves and are thus added to increase copper production. In the case of the Upper Shale mining unit, bed 47 is sampled and assayed at 6 in intervals so that the geologist can select the optimum column.

Once each block has been examined in this way, the limits of four mining areas are defined and digitized. The program then computes the overall grade and volume of these areas, and then tonnage, using a fixed tonnage factor of 11.5 ft^3/ton. There is, however, the facility should it be required, to assign a different factor to each bed.

The 100 × 100 ft block model of each unit in the orebody is also used for grade control purposes for underground sampling is no longer undertaken. The geologist measures the thickness of each bed exposed at the face (l_1 – l_6 in Figure 8.4) together with the face advance and the face width (W). In the text figure, the Parting Shale Mining Unit is shown (units 21–29 inclusive). Also measured is the amount of overbreak (OB) and underbreak (UB) so that an estimate can be made of the percentages these represent of the area of the relevant unit in the face (e.g. 70% overbreak of 30 and 60% underbreak of 21). All this information is then input to the computer together with the coordinates of the centre of the nearest 100 × 100 ft ore-block taken from the mine plan. The program is then able to compute the diluted grade for the current round at the face using the block slice grades and the number of tons broken. These can be accumulated for all advancing faces or all stopes over a monthly period and then the contained copper compared with the mill recoveries for the same period. Typical mill feed grades 23 lb/ton copper and 0.22–0.26 oz Ag/ton. The two are remarkably consistent proving that the existing level of sampling is adequate taking into account the predictable and uniform nature of the mineralization. Where problems of overbreak in the back

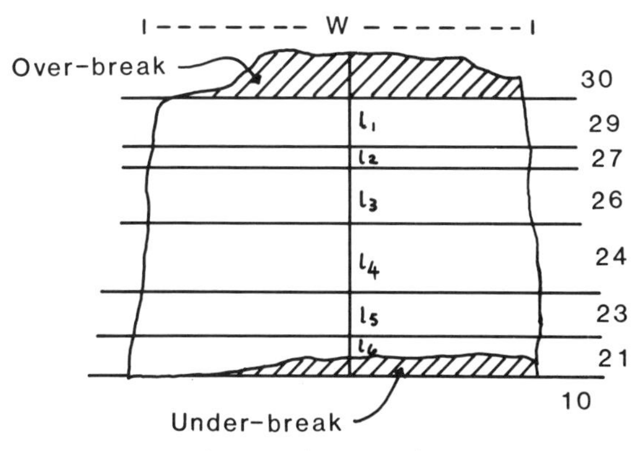

Fig. 8.4 Stope face at White Pine showing subdivision of the Parting Shale for grade control purposes.

*Native Copper in the chloritic facies of the 'Lower Sandstone' may require the addition of 0.5 to several metres of bed 10 to the base of the mining interval.

are occurring the sampler sprays on the correct profile of the face ready for the drilling of the next round.

BIBLIOGRAPHY

Brown, A. C. (1971) Zoning in the White Pine copper deposit, Ontonagon County, Michigan. *Econ. Geol.*, **66**, 543–573.

Brown, A. C. (1974) An epigenetic origin for strati-form Cd-Pb-Zn sulfides in the lower Nonesuch Shale, White Pine, Michigan. *Econ. Geol.*, **69**, 271–274.

Burnie, S. W., Schwarcz, H. P. and Crocket, J. H. (1972) A sulfur isotopic study of the White Pine mine, Michigan. *Econ. Geol.*, **67**, 895–914.

Ensign, C. O. and Patrick, J. L. (1968) Ore reserve computation and ore dilution control at the White Pine Mine, White Pine, Michigan. *CIM*, **9** (Spec. Vol.: Ore Reserve Estimation and Grade Control).

White, W. S. (1971) A paleohydrologic model for mineralization of the White Pine copper deposit, northern Michigan. *Econ. Geol.*, **66**, 1–13.

White, W. S. and Wright, J. C. (1954) The White Pine copper deposit, Ontonagon County, Michigan. *Econ. Geol.*, **49**, 675–716.

8.3 CASE HISTORY – EVALUATION OF THE J-M Pt-Pd REEF, STILLWATER, MONTANA

8.3.1 Location

The Stillwater Mining Company, jointly owned by Chevron Resources and the Manville Corporation, is recovering platinum and palladium from the J-M Reef which is a sulphide-bearing zone within the 43-km-long Stillwater Complex. This reef is discontinuously exposed over almost the entire length of the Complex along the north-eastern margin of the Beartooth Mountains, NW of Red Lodge, Montana.

8.3.2 Geology

The Stillwater Complex is a layered mafic/ultramafic body whose original thickness probably exceeded 6 km and which is dated as early Pre-Cambrian at 2700 Ma (Todd *et al.*, 1982). The Complex commences with the Basal Series which consists of norites and bronzite-rich, olivine-poor, cumulates. These rocks are of economic significance in that they contain pods of copper and nickel sulphides. The Basal Series is overlain by the Ultra-Mafic Series which consists of a lower 'Peridotite Zone' and an upper 'Bronzitite Zone'. In the past, chromite bands have been mined from these units, especially during the Second World War. The Lower Banded Series is the next unit in the stratigraphy which begins with 'Norite Zone 1' and which passes up into 'Gabbro Zone 1'. At the top of this series is the J-M Reef Package which directly overlies a series of plagioclase and plagio-clase-bronzite-augite cumulates and which generally fluctuates between 1 m and 3 m in thickness. The Reef Package commences with olivine-bronzite pegmatoids and then passes up into sulphide-bearing plagioclase-olivine cumulates and then into plagioclase cumulates. The hangingwall formations consist of norites ('Norite Zone II') and gabbros ('Gabbro Zone II'). A more detailed description of the geology can be found in Turner *et al.* (1985) and Todd *et al.* (1982). Figure 8.5, however, summarizes the main elements of the geology in the proximity to the exploited section of the J-M Reef.

An exploration programme on the J-M Reef was commenced in 1967 by the Johns-Manville Corporation. This revealed that the mineralized zone within the Reef Package strikes at 290° and dips between 55° and 90° to the NE. Generally, however, the dip lies between 65° and 75°. The zone was also found to pinch and swell along the explored strike length but, although it locally attained thicknesses of 10 m, it was generally less than 1.2 m thick. Locally, transgressive extensions of the mineralization penetrate the footwall formations and in other areas, a separate footwall zone is recognized and mined. The mineralization is the product of accumulation of immiscible droplets of sulphide during crystallization of the magma and consists largely of pyrrhotite, pentlandite and chalco-pyrite, together with secondary pyrite. The volume percentage of sulphide is low at between 0.5% and 1%, but a 5.5 km length of the reef, in the vicinity of the Minneapolis Adit (Figure 8.5), was found to grade 22.3 g combined Pt + Pd per short ton over an average width of 2.1 m.

Turner *et al.* (1985) state that a normalized sulphide fraction from the reef would contain 9.3 ± 2.9% Ni and 6.9 ± 3.0% Cu. The sulphides are strongly associated with olivine-rich cumulates which are now heavily serpentinized and sheared. Other mineralized rocks include bronzite-rich ultramafics and anorthosites. The host rocks are generally soft and incompetent and the present plunging podiform nature of the ore-lenses is in part a primary feature and in part the product of shearing and attenuation by both faults and shears. The hangingwall of the J-M Reef is closely followed by the South Prairie Fault, a Lamaride reverse fault, and later southwesterly dipping reverse faults further disrupt the dip continuity of the mineralized zone. Figure 8.6, a simplified VLP showing the distribution of stopes and ore-lenses, highlights the discontinuous nature of the ore-bearing J-M Reef.

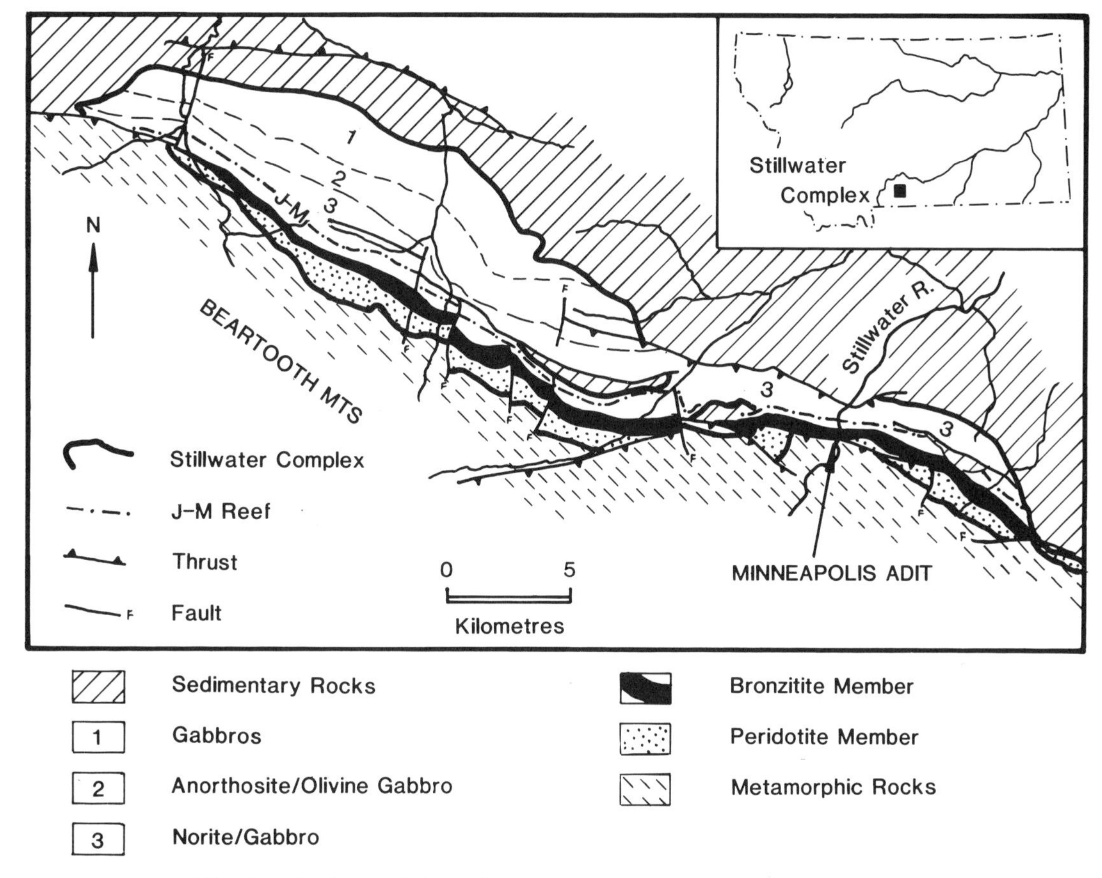

Fig. 8.5 Geology of the Stillwater Complex, the host to the J-M Reef.

Turner *et al.* (1985) contend that platinum group metals (PGMs) were enhanced in the J-M Reef by volatile streaming during the final phases of crystallization and compaction of the magma. They point out the strong produces 3 t of PGMs per annum from ore grading close review of PGM deposits around the world, the *Northern Miner Magazine* (1988) states that the Stillwater mine pro-duces 3 t of PGMs per annum from ore grading close to 22 g/t. The contained metal value is, however, only 133 US$/t because it consists of 77% palladium (price 124 US$, September 1988) and only 20% platinum (price 531 US$, September 1988) by weight. The Merensky Reef in South Africa has an average grade of only 9 g/t but of this, 60% is platinum and 26% palladium giving a contained metal value of 127 US$/t. Within the J-M Reef 50% of the palladium is in solid solution in pentlandite while the remainder is in braggite ((Pt,Pd,Ni)S). Platinum largely occurs in braggite and as a Pt-Fe alloy.

8.3.3 Mining history and methods

Permission to develop the present mine was gained in 1986 and the mill commenced operations in March 1987. Mining is via cut and fill methods and by June 1989, the mill through-put was running close to 750 tons per day. The ore is ground to -200 mesh followed by a bulk sulphide flotation using a talc depressant. In this way, 90% recovery of the PGMs is achieved allowing 10 tons of concentrate to be shipped to Belgium per day.

Although the initial Anaconda plan was to develop a constant 3 m width stope on ore, once Chevron Resources took over the management of the mine, they decided to follow the sulphide contacts as defined by the geologist so that development became variable in width and direction. A horizontal minimum stope width of 1.2 m was, however, applied and the average stope width is presently close to 2.1 m. Figure 8.7 shows, in

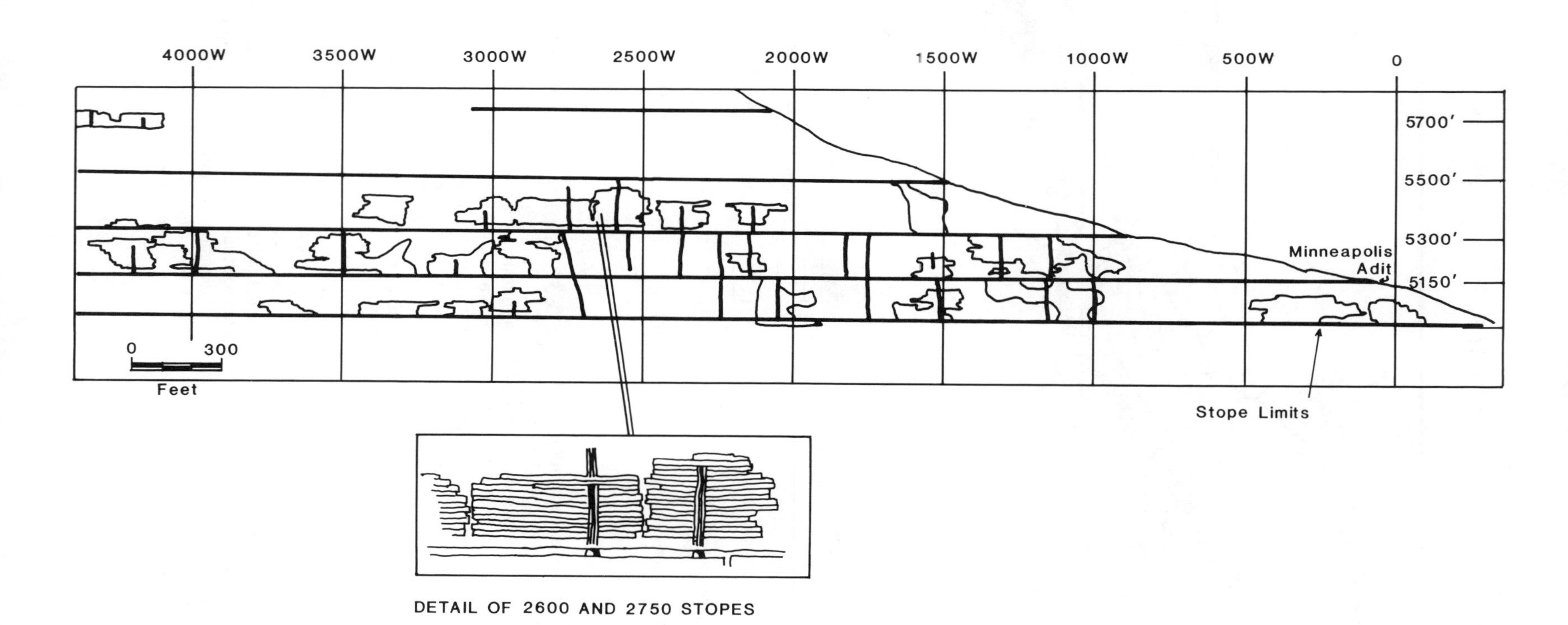

Fig. 8.6 A VLP of the J-M Reef, Stillwater Pt-Pd Mine, Montana, showing distribution of adit levels, raises and cut and fill stopes.

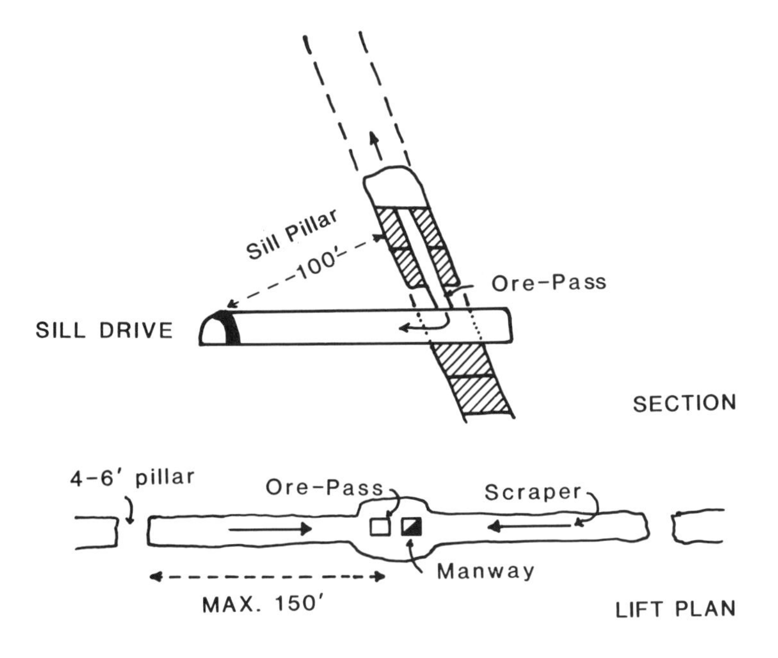

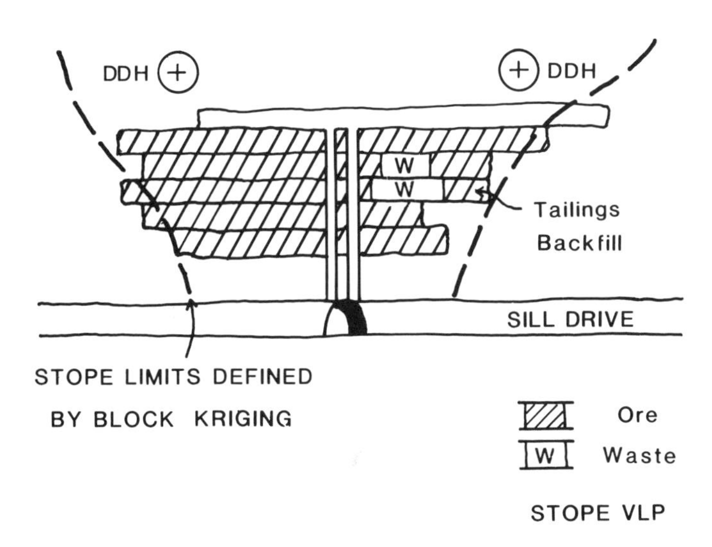

Fig. 8.7 Section and VLP showing lifts and development associated with J-M Reef stopes.

diagrammatic form, a typical stope layout with lifts of 2.4 m. Each lift is backfilled with uncemented tailings as the stope progresses up the dip from the underlying sill drive. Each stope is numbered on the basis of the distance W from the 5150 ft (1570 m level adit portal (Minneapolis adit) to the stope access man-way or drawpoint cross-cut, and on the draw-point level (e.g. 3500 stope, 5150L). Drawpoint cross-cuts ares spaced at intervals of 30–60 m along the sill drives. During stoping the current lift back is supported using close-spaced tubular rock bolts and blocks of wood. These are removed as scrap during the mining of the succeeding lift.

8.3.4 Ore reserves

Fan drilling (BQ size) from the sill drives on 15 m sections allows the intersection of the reef at approximately 15 m vertical centres. Four holes are drilled upwards from each drive and sometimes a down-hole is drilled to evaluate the underlying block. The cores are logged in considerable detail and the lithologies are then plotted on sections using a colour scheme. This is deemed worthwhile for it allows correlation of lithological types and hence the determination of the throw on faults or the thickness of post-ore dykes. It also allows the construction of a detailed stratigraphy which aids future exploration drilling. A red dot adjacent to a lithological unit indicates the presence of sulphides.

Once the sections are drawn, the drill-hole intersections of the orebody are projected on to VLP. As holes generally remain fairly straight, they are no longer surveyed internally and the collar coordinates and inclination only are used by the computer to determine the positions of the orebody mid-points on sections and VLPs. However, to allow for possible drift, a 1.5–1.8 m radius circle is drawn around each point on the VLP.

Each drill-hole is sampled at 0.15–0.45 m intervals over the visible sulphide zone and the assays for Pt and Pd are quoted separately for each. The assays are then combined and if they exceed 3.5 oz/ton (120 g/t) they are cut to this level. This figure is based on experience and on the fact that bulk samples rarely, if ever, exceed 3.5 oz/ton. Individually Pd is cut to 2.72 oz/ton (93 g/t) and Pt to 0.78 oz/t (27 g/t) giving a combined cut grade of 3.5 oz/ton. The level of cut for each metal, however, varies with the current relative price. A cut-off grade is 0.3 oz/ton (10 g/t) is applied in the calculation of ore-zones but it has been found in practice that, if visible sulphides are present, then the grade exceeds this level. For economic evaluation purposes, the combined grade is converted to a dollar value. As the current Pd/Pt ratio is fairly constant at 3.5 : 1, the relative proportions of Pd and Pt are 77.2% and 22.8% respectively.

At September 1988 prices of 124 $/oz for Pd and 531 $/oz for Pt, each ounce of combined metal is worth $(0.228 \times 531) + (0.772 \times 124) = 217$ US$. Hence if the weighted grade of an intersection is 1.05 oz/t (combined), this is quoted as 227.9 $ ore (227.9 $/ton). This figure has, however, to be abated to allow for metallurgical losses. The final value can now be compared with the current cost of mining, milling and concentration of each ton of ore.

All drill-hole thicknesses are converted to horizontal equivalents and the horizontal metal accumulation calculated. These values are then kriged using a 25 × 25 ft (7.6 × 7.6 m) block size on the VLP and a search radius of 75 ft (22.9 m). The semi-variogram parameters used are based on experimental semi-variograms produced from combined drill-hole and chip sample data from several mined-out test stopes. The drill-hole data alone were insufficient to produce reliable semi-variograms (they were erratic) but the combined data gave good simple spherical semi-variograms with ranges of 19–23 m with weak directional anisotropism. The mixing of sample types is not normally acceptable but, perhaps, in this instance the similarity of the sample variance, and perhaps nugget variance, enabled the company to overlook this aspect. Although at present an isotropic case is assumed, future revisions of the method may take into account both zonal and directional anisotropism.

The kriged horizontal thicknesses and kriged block grades for each element, together with the relevant kriging standard deviations, are produced using 'Geology Data Manager Software' originally produced by the BRGM and marketed in the USA by Geomath Inc. of Wheat Ridge, Colorado. This information is then used by the AutoCad software to produce the final VLP plan. Where the kriged thickness of a block is less than the 1.2 m MSW, the program recalculates the kriged grade to a diluted equivalent over this width. The geologist then marks with a red dot all those blocks whose combined Pt + Pd kriged grades are >0.6 oz/ton (20 g/t), the currently accepted minimum mining grade (MMG). The stope area is then defined on the basis of both the MMG and on the kriging standard deviation taking into account both the distribution of ore and non-ore blocks and mining practicability. To date, about 40% of the reef can be stoped. The average kriged thickness in the stope is then calculated and then diluted using a dilution factor (DF) calculated from a power-curve equation. This was derived from the mathematical modelling of data from mined-out stopes and is as follows:

$$DF = 1 + 25.788\ (\text{thickness in feet})^{-2.802}$$

Hence a 4 ft or 1.2 m (MSW) stope would be assigned a dilution of 53% (DF = 1.53) while an 8 ft or 2.4 m stope would be diluted by only 8%. In other words, the thicker the ore the lower the percentage dilutions. The stope grade is then recalculated on the basis of this dilution and the diluted stope tonnage calculated using:

$$N \times T \times DF \times 25 \times 25/11.7$$

where:

N = the number of blocks in the stope
T = the average kriged thickness
11.7 = cubic foot factor (ft^3/ton) to convert ft^3 to tons

The block grades and overall stope tonnage and grade are later compared with those calculated from stope sampling and planimetry and from the mill feed. Aspects of grade control at this mine are described in section 7.2.2, Case history III.

REFERENCES

Northern Miner Magazine (1988) **3**, (9), September.

Todd, S. G., Keith, D. W., Le Roy, L. W. *et al.* (1982) The J-M platinum-palladium reef of the Stillwater Complex, Montana: I stratigraphy and petrology. *Econ. Geol.*, **77**, 1454–1480.

Turner, A. R., Wolfgram, D. and Barnes, S. J. (1985) Geology of the Stillwater County Sector of the J-M Reef, including the Minneapolis Adit. The Stillwater Complex. Montana Bureau of Mines and Geology. **92**, (Spec. Pub.), 210–231.

8.4 CASE HISTORY – EAST ORE ZONE, TECK-CORONA GOLD MINE, HEMLO, CANADA

8.4.1 Location

The Hemlo gold deposit lies close to the NE shore of Lake Superior in Canada, within the Abitibi-Wawa greenstone belt. It is approximately half-way between the towns of Thunder Bay and Sault Ste Marie on the Trans-Canada Highway. The mineralized zone strikes E–W and dips northwards at approximately 65°. It has a proven strike-length of 2 km and a dip-length of 1.5 km and is exploited by three separate operations; Teck-Corona joint venture in the E and central sections, Noranda's Golden Giant in the deeper part of the central section and Page Williams (originally a Lac Minerals operation) in the W.

8.4.2 Geology

Hemlo is a stratiform to stratabound gold-molybdenum deposit which has a total reserve of 75 million tonnes at 7.2 g Au/t. It occurs in amphibolite facies rocks dated at 2.6 billion years. Figure 8.8 represents a diagrammatic stratigraphic column for the Hemlo deposit indicating the presence of three potential ore horizons, 5A, 5B and 5C. The East Ore Zone, the subject of this case history, is 200–300 m wide and averages close to 2.5 m in thickness, although locally it may attain 5 m (Figure 7.27). It is within horizon 5A which is microcline rich and which contains quartz, barytes, barian microcline, sericite, vanadian muscovite, tourmaline and rare calcite. It is underlain by mafic lapilli tuffs (4a) or interbedded quartz-feldspar and quartz-biotite tuffite (3e), and overlain by meta-siltstones of unit 7. To the west, in the central section, the ore-zone thickens considerably to 10 m, although the grade declines.

The principal ore minerals are pyrite and molybdenite but minor amounts of stibnite, realgar, cinnebar, sphalerite and tennantite are also present. The ore-zone is marked by the disappearance of kyanite and the appearance of muscovite and molybdenite, the latter giving the rock a definite grey colouration. It is enriched in Ag, Mo, Ba, V, B, As, Sb and Hg, while the gold is in solid solution with Hg and occurs as clots closely associated with the molybdenite. The reserves of the East Ore Zone (5A) are estimated as 7.6 million tonnes at 12 g Au/t.

Although the gold is associated with molybdenite, it is also clearly related to the presence of pyrite which is concentrated along bands reflecting the original bedding on the host rock. The overall impression is thus of a syn-diagenetic stratiform deposit. However, at depth to the west, the zone is heavily invaded by barytes, which initially picks out the compositional banding, and then forms transgressive veins. This is accompanied by the appearance of quartz veins bearing visible gold.

8.4.3 Mining

The Hemlo deposit was first discovered (or perhaps rediscovered) in 1981 and was brought into production in a remarkably short time after the expenditure of $C80 million. By 1986 the first three levels (L1, L2 and L3 on Figure 7.27) in the East Ore Zone had been developed and stoping, by cut-and-fill methods, was well advanced above level 1 (130 m below the surface). Production in May 1986 had reached 20 000 t/month with an ultimate target of 30 000 t/month. Below level 1 the level spacing is at 100 m intervals and mining blocks are 100 m long. The latter are numbered on the basis of an E to W progression and on the level on which they are based. Thus block C3 lies above L3 and between 11 400E and 11 500E.

All mine development and the main elements of the geology (faults, orebody fringes, diabase dykes, etc.) are stored on hard disk and thus enlargements of specific areas of the mine can be displayed on a computer monitor for detailed mine planning purposes or to allow

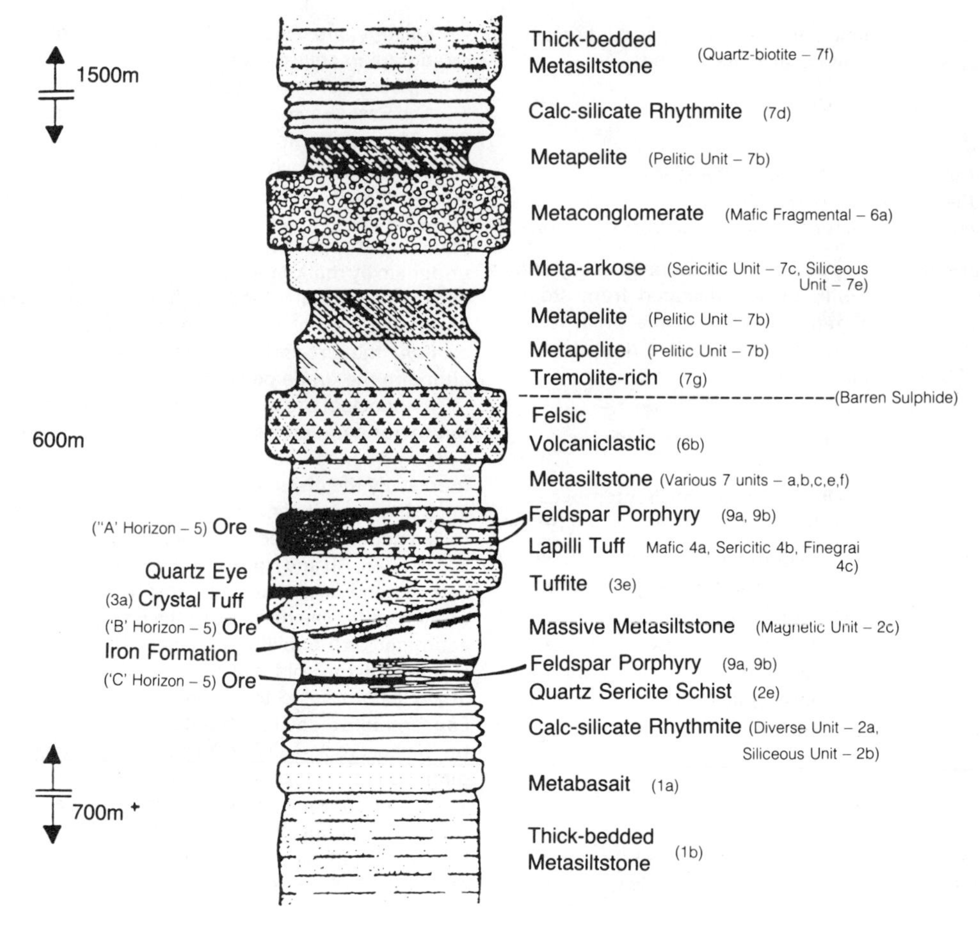

Fig. 8.8 The relationship of mineralization and stratigraphy in the East Ore Zone, Hemlo. (Diagrammatic: not to scale.)

the production of sections, composite level plans, etc. using the AutoCad software.

All sill drives on ore are separated from the first lift in the cut-and-fill stope by a 20 ft (6 m) pillar. An 8 ft (2.44 m) minimum width is re-quired for the drifter and scoop tram in the stope and thus the MSW is set to this width. The stope length is typically 1000–1500 ft and each lift is 10 ft high. Mill-holes are at 30 m intervals along the stope length and fill is pumped into the stopes via a special bore-hole. The fill is mainly sand from a surface quarry as only 10% of the tailings is suitable for this purpose due to the fine grain-size of the remainder.

8.4.4 Ore reserves

The deposit was originally drilled so that each intersection was at 100 × 100 ft centres on the plane of the lode. Once it was realized that the mineralization was remarkably uniform, this spacing was widened to 300 × 300 ft. The core was split and sampled over 1–1.5 m lengths depending on the geology. The samples were then assayed for Au and Mo and occasionally Ag also. Over the top three levels gold grades were cut to 20 g/t whereas grades from the lower levels were cut to 1 oz/metric ton. (**Note.** At the time of the author's visit to the mine in 1986 the mine had taken a tentative step towards metrication by quoting grades in mixed units and using both feet and metres in technical documents.) The weighted grade of each intersection was calculated by applying a 0.088 oz/metric ton cut-off grade (0.1 oz/short ton).

The global reserves of the A zone mineralization are calculated using manual polygon methods (perpendicular bisector) on an unrolled plan (Figure 8.9) or on a VLP, but the dimensions of the polygons are governed by

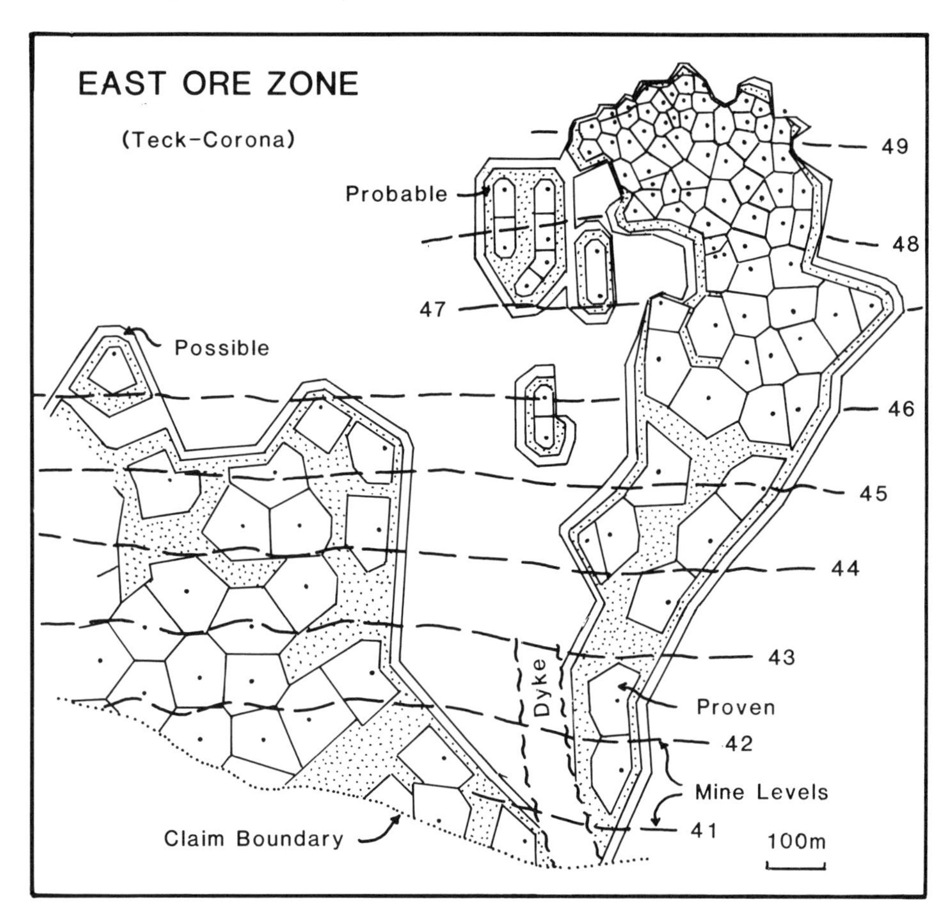

Fig. 8.9 Unrolled plan of the East Ore Zone at Hemlo (Teck-Corona) showing ore reserve polygons and ore classifications.

mine-reserve classifications. Those polygons which lie within the orebody, away from the fringes, and which contain positive holes less than 100 m apart, will have mutual boundaries (Figure 8.10). Where the hole spacing is greater than this, then internal zones will exist which will not be categorized as Proven Ore at this stage. This method thus constrains the dimensions of polygons. The areas of those polygons whose sides are less than 50 m from the contained drill-hole are classified as 'drill indicated' reserves, while those areas between 50 m and 80 m from drill-holes are classified as 'probable' reserves. Any areas outside the 80 m limit are defined as 'possible' reserves. Peripheral holes have their polygons extended outwards for 15 m for 'Drill Indicated' reserves, for 30 m for 'Probable' reserves and for 50 m for 'Possible' reserves (Figure 8.10). The latter thus defines the limit of ore based on the existing density of drilling, i.e. ore is only extrapolated a distance of 50 m from positive holes towards negative (subeconomic holes).

The geological reserves in each category are calculated by planimetry and by the application of a tonnage factor of 2.8 t/m^3 for the East Ore Zone and 3.0 t/m^3 for the higher grade Central Ore Zone. Grades of each polygon are tonnage weighted. No minimum stoping width (MSW) is applied at this stage. The mine reserve is thus the sum of the 'drill indicated' and 'probable' reserves. Mineable reserves are, however, calculated using a 2.44 m (8 ft) MSW but using the geological reserve fringe. Mineable-diluted reserves are calculated applying a variable dilution factor as below:

Ore thickness	% Dilution
<3 m	30
3–5 m	20
>5 m	10

In this case dilution is expressed as a percentage of the ore width: mineable-diluted-recoverable reserves

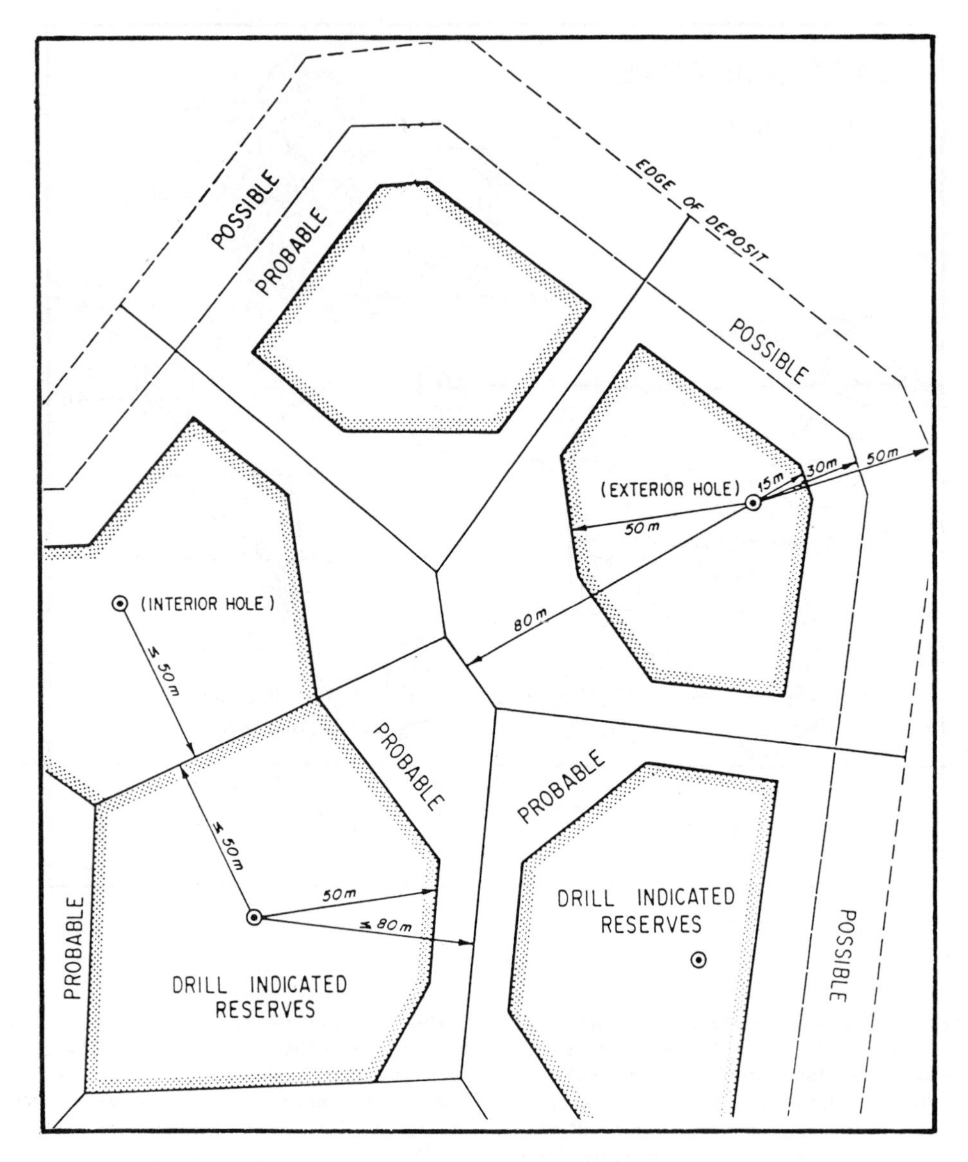

Fig. 8.10 Classification of ore reserves on the basis of polygon size.

assume that a 90% recovery of the broken reserve will be achieved.

Although the polygon method is deemed satisfactory for the global reserves at Teck-Corona, they are unsuitable for mining purposes and are thus recalculated on the basis of 100 × 100 m mining blocks (Figure 8.11). The area of each polygon lying within the mining block (A_1–A_4) is planimetered and the tonnage calculated based on:

$$\Sigma_{i=1}(A_i \times T_i) \times SG$$

where T_i represents the horizontal thickness assigned to each polygon. The block grade is then calculated from:

$$\Sigma(A_i \times T_i \times G_i)/\Sigma(A_i \times T_i)$$

8.5 CASE HISTORY – OPENCAST COAL MINING IN SOUTH WALES

(Mr R. MacCallum, British Coal – Opencast Executive, South Wales Region)

8.5.1 History

Coal was first mined in South Wales in the Bronze Age and mining continued, on a limited scale, through Roman and Mediaeval times. Extensive mining began in the mid-eighteenth century when coal and ironstone was exploited via shallow opencast workings known as 'patch workings'. Modern opencast mining began in 1942 under the Defence of the Realm Regulations as a wartime contingency. The Opencast Coal Act of 1958 passed control of the industry to the Open-cast Executive of the National Coal Board which had been established at nationalization in 1947. About 3000 million tonnes of bituminous and anthracitic coal are thought to have been won from the coal measures of South Wales to date. Of this, about 71 million tonnes have been mined by opencast methods since the Second World War.

Up to the 1950s, sites were generally small and restricted to depths of about 18 m, largely because of the limitations of the plant and machinery available at the time, together with the economic stripping ratio at which the coal could be worked. Some improvements to plant and machinery for both exploration and extraction purposes, plus an increase in the economic stripping ratio, have led to a substantial increase in the size and depths of opencast mines in the region. Sites are currently being worked to depths of 150 m. All sites are worked on a contract basis for the Opencast Executive by large companies who specialize in, or who have specialist branches in, opencast coal mining. Figure 8.12 shows the distribution of open-cast coal sites in South Wales which are referred to later in this case history.

8.5.2 Exploration and sampling

Sites are selected for exploration after basic research has been carried out into the potential viability of an area. This involves investigating the geology from published sources, abandoned deep mine records, and from records and maps held by the Executive. This preliminary investigation also involves the potential marketability of the type of coal anticipated in the area, geographical location and access and anticipated planning controls. If the above conditions seem favourable, then detailed exploration by drilling and sampling is carried out.

The drilling is done on a contract basis by specialist drilling companies, but British Coal geologists maintain close control of the drilling operation. Mobile, tractor-mounted, rotary air-flush rigs are normally employed (Figure 8.13), but occasionally, specialized rigs are used where ground conditions are very soft or where depths in excess of 200 m are required. Most of the boreholes are 'open-holes', drilled with tricone bits which produce a 120 mm borehole through both superficial deposits and solid bedrock. Selective coring is undertaken using a 412

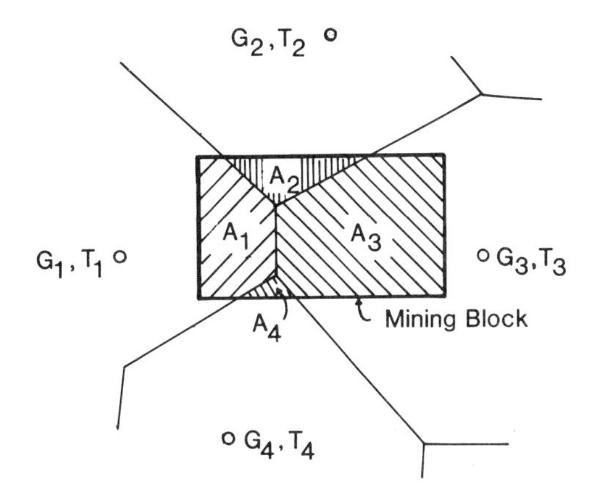

Fig. 8.11 Computation of mining block reserves from polygons.

core barrel which can be fitted with a variety of bits depending on strata conditions. All coal horizons are sampled to determine accurate seam thicknesses and to provide material for chemical analysis. In addition, as many as ten continuous cores are taken of the whole stratigraphic sequence on every site for geotechnical and engineering evaluation.

Borehole spacing on most sites is at 30 m centres, and few sites are drilled with a bore-hole spacing in excess of 50 m. This close spacing is controlled by the need to establish the extent, and percentage extraction, of abandoned underground workings in the seams, and to elucidate the often complex geological structure. In extreme cases, particularly in the southern and western parts of the South Wales Coalfield, borehole spacing may be reduced to 10 m in order to more accurately define complex geological structures.

Nearly all boreholes drilled are now geophysically logged. A pilot scheme using the nuclear geophysical suite of wireline logging tools was carried out in the early 1970s and such was the success of this that geophyysical logging has become one of the most important routine exploration techniques. The logging unit can be mounted in a rough-terrain vehicle and run by a logging engineer but the system currently in use in South Wales employs driller-operated geologging units (Figure 8.14). These units consist of a steel box which can be mounted on the drilling rig and encloses a completely self contained geologging system comprising sonde, cable and winch, electronics package and data recorder. An easy to use control panel on the front of the unit enables the driller to log each borehole immediately on completion, the data being recorded on cassette tape. The tapes are processed by computer at the Opencast Executive Regional Headquarters and a graphic log is produced on an electrostatic printer/plotter. The logs generated

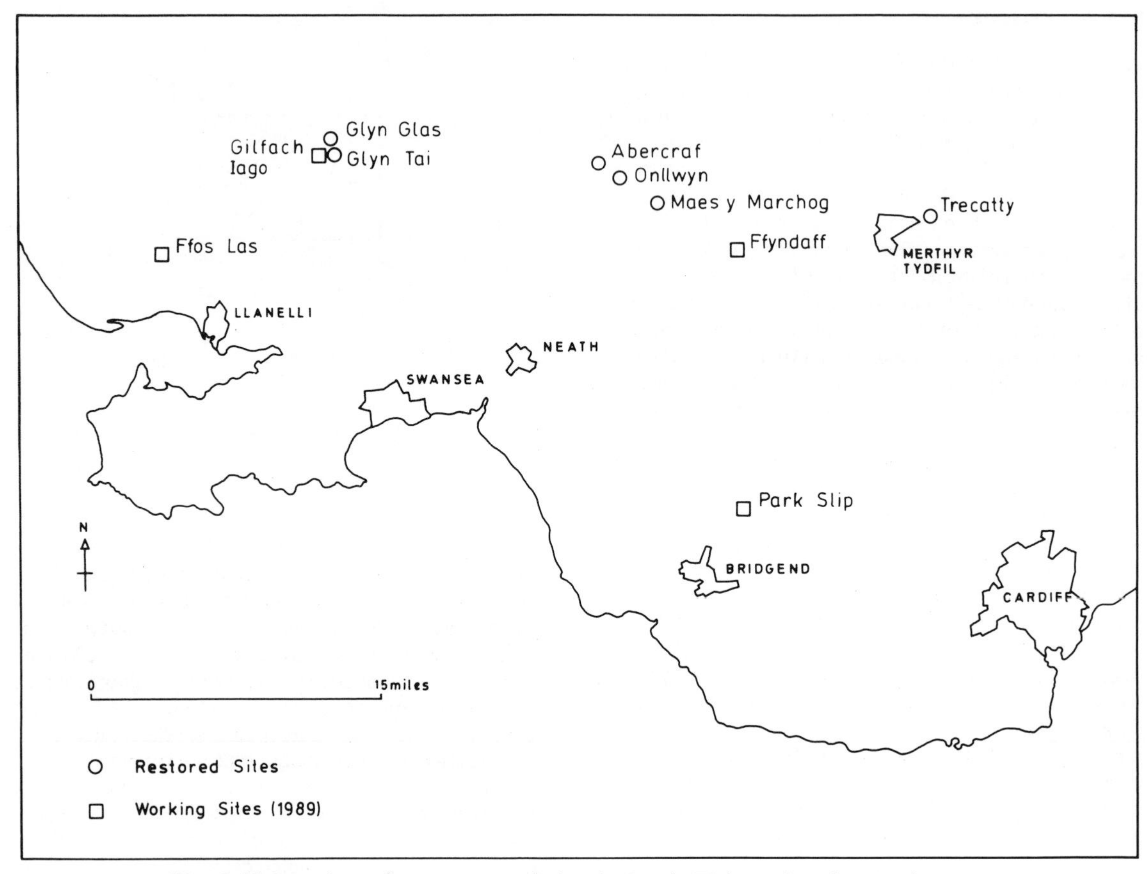

Fig. 8.12 Location of open-cast coal sites in South Wales referred to in the text.

comprise a natural gamma trace, together with long spaced and high resolution density traces (Figure 8.15).

All borehole data gathered during the exploration phase on a site is input to a computer. British Coal's Opencast Executive have their own computer system known as GEOMODEL (registered trademark of British Coal) which was programmed 'in house' by Compower Ltd, a British Coal subsidiary. This system anables the large volume of data produced during drilling to be conveniently stored, and in addition has a large set of extraction programs which run against this data base in order to speed up the elucidation of geological structure and stratigraphy, and to assist in the evaluation of the reserve.

8.5.3 Preliminary evaluation

During the initial stages of site exploration, enough borehole data are gathered to indicate viability. Preliminary exercises are thus carried out in order to establish the economic potential of the site.

The ratio of the tonnages of coal to over-burden has to remain within acceptable limits. Currently, sites in South Wales are being worked at ratios of up to 25 : 1. Estimates of the ratio can be provided by GEOMODEL which examines the proportion of coal to non-coal strata in every borehole for a given area. Variations in this ratio are often due to the incidence of old workings rather than hole depth. This method provides an indication as to whether the ratio on the site is within acceptable limits.

Chemical analyses of cored samples of the seams are required to confirm the marketability of the coal. At a preliminary stage, the most important aspects of these analyses are the ash and sulphur content. Generally the coals of South Wales are of good quality (i.e. they have ash values $<10\%$ and sulphur values $<1.5\%$) and those seams notable for either high ash or high sulphur values tend not to be grouped together stratigraphically. As a result, it is often acceptable to have one or two seams on

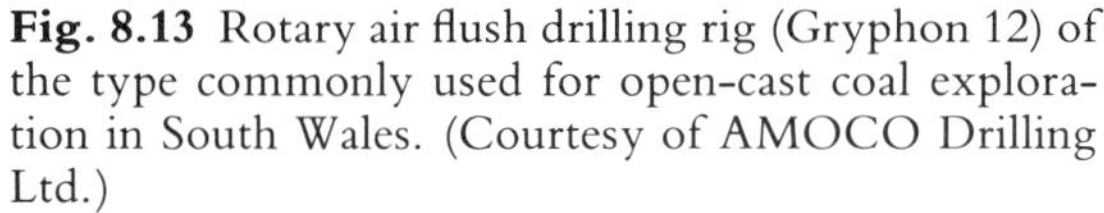

Fig. 8.13 Rotary air flush drilling rig (Gryphon 12) of the type commonly used for open-cast coal exploration in South Wales. (Courtesy of AMOCO Drilling Ltd.)

Fig. 8.14 Driller-operated geologging unit in place over a borehole. (Courtesy of Brown Electronics Ltd.)

a site, coaling say twelve seams, to have high ash and sulphur values, since the product from these seams can be blended with other coals at the coal preparation plant to produce an acceptable product. Where seams are less than 30 cm thick, and are of inferior quality, they are frequently excluded from a contract. Table 8.1 illustrates some typical analyses from some seams from open-cast mines in South Wales whereas Table 8.2 lists typical specifications for a variety of markets.

If an exploration site is shown to contain coals which can be mined with an acceptable stripping ratio and quality then preliminary estimates of tonnages are calculated. This can be done using GEOMODEL which can evaluate volumes or provide average seam thicknesses within a given area (Table 8.3). Volumes are converted to tonnages using conversion factors of 1.33 to convert cubic metres of bituminous coal to tonnes and 1.40 for anthracite.

Planning considerations are closely examined at all stages in site development. Timely liaison with the relevant Mineral Planning Authority may enable important planning aspects to be incorporated into the site design at an early stage. This may avoid major modifications, a planning refusal or a Public Inquiry at a later stage.

If the above preliminary investigations show the viability of a prospect to be in serious doubt, it may be shelved at this stage. On the other hand, if preliminary work has firmly established the viability of a site, then more detailed exploration and sampling will continue in order to provide all the data necessary to carry out a final site design and evaluation. This will involve reducing the borehole spacing to gain more information on geological structure, seam thickness and continuity, and the incidence of old workings, together with increased coring and sampling of coals for analysis. Also at this stage, a comprehensive geotechnical investigation will be carried

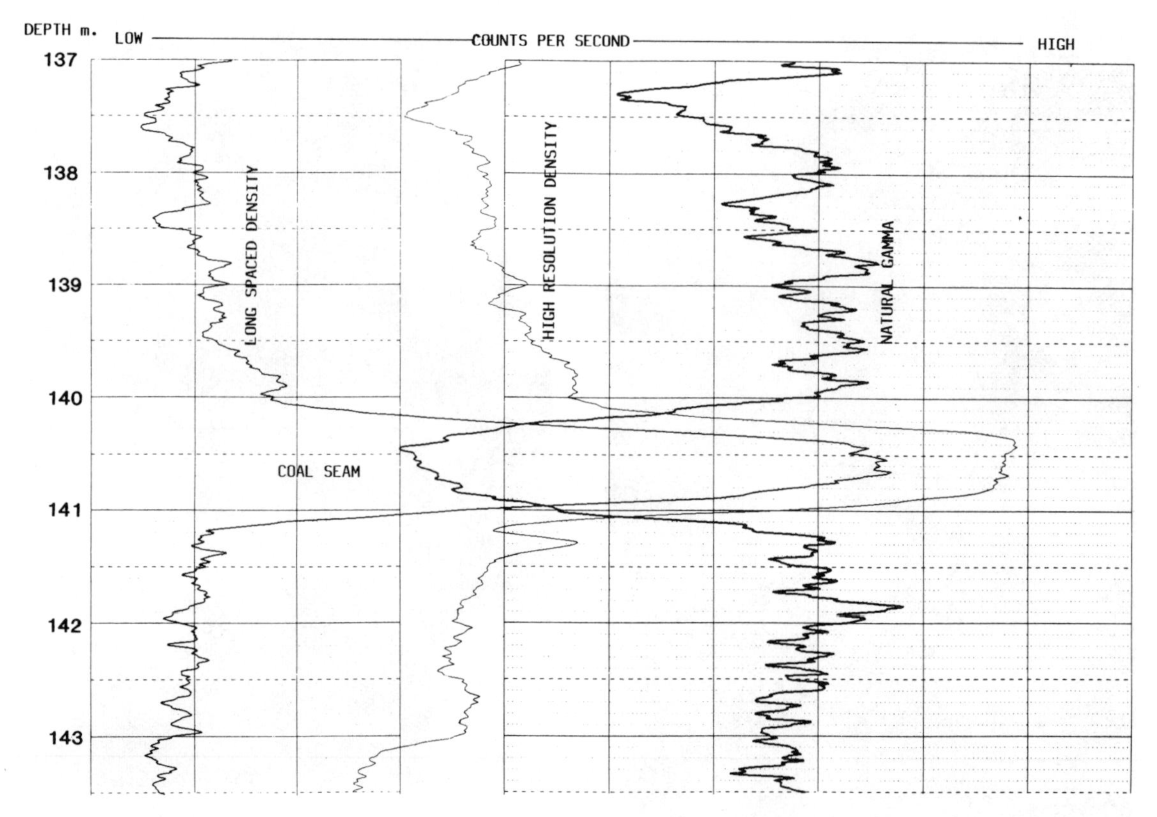

Fig. 8.15 Typical natural gamma and density responses produced by coal and associated strata at Trecatty (Little Vein seam). High counts per second (c.p.s.) on the density traces indicate the low bulk density of coal compared to the mudstone above and below. Low counts on the natural gamma trace are typical of coal (high organic content, low K^{40} compared with mudstone).

out to provide data essential to the final site design.

Areas of the site, in which detailed geotechnical investigation is necessary, will have been identified during the preliminary exploration phase. Geotechnical investigations on all sites involve additional coring and the production of an engineering log for all solid strata likely to be encountered in the excavation, and also strata surrounding the excavation which are likely to influence the stability of the walls. Particular attention is focussed on faulted zones, or other structurally complex areas, so that the excavation can be designed to minimize the chance of failure. Superficial deposits are sampled using U4 sample tubes, the samples being sent to the laboratory for shear-strength testing. In addition, superficial deposits may be sampled and logged using shell and auger drilling equipment or in trial pits and trenches. Soil sampling is frequently carried out at the same time and is often augmented by hand augering. Particular attention is paid to the superficial deposits around the margins of the proposed excavation area and in areas which will be used for overburden storage mounds. Groundwater is investigated by the placing of piezometers in those horizons in the superficial deposits and solid strata which appear from the drilling and geophysical logging to have some hydrogeological significance. Of particular importance for open-cast mining is the existence of abandoned underground workings which may contain, or connect with, large volumes of water, sometimes with a considerable head. A large number of piezometers may have to be set in such old workings in order to establish the water regime as this will be critical to the working of the site.

8.5.4 Final site design

When all the detailed exploration is complete, a comprehensive final site design can be produced. First, the working site boundary has to be established since this

Table 8.1 Analyses of coals from various opencast sites in South Wales

Site + seam	Thickness (m)	Moisture % air dried	Ash (%)	Volatiles (%)	Sulphur (%)	Calorific value (kJ/kg)
Trecatty, Big	1.05	2.0	6.5	12.9	0.89	27 960
Park Slip, 9 ft	2.25	1.7	5.3	27.2	0.78	31 890
Glyn Tai, Stanllyd	1.95	2.4	2.3	5.8	0.94	34 610

Table 8.2 Specification of coals for various markets in South Wales

Market	Moisture % as received	Ash (%)	Volatiles (%)	Sulphur (%)	Calorific value (kJ/kg)
Power station	10.0	17.0	12.2	1.0	25 960
Coking coal	12.0	6.4	17.5–28.0	0.65	29 720
Domestic anthracite	3.0–4.0	2.0–5.0	5.0–7.0	1.0	33 330

Table 8.3 Summary of the evaluation of the upper leaf of the Lower Four Feet Seam on Trecatty Site, using 'GEOMODEL'. (For details of method see section 8.5.5)

	Method of evaluation*		
	1	2	3
Area of seam (m^2)	204 561	204 561	204 561
Boreholes (no.)	98	—	—
Average seam thickness (m)	0.65	—	—
Derived mean seam thickness (m)	—	0.66	—
Mean seam thickness from grid values (m)	—	0.66	—
Volume of seam (m^3)	132 965	135 010	135 010
Tonnes	176 843	179 563	179 563

*1, Using simple arithmetic mean of borehole data; 2, using values from triangulated borehole data (triangle prisms); 3, using interpolated grid values from triangulated borehole data.

may vary from the boundary used in the exploration phase. The location of the working site boundary will control the area within which the coal and overburden can be excavated, overburden mounds and baffle banks established, drainage works and water treatment areas located, together with areas for offices and workshops essential to the working of the site. Figure 8.16 shows the basic site design for the Glyn Tai Site in Dyfed, Wales, which was worked over the period 1983–1986.

The limit of excavation, although initially controlled by the location of the site boundary, is usually subject to other factors both inside and outside the site. Minimum clearances have to be allowed between the limit of excavation and both natural features, such as rivers or

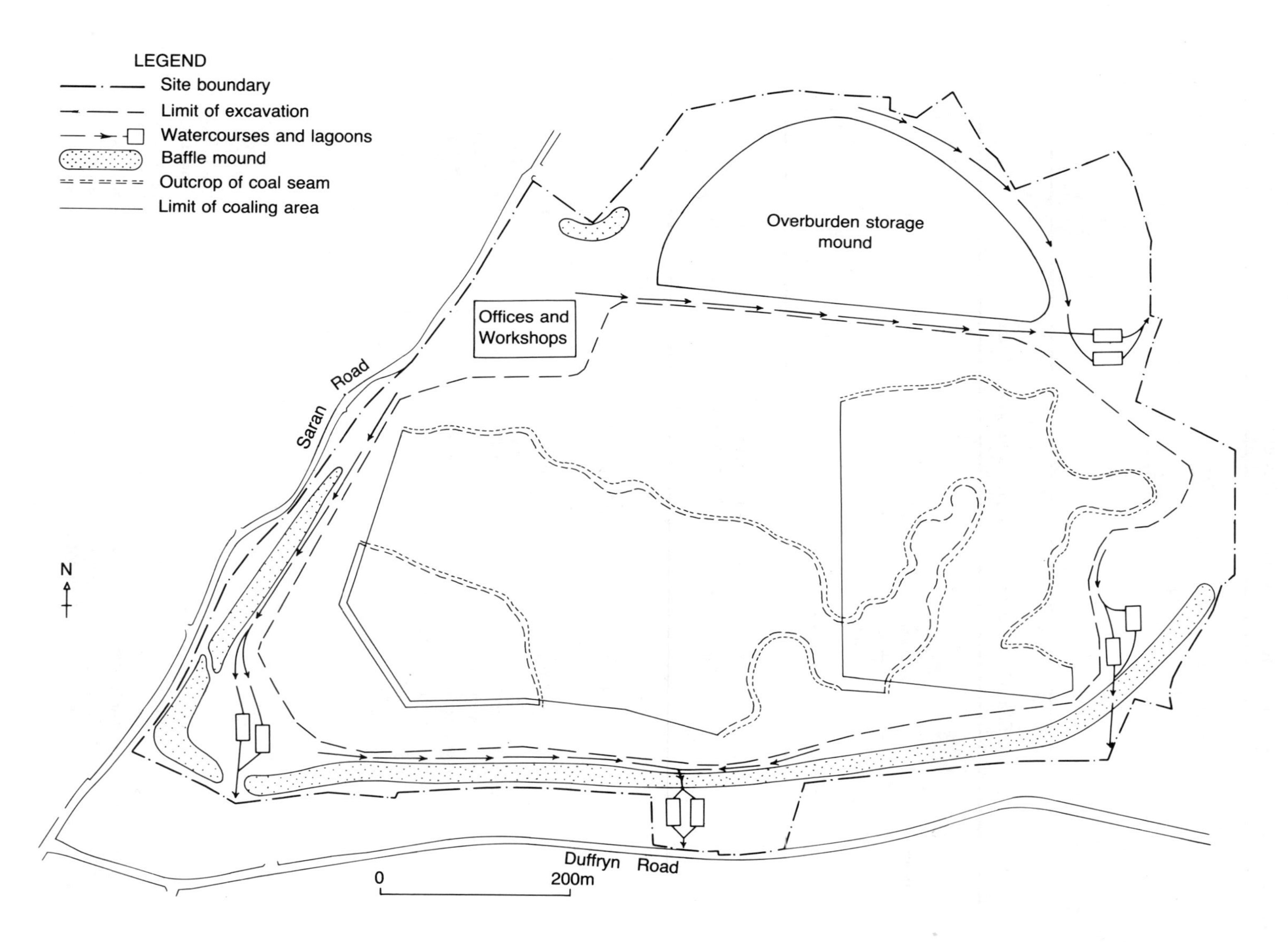

Fig. 8.16 Outline plan of the Glyn Tai Site illustrating the major elements of the site design.

woodland which is to be preserved, and man-made features, such as roads, housing, railways, power lines and gas and water mains. Some natural or man-made features may be relocated or diverted during the development of the site but those which remain have to be given adequate clearance from the limit of excavation to ensure their safety, and/or to provide for the establishment of baffle embankments between the excavation and, for example, housing or roads. Clearances from the limit of excavation can vary considerably from site to site depending on topography and geology but are normally kept to at least 40 m.

Drainage works on every site are carefully planned, not just to accommodate the high rainfall expected in South Wales, but to cater for the large volumes of groundwater which may be encountered during the course of excavation. Also, since abandoned and flooded underground workings are intersected on many sites, considerable volumes of mine water may be encountered. This is frequently acid and may need special treatment before discharge. Strict controls on the quality of water which may be discharged from a site are maintained. To this end, water treatment facilities are provided on all sites and the quality of water discharge is carefully monitored. Attenuation basins are constructed to cope with maximum predicted flood conditions on the site, ranging from 10- to 50-year flood predictions.

Temporary baffle embankments and overburden mounds are designed to ensure their stability for the length of time during which they have to stand. Baffle embankments last for the life of the site and are usually constructed from soil and subsoil which will eventually be used in the final stages of restoration of the site. They are typically 5 m high on a base of 25 m, with batter slopes of 1 in 2. Temporary overburden mounds are constructed to a maximum height of about 40 m, with slopes of 1 in 2. Temporary overburden storage rarely involves more than 35% of the total excavation. Both baffle embankments and overburden mounds which are to stand for any length of time are grassed to minimize the visual impact.

Within the limit of excavation, the areas from which the coal is to be won from each seam have to be defined. In the first instance, safe batter slopes have to be constructed so that the stability of the walls of the excavation can be maintained for the length of time they are exposed. Experience over more than 40 years in South Wales has shown that overall batter slopes of 1 in 2 in the superficial deposits and 1 in 1 in solid strata with a suitable bench at rockhead, frequently provide the basis for a safe design. However, every site is examined in detail and each has a geotechnical analysis carried out to ensure an adequate factor of safety. This analysis involves a detailed investigation of the nature of the strata, geological structure, groundwater conditions, and the incidence of mine workings. Linked to the geotechnical factors which control the detailed site design, are such factors as the working method for the site and the maximum depth to which the site can be worked. This latter consideration is usually controlled by the maximum gradient for the main haul roads from the lowest point in the excavation. As a general rule, haul roads should not be steeper than 1 in 10 and accordingly, in an ideal situation, the maximum horizontal dimension of an excavation area should be ten times the maximum depth.

Once the geotechnical and working method aspects of site design have been determined and safe batter slopes have been calculated, the intersection of the walls of the excavation with the seams to be coaled are defined. Benches in the walls of the excavation are frequently designed to coincide with those horizons which contain the coal seams where this is practical. Where seams are widely spaced, however, intermediate benches may be necessary, at a maximum interval of 30 m.

Characteristics of the seams to be coaled may influence the delineation of coaling areas within the excavation. 'Washout' conditions, or a high incidence of old workings in a basal seam, may lead to wide benches being established on higher seams which are not so affected. Also, complex geological stuctures including folding, faulting and overthrusting may influence the delineation of coaling areas, sometimes leading to a complex configuration of internal walls at certain stages of site development.

When the coaling limits have been delineated, a precise evaluation of coal volumes and tonnages can be carried out for each seam. In addition, the overburden volume lying vertically above the coal can be calculated and an accurate ratio of coal to overburden determined.

8.5.5 Detailed evaluation of seam thicknesses

The estimation of an average thickness for each seam, within its coaling limits, is critical to the precise evaluation of quantities of coal available for extraction. The seams of the South Wales Coalfield vary from simple, single-leaf seams showing little variation in thickness, to complex multi-leaf seams or seams showing extreme variation in thickness.

The data used in determining seam thickness comprises open-hole drilling records, coring records and geophysical logs, gathered during the exploration phase and stored on the computer. Open-hole data have been largely superseded by geophysical data for the evaluation of seam thickness, but coring information is essential since it provides a precise measurement of seam and parting thickness and is necessary for the calibration of

Table 8.4 Data set of thicknesses of the Lower Brass Seam on Maes-y-Marchog Site, from cores and equivalent geophysical logs

	Thickness (m)	
Borehole no.	Cores	Geophysical logs
1016	0.56	0.55
1041	0.63	0.64
1043	0.70	0.65
1072	0.54	0.60
1076	0.45	0.58
1170	0.60	0.61
Analysis of data set	**Cores**	**Geophysical logs**
Total thickness (m)	3.48	3.63
Boreholes (no.)	6	6
Max. thickness (m)	0.70	0.65
Min. thickness (m)	0.45	0.55
Average thickness (m)	0.58	0.60
Coefficient of variability (%)	17.03	6.16
Standard deviation	0.10	0.04

geophysical data. The geophysical data have been found to provide very accurate estimates of seam thicknesses, but the resolution is such that it does not allow accurate measurements for leaves of coal less than 0.15 m thick, or for shale partings within a seam less than 0.10 m thick. Comparison of geophysical logs with coring data is therefore vital. Extraction programs enable the computer to list the seam thicknesses for those bore-holes penetrating the seam within, or adjacent to, the coaling area.

The Lower Brass Seam on the Maes-y-Marchog Site (Figure 8.12) is a single-leaf seam showing little variation in thickness. Table 8.4 shows a listing of thickness information from cored holes and the equivalent values determined from geophysical logs, together with an analysis of the data for this seam. There is a good correlation between the two sources of data, indicating that the geophysical logs are providing an accurate picture of seam thickness, with an average value of 0.60 m compared with 0.58 m from the cores. A listing of all 92 geophysical logs taken through the seam reveals that the average value (Table 8.5) also compares well with the average from the cored holes and that the coefficient of variation is only 17.36%. With such a seam, providing there is comprehensive borehole coverage, a simple arithmetic mean will give an adequate estimate of the average seam thickness.

The Geomodel system provides for automatic triangulation of data held on computer. The thickness of a seam can be triangulated between bore-holes to form a triangle net with the seam thickness in each bore-hole forming the apices of the triangles. The area of each triangle is multiplied by the mean of the thickness values lying at the apices to generate a volume for each triangle. Summation of the individual triangle volumes will provide a total seam volume for a given area, and the average seam thickness over that area is derived by dividing the volume by the area. This method of estimation is in effect using a set of triangular prisms to calculate the volume of a seam in a known area and hence derive a thickness. In addition to this triangular prism method, a calculation based on grid points superimposed on the triangulated data can be used. A grid origin can be selected manually or is fixed automatically in the SW corner of the plan. A grid spacing is then selected and the seam thickness values at the grid points are interpolated from the triangle data depending on

Table 8.5 Summary of thickness estimates, Maes-y-Marchog and Abercrâf Sites

	Thickness (m)	
Data source	Maes-y-Marchog Site Lower Brass Seam	Abercrâf Site White Four Feet Seam
Cores	0.58	0.47
Geologs of cores	0.60	0.43
All geologs	0.58	0.72
Grid values	0.60	0.79
Triangle prisms	0.59	0.80

where these grid points lie in each triangle (Figure 8.17). The average of all the grid point values provides the average seam thickness and the volume is then generated by simply multiplying this thickness by the area involved.

A good borehole coverage for the Maes-y-Marchog Site and the uniformity of thickness values, ensures that the results obtained by the two triangulation methods compare well with those derived by applying an arithmetic mean to the raw data (Table 8.5).

The White Four Feet Seam on the Abercrâf Site (Figure 8.12) is a simple single-leaf seam but it varies considerably in thickness, and ground conditions precluded the drilling of evenly spaced holes. Table 8.5 presents a summary of the various estimates of thickness for this site. In this instance, a simple arithmetic mean value for the seam thickness has to be viewed with suspicion since it does not take into account the spatial distribution of the very variable (coefficient of variation 62.77% for Geologs, 104.11% for cores) seam thickness values obtained from the bore-holes. This problem is to some extent overcome by the use of triangulated data. As can be seen from Table 8.5, thickness values derived from grids and triangular prisms show a considerable difference from the simple arithmetic mean values but are considered to be a better representation of the average seam thickness.

The Lower Four Feet Seam on the Trecatty Site (Figure 8.12) is a complex multi-leaf seam consisting of up to six leaves of coal with intervening partings. The seam can be divided into three main sections but only the middle section will be considered here for it illustrates some of the problems encountered in the evaluation of multi-leaf seams. Coring data shows that the middle section of this seam consists of a main upper leaf of coal with two thinner leaves below. These thinner leaves, and the intervening partings, cannot be fully resolved by geophysical means on their own and are thus grouped as 'shaly coal' (Figure 8.18, Table 8.6). Core data have to be used to assist in the interpretation of the geologs. The total seam section (roof to floor) based on the geophysical data compares well with the coring data. The cores show, however, that this thickness includes a combined parting thickness of 0.13 m which is thus deducted prior to the calculation of the average coal thickness.

An alternative to the above approach would be to derive a thickness for the main upper leaf from the overall geolog coverage of the seam, and then to rely solely on cores for the thicknesses of the thin lower leaves. This would rely heavily on good core coverage and would still need the more numerous geophysically logged boreholes to confirm the presence, if not the thickness, of the thin lower leaves across the whole coaling area.

Grid and prism methods of evaluation using triangulated data, applied to the whole seam section, produce similar results to those produced by the arithmetic mean (Table 8.6). This indicates that, although the seam has a complex section, the variation in thickness and the spatial distribution of the boreholes has little influence. Where these factors are significant, however, total seam thickness derived from either grids or prism methods would be accepted and the necessary reduction for included partings, derived from coring data, would be applied to this figure.

When evaluating coal seams containing partings it may be considered impractical to remove some of the partings during the coaling operation. In this situation, the parting thickness would be included in the average thickness for the seam. In the example cited above, the middle section of the seam was coaled in two lifts with the main upper leaf being lifted separately as clean coal. The parting below this leaf was discarded during the coaling operation and the two thinner, lower leaves were lifted together with the intervening parting. Coal and parting material were then separated in the coal preparation plant.

8.5.6 Evaluation of reserves in structurally complex terrains

In some parts of the South Wales Coalfield the coal measures have been intensely folded and faulted and the structures in such areas pose special problems in evaluation. Complex structures have not proven a barrier to exploration or evaluation, but rather the converse, since in such areas the amount of extraction by previous underground mining is frequently less.

Complex, and sometimes multiple overthrusting has been a feature of many sites in South Wales, e.g. Ffyndaff, Park Slip, Glyn Glas and Ffos Las. It is critical that an accurate picture of the structural geology is built up from the borehole data, prior to evaluation, and that repetition of coal seams is recognized. The intersections of thrust planes with coal seams have to be carefully plotted so that overlapping panels of coal on the same seam can be readily identified. Once this has been done, the coal in each panel can be evaluated separately. As a result, within a single site, one seam may have a number of overlapping coaling areas bounded by the intersection of the seam with thrust planes. The seam thickness would be separately evaluated for each overlapping area. On Ffyndaff Site, currently in production, four identifiable seams are being worked as 11, separately assessed, panels of coal bounded by thrust faults.

On sites where the strata have been intensely folded, e.g. Ffos Las and Gilfach Iago, a different approach to evaluation is required. In particular, the coal contained in monoclinal folds with vertical limbs cannot be evaluated

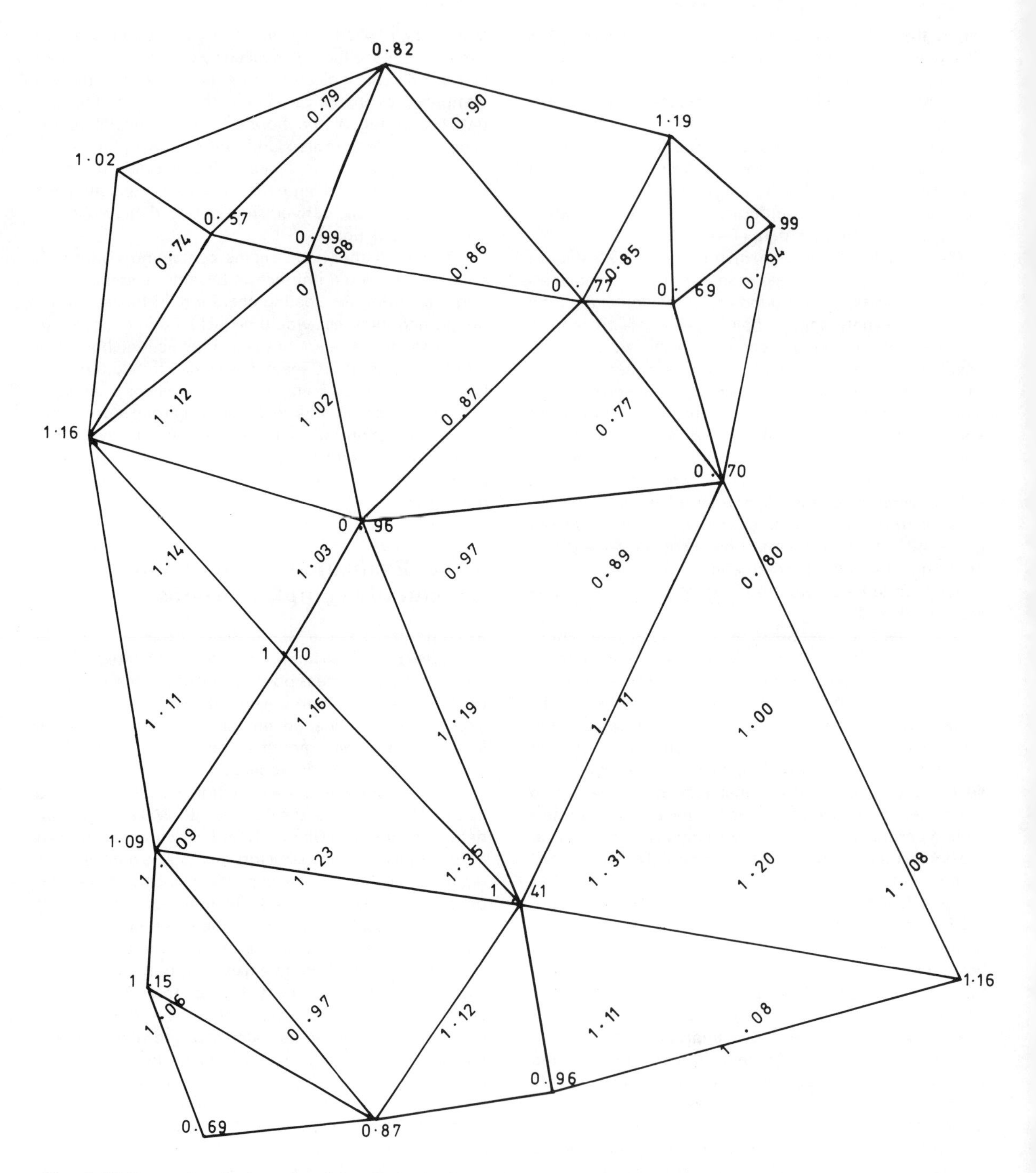

Fig. 8.17 Example of triangulated borehole thickness data. The boreholes are situated at the triangle nodes and the adjacent figures are seam thickness values. The sloping figures are interpolated values at a grid spacing of 25 m.

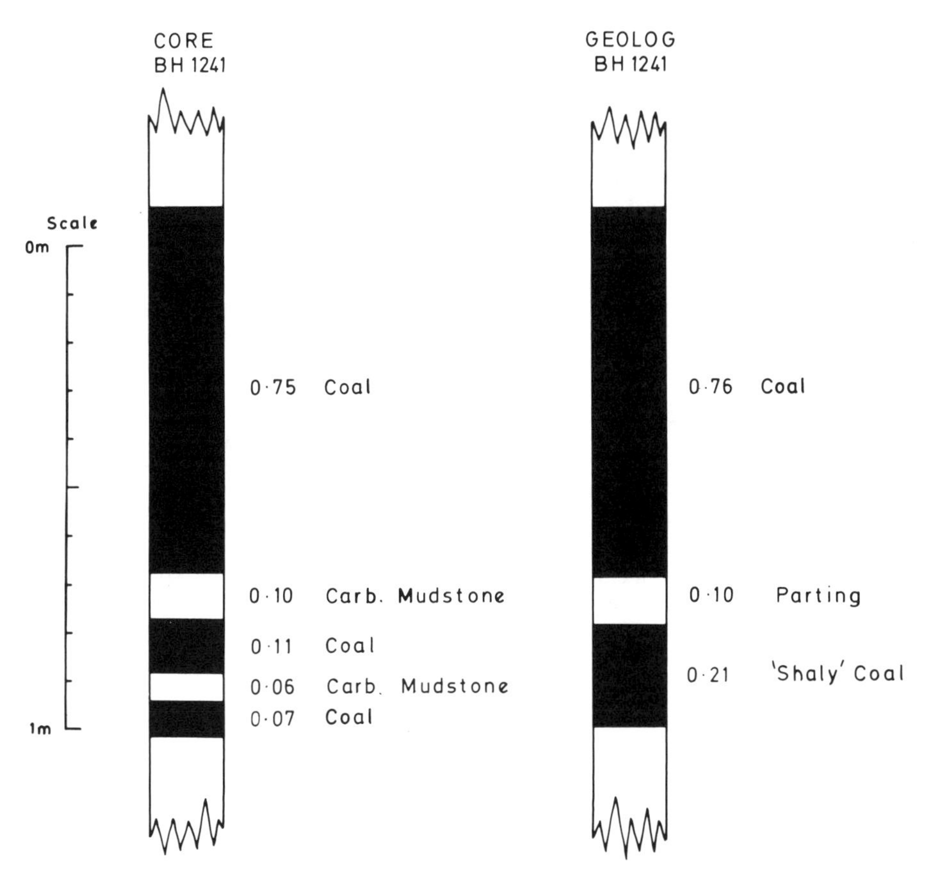

Fig. 8.18 Comparison of a core and geolog from the middle section of the Lower Four Feet seam on the Trecatty Site. The thin parting near the base of the seam is not fully resolved by geophysical means so cores are essential to establish the fine detail of the seam section.

Table 8.6 Summary of thickness estimates, Lower Four Feet Seam, middle section, Trecatty Site

	Thickness (m)		
	Cores	Geologs of cores	All geologs
Top leaf	0.72	0.69	0.68
Parting	0.08	0.12	0.13
Middle leaf	0.15 }		
Parting	0.05 }	0.21 'Shaly Coal'	0.23 'Shaly Coal'
Lower leaf	0.08 }		
Roof to floor	1.03	1.01	1.04
Grid values, roof to floor			1.03
Triangle prisms, roof to floor			1.01

using the vertical thickness derived from bore-hole data. In such situations, the coal contained in vertical fold limbs is separately assessed. Precise quantification of the coal is difficult and potentially has the largest margin of error in coal evaluation. The method usually employed is to first calculate the area of coal contained in a vertical limb as if viewed horizontally. This area is multiplied by the thickness to generate a volume for the coal. The thickness used is the true thickness of the seam as calculated from the bore-hole records of seam thickness in adjacent areas where the seam is not vertical, using the formula $T = VT.\cos\alpha$, where T is a true thickness, VT is the average vertical thickness and α is the average dip angle.

Experience has shown that monoclinal folds often have minor folding associated with the vertical limbs which increases the amount of coal available. However, there is no satisfactory method of precisely evaluating the coal in such small-scale structures. Occasionally, totally unexpected structures are exposed during the working of a site. On the Onllwyn Site the drilling indicated that the Bluers Seam consisted of a main top leaf of coal underlain by up to six very variable thinner leaves. During excavation, however, it became apparent that these lower leaves were in fact part of the main leaf which had undergone considerable movement in relation to its floor, resulting in the generation of an imbricate structure at the base of the seam.

8.5.7 Estimation of allowances

Frequently, the coaling areas on a site need to have allowances set against them in order to account for losses which it can be predicted will occur during the coaling operation. These losses fall into a number of clearly defined categories, described below.

Old workings

Few open-cast coal sites in South Wales have seams which are unaffected by old underground workings. Accordingly, all such seams have to have an estimate made of the proportion of the seam which has been removed by previous mining. Such estimates are rarely straightforward for a variety of reasons. Since 1871, it has been a legal requirement for all mine workings to be recorded on plan. Unfortunately, these abandoned mine records are far from complete since a great deal of mining was carried out prior to this time. However, some records were kept before this date and plans are available back to about 1812. The incomplete nature of the abandonment plans means that drilling for unrecorded old workings is frequently one of the most important aspects of the exploration phase and the drilling results are vital in estimating the allowance which has to be made for old workings in a seam. Occasionally, there are no abandonment plans available for old workings on a prospective open-cast site and the estimates for old workings have to be made entirely on the drilling information. More usually, however, some records are available but the drilling shows them to be incomplete. Only rarely is a complete set of abandonment plans available for a site.

The Bluers Seam on the Maes-y-Marchog Site forms a good example of a seam containing extensive old workings for which an estimate had to be made. Figure 8.19 shows the boreholes drilled in part of the site and illustrates which recorded solid coal and which recorded old workings in the Bluers Seam. The mid-points between the boreholes which contained coal and those which contained old workings have been linked to form a mosaic of coal and old workings. The area of coal remaining as indicated by the drilling can thus be measured. This is a somewhat empirical approach but it has been shown to provide good results in most circumstances, providing the borehole spacing is not greater than about 30 m. The abandonment plan for this same area (Figure 8.20) shows both pillars and stalls and measurement of the pillars remaining gives an extraction percentage of 34%. Many abandonment plans, however, do not show details of the pillars and simply depict the main drivages in the seam. Experience has shown that old mine records often show less coal to have been extracted than is actually the case. This usually results from pillars of coal being removed after the last survey has been taken in the mine. Exploitation of the Bluers Seam on the Maes-y-Marchog Site showed that the actual extraction ratio was 50%, which lies between the 60% predicted from the drilling and that shown on the abandonment plan.

Drilling results are the most frequently used means of estimating the percentage of the seam worked out but sometimes these estimates are modified on the basis of data from old plans (perhaps from adjacent areas), providing that the method of extraction is thought to be the same throughout the area.

The estimation of allowances for old workings is the aspect of coal evaluation which is subject to the greatest margin of error. On multi-seam sites, where many of the seams contain old workings, it is often found that the estimates, based mainly on the drilling results, lie within 10% of the actual extaction ratios (Table 8.7).

Washouts

Where drilling indicates that a seam is locally thin or absent within a coaling area, a 'washout' allowance is made which takes the form of a deduction from the gross coaling area of the seam. Frequently, these are not true washouts but may be structural in origin or may result

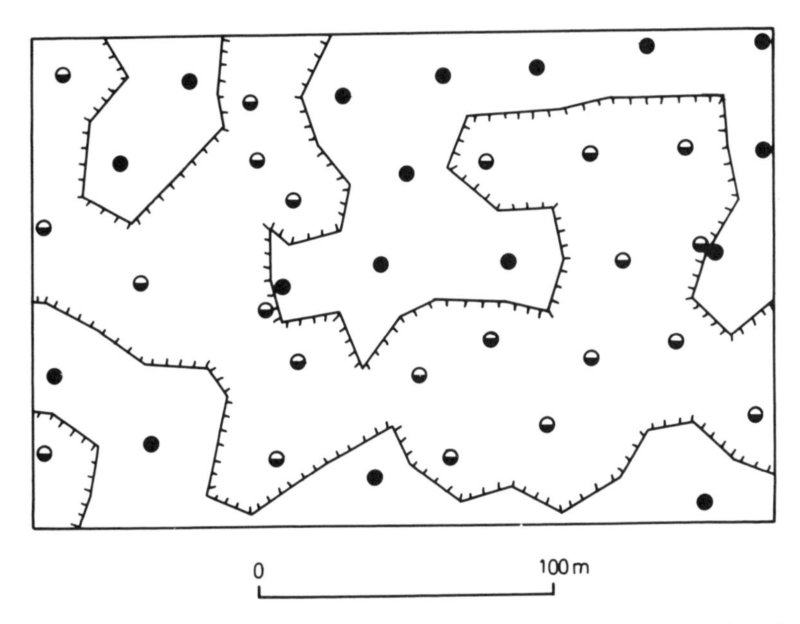

Fig. 8.19 Incidence of old workings in the Bluers Seam on the Maes-y-Marchog Site based on drilling results. Black circles show boreholes which penetrated coal, half black circles show boreholes which penetrated old workings.

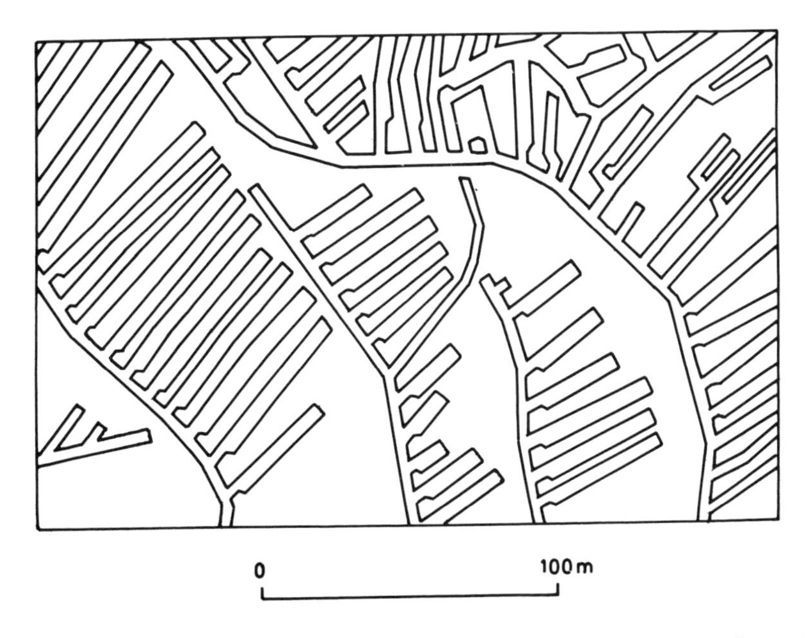

Fig. 8.20 Abandonment plan for part of the Bluers Seam on the Maes-y-Marchog Site illustrating pillar and stall workings.

Table 8.7 Estimated and actual extraction rates resulting from previous undeground mining in some of the seams on Glyn Tai and Trecatty Sites

		% Old workings	
Site	Seam	Estimated	Actual
Glyn Tai	M. Pumpquart	17	12
Trecatty	Black	0	1.9
Trecatty	Big	55	45
Trecatty	Bute	5	3
Trecatty	Three Coals	12.5	11.5
Trecatty	Upper Four Feet	90	91

from depositional thinning or non-deposition of the seam. A seam thickness below which it is considered impractical to lift the coal is decided upon and any areas shown by the drilling to be below this value are delimited on the plan. The area so delineated is accurately measured and deducted from the coal area. In practice, where the seams decrease to below about 15 cm they are difficult to coal and this thickness often forms the basis for the delineation of a 'washout' area. Occasionally, however, hard coals which separate cleanly from their roof and floor can be coaled down to a thickness of 10 cm.

Faulting

Although close-spaced drilling will identify faults with a throw down to a few metres, it is rare that the precise hade of faults can be determined. As a result, a precise area of fault 'want' cannot usually be calculated. In any case, the disturbance of coal close to faults often leads to losses greater than would be predicted by an accurate measurement of the 'want' area. Accordingly, a purely empirical rule is normally used in the calculation of allowances for faulting. Most fault allowances are based on multiplying the length of the fault by the average throw and the resulting area is deducted from the area of the seam to be coaled. This method corresponds to a fault which has a constant 45° hade and, although experience has shown that most faults which intersect the coal measures in South Wales are steeper than this, it takes account of the fact that increased losses may occur as a result of disturbance of the coal adjacent to the fault. Normal faults with a high hade angle (lag faults) have large areas of 'want'. Occasionally these may be proven by drilling and thus it is possible to measure the area of 'want' on plan and this value used instead of that based on throw as described above.

Dislocation

Minor faulting of seams is commonplace in many parts of the South Wales Coalfield. Frequently, small faults with throws of about 2 metres or less may be present and it is thus impossible to detect these during drilling. The only indication of their presence may be occasional random absences of the seam. If a seam has been subject to previous underground mining, the abandoned mine plans may be invaluable if they show detail of any minor faults encountered. However, it is not possible to accurately quantify the losses which may occur as a result of such small scale faulting, and accordingly a nominal percentage allowance is applied. Such allowances are usually below 5% and experience has shown them to be adequate. Formerly, a dislocation allowance was also applied to complex sites in which overthrusting had occurred. However, it was found that the increased volume of coal which was often available in such areas, as a result of minor repetition of seams or localized thickening of the coal (undetected in the drilling), more than compensated for losses resulting from disturbance of the coal. As a result, dislocation allowances in such circumstances are now rarely applied.

8.5.8 Calculation of overburden volumes

Calculation of the overburden volume for a site is one of the most important aspects of evaluation since it governs the overburden to coal ratio which in turn influences the profitability of the operation. The calculation will determine the volume of overburden which lies vertically above the coal extraction areas as defined on plan. This approach is adopted since it enables an accurate measurement to be made related to clearly defined limits. Thus, it is not subject to variables such as the batter slopes which can be achieved or the number and dimension of interseam benches which may be required during the excavation of the site. These are the aspects which vary most from the site specification, depending on the actual conditions found during excavation. By using the vertical excavation above coal, a precise comparison can be made between the excavation volumes and the ratio predicted at the design stage, and that which is actually achieved on completion of working the site. Ratios calculated in this way are termed in situ ratios.

There are various ways in which the overburden volume can be calculated. Experience has shown that the most precise method is based on the construction of an isopachyte map for the site. The isopachs represent the thickness of the strata from ground surface to the base of the excavation, within the coaling limits. The isopachyte lines are constructed by comparing the values of surface level contours with structure contours of the base of the coal seam. Construction of such an isopachyte map is essentially a manual task, although it can be computer aided. When the contour map has been completed, the areas lying between each isopach can be measured using a digitizer. These areas are multiplied by contour interval to give a set of volumes. Summation of these volumes gives the total vertical excavation for the site. The ratio is calculated using the formula: $(x - y)/y$, where x is the total excavation volume including coal and y is the *in situ* volume of coal.

The GEOMODEL system is now used to calculate excavation volumes using the gridding and triangular prism methods. In the former case, excavation volumes are calculated by subtracting grid values derived from two sets of triangulated data, each representing the elevation of a coded horizon in the boreholes. These horizons could, for example, be the base of a coal seam and the ground surface or two different coal seams. The latter approach is particularly useful in that it enables the excavation to be broken down on a seam by seam basis, and also enables the excavation to be built up on a 'layer-cake' principle. Alternatives are provided in the programming so that, where an excavation volume is being evaluated between two coal seams and the higher seam has outcropped, then the ground surface can be substituted for the area outside the outcrop of the higher seam. The plotted output provides a graphic illustration of how the volume between any two horizons changes across a site and also enables a rapid check to be made for any anomalous values.

The triangular prism method uses triangulated data from the two horizons between which the overburden volume is required. For each horizon, a mean thickness ≡ depth) value is calculated for each triangle down to a given datum. These thickness values are multiplied by each triangle area to generate a volume and summation of these triangular prism volumes will produce the total volume contained in a given area, between each horizon

Table 8.8 Schedule of estimated quantities

		Coal *in situ*		Recoverable coal		
Name of seam	Seam area (m^2)	Thickness (m)	Volume (m^3)	Thickness (m)	Volume (m^3)	Tonnes
Trichwart	124 400	1.64	204 016	1.54	191 576	254 796
Black	97 680					
Less 'Washout'	65 360					
	32 320	0.54	17 453	0.39	12 605	16 765
Little Vein	64 480					
Less 'Washout'	17 360					
	47 120					
Less Faulting	570					
	46 550	0.40	18 620	0.25	11 638	15 479
Nine Feet	85 426					
Less 'Washout'	786					
	84 640					
Less Faulting	910					
	83 730					
Less 55% O.W.'s	46 052					
	37 678	2.45	92 311	2.30	86 659	115 256
Totals			332 400			402 296

Overburden volume vertically above coal: 5826 972 m^3.

$$\text{Ratio: } \frac{\text{Overburden volume}}{\textit{in situ}\text{ coal volume}} = \frac{5826\,972}{332\,400} = 17.53:1.$$

and the datum. The volume between the two horizons is obtained by subtracting the smaller volume from the larger.

The flexibility of both the grid and the triangular prism methods is a major advantage. Not only can they be used to calculate volumes between selected coal seams, or between a coal seam and ground surface, but volumes can be determined for hard rock (e.g. sandstone) and for unconsolidated material (e.g. glacial drift).

Such is the importance of obtaining an accurate vertical excavation volume for a site that at least two methods are employed as a cross-check. Any discrepancies between the two results are thoroughly investigated to ensure that the best possible prediction of excavation volume is made.

The methods considered so far are used to calculated the vertical excavation above the coal and hence an *in situ* ratio of coal to overburden. It is now necessary to estimate the impact of applying batters to the pit-walls which is normally done by measuring the cross-sectional areas of the excavation between these walls and the coaling limit at intervals around the excavation. Each cross-sectional area is then multiplied by the length of the excavation with which it is associated. The sum of these volumes is then added to the vertical excavation volume to determine the total pit volume and hence, knowing the volume of coal, the overall working ratio for the site.

8.5.9 Documentation

The evaluated coal quantities and over-burden volumes for a site are summarized on a schedule of quantities (Table 8.8). This is accompanied by a schedule of the drilling and geophysical data, a geological plan, report and cross-sections. In addition, plans of rockhead contours, drift thickness and abandoned underground workings are usually prepared, together with detailed drawings of selected cored boreholes. These documents accompany the contract presented to contractors prior to the submission of a tender to work the site.

BIBLIOGRAPHY

Atkinson, K. and Brassington, R. (eds) (1983) *Prospecting and Evaluation of Non-metallic Rocks and Minerals*, Institution of Geologists, London.

Hoare, R. H. (1979) Exploration 2000. *Mining Eng.*, **138**, 131–40.

Kloosterman, R. A. and Brom, R. W. C. (1977) *Application of Wireline Logging Techniques in the Assessment of Surface Mineable Coal*, Shell Mijnbouw NV, Indonesia, June 1977.

Reeves, D. R. (1971) In-situ analysis of coal by borehole logging techniques. *Canad. Min. Met. Bull.*, (February).

Ward, C. R. (ed.) (1984) *Coal Geology and Coal Technology*, Blackwell Scientific, Oxford, 345 pp.

Whincup, G. T. (1972) Some Aspects of Opencast Mining in South Wales. *Ext. Proc. S. Wales Inst. Eng.*, **LXXXVI**, (1), 15–34.

8.6 CASE HISTORY – BOULBY POTASH MINE, CLEVELAND, UK

8.6.1 Location

Cleveland Potash Ltd has been mining potash at its Boulby operation since 1974. The mine is located on the NE coast of England in the County of Cleveland, and approximately 16 km NW of the town of Whitby. Current workings cover an area of approximately 20 km^2 and extend beneath the North Sea for a distance of 2.5 km.

8.6.2 Regional geology

The potash seam under exploitation lies in Upper Permian rocks at the transition between the English Zechstein cycles 3 and 4 (Teeside and Staintondale Groups respectively). The stratigraphic units of the Upper Permian are presented in Table 8.9 and the sequence cut by the No. 1 and 2 shafts is shown in Figure 8.21. The latter shows that the potash seam lies at a depth of slightly over 1100 m and is overlain by strata of Triassic (Bunter Mudstone and Sandstone and Keuper Marl) and basal Jurassic (Rhaetic and Lias) age. Figure 8.22, a footwall structure contour map of the base of the potash seam, reveals that in the area covered by the mine workings, the dips are generally to the SE at approximately 3°. However, various WNW to NW trending tranverse structural features exist which are related to faulting in the underlying Upper Magnesian Limestone. More information on the English Zechstein Basin can be obtained from Smith and Crosby (1979). Figure 8.23 represents a detailed section of the strata immediately underlying and overlying the potash seam as intersected in the No. 2 shaft. A detailed description of the mine geology appears in Woods (1979) and only a summary of the main characteristics of these strata will be given here.

The most critical horizon, from the mining point of view, is the Carnallitic or 'Rotten' Marl which is very weak and which consists of a mixture of red-brown dehydrated clays and salt (up to 60% halite locally). It is also cut by slickensided halite-silvite veins which further weaken this horizon. The Carnallitic Marl grades down via a

Table 8.9 Stratigraphy of the English Zechstein (Cycles EZ3 and EZ4)

Group	Formation	Cycle	Shaft Section
Staintondale	Upper Halite Upper Anhydrite Upgang Formation Carnallitic Marl	EZ4	
Teeside	(Boulby Potash) Boulby Halite Billingham Main Anhydrite Upper Magnesian Limestone	EZ3	

Woods and Powell (1979).

transition zone, consisting of halite, dark grey clay and anhydrite, into the Boulby Potash Horizon which consists of a variable mixture of halite and sylvite ('sylvinite'). This seam varies in thickness from zero to in excess of 20 m, and in composition from 10% to 60% KCl and 2% to 50% impurities. This thickness variation is partly due to sedimentological reasons but tectonic thickening and thinning (mass flow and thrusting) have also played a role. The sylvinite varies from greyish-pink to orange or bright red in colour (the colouration being due to tiny platelets of haematite), while its texture varies from gneissic, with flow banding, to bedded. Sylvinite veins and intergrowths also exist within shale-rich areas. Impurities (insolubles), mostly anhydrite and clay, average close to 12%, while the sylvite content gives the ore a typical mill-head grade of 34–38% KCl (lower grade areas are not mined). In some areas carnallite ($KCl.MgCl_2.6H_2O$) is also present, mixed with halite, as a separate unit at the base of, or within, the potash seam. Nodules of 'boracite' (a B,Cl,O,Mg,Ca,Fe bearing mineral) may be present which vary from egg-sized to almost 1 m in diameter. These nodules present a problem during mining as they are very hard and cause excessive wear to the tungsten-carbide teeth on drills and heliminers (Figure 8.24). The 'sylvinite' usually has a sharp contact with the underlying Boulby Halite, the top 2–3 m of which is a brownish-grey halite often containing red sylvite grains. This passes down into 1–2 m of pure orange-pink halite and then into a coarse orange-brown halite with interstitial grey clay. The latter becomes increasingly colourless and anhydritic downwards towards the Billingham Main Anhydrite. The Boulby Halite shows marked lateral changes in thickness due to flowage over faults in the Upper Magnesian Limestone. As a result, few faults are seen in the mine and their presence can only be inferred from ramp structures and sub-horizontal thrusts affecting the potash seam, and from reflection seismic surveys on surface.

8.6.3 Mining

Shaft sinking began at Boulby in 1969 and by 1973 the first of the two shafts had reached the potash seam. Mining of the ore was thus able to commence in 1974. By 1988 ore production had reached 2.5 million tonnes per annum from an area of 1.2 km^2, making the mine the largest single underground mine in the UK from the point of view of tonnage, as well as being the deepest mine in the UK. Lesser amounts of salt are also produced (derived from development in the underlying Boulby Halite). Initially, mining was based on the 'room and pillar' method in which 24 m square pillars were retained at 30 m centres (Figure 8.25). However, roadways suffered from collapse due to the weak nature of the potash itself and also the overlying Carnallitic Marl, and high overburden pressures. The problem with the latter horizon was reduced by leaving 2 m of potash in the roof; however, where the potash seam is thin, this does require mining down into the underlying halite and accepting the resulting dilution. Mining heights are dictated by the equipment used and are fixed between narrow limits of 3–3.8 m. In 1979, continued ground control problems forced the mine to change its method of working to one in which groups of four, five or six, parallel roadways are mined together, as in the northern and western areas of Figure 8.25 (such a group of roadways is referred to as a panel). The outer roadways are mined so that they advance 5–10 m ahead of the two inner roadways. Yielding pillars, 2.5–5.0 m wide, separate the 3–3.8 m high roadways. Each group of roadways is then separated

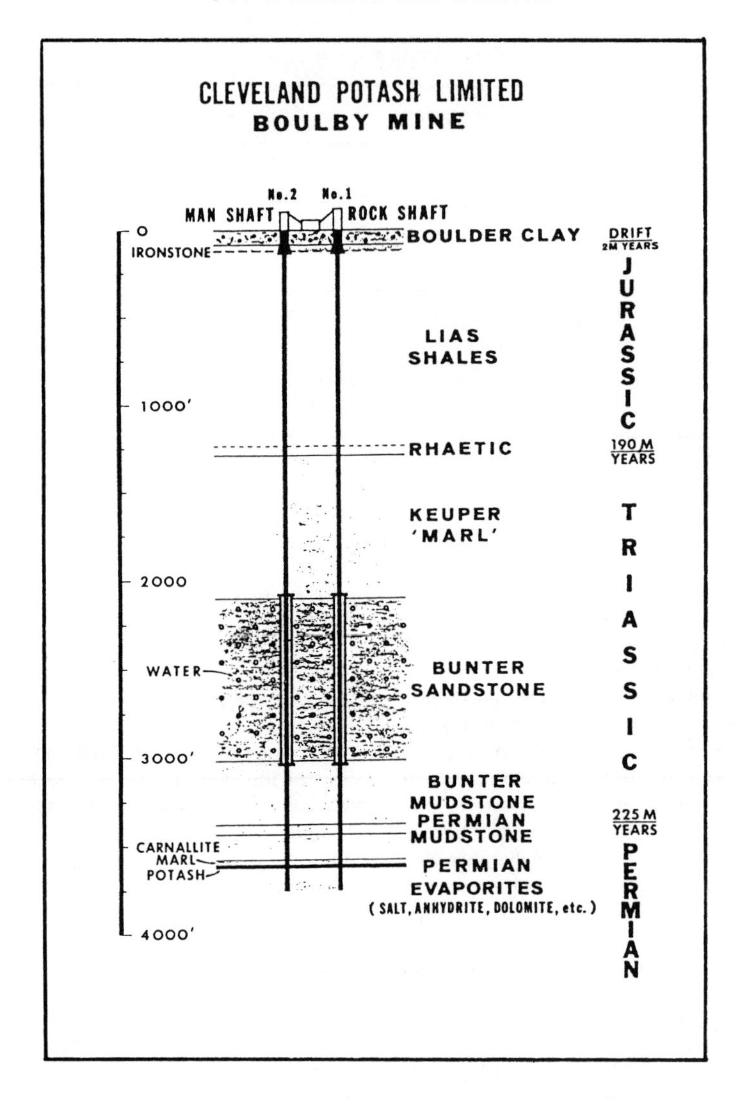

Fig. 8.21 Stratigraphic sequence intersected by No. 1 and No. 2 shafts at Boulby Mine, Cleveland Potash Ltd, Cleveland, UK

by barrier pillars, 100 m wide, as in Figure 8.25. This mining configuration was thus designed as a stress-relief system. The outer roadways suffer heavy damage due to roof break against the barrier pillars but the inner roadways (which are required for access, conveying and ventilation) remain accessible because the yielding pillars allow the roof beam to flex in a controlled manner.

Gas (N_2: 60–80%; CH_4: 20–40%) is present throughout the potash seam, and in some areas where there are significant concentrations of shale, outbursts of rock, propelled by the high gas pressures, occur during mining. Where no gas outbursts are anticipated, continuous 'full-face' mining methods are employed using a Jeffrey's Heliminer (Figure 8.24). Where shale is present, the heading is undercut, drilled and blasted. Blasting releases the gas and the heliminer is then used to remove the broken rock and to trim the roadway to shape. After each advance during mining, the backwall and sidewalls are secured by roof bolting on a 1.4 m grid. When 'full face' mining is used, groups of four safety cover holes

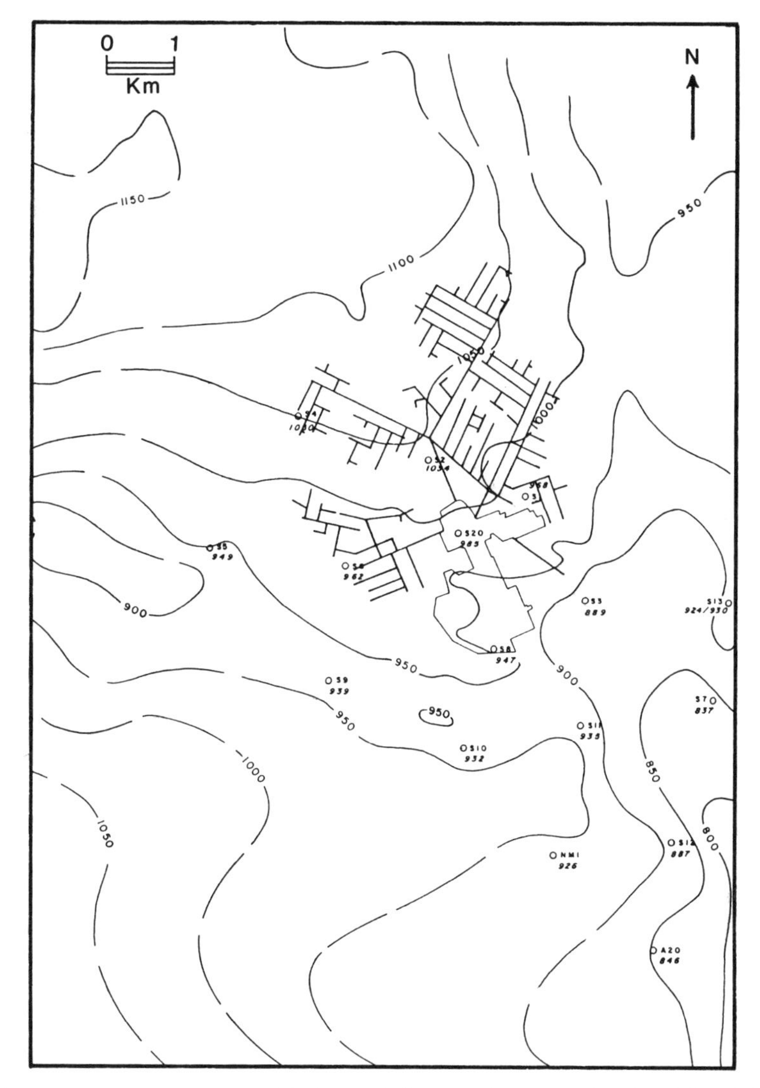

Fig. 8.22 Footwall structure contour map of the base of the potash seam at Boulby Mine. Elevations in metres above mine datum (2000 m below Ordnance Datum.)

are drilled for distances of 8 m ahead of the current face to check for the presence of shaley potash, permitting 5 m of advance.

As mining proceeds, the roadways are mapped on a daily basis to define the different categories of potash ore, and the underlying halite, and also to locate the mined roof and floor. The mapping is done relative to a level datum at 1 m intervals from a survey peg. This information is plotted on longitudinal roadway sections at a scale of 1:250 (Figure 8.26). Short (20 m) inclined rotary holes are drilled in the face (Figure 8.27) to locate the top of the potash seam and the base of the Carnallitic Marl and allow the geologist to ensure that a 2 m roof beam of potash is maintained. This is accomplished using a gamma probe (section 2.14.1 and Figure 8.28) which is inserted into the holes by means of fibre-glass rods. Measurements are taken, at 1–2 m intervals along the holes, and over 1 minute periods, of ^{40}K-derived gamma radiation. Calibration standards allow the bulk KCl composition, over a radius of 0.5 m from the hole, to

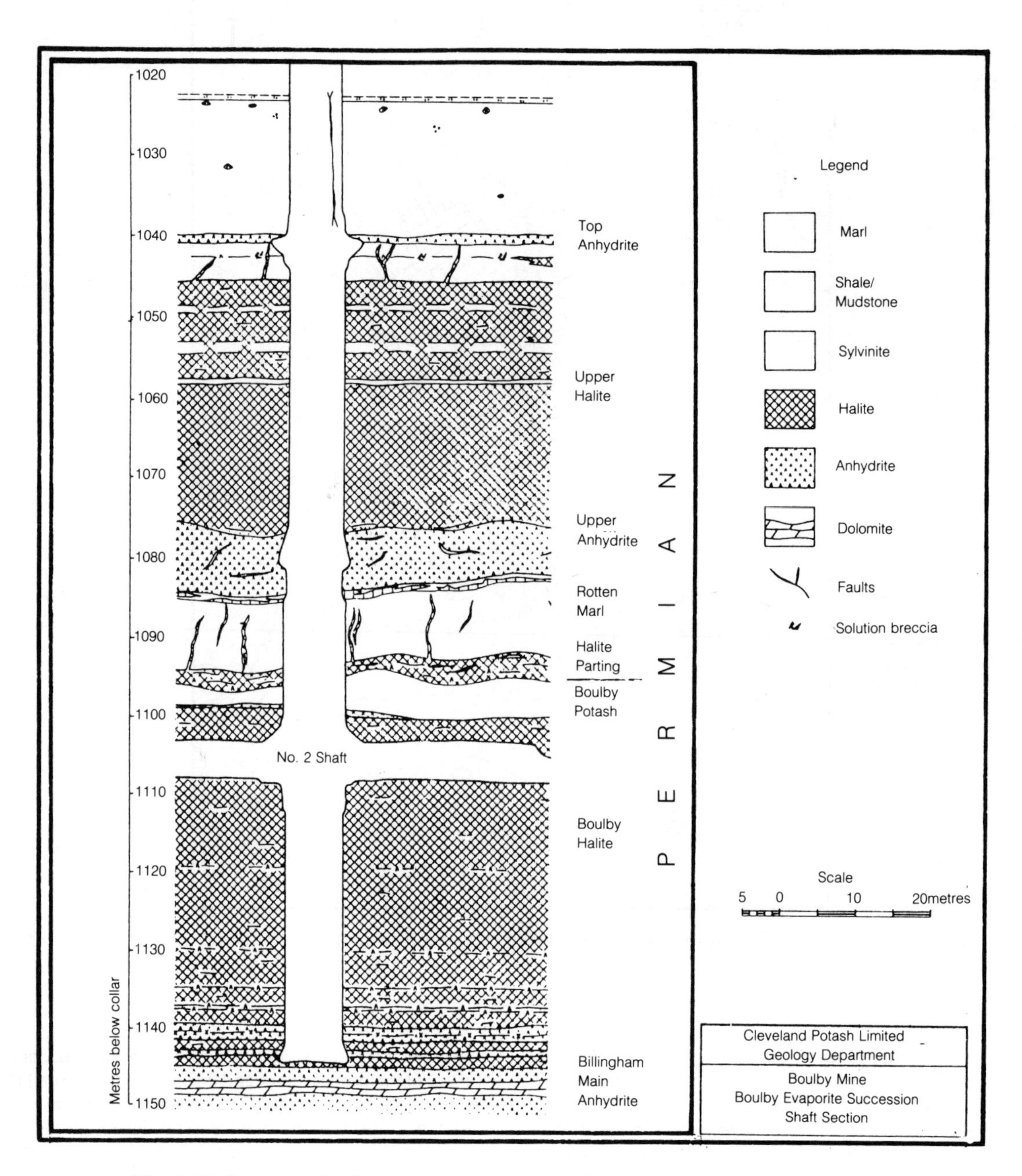

Fig. 8.23 Seam proximal stratigraphy as exposed in the No. 2 shaft, Boulby Mine.

Fig. 8.24 Heliminer cutting the potash seam at Boulby Mine. (Photograph courtesy of Cleveland Potash Ltd and Marcon Ltd.)

be determined. The geologist is thus able to mark on the sidewall a grade line, with the required gradient and the roof and floor positions (Figure 8.27), using a clino-rule and spray paint. This serves to guide the blast-hole driller and the heliminer driver in the next mining advance.

8.6.4 Sampling for mine ore reserves

Initial exploration drilling in Cleveland Potash's lease area was on the basis of a 1.5–2 km spacing and was aimed at proving the continuity of the seam. As only 15 holes (including the shaft pilot hole), and five deflections, were drilled, once mining began it was necessary to develop a method to obtain advance information on seam quality and thickness. Long horizontal, continuously cored, boreholes are drilled by reverse-circulation techniques using PCD (polycrystalline diamond) bits (Woods and Hopley, 1979). Drilling fluids used are KCl- and NaCl-saturated brines to prevent dissolution of the cores during drilling. Fans of 86 mm diameter drill-holes are thus drilled outwards from roadways and 20–30 m below the footwall of the potash in the Boulby Halite. These holes normally extend for distances

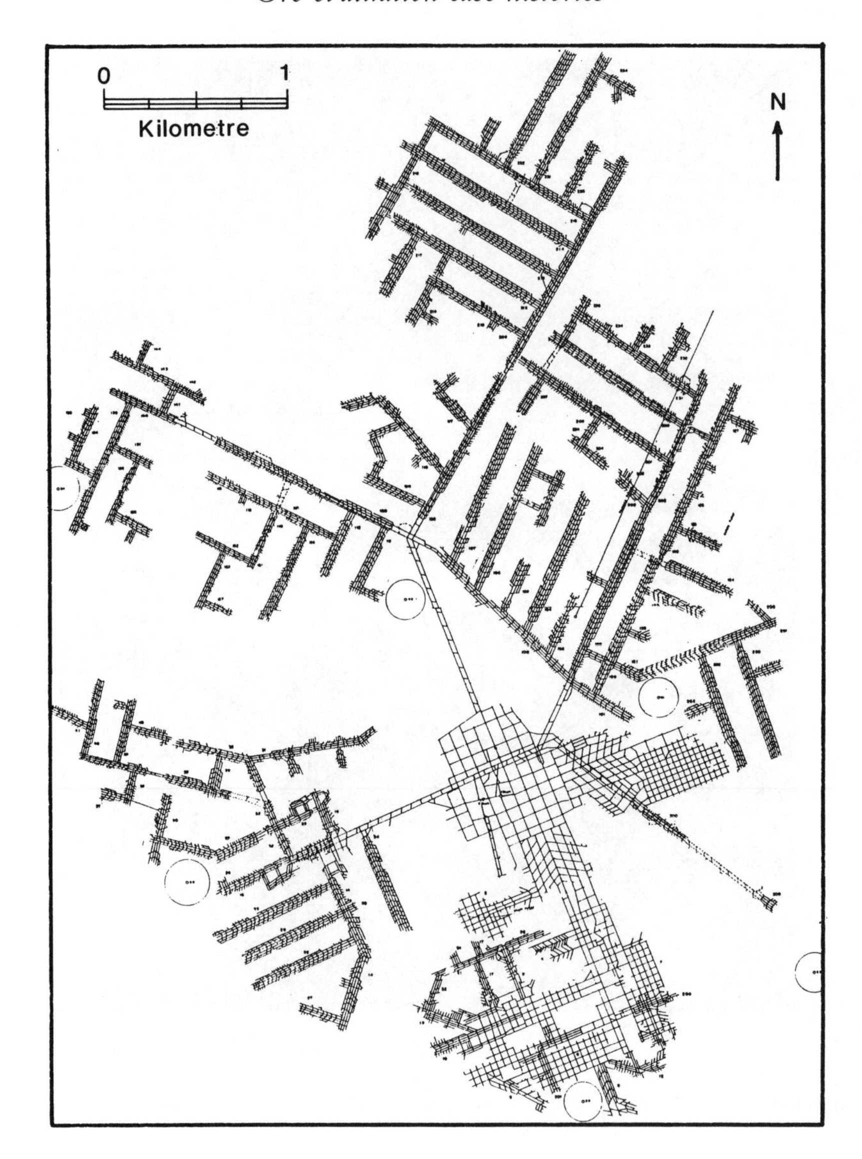

Fig. 8.25 Plan of the mine workings at Boulby Mine showing mining layouts and surface drill-hole locations.

of 1300 m, with upward deflections into the overlying potash at intervals of approximately 200 m (Figure 8.29), although locally this may be reduced to 50 m intervals. In the peripheral areas, or in marginally economic areas, the spacing of intersections is reduced to a 120 × 120 m grid whereas, in unpromising areas, the grid is expanded to 400 × 1000 m and in some cases the holes are extended to distances of 2000 m. As the holes progress, they are internally surveyed at intervals of 30 m using a Sperry-Sun or Eastman single-shot instrument. Thus the progress of the hole is monitored and the 3D coordinates of the footwall and hangingwall of the potash are calculated. In 1988, Cleveland Potash drilled in excess of 30 km and it is anticipated that this figure will exceed 40 km in 1989.

A summary log of the core is made except for the

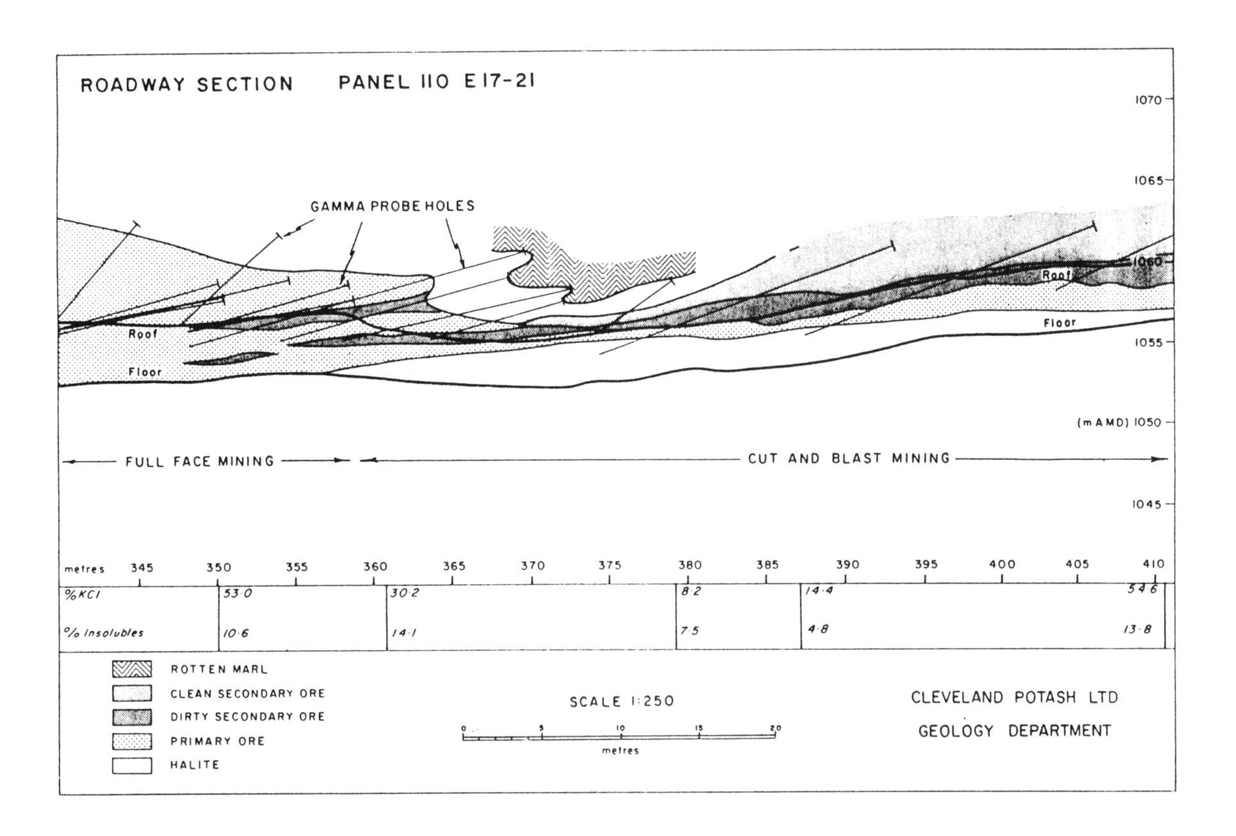

Fig. 8.26 Longitudinal roadway section showing seam geology relative to mine roof and floor, ore types, cover drilling and sample grades.

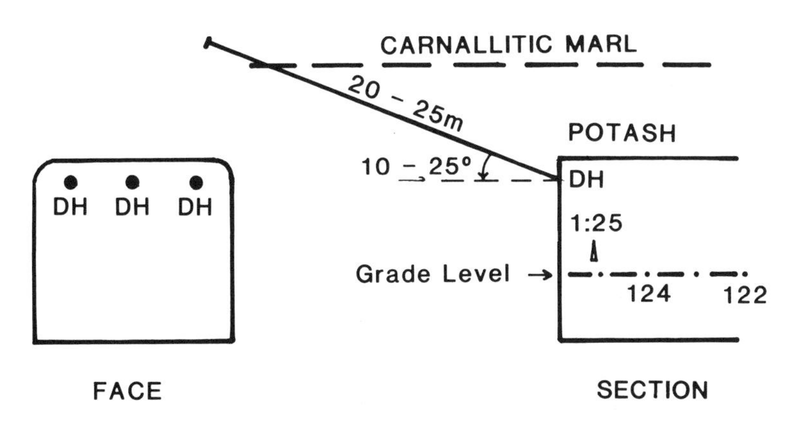

Fig. 8.27 Roadway guidance holes.

Fig. 8.28 Geologist inserting gamma probe into rotary hole to locate the hangingwall of the potash seam. (Photograph courtesy of Cleveland Potash Ltd and Marcon Ltd.)

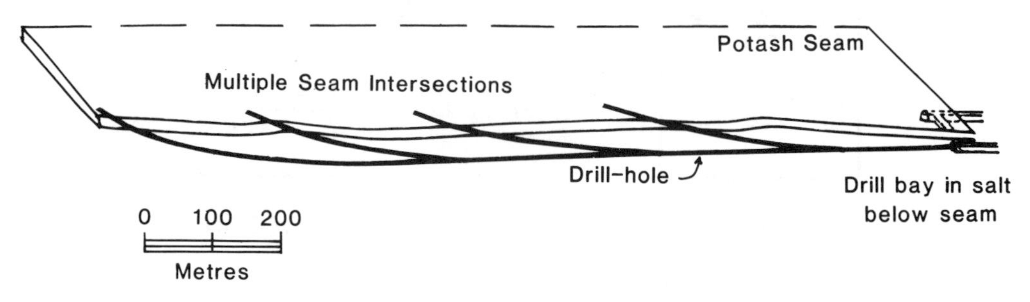

Fig. 8.29 Longhole drilling section showing multiple seam intersections.

potash seam itself, where more detail is recorded. A series of codes are used to represent core recovery levels and whether the core has suffered stress-relief disking. Generally, the geologist records the colour, grain size, the existence of anhydrite bands or shale, foliation and whether the grains are anhedral to euhedral. Three main types of ore are recognized using the following mine terms and these are colour-coded for plotting on the geological plans (ore classifications are similar to those used in the geological mapping and gamma probing described above).

Primary Ore (P/O)	a speckled gneissose ore at the base of the seam; usually high grade and with a low insolubles content.
Dirty Secondary Ore (DSO)	grey- to red-coloured shaley ore of widely variable grade.
Clean Secondary Ore (CSO)	banded/bedded ore at the top of the seam; low in insolubles and with widely variable potash content. Interbanded high and low grade clean ore also exists, the latter containing clay and marl partings.

The presence of cavities in the core usually indicates that carnallite has been leached away because, at the moment, the drilling brine is not saturated in $MgCl_2$.

For sampling purposes, the core is divided into the above potash units, as in Figure 8.30, where three types of CSO have also been recognized. Sampling will then commence from the base of the Carnallitic Marl – usually at a marker horizon called the Halite Parting (H/P) which is very weak, contains little potash, and which is a shaley and anhydritic halite. Sampling will then extend from this horizon, through the potash, into the halite where at least one sample is taken. Sample lengths are controlled by the lithological units and by the requirement that each sample represents a true thickness of not more than 1 m. As the intersection angle is close to 20°, the actual maximum length of core represented by one sample is approximately 3 m. From this length, 8–12 core pieces (discs) are taken so that their combined length is approximately 10% of the interval. These are then bagged, assigned an assay number and sent for assay by AAS and XRD.

The sampling log sheet used will eventually contain the following information: unit code, sample number, from, to, intersection angle, total KCl, KCl as sylvite, NaCl, $CaSO_4$, Al_2O_3, SiO_2, total MgO. Water-soluble Mg, total chloride and % carnallite will also be shown where carnallite is visibly present. The percentage of carnallite is first calculated from the water-soluble Mg value (using a factor of 11.43). The amount of KCl in this carnallite is then calculated and the result deducted from the total KCl to give the amount of KCl as sylvite. By deducting the amount of Cl combined with K and Na from the total Cl value the amount of $MgCl_2$ can be calculated, thus a second estimate of the amount of carnallite present can be made as a cross-check. From this sheet, the following information can be input to the computer: sample number, unit name, from, to, intersection angle, seam dip (in direction of the hole), total KCl, KCl as sylvite, carnallite and insolubles. Rarely can the true intersection angle be seen in core due to recrystallization and flowage and it is thus calculated from $\alpha + \beta$, where α is the dip of the borehole to the horizontal and β is the dip of the seam on the drill section. Where carnallite-rich horizons are intersected, dissolution of the walls of the hole causes the hole to droop. In this instance, the hole is surveyed on either side of the carnallite zone. The dips at the HW and FW are then calculated by extrapolating downwards and upwards from the survey points respectively. This allows the dip α of the hole over the zone to be calculated and thus the intersection angle.

Using the data file created above, various units can be combined to produce overall thicknesses and weighted grades. One of these will be the 'most likely mining-cut' based on a 3–3.8 m true thickness. This cut, which is defined manually and individually for each seam intersection, is usually placed at the base of the potash seam, although if higher grade zones exist above this cut, the base of the revised cut may be set at a higher level. The computer starts at the specified base, or top, and

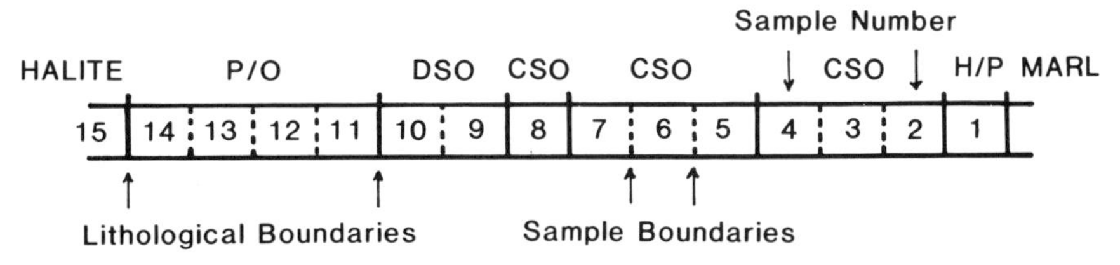

Fig. 8.30 Subdivision of core for sampling purposes. P/O, Primary Ore; DSO, Dirty Secondary Ore; CSO, Clean Secondary Ore; H/P, Halite Parting.

accumulates the true thicknesses for each successive sample until it has the specified thickness (the last sample may have to be subdivided) when it calculates the weighted grades of all the significant variables. At this stage, it is necessary to check whether a 2 m roof beam of potash exists beneath the top of the seam or whether a clay band, or other weak zone, lies immediately above the proposed roof. If the latter is the case, then the mining cut is moved upwards to take out this band and thus prevent potential instability problems. Where the roof beam is too thin, then the base of the mining cut is moved into the halite resulting in dilution. Calculations can also be made to reveal the impact of changing the mining height and beam thickness parameters.

8.6.6 Sampling for grade control and mining statistics

Grade control is maintained by vertical chip sampling of the faces of the advancing roadways every 10–15 m. The sampling is controlled by the contacts between the different types of potash and also the halite. The samples are analysed for KCl, NaCl and the difference is assigned to 'insolubles' (unless carnallite is visibly present). Each is then weighted by the respective areas (A_1–A_3) to calculate a weighted average for the whole face, which can then be plotted on longitudinal sections (Figure 8.26).

Every week the Survey Department measure the roadways for width, height and advance, from which the tonnage mined during the week can be calculated using a tonnage factor of 2.1 t/m^3. Geology determines the overall grades of each roadway advance using the area-weighted chip grades for each face sampled. From this the contained KCl can be determined. All this information is written to a computer file. The grades calculated from the chip samples can then be compared with the predicted grades from core drilling and the plant feed grades. It is also used in the determination of royalty payments as the lease area is owned by several different parties. For this purpose, royalty codes can be assigned to panels, or part panels, so that data can be resorted by the computer on this basis. The tonnage broken is also subdivided into different categories as below:

Potash roads	potash recovered by the heliminer in the first cut.
Potash milling	floor milling to increase potash recovery from the seam.
Potash beam blasting	recovery of potash from the roof.

From these a total potash tonnage is calculated.

Salt roadways	long-term development wholly in halite.
Salt milling	floor milling associated with salt roadway mining.

From these a total salt tonnage is calculated.

Roadway maintenance (ripping)	ripping of potash or salt to maintain safe roadway conditions.
Roadway maintenance (milling)	additional cut into the floor to maintain roadway height after roof sag.

Thus, each week an overall total of 'rock broken' can be calculated from the above which can then be subdivided into tonnes stowed underground, tonnes trammed as potash ore, and tonnes trammed as salt (salt trammed and hoisted separately from potash is sold as road-salt). These results can then be further combined on a monthly, 6-monthly or annual basis.

Panels, defined by groups of roadways, may be subdivided into lengths of 100–200 m so that each subdivision, or block, contains 50–150 chip samples. These are averaged, as described above, and the result assigned to the centre of gravity of the block. Each value is thus taken to represent the tonnage mined (usually between 50 000 and 150 000 tonnes). These values are used in conjunction with core drilling data to predict grades of future mining (see next section) and comparisons can be made between the mined grade of KCl and insolubles, and previous predictions.

8.6.7 Ore reserves

Annual review of global reserves

This involves a statement as to the gross reserves before mining, the amount mined during the last year, the total mined in previous years and the reserves remaining. The latter are classified on the basis of various confidence categories as defined below.

Indicated or Class A reserves – these are based on long-hole underground drilling (and mined panels in the case of the 'most likely mining cut'):

A1 – areas where drilling is complete and there is confidence in grade estimates.

A2 – these represent peripheral blocks to the A1 area where further drilling is planned or where the drilling is widely spaced because grades are low and further drilling is not anticipated (Figure 8.31).

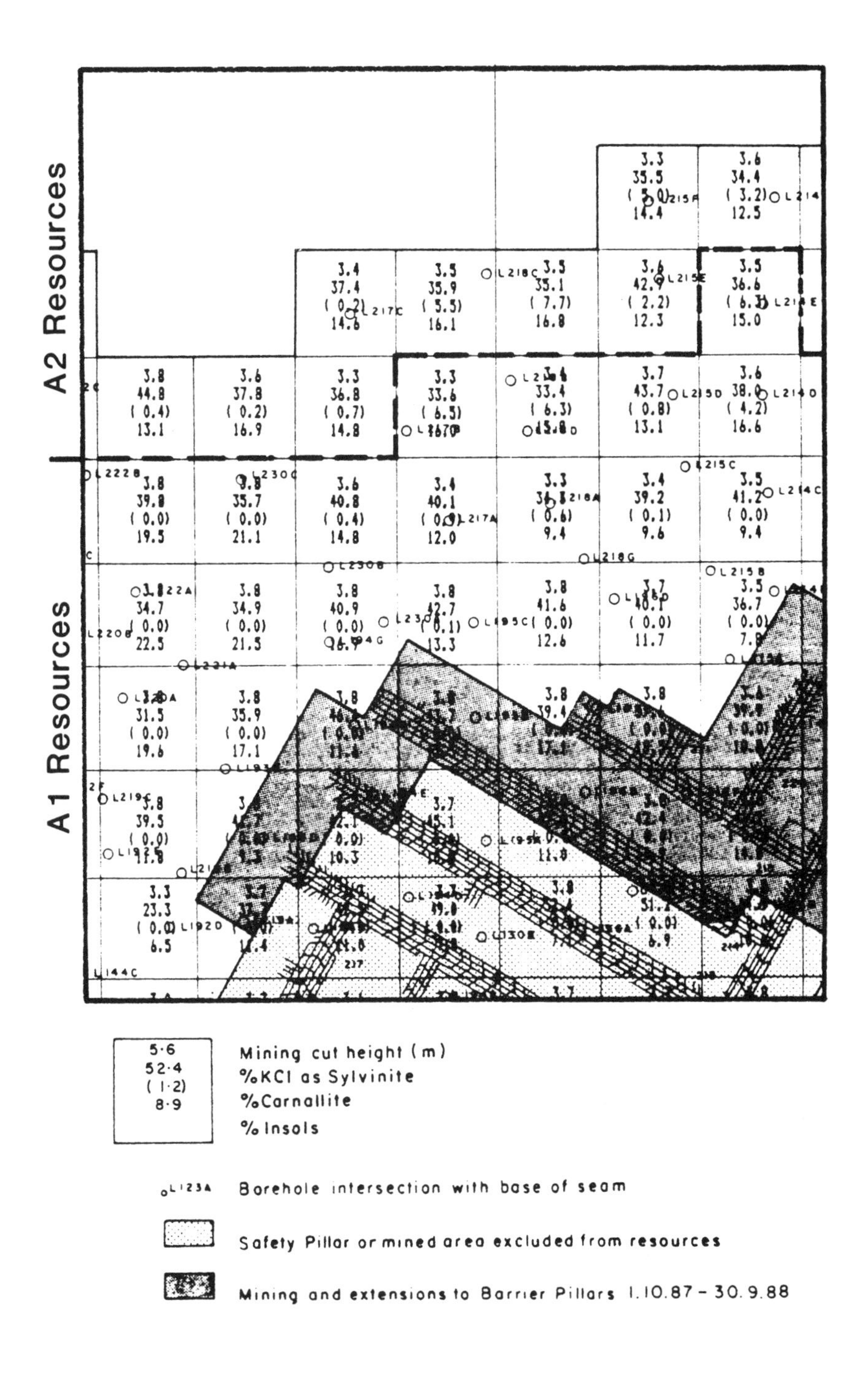

Fig. 8.31 Part of the mine reserve plan showing 200 m square ore blocks and reserve classifications.

Class B reserves – these are based on surface drill-holes (2 km spacing), and the outermost underground boreholes, and extend to 1 km beyond the outermost borehole (either underground or surface).

Possible reserves – these are basically inferred reserves beyond 1 km from the outermost boreholes but within the lease area boundary, either onshore or offshore. The grade is a 'guesstimate'.

Calculation of Class A reserves

These reserves are calculated using 200 m^2 blocks based on the National Grid system. Each block is stored on the computer together with its various codes, areas, and cut grades (see below). The grade of each is calculated using inverse distance weighting ($1/d$) and a search radius (from the centre point of the block) which is specified manually for each block, and which could lie between 100 m and 500 m but is usually 200–300 m. The value used depends on the density of data and the variability of the seam in the area. The user tries to capture at least four holes in the circular search area and the computer program allows him/her to delete any hole considered dubious or irrelevant to the block under evaluation, or to drag in outlying holes. As the grade and the thickness of the potash seam are very much controlled by structure (faults), the seam can be divided into structural domains in which conditions are more uniform. Only holes from one domain would be used to evaluate a block in this same domain. Inverse square distance ($1/d^2$) weighting was rejected because it was thought that too much weighting was often given to a single hole close to the centre of each block. Also, if the distance 'd' is less than 50 m, then the program sets 'd' to 50 m.

Various cuts are calculated for each block:

1. a total seam or geological reserve;
2. a 'most likely mining-cut' reserve (3.0–3.8 m + 2 m roof beam);
3. alternative mining-cut reserves (1.5–3.5 m + 2.0 m roof beam).

Other cuts can be calculated as required.

Each of these can then be subdivided on the basis of the confidence levels described earlier (A1, A2), grade (see below) and area code (see below).

Ore-reserve plans (Figure 8.31) are drawn for each cut and each block is colour coded as follows:

purple ⩾ 40% KCl
red = 35–39.9% KCl
orange = 30–34.9% KCl
yellow < 30% KCl

and contains the following values:

Height of cut (e.g. 3.8 m)
% KCl (e.g. 30.2%)
% Carnallite (e.g. 1.2%)
% Insolubles (e.g. 17.6%)

Each block has an area code (0–9) as defined below:

0 – still to be mined
9 – ore in shaft pillar (600 m radius)
8 – ore in surface borehole protection pillar (100 m radius)
7 – ore mined previously (includes barrier pillars between panels)
6 – ore mined last year (includes barrier pillars between panels)
5 – barrier pillar around brine-filled old workings
4 – area under the town of Loftus
3, 2, 1 – not used at present

Each block can be split into subareas on the basis of area codes, e.g. code 6 given 25 units (25 000 m^2) and code 7 given 15 units (15 000 m^2), representing a total of 40 000 m^2, the area of one block.

Calculation of Class B reserves

These are calculated using polygons constructed around surface boreholes. The boundaries of these polygons are based on the boundaries of 200 m blocks and are thus stepped. This allows the polygons to fit against adjacent Class A areas and allows the Class B blocks to be stored and manipulated in the same way as the Class A blocks. The grade of each central hole is assigned to its associated polygon, except in a 0.5 km wide zone adjacent to Class A areas, where the polygon is given a grade based on the average of adjacent Class A blocks. Where no surface boreholes exist (e.g. offshore) averages of the peripheral Class A blocks are used instead. In both instances, the total seam thickness and grade are used and not the 'most likely mining-cut' values. Tonnages and areas are quoted for the reserves but it is the areas which are most relevant and useful to the mine.

Detailed reserves (panel grade predictions)

These reserves are calculated to predict grades and tonnages that will actually be mined and are thus based on the panel (roadway) design. 100–200 m long blocks (representing 2–3 months mining) are thus fitted to the proposed mining panels as in Figure 8.32. Each is divided into two sub-blocks and the centre of each is evaluated, by inverse distance weighting of drill-hole and mined panel data (section 8.6.6). Data for the 'most likely mining-cut' (including dilution from underlying halite, if necessary) are used and the grades for each sub-block

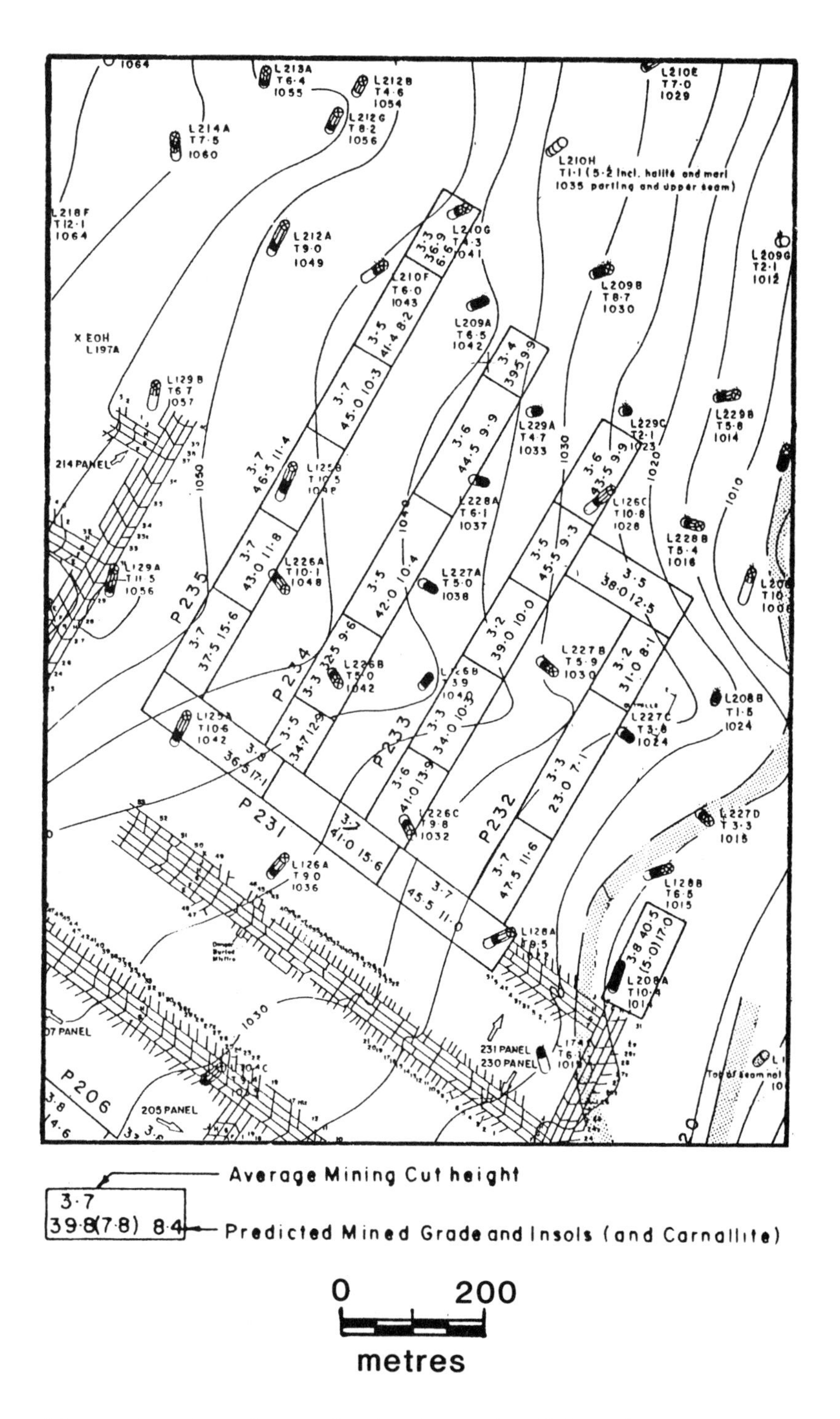

Fig. 8.32 Part of the panel grade prediction plan showing grades of planned panels over an average mining cut height.

are averaged. An estimate of additional unplanned dilution has to be added to allow for mining errors and seam irregularities beyond the flexibility of the mining equipment – this can be considerable. Although the extraction ratio within a panel is 80–85%, if the barrier pillar between adjacent groups of roadways is taken into account, this drops to 30%. It must be realized, however, that this is not the true extraction ratio of the seam because no allowance is made for the potash lost in the roof beam, or low grade/thin seam areas which are not mined.

REFERENCES

Smith, D. B. and Crosby, A. (1979) The regional and stratigraphic context of Zechstein 3 and 4 potash deposits in the British sector of the southern North Sea and adjoining land areas. *Econ. Geol.*, **74**, (2), 397–408.

Woods, P. J. E. (1979) The geology of Boulby Mine. *Econ. Geol.*, **74**, (2), 409–418.

Woods, P. and Hopley, R. (1979) Horizontal long hole drilling underground and Boulby Mine, Cleveland Potash Ltd. *Mining Engineer* (Journal of the Inst. of Mining Engineers), **139**, 585–90.

BIBLIOGRAPHY

Cleasby, J. V., Pearse, G. E., Grieves, M. and Thorburn, G. (1975) Shaft sinking at Boulby mine, Cleveland Potash Ltd. *Trans. A IMM*, **84**, 7–28.

Woods, P. J. E. (1973) Potash exploration in Yorkshire: Boulby mine pilot borehole. *Trans. B IMM*, **82**, (801), 99–106.

8.7 CASE HISTORY – EXPLORATION AND EVALUATION OF A GLACIAL SAND AND GRAVEL DEPOSIT

(P. Brewer and P. Morse, Tarmac Roadstone, Northwest Limited)

8.7.1 Target selection

It is essential for a source of bulk materials to be located close to the potential market-place because of the relatively high cost of transportation in relation to the final ex-works selling price. Hence, the identification of a specific market-place was the starting point for this case study; the area of search for a viable sand and gravel deposit was defined as being within a few tens of miles of that market-place, i.e. within a distance at which transport costs could be sustained.

The first stage of the geological investigation involved a preliminary study of readily available published geological information covering the area of search. However, the prime source of geological information, the 1 in to 1 mile and 1 : 50 000 geological drift maps produced by the British Geological Survey, indicated only a sparse distribution of potential sand and gravel resources. It thus became essential to undertake a very detailed literature study to build up a thorough understanding of the basic geology and, in particular, of the Pleistocene geological history of the search area. This work enabled specific target areas within the main area of search to be identified where there was a likelihood that sand and gravel could be present.

These target areas were then assessed on the basis of the feasibility of obtaining a valid planning consent to extract the minerals. Each Local Planning Authority in England and Wales has its own policies and guidelines on mineral extraction. In particular, certain counties have a presumption against mineral workings on good quality agricultural land (grades 1 and 2); Areas of Outstanding Natural Beauty; Sites of Special Scientific Interest; Green Belt Land, etc. Each of the target areas was therefore specifically assessed against these basic planning criteria. In addition, factors such as ease of access, possible environmental/visual impact and availability of water for processing needs were also taken into consideration.

Having thus identified those target areas without obvious planning constraints, approaches had to be made to landowners for permission to undertake a more detailed evaluation of these sites. Consent was eventually obtained from a number of estates with tenant farms and also from several individual farmers.

8.7.2 Initial exploration

To keep exploration costs to a minimum, a phased approach to exploration was adopted. A reconnaissance exploration programme of the most promising target areas was undertaken to determine whether any sand and gravel was in fact present. This work consisted of drilling widely spaced (500 m) boreholes using continuous flight augers, and an intermittent drilling technique (section 2.12.1).

The vast majority of boreholes penetrated thick sequences of glacial till or thin, silty, fine-grained sands interbedded with thicker silt and clay bands. However, on the site of this specific case history, two adjacent boreholes (although widely spaced) encountered sand and gravel of both a reasonable quality and thickness. The sand and gravel occurred beneath up to 4.7 m of glacial

Table 8.10 British Standard sand specifications for BS 882/1983 – Concrete Sand

Sieve size	Overall limits	% by mass passing BS sieve		
		C	M	F
10.00 mm	100			
5.00 mm	89–100			
2.36 mm	60–100	60–100	65–100	80–100
1.18 mm	30–100	30–90	45–100	70–100
600 μm	15–100	15–54	25–80	55–100
300 μm	5–70	5–40	5–48	5–70
150 μm	0–15			
75 μm	0–3			

Table 8.11 British Standard sand specification for BS 1200/1983 – Building Sand/Mortar Sand

Sieve size	% by mass passing BS sieve	
	Type S	Type G
6.30 mm	100	100
5.00 mm	98–100	98–100
2.36 mm	90–100	90–100
1.18 mm	70–100	70–100
600 μm	40–100	40–100
300 μm	5–70	20–90
150 μ	0–15	0–25
75 μ	0–5	0–8

Table 8.12 British Standard Sand specification for BS 594 – Asphalt Sand

Sieve size	% by mass passing BS sieve
5.00 mm	100
2.36 mm	95–100
600 μm	75–100
212 μm	15–60
75 μm	0–8

till but the overburden to mineral ratio was not excessive at 1 : 2.

To determine the possible extent of sand and gravel present, a geophysical survey was undertaken. This survey consisted of 12 resistivity soundings using the ABEM SAS300 Terrameter and an Offset Wenner Multi-core sounding cable. This equipment was used to produce standard Wenner resistivity sounding curves from which an interpretation of material type was made. If resistivity values were high enough to suggest the presence of sand and gravel, an interpretation of overburden thickness was made using a computer curve matching program. As a control, soundings were carried out at the locations of two boreholes where sand and gravel were present. The interpretation of these correlated well with the drilling results, with a marked contrast recorded between the resistivity of the clay (less than 50 Ω-metres) and the resistivity of the sand and gravel deposit (more than 100 Ω-metres). The remaining soundings were used to delineate the irregular-shaped deposit. This was essentially in qualitative form by simply identifying the presence or absence of sand and gravel. An initial reserve estimation could therefore be made of this potential deposit to determine whether its size justified further detailed investigation. In addition, samples from the boreholes encountering the sand and gravel were submitted to the laboratory for a washed particle size analysis (grading). These samples are weighed and washed through a nest of British Standard sieves of the following sizes, 75 mm, 63 mm, 50 mm, 37.5 mm, 28.0 mm, 20.0 mm, 14.0 mm, 10.0 mm, 6.3 mm, 5.0 mm, 3.35 mm, 2.36 mm, 1.18 mm, 600 μm, 425 μm, 300 μm, 212 μm, 150 μm, 75 μm and 63 μm. The retained weight in grams for each sieve is then determined to produce cumulative percentages by weight passing each of the sieves – a grading curve. These results are compared to the British Standard Specifications (Collis and Fox, 1985) for different types of sand (Tables 8.10–8.12). These technical results proved to be encouraging and an initial geological report on the overall findings was written with recommendations for further, more intensive, exploration which were subsequently accepted.

8.7.3 Detailed exploration and evaluation

Having identified the potential size of the deposit, it was necessary to carry out a detailed exploration drilling project to establish:

1. overburden thickness and its nature;
2. mineral thickness and its nature;
3. confirmation of the boundary of the deposit;

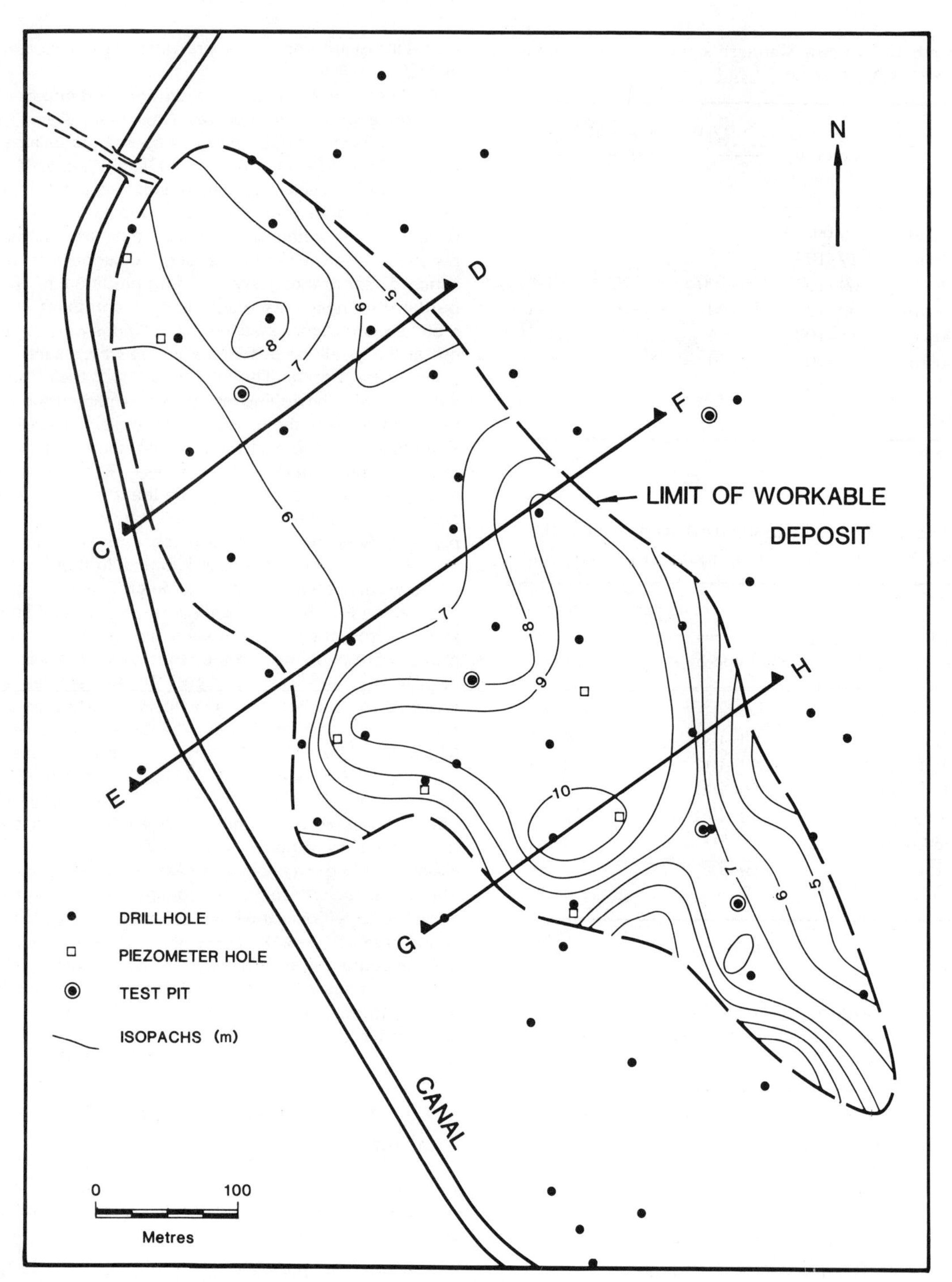

Fig. 8.33 Isopachyte map of a sand and gravel deposit in Lancashire showing boundary limits, drillhole layout and transverse section lines.

4. the quality of the sand in terms of grain size distribution to determine its potential end use;
5. the quality of the gravel in terms of size distribution and lithologies;
6. the presence of deleterious materials such as clay, silt or coal;
7. the water-table.

A further 35 intermittent flight auger boreholes were drilled with bulk samples being taken on the basis of 2 m intervals or lithological units. These samples were submitted to the laboratory for subsequent washed grading analysis. The boreholes were located (Figure 8.33) mainly within the boundary identified by the geophysical survey, but a few holes were drilled outside this boundary to confirm the geophysical interpretation. As the water-table was found to be above the base of the deposit, five piezometers were installed to monitor the seasonal fluctuations of the water-table (Figure 8.33).

The drilling proved an elongate lenticular deposit (Figures 8.33, 8.34) trending NW to SE, parallel to the regional fluvio-glacial drainage direction. An overburden of soils and boulder clay, averaging 3.7 m in thickness was found to overlay a fine glacial sand, averaging 2.8 m in thickness, which in turn overlay a sandy glacial gravel, averaging 6.3 m in thickness.

Isopachyte maps were produced for both layers in the sand and gravel and also for the overburden. Volumes were then calculated by planimetry of the isopachs and then converted to tonnages using conversion factors of 1.6 t/m^3 for the sand, and 1.75 t/m^3 for the sandy gravel. Adjustments were subsequently made for silt wastage, following the detailed assessment of the grading data received from the laboratory. This silt wastage is that considered to be removed by the washing plant. For this purpose, it is assumed that the percentage passing the 75 μm sieve size only represents 85% of the total removed by washing. On average, this amounted to 39.4% for the silty sand, but only 16.8% for the gravelly sands, resulting in an overall waste factor of 23.6% for the whole deposit.

Grading analyses returned from the laboratory indicated that the deposit was rather silty in nature, and would therefore require extensive processing to produce suitable construction materials. A theoretical washing process was carried out on these individual gradings – this assumed that all of the 75 μm sieve size material and 15% of the 150 μm sieve size material, would be removed in a washing plant. The original laboratory grading was therefore recalculated to produce a theoretically washed particle size distribution curve for further comparison with the relevant British Standard Specifications. The most suitable end use has a grading envelope which totally encloses the processed grading curve (two examples of BS grading envelopes are shown in Figure 8.35).

For every borehole, weighted averages of the sand gradings were produced for each of the major identified lithological units and for the entire borehole. This weighting takes into account both the different sample lengths and the varying proportions of sand to gravel within each sample. The individual sample gradings were therefore multiplied by the sample length and the percentage of sand within that sample. These figures were totalled for the entire lithological unit or for the borehole and re-expressed as percentages. Figures 8.36(a) and (b) illustrate the changes in size distribution between the raw grading data and the theoretical washed grading data for the gravelly sand and the silty sand respectively. By comparison with the British Standard curves and the specifications listed in Table 8.10, it was established that the upper fine sands could be washed to produce a sand suitable for building, mortar or asphalt purposes, whilst the lower sandy gravel sequence could produce, after washing, material for use in concrete.

However, during the drilling process, it was apparent that the proportion of gravel being returned on the augers was not truly representative of *in situ* materials – large cobbles and boulders were being encountered, but were being either broken down or pushed to one side by the drilling. In addition, some contamination of the lower sandy gravel sequence was possible by the overlying finer sands and the boulder clay.

The final evaluation phase of the deposit therefore involved 'test pitting' (Figure 8.33) to obtain a better assessment of the percentage, size and quality of the gravel fraction and to clarify the quality of the sand portion. This information is crucial to the final processing plant design. Due to the depth of overburden and then fine sand, it was necessary to employ a track-mounted back-actor to excavate a pit to a depth of at least 6 m. A total of five test pits were dug, each approximately 6 × 10 m across, and bulk samples of the overburden, fine sand and sandy gravel were taken. In addition, a 15 t load was taken to an established sand and gravel processing plant in the area for full-scale plant trials. Further wet grading analyses were carried out on subsamples and a lithological classification of the gravel fraction undertaken. It was found that the samples of fine sand obtained by drilling were in fact representative, but that the samples of sandy gravel had been contaminated, as anticipated, during drilling. In addition, the gravel fraction was found to be very cobbly and, as had been suspected, many of the larger cobbles were not recovered during drilling. In consequence, final adjustments were made on the proportions of gravel, sand and silt within the deposit, together with a reserve estimate.

A final geological report was therefore produced, together with a reserve estimate, which identified the existence of in excess of 1 million tonnes of sand and

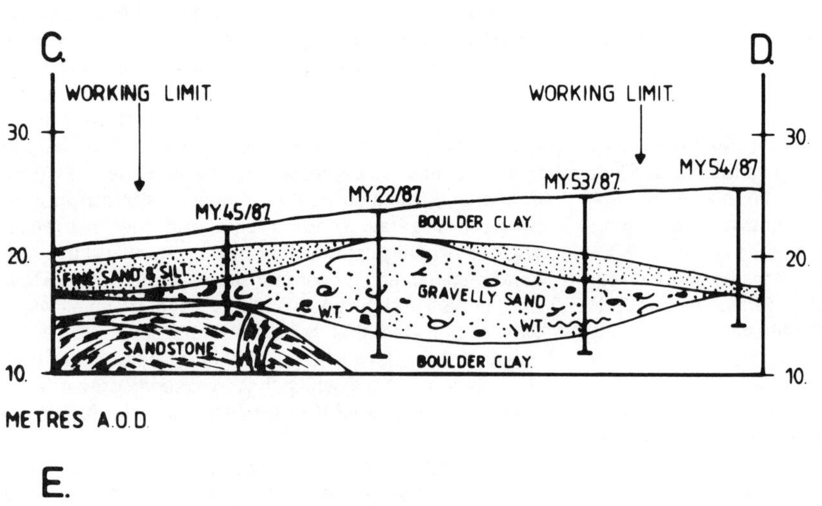

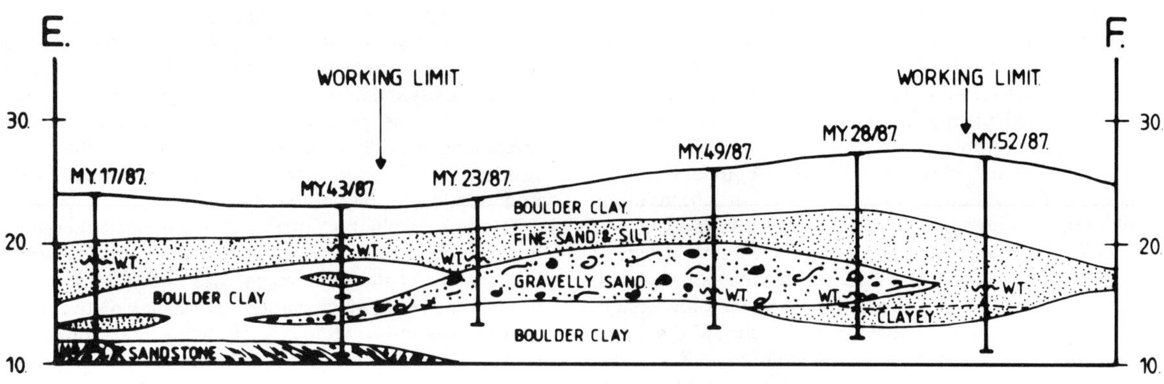

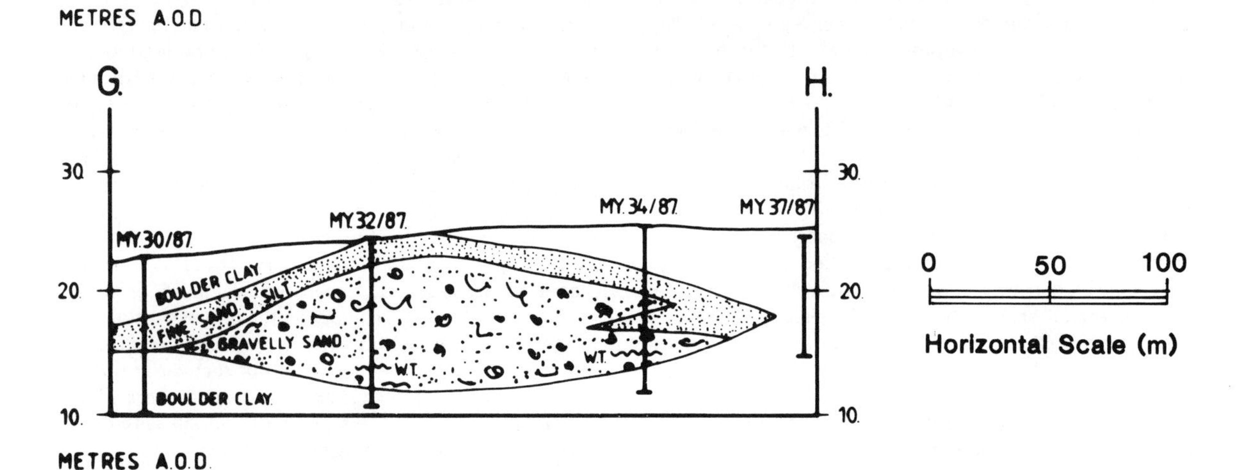

Fig. 8.34 Transverse sections of the sand and gravel deposit in Fig. 8.33.

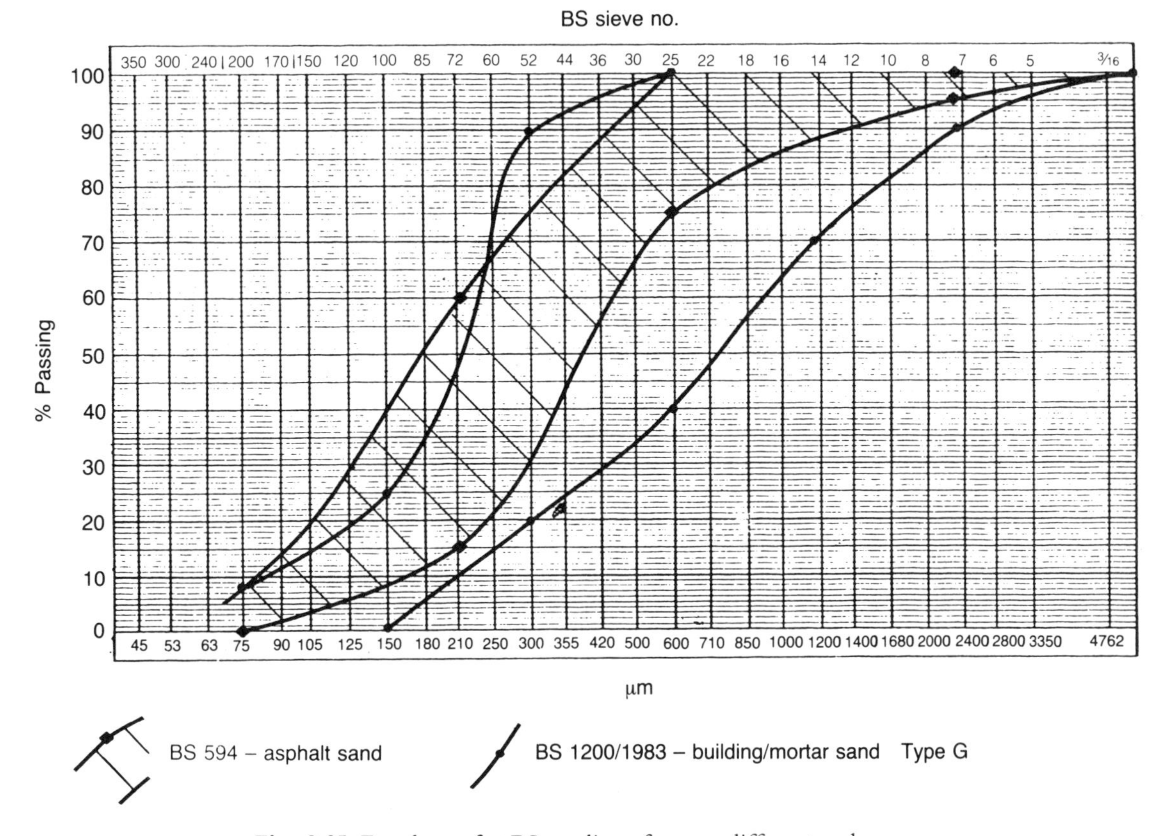

Fig. 8.35 Envelopes for BS gradings for two different end-uses.

gravel for use as concrete aggregates, general purpose mortar sands, sand for use in the manufacture of hot-rolled asphalt, and 'grits' for horticultural purposes. Whilst this is not regarded in the industry as a large reserve, resources of sand and gravel in the search area were so limited that it was considered to be an important and economically viable deposit. In conjunction with the Estates, Production, Engineering, Technical, Marketing, Sales and Landscape Architects Departments, a planning consent was obtained in September 1989 for the extraction, processing, and sale of the sand and gravel and subsequent restoration of the land.

REFERENCE

Collis, L. and Fox, R. A. (eds) (1985) *Aggregates: Sand, Gravel and Crushed Rock Aggregates for Construction Purposes* (Engineering Geology Special Publication, No. 1), The Geological Society, London, 220 pp.

8.8 CASE HISTORY – LIMESTONE AGGREGATES – THE TYTHERINGTON LIMESTONE QUARRIES (ARC LTD)

8.8.1 Introduction

The Tytherington Quarries are three separate but adjacent limestone quarries which lie just under 1 km NW of the village of this name and 16 km N of Bristol in the county of Avon, UK (Figure 8.37). The central quarry (Grovesend), which began in 1928, was finally exhausted in 1960 by which time the northeastern quarry had commenced. The new quarry (Hetherington) was later accessed by way of a tunnel beneath the public road separating the two quarry areas.

Production in this quarry has continued progressively since the early 1960s and it reached a peak in 1983/84 at 1.5 million tonnes per annum. The total resource that can be recovered from the northeastern quarry is constrained

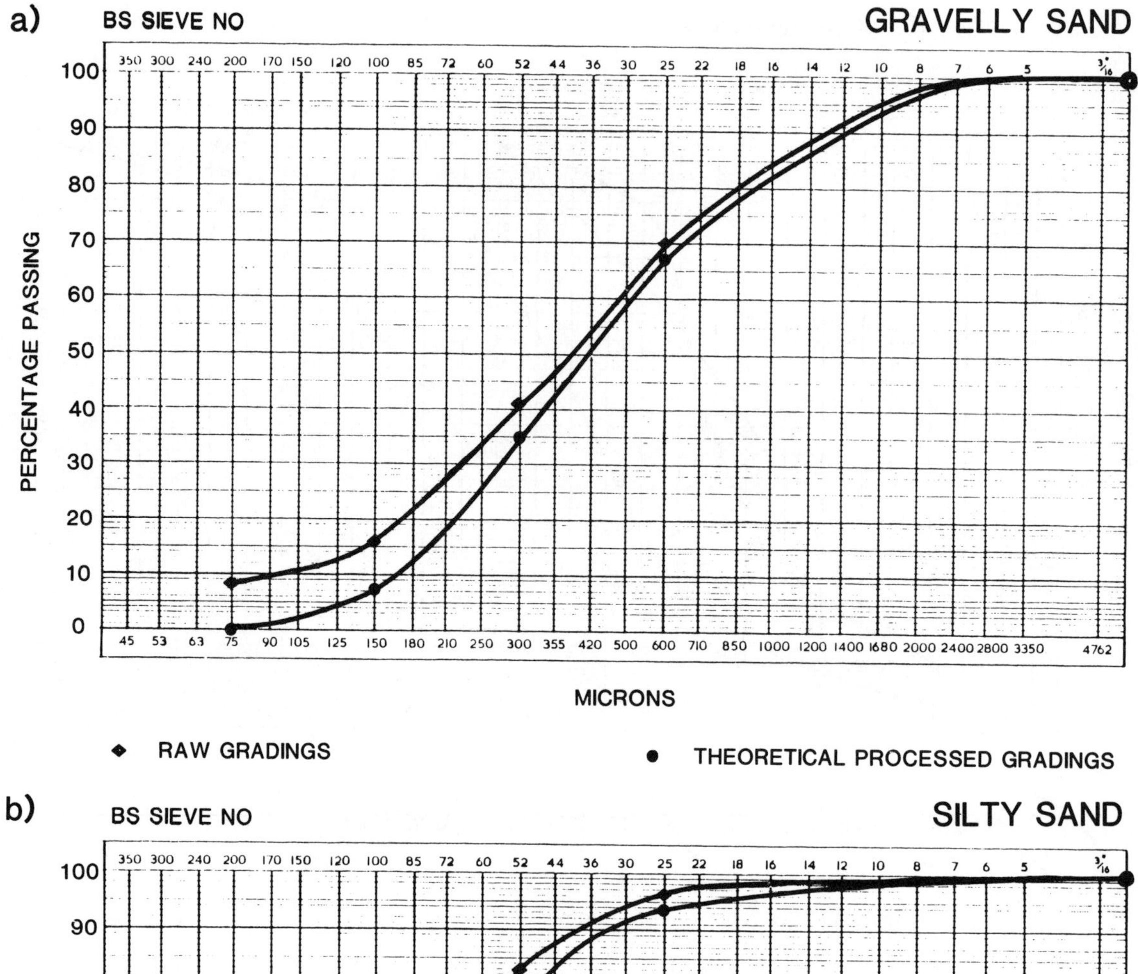

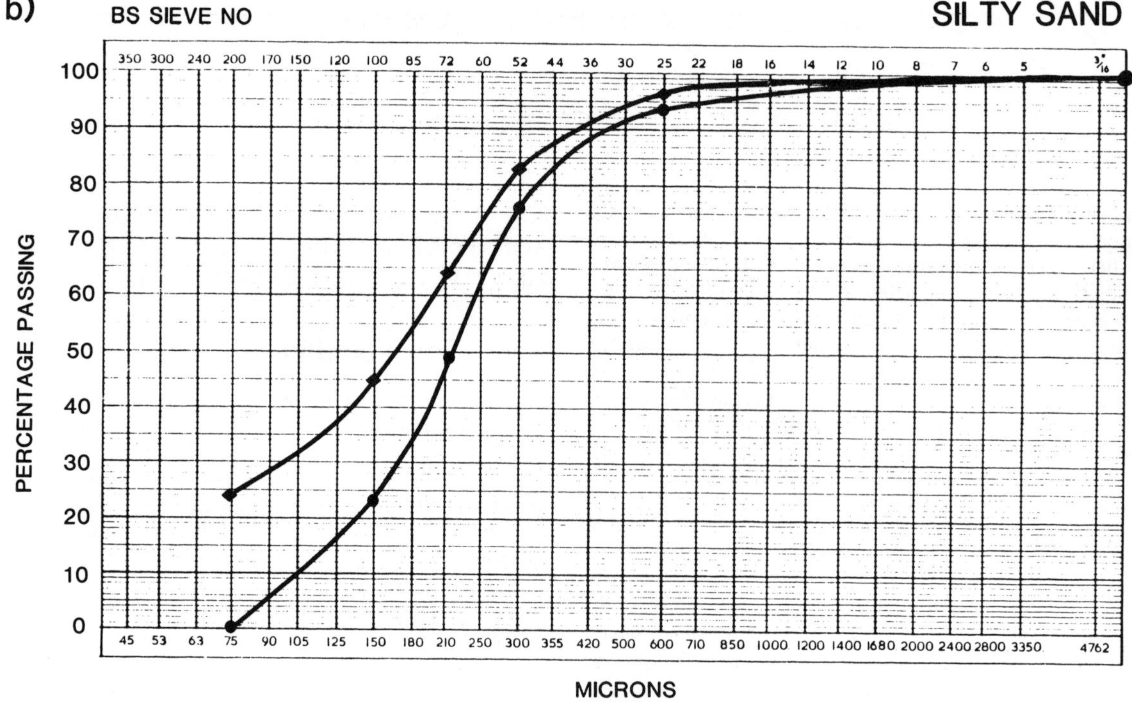

Fig. 8.36 Raw (◆) and theoretical processed (●) grading curves for: (a) the gravelly sand and (b) for the silty sand.

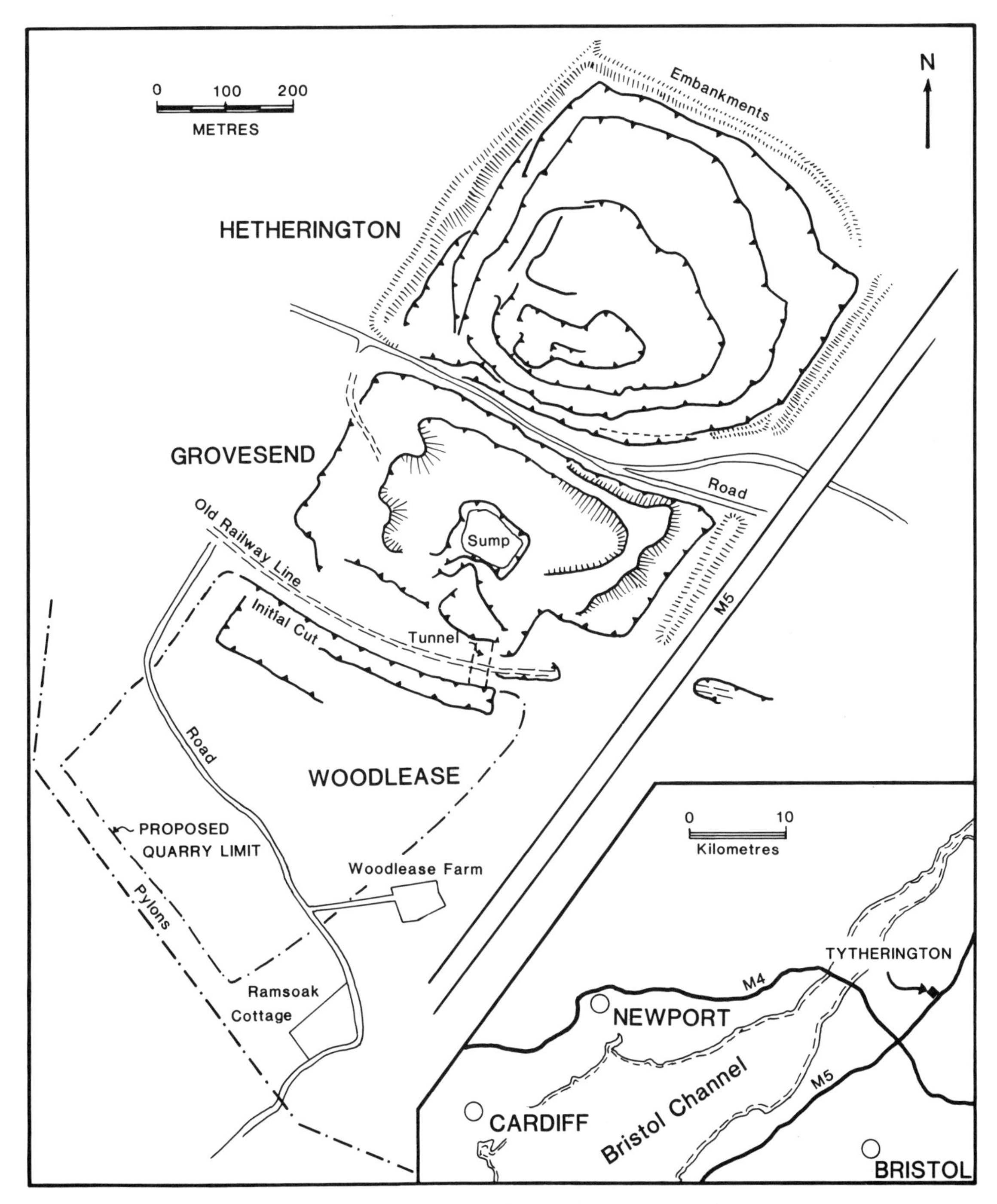

Fig. 8.37 The Tytherington quarries with inset showing their location near Bristol.

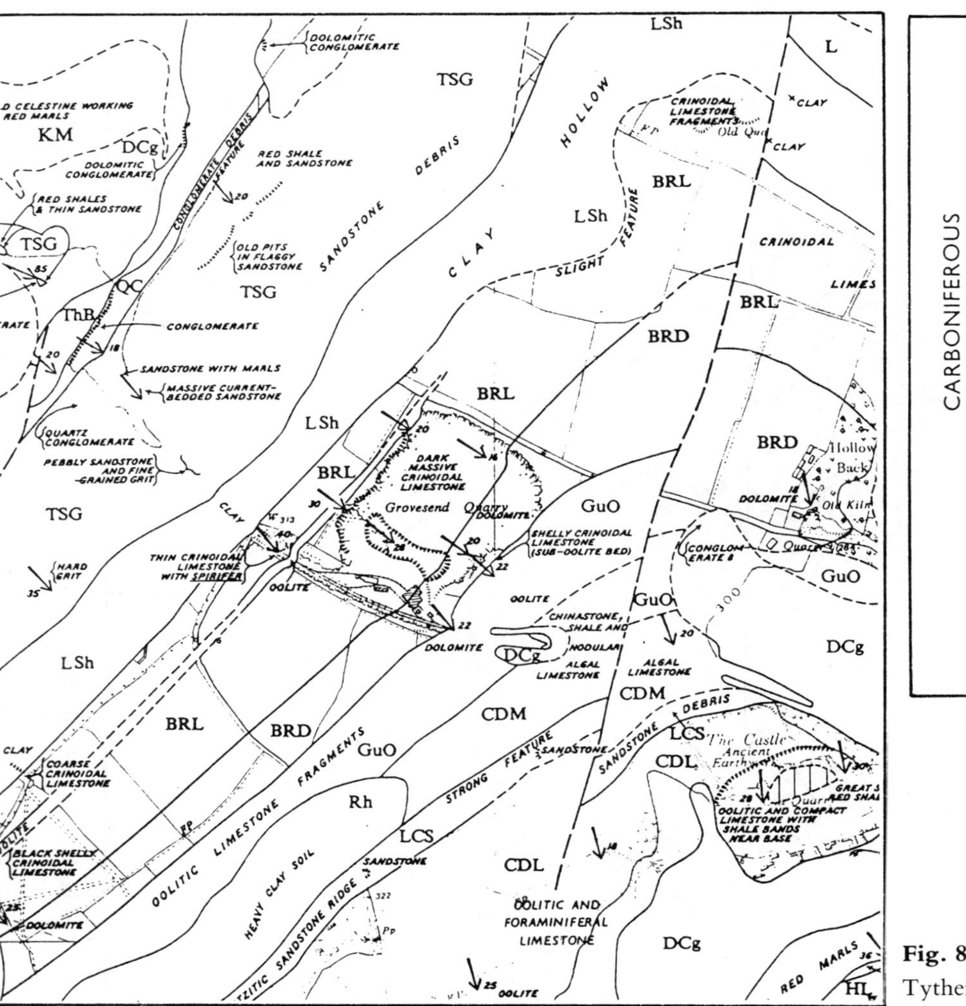

Fig. 8.38 Geological plan and stratigraphic section of the Tytherington area.

by both stratigraphy and synclinal folding. By 1989/90, when production will cease, a total of approximately 25 million tonnes will have been extracted.

Production demand beyond 1990 is forecast to be more than 2.0 million tonnes per annum, most of which will be distributed by rail to the Midlands and SE England. The quarry was, in fact, connected to the rail network in 1968 to service the older quarries and trains can be loaded at the rate of 1000 tonnes per hour. The demand forecast requires that continuous operation be maintained and that, by the time the northeastern quarry is exhausted, a new quarry area to the S of the Grovesend property would have to be in operation and able to maintain a supply rate of 2 million tonnes per annum.

This present case history deals specifically with this new quarry (Woodleaze) and examines the methods used to explore and evaluate the site and to produce a final pit design and reserve which would satisfy the high output requirement.

The quarries lie on a prominent NE–SW trending ridge rising to an elevation of 100 m above ordnance datum (AOD) and are immediately adjacent to, and NW of, the M5 motorway. The Woodleaze site covers an area of 35 hectares, ranges in altitude from 94 m to 98 m (AOD), and is bounded to the N by a railway line, to the E by the motorway and to the S by a line of electricity pylons. The western limit is represented by the outcrop of the base of those Lower Carboniferous limestone horizons which are considered suitable for road aggregate. Quarry design is further constrained by the existence of a small road which passes over its central section and by a cottage in its SE corner.

8.8.2 Geology

The quarries lie close to the nose of a gently plunging syncline which is a minor structure on the western limb of the Bristol Coalfield syncline. The Carboniferous strata thus dip to the SSE at 20–30° but the strike changes to a more easterly direction, and the dip flattens, as one proceeds northwards towards the nose of the syncline. The stratigraphic section (Figure 8.38) begins in the NE with the Lower Limestone Shales, which consist of interbedded impure limestones with thin dark grey to black bands of fossiliferous shaly mudstone up to 0.5 m thick. These rocks grade up into the overlying Blackrock Limestone whose base is located at the point at which the last significant (>10 cm) mudstone band disappears. The Blackrock Limestone is a fossiliferous (crinoidal), fine to medium grained, massive, light grey limestone whose upper section has been subjected to secondary dolomitization and hence is referred to as the Blackrock Dolomite. The total thickness of the limestone and dolomite is approximately 145 m. The top of the Blackrock Dolomite is placed at the Tournaisian–Visean boundary. It is overlain by the Gully Oolite, a massive, cross-bedded oolite which is generally light grey and unfossiliferous and 35–40 m thick. The eroded upper surface of this oolite is overlain by the Clifton Down Mudstone. This contact represents the upper limit of the available limestone resource.

One to two metres of Rhaetic marls and Recent drift obscure the underlying Carboniferous rocks and may also infill solution cavities and joints in the limestones. These younger sediments must be removed during quarrying of the limestones to avoid a depreciation in the quality of the product. Moderate weathering of the limestones has occurred to depths of 1.5–4 m below the surface and this zone (the 'Top Rock') is also treated separately.

Although the water-table is relatively deep, and fluctuates between 55 m and 65 m (AOD), flooding of the lower benches in the quarries is a major problem, especially in the winter months.

8.8.3 Site exploration

Rotary water-flush core drilling of the Woodleaze site began in April 1982 and continued spasmodically until May 1985, by which time, a total of 13 holes had been completed. The holes were drilled, using a Dando 250 rig, on a random grid but with an average spacing of approximately 200 m. All the holes were drilled vertically, at 76 mm size, and were located in such a way as to cover the stratigraphic interval from the base of the Clifton Down Mudstones down to the top of the Lower Limestone Shales. These holes were carefully logged to determine the location of all lithological subdivisions, especially mudstone bands. At the same time, core recoveries, RQDs, fracture indices and weathering grades (I = fresh; II = slightly weathered; III = moderately weathered) were assigned to drill runs (see example log sheet, Figure 8.39).

The cores were then subjected to routine petrographic analyses and crushed aggregates tested to determine their physical properties. The preliminary results of the physical testing are listed in Table 8.13 and the methods are briefly described below. More detailed information on the testing of aggregates can be gained from Collis and Fox (1985).

Relative density

Three values for relative density, quoted in the table, were determined as follows:

$$\text{Oven dried (OD)} = \text{ODW}/(\text{SDW} - \text{WW})$$

$$\text{Saturated and surface dry (SSD)} = \text{SDW}/(\text{SDW} - \text{WW})$$

$$\text{Apparent} = \text{ODW}/(\text{ODW} - \text{WW})$$

Location TYTHERINGTON QUARRY

Drillhole no. 30

Sheet 2 **of** 21

Type of drilling ROTARY CORE W/FLUSH Angle from horizontal 90° (vertical) Co-ordinates E 365957

Rig DANDO 250 Bearing N ____ E ____ N 188404

Bit DIAMY SURFACE SET 20/40 SPC Ground level 94.42m aod Water table level ____

Drilling progress	Casing depth, size	Water level	Notes e.g. colour water return, caving instrumentation, sampling disc.	Depth and diameter (metres)	Reduced level	Core recovery %	Rate of penetration	R.Q.D.	Fracture index	Legend	Description	Weathering grade
10.05				10, 11, 12		T100 S100		44	12		10.35 2cm band of pale grey weak mudstone. 11.95-12.10m Dark grey band of mud-stone. 12.10-12.15 Dark grey weak mudstone partially unconsolidated.	
12.35				13		T100 S 87		45	14		13.06-13.24 Dark grey weak and unconsolidated mudstone band. 13.50m 1cm band of weak brown grey mudstone.	
13.78				14, 15		T100 S100		40	11		14.24. 2cm band of weak brown mud-stone. 14.35 Dark grey mudstone with reddy brown patches. 15.25 5cm pale grey mudstone band.	
15.35			TOP OF GULLY OOLITE	16, 17, 18		T100 S100		53	8		15.30-15.70m Pale grey oolitic lime-stone with purply-red streaks. 15.70m Pale grey fine grained oolitic limestone. 16.00-16.50m Oblique fracture zone patches of orange fe staining along fracture surface. 18.0-18.35 Oblique fracture zone.	
18.35				19, 20		T100 S100		51	18			

Legend

Remarks

Scale 1:50

Logged by D.M.L.L

Contractor A.R.C.

Date started 12.3.85

Date finished 26.3.85

ARC

Date May 1985

Fig. no.

Fig. 8.39 Typical page from geologist's log of a borehole from the Woodleaze site.

where ODW = the weight of the sample after oven drying for 24 h at 100–110°C; WW = the weight in water of a thoroughly saturated sample; SDW = the surface dried weight of this sample. From the above information, the percentage water absorption can also be calculated from:

$$100 \times (SDW - ODW)/ODW$$

Mechanical tests

The five values obtained from standard mechanical tests are shown in Table 8.13 and each is briefly explained below. The strength tests are defined by British Standad 812 (1975).

Aggregate impact value (AIV)

This represents the amount of degradation of a sample, expressed as the weight percentage passing through a BS 2.40 mm sieve, after it has been subjected to 15 blows with a piston hammer. The hammer should weigh between 13.5 kg and 14.1 kg and should fall through 381 ± 6 mm. The initial sample should also be in the size range 9.5–12.25 mm. AIV thus measures the resistance to granulation and, hence, the lower the value the more resistant is the rock.

Aggregate crushing value (ACV)

The weight percentage of fines passing through a BS 2.40 mm sieve is determined after a 2 kg sample has been subjected to a piston loading of 400 kN over a period of 10 minutes. Again, the lower the percentage, the higher the resistance of the rock.

10% fines value

This is a variant of the ACV test in that it is based on the force required to produce approximately 10% fines (<2.40 mm) after uniform loading for 10 minutes and a piston penetration of 20 mm. The value is calculated from:

$$(14 \times F/P) + 4$$

where F is the maximum force in kN and P is the mean percentage of fines from two tests at force F.

Aggregate abrasion value (AAV)

This value is based on the percentage loss in weight during the test which is performed on 33 cm^3 of clean 10–14 mm non-flakey aggregate mounted in resin so that 6 mm protrudes above the surface of the resin. This sample is held against a lap rotating at 28–30 r.p.m. by a force of 2 kg for a period representing 500 revolutions, while sand (Leighton Buzzard Sand specified by BS 812, 1975) is fed over the sample at 0.7–0.9 kg/min.

Polished stone value (PSV)

In this test, the sample is set in resin, as above, and is then subjected to a rotating pneumatic tyre under a 40 kg load. Corn emery and water are fed to the surface of the tyre and the polish induced on the stone measured by a standard pendulum arc friction tester. The test is repeated using emery flour as the abrasive. The PSV is thus the coefficient of friction expressed as a percentage. A high value implies a high resistance to polishing.

Evaluation of the Woodleaze results

Coarse aggregates for pavement wearing surfaces are covered by British Standard 882, 1983. The requirements are:

10% Value > 100 kN
AIV < 30
AAV < 15

Table 8.13 reveals that the Woodleaze aggregate meets these specifications.

Chemical analysis

Although the bulk chemical analysis is not particularly relevant when the end-use is aggregate for concrete or roads, the results for +10 mm material are presented in Table 8.14. These average values do, however, mask a marked variability in CaO/MgO ratios.

Table 8.13 Physical test for aggregates from the Woodleaze Site

Relative density (OD)	= 2.69
Relative density (SSD)	= 2.72
Relative density (Apparent)	= 2.76
Water absorption (%)	= 0.86
Aggregate impact value (AIV)	= 23
Aggregate crushing value	= 22
10% fines value	= 180 kN
Aggregate abrasion value (AAV)	= 10.2
Polished stone value (PSV)	= 43

8.8.4 Geotechnical studies

In order to gain an insight into what stability problems might occur in the proposed quarry, a geotechnical study was undertaken of the adjacent Grovesend quarry using the methods described in section 5.3.5. Figure 8.40 shows a geologist using a stratum compass to measure

Table 8.14 Chemical analyses for Woodleaze aggregate

CaO	43.70%	($CaCO_3$ = 78.0%)
MgO	8.08%	($MgCO_3$ = 17.0%)
SiO_2	2.81%	
Al_2O_3	0.63%	
Fe_2O_3	0.52%	
LOI	43.70%	

discontinuities in limestones. Stereographic plots of the poles of all the joint and bedding planes measured revealed that most of the joints are subvertical and would thus provide release surfaces for planar sliding on those mudstone horizons which daylighted into the quarry. The spacing of the joint sets within the limestones were found to be less than 3 m and two of these had values less than 1 m. Bedding is well developed in both the limestone and mudstone, and open planes can be traced for the full length of individual faces. Units defined in the limestone by these bedding planes vary in thickness from 0.3 m to 3 m. Geological mapping also located a series of minor thrust faults with dips of 20–40° to the SE and with throws generally less than 2 m. Stereographic plots indicated that there is a possibility of wedge failure when such thrusts intersect bedding. Additional, but minor, wedge failures could also affect the other faces due to conjugate joints sets in the limestone.

Laboratory shear box testing of 60 mm cores indicated peak and residual angles of friction for mudstone surfaces of 31° and 19° respectively, and natural cohesions of 11 kN/m^2 and 0 kN/m^2 respectively. Instability can thus be expected in the west wall of the quarry which would be following the footwall of the limestones at a dip of approximately 26°, especially if this wall were to be undercut in any of the lower benches and if joints, etc., remained waterlogged. Both the geological mapping and study of drill cores confirmed the existence of clay infilled joints and bedding planes which will not only impede water drainage through the rock mass into the proposed pit, but will, in conjunction with the expected water saturation, significantly decrease the shear resistance of these planes. Eadie (1982) has calculated the safety factors for potential block sliding on the bedding using the methods described in section 5.3.5 and in Stimpson (1979). He used friction angles varying between 15° and 19°, natural cohesion values varying between 0 kN/m^2 and 50 kN/m^2 and release surfaces 14 m and 28 m back from the quarry face. Where a water saturation of 25% was used, safety factors ranged from 1 to 4, but a saturation of 50% resulted in a marked drop in the safety factors. When low natural cohesion values were applied, the safety factors dropped to below the critical level of 1. Thus good drainage conditions will have to be maintained to assure a reasonable degree of safety.

Fig. 8.40 Geologist undertaking a geotechnical survey of a limestone quarry.

8.8.5 Preliminary pit design/evaluation

In order to assess what problems might be encountered during quarrying operations on the Woodleaze site, a manual design and evaluation exercise was undertaken. The first phase of this involved the production of dip sections on which boreholes were plotted (projected) together with lithological units. This allowed stratigraphic correlations to be made and also the determination of the dips of these units. This work revealed a discrepancy between the dip of the Lower Limestone Shale–Blackrock Limestone contact and that of the Gully Oolite–Clifton Down Mudstone contact, i.e. 27° versus 22° respectively. Structure contour maps of these horizons also revealed a slight discrepancy between the two and an easterly curvature towards the N, as the nose of the syncline was approached.

From the above information, the outcrop traces of the two key horizons were plotted by trigometric extrapolation of the borehole information. The next step was to set the limits of the proposed quarry. This was done using the following criteria:

1. A 50 m wide barrier from the pylons in the SW.
2. A 50 m wide barrier from the railway line in the NE.
3. The projected outcrop of the base of the Blackrock Limestone in the E.
4. An eastern perimeter based on the point at which a 60° face would cut back into the Clifton Down Mudstone from the point at which the first bench cut the base of this horizon (Figure 8.41). For a dip of 22° and a bench height of 15 m, this would be approximately 45 m E of the outcrop trace. At the same time, the proximity of the motorway had to be taken into account.

The next stage was to set the design parameters for the quarry itself as follows:

1. Five benches each 15 m high (80, 65, 50, 35, 20 m AOD) plus a sump.
2. Face of first bench cut to 60°, but thereafter 70° rock faces.
3. Berm width 7 m.
4. Haul roads 15 m with 1:10 maximum gradient.
5. West face to be planed off at 27°.

Items 2 and 3 are based on the requirement for a batter of 50° for the N, E and S slopes as indicated by geotechnical studies. The base of the Blackrock Limestone was

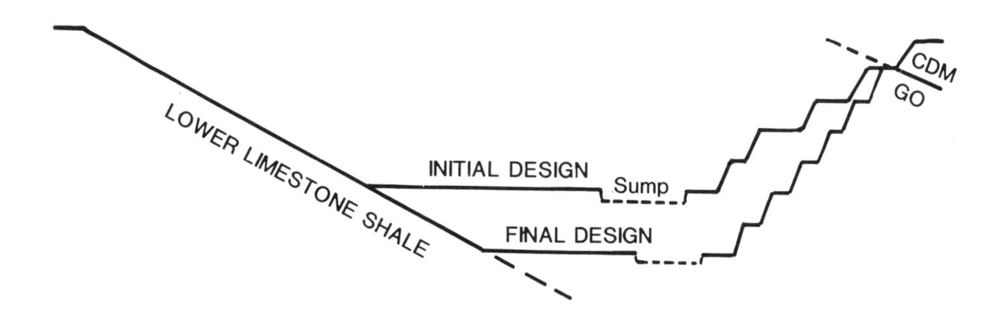

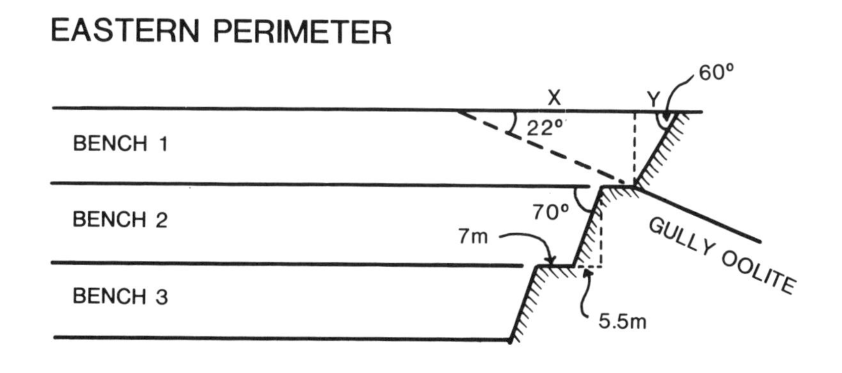

Fig. 8.41 Profile of initial quarry design for the Woodleaze site showing the method used to determine the eastern perimeter.

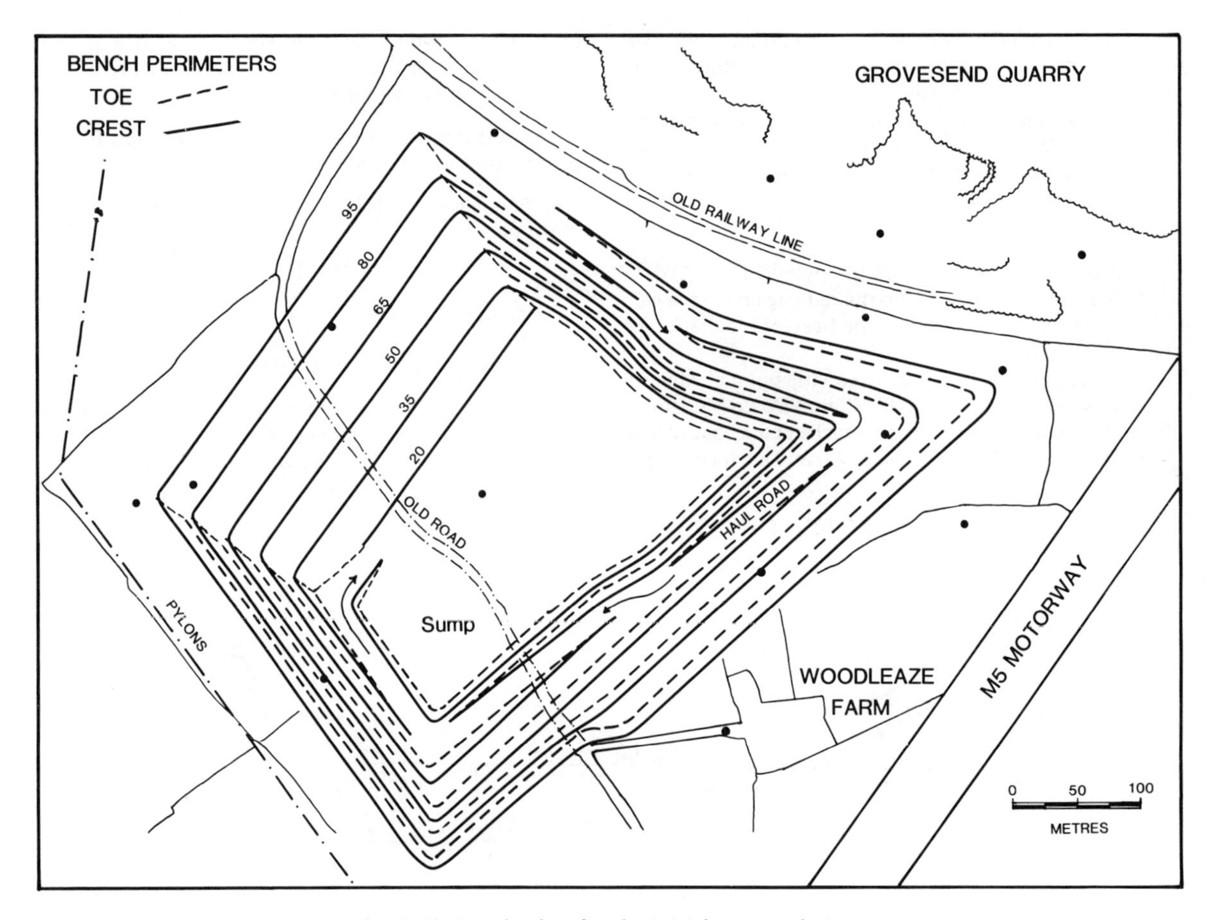

Fig. 8.42 Bench plan for the initial quarry design.

then contoured at the above bench levels and the other pit-walls were then drawn in using horizontal spacings of 8.7 m (bench 1) and 5.5 m (benches 2–5) between bench crests and toes, and 7 m berm widths. This contour plan was then modified to allow for the haul roads (Figure 8.42) and for the sump in the south-eastern corner. Also, the overall pit slope was reduced on the eastern side by reducing the berm width, for the 50, 65 and 80 m benches, to 15 m. An overlay showing the mid-bench contours was then produced and planimetered to determine bench areas. From these areas, volumes could be calculated making allowance for the reduced thicknesses on bench 1 due to the surface topography. A tonnage factor of 2.6 t/m^3 was then applied. The tonnage of overburden to be stripped was also estimated and the proportion which would be required for baffle embankments. Within each bench, the proportion of waste (cavity/joint fill and Clifton Down Mudstone – the former from experience in the other quarries) was also determined. Finally, estimates were made of the percentage of 'scalpings' (i.e. the oversize material which would have to be removed for recrushing after a first pass through the primary crusher, together with the undersize material). These were set at 30% for bench 1, 25% for bench 2 and 20% for the remaining benches.

8.8.6 Final pit design/evaluation

This exercise was undertaken using MICL's DATAMINE software. The data base created contained hole numbers, collar coordinates and the elevations, and finally the elevations of the bases of the Clifton Down Mudstone and the Blackrock Limestone. Where either horizon was not intersected, its elevation was estimated from sections (hence some are above the present surface).

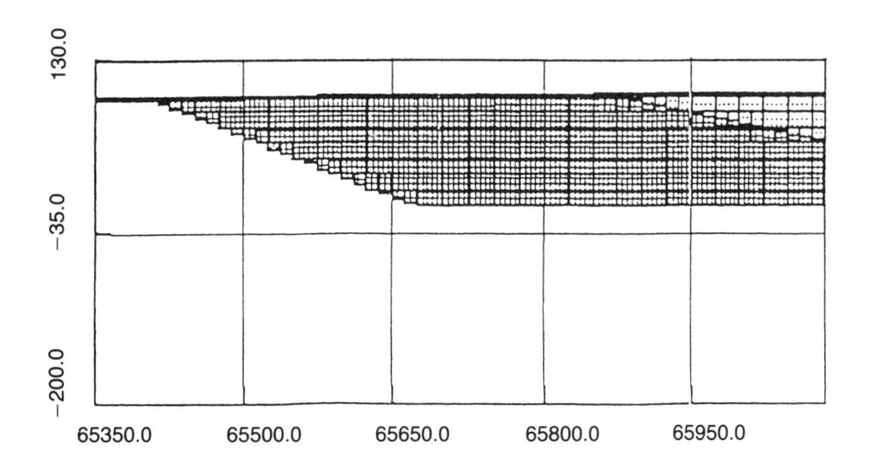

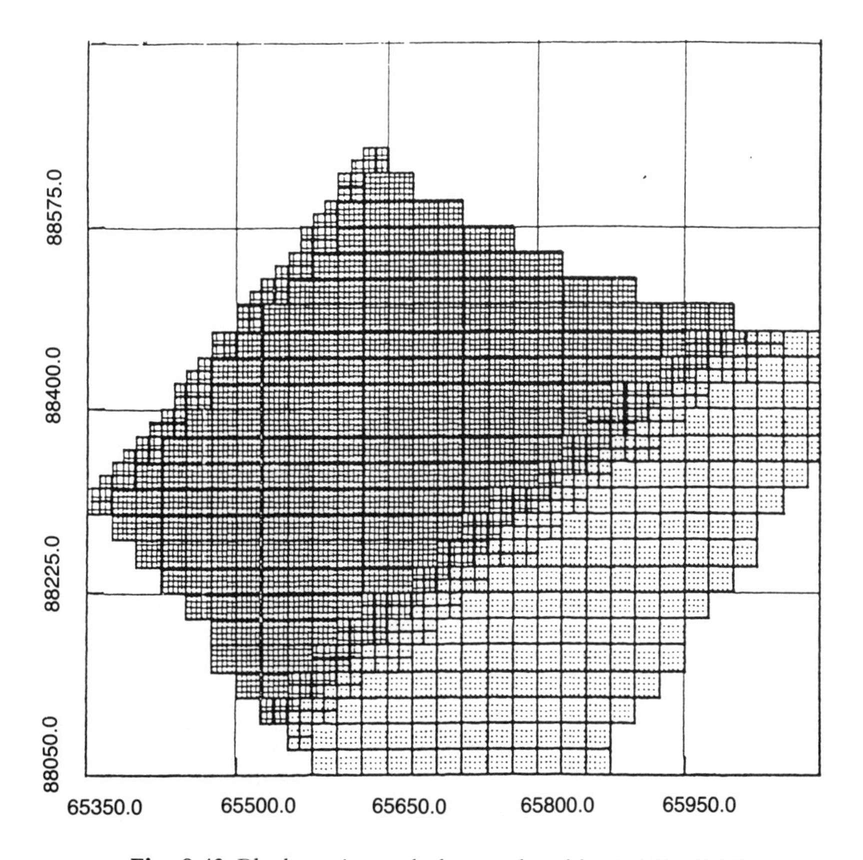

Fig. 8.43 Block section and plan produced by DATAMINE.

Dummy holes were also added at the boundaries of the area to assist in the modelling of the deposit. From this information, DATAMINE produced surface plans of the boreholes and also borehole sections. The limits of the quarry, as defined earlier, and other geographical features, were digitized from a plan. The program then calculated 3D block models (using 25 × 25 × 15 m cells, locally subdivided into subcells to gain better definition) of the following horizons over an 800 × 800 m area:

1. surface topography;
2. base of the overburden (1 m below surface);
3. base of the 'top rock' (3 m below surface);
4. base of mudstone;
5. base of limestone.

The base of the block model was set at − 10 m (AOD), as the lowest possible bench thought possible would be −10–+5 m (AOD), while the top was set at +110 m (AOD), allowing for a bench between +95 and +110 m (AOD). A planar trend was estimated separately for each surface using DATAMINE's trend fitting facility. The results were then used in a surface interpolation process which removes the trend from the data points and determines the residuals by inverse weighting methods. The trend is then added back to the interpolated residuals to produce a model of the surface based on blocks and sub-blocks. Each block or sub-block in the 3D model is then assigned a rock code and sections and level plans produced (Figure 8.43). Detailed blow-ups of these sections and plans (Figure 8.44) can be produced, together with structure contour plans of any of the surfaces modelled.

Three alternative quarry designs were considered:

A. One large quarry involving the removal of the road but not the cottage in the SE.

B. One large quarry as above, but removing the cottage.

C. Two separate quarries either side of the road.

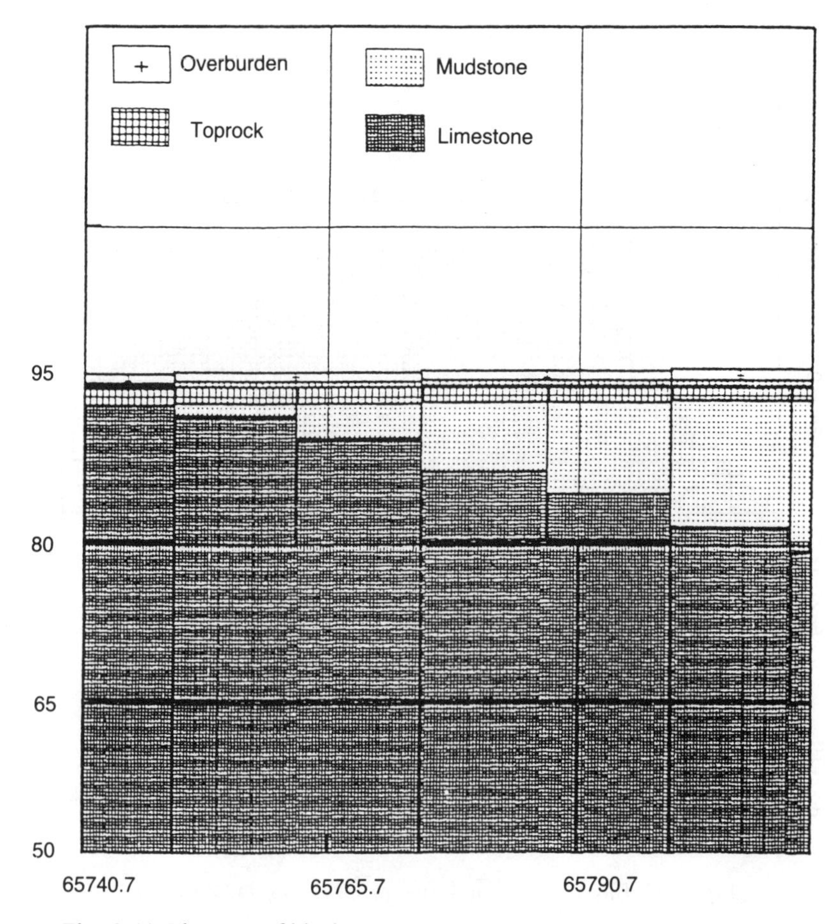

Fig. 8.44 Blow-up of block section showing the geology in more detail.

Figure 8.45 represents the design produced for case B in which each bench is represented by the mid-bench contour. As bench 1 had little material in it, the initial quarry perimeter was taken at the bench 2 mid-level and the walls projected up to bench 1 and down to bench 8 at a slope angle of 60°. In the case of the N wall, this slope was set at 50° to give a greater factor of safety for the railway line along that edge. The steeper walls used in this final stage were to obtain an increased recovery of limestone over that achieved in the initial study. The design was then modified by insertion of haul roads and was then digitized for computer storage.

The *in situ* reserves of the quarry were then calculated by applying different density values to each rock type (overburden 2.0, 'top rock' 2.5, mudstone 2.6, limestone 2.7 t/m^3). Tonnages could then be calculated for each bench, and each rock type, in the three quarry designs as in Table 8.15. At the time of writing no final decision had been made as to which option would be chosen. However, an initial bench cut had been made adjacent to the disused railway line as shown in Figure 8.33.

REFERENCE

Collis, L. and Fox, R. A. (eds) (1985) *Aggregates: Sand, Gravel and Crushed Rock Aggregates for Construction Purposes* (Engineering Geology Special Publication, No. 1), The Geological Society, London, 220 pp.

Eadie, A. (1982) An assessment of the parameters which influence slope stability and their application to the northern section of Tytherington Quarry.

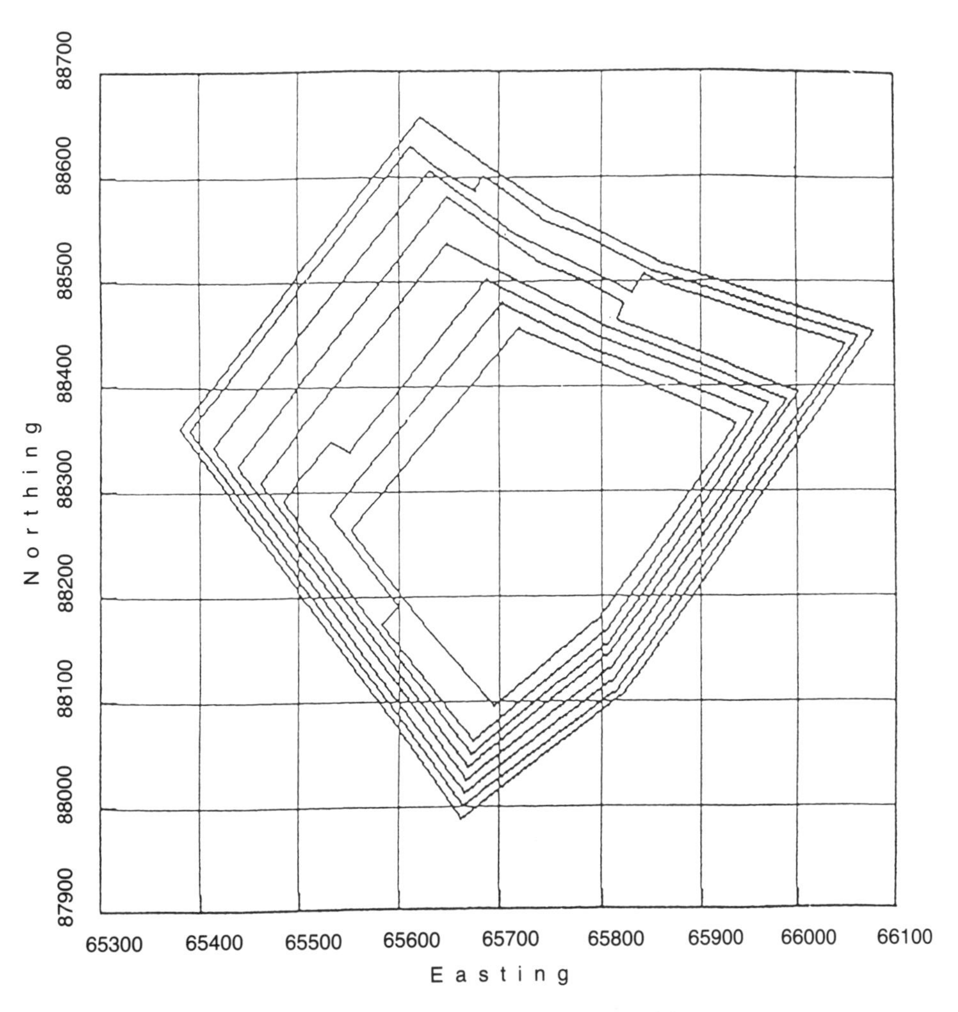

Fig. 8.45 An example of a pit design produced by DATAMINE.

Unpublished BSc dissertation, Royal School of Mines, Imperial College of Science, Technology and Medicine, London, 93 pp.

Stimpson, B. (1979) Simple equations for determin-ing the factor of safety of a planar wedge under various groundwater conditions, *Quarterly Journal of Engineering Geology*, **12** (1), 3–7.

8.9 CASE HISTORY – CEMENT QUALITY LIMESTONES AT LOS CEDROS, VENEZUELA (BLUE CIRCLE INDUSTRIES PLC)

8.9.1 Introduction

Following an initial reconnaissance of the Torococo–Cuicas area of Trujillo State, Western Venezuela (Figure 8.46) by C.A. de Estudios de Factibilidad y Proyectos (EFP), which indicated that several areas contained both limestone and shale suitable for cement production, Blue Circle's Technical Services Division (BCT) were asked by Cemento Andino to undertake a follow-up survey. The intention was to locate a potential quarry site which could supply a plant producing at least 500 000 t of cement clinker per annum.

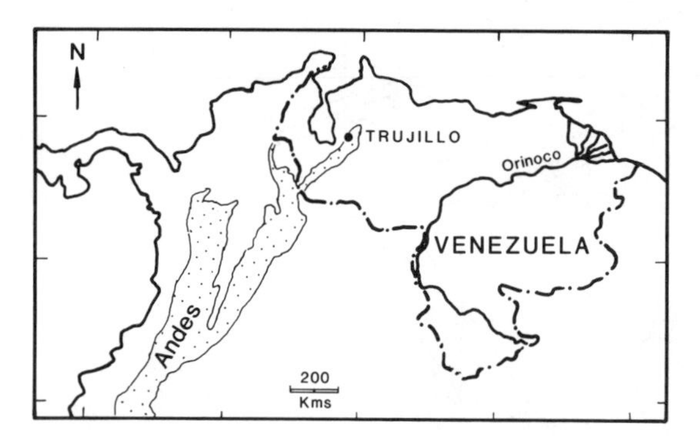

Fig. 8.46 Location of the Torococo–Cuicas area, Venezuela.

An initial reconnaissance programme by BCT in 1976 revealed that the most suitable potential site was Los Cedros Hill near the town of Torococo (Figure 8.47). This hill lies on the eastern margin of the Llanos de Monay, flat-lying plains which are approximately 38 km by road N of Trujillo, the capital of Trujillo State. The area outlined for detailed study (Figure 8.48) is approximately 1800 m E/W and 2000 m N/S and it lies immediately N of the village of San Pedro on a NNE striking ridge which rises to 660 m. The adjacent plains lie between 260 m and 300 m above sea-level and the mountainous areas to the E and S of Los Cedros rise to an elevation of 900 m. A site suitable for the cement plant was also identified at Silos de Monay, 2.5 km to the NW.

The reconnaissance survey on Los Cedros was undertaken using compass and tape traverses and allowed the elucidation of the local stratigraphy and structure, together with a programme of outcrop sampling. This work revealed that, within a 280 m thick stratigraphic section, there were potential reserves of cement quality limestone and shale which exceeded 35 million tonnes. High grade limestones were also located containing between 58% and 99% $CaCO_3$ and generally less than 1% $MgCO_3$. Significant amounts of potentially deleterious elements (P, S and C) were detected in the limestone and shale but it was thought that blending with higher grade, slightly siliceous, limestones would dilute these sufficiently to produce an acceptable plant feed. A drilling programme was thus proposed to further define the local stratigraphy, structure and available reserves of suitable limestone. This study would also further examine the quality of each stratigraphic unit in the area and the associated contaminants. The final phase of the exercise would be the design of a quarry suitable for the exploitation of this reserve.

8.9.2 Regional geology

The Torococo–Cuicas area contains rocks which range from Lower Cretaceous to Recent in age. Cretaceous rocks are dominant in the upland areas to the E whereas

Table 8.15 Reserve estimates for quarry design options

Design	Tonnages × 10^6			
	Limestone	Mudstone	Top rock	Overburden
A	31.0	3.1	1.0	0.4
B	33.9	4.9	1.1	0.5
C1	17.5	2.7	0.7	0.3
C2	3.7	0.3	0.2	—
C	21.2	2.9	0.9	0.4

Fig. 8.47 Photograph of Los Cedros hill looking east. The site of the potential high grade limestone quarry is on the left of the photograph with the main quarry in the central section.

Tertiary and Quaternary rocks underlie the plains to the W (Figure 8.48). The older rocks have been heavily disrupted by a series of NNW and NNE striking faults, the latter being strike faults in that they approximately follow the trend of the local folding. The most important formation from the point of view of cement production is the La Luna Formation (Table 8.16) which is Upper Cretaceous in age and which consists of thinly bedded laminated carbonaceous limestones with interbedded non-calcareous and calcareous shales. The underlying Maraca Formation contains massive limestones and sandstones, the former being a source of high quality feed whose only significant contaminant is silica (usually 4–13%).

8.9.3 Detailed exploration programme

Fieldwork on Los Cedros commenced early in 1977 and was completed in January 1978. Detailed geological mapping was accomplished by cutting 16 E–W traverse lines at approximately 200 m spacings and marking all lithological contacts with pegs which were later surveyed. This allowed the production of a geological map as illustrated in Figure 8.49. Table 8.16 represents a summary of the stratigraphic information gained from this exercise. Ten units are recognized, with Unit 3 marking the top of the La Luna Formation. This unit consists of non-calcareous laminated black shale with interbeds of calcareous shale and limestone up to 10 m

Table 8.16 Stratigraphic summary of the Los Cedros Site

Unit	Lithology	Thickness (m)	Formation
1	Sandstone, red/orange-brown	>30	
2	Variegated shale, black to red/brown	25–40	
3	Black shale with minor limestone beds	60–80	La Luna
4	Interbedded limestone, argillaceous limestone and shale	60–70	
5	Massive limestone and sandy limestone	30	Maraca
6	Sandstone	30–33	
7	Massive limestone, sandy at base	10–18	
8	Sandstone	8–10	
9	Massive limestone, with sandstone interbeds (3 m thick)	35–45	
10	Sandstone, with limestone beds	24	

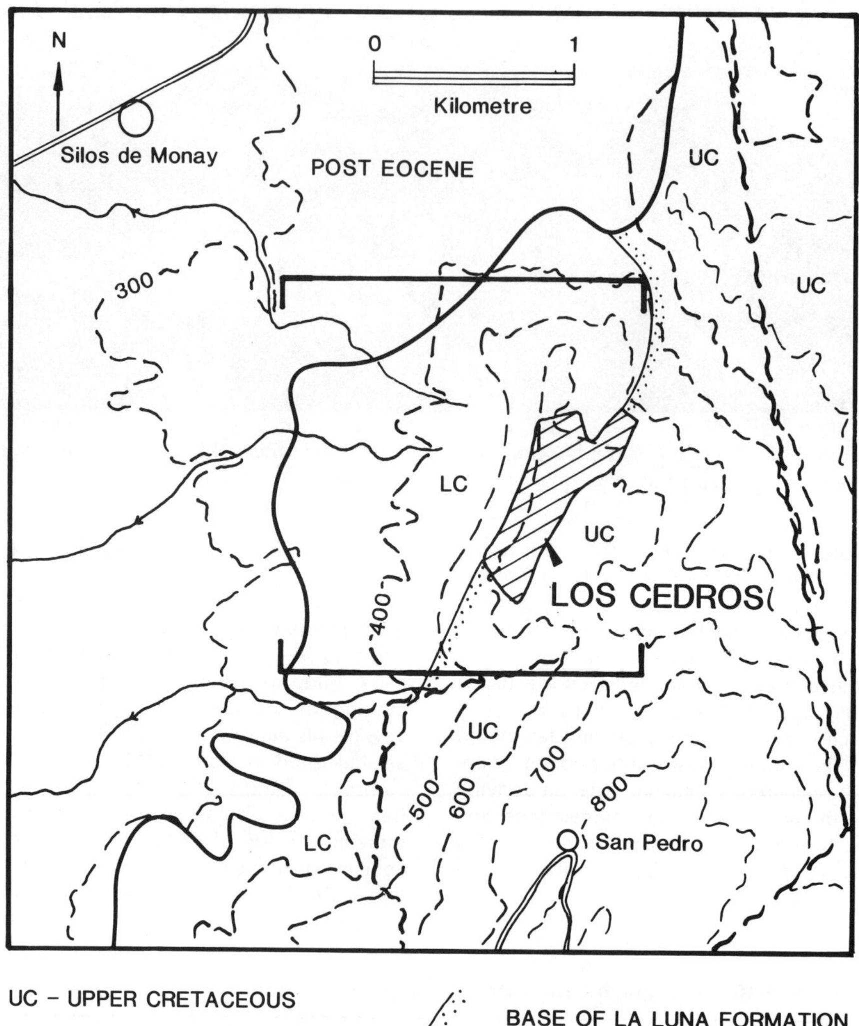

Fig. 8.48 Location of the study area and the proposed Los Cedros quarry.

Table 8.17 Weighted average analyses for potential sources of limestone and shale based on core sampling

Unit	SiO_2	Al_2O_3	Fe_2O_3	$CaCO_3$	$MgCO_3$	K_2O	Na_2O	P_2O_5	S	C
3a*	49.2	12.1	3.5	25.6	1.6	1.14	0.11	0.28	1.42	2.3
3b*	51.4	12.5	3.2	24.8	1.2	1.15	0.11	0.31	0.23	1.4
4a	11.8	3.8	1.2	78.7	1.1	0.50	0.05	0.59	0.51	1.1
7	5.8	1.4	1.2	88.6	0.6	0.32	0.03	—	—	—
9a	9.9	1.4	0.9	85.6	0.6	0.28	0.04	—	—	—
9c	9.9	0.6	1.0	85.7	0.9	0.11	0.02	—	—	—
9e	3.2	0.4	0.8	92.2	1.4	0.06	0.04	—	—	—
9g	3.7	0.6	1.2	88.1	4.5	0.11	0.11	—	—	—

*3a, Unweathered; 3b, leached.

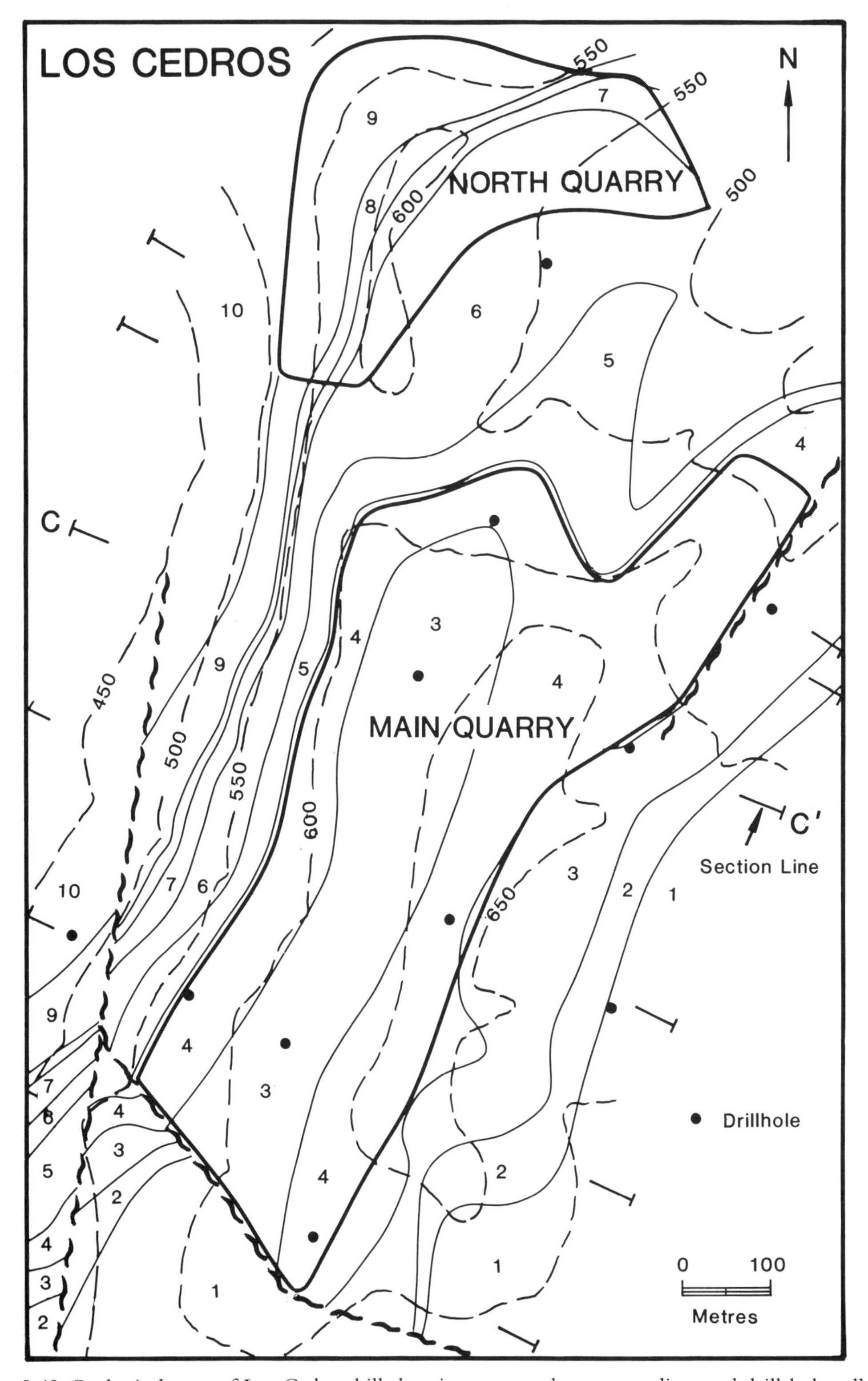

Fig. 8.49 Geological map of Los Cedros hill showing proposed quarry outlines and drill-hole collars.

Fig. 8.50 Drilling of limestones to determine their suitability as kiln feed.

thick throughout the sequence. Although disseminated and nodular pyrite is present in 'fresh' rock, oxidation and leaching have substantially reduced sulphur concentrations to depths of 15–20 m below surface. Carbon has also been selectively removed in the weathered zone.

Unit 4 is subdivided into 4a and 4b, the former is a limestone containing variable amounts of pyrite, carbonaceous material and thin phosphatic bone beds, whereas the latter is a 3 m thick black carbonaceous shale at the base of the limestones.

Additional limestone resources are available in Unit 7 and in Unit 9. The limestones in the latter are separated by three sandstone interbeds and are numbered 9a, 9c, 9e and 9g.

The surface geological mapping was augmented by information gained from the drilling programme in which 19 holes were completed at 12 sites (Figures 8.49, 8.50). Six of these were inclined and the rest were vertical and together they provided 1906 m of core. Rotary diamond drilling, using conventional and wireline methods, was undertaken, the latter using air-flush techniques because of the paucity of water supplies on the hill. Initially the holes were drilled NX size, but in some cases a reduction was necessary to AX where ground conditions were poor. Where recoveries fell beneath the required 90%, the hole was redrilled. The core produced was then cut longitudinally with a diamond saw so that one-half could be used for bulk testing and the other split, on the basis of lithology, into samples each less than 3 m long in preparation for chemical analysis. The core was logged on site to allow the production of detailed cross-sections (a typical example is presented in Figure 8.51) showing both lithological and structural information. Sampling and analysis were, however, done in the UK by BCT's Geology and Research Divisions respectively.

The drilling confirmed the interpretation of the structure gained from the geological mapping. A symmetrical synclinal fold (limbs 40–45°) occupies the western flanks of Los Cedros whereas an anticlinal axis follows the crest of the ridge in a north-northeasterly direction. The eastern limb of this latter structure is, however, subvertical. Both folds plunge to the SSW at between 1° and 5°. The structure is further complicated by dip and strike faulting.

Table 8.17 summarizes the results of the chemical analyses of the most important units. Unit 3 is very variable in quality, a fact hidden by the averaging of the analyses. The high S and C is, however, evident as is the reduction in these contaminants in the leached portion of the unit. This leached material is thus suitable as a secondary raw material whereas the unweathered material will have to be discarded as waste.

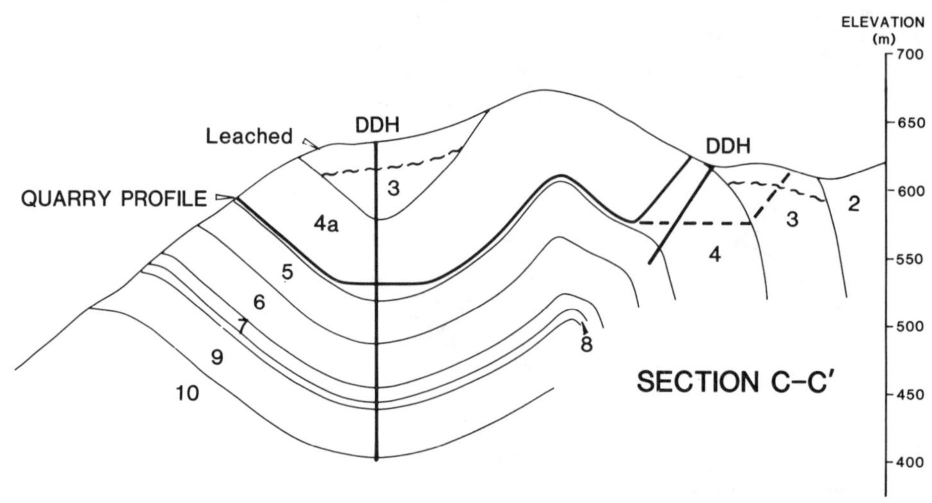

Fig. 8.51 Cross-section along line C-C′ (Fig. 8.49) showing quarry profile.

Unit 4b is not suitable as feed and will not be mined. The upper portion, Unit 4a is, however, marginally above kiln feed grade but is also heterogeneous vertically and laterally due to variable amounts of interbedded argillaceous limestone and shale. The high phosphate, especially in the upper 30 m of the unit, is also a concern. Multiface quarrying and blending should ensure that the average kiln feed will be close to that shown in Table 8.17. Units 7 and 9 contain high grade limestones with low levels of contaminants other than silica. The high levels of S and C in both Units 3 and 4a will almost certainly require their dilution by the addition of limited amounts of this high grade material.

In order to investigate how best to blend the different materials, bulk samples (40–140 kg) of the respective raw materials were produced from selected intervals from selected drill-holes. The results are summarized in Table 8.18. Two possible scenarios are presented in an attempt to obtain a feed which approximates that of an ideal mix (see below). Mix 1 combines 4a, 7+9 and 3 in the ratio 59.37%, 30% and 10.63% respectively, while Mix 2 uses only Unit 4a (95.25%) and Unit 3 (4.75%).

Table 8.19 shows the ideal composition of the raw feed (kiln feed) for Ordinary Portland Cement but the raw mix may vary between set limits without seriously affecting the end product. $CaCO_3$ (calculated from CaO × 1.7848) may range between 76% and 77%, the MgO content should be less than 2.6% whereas the SO_3 content (total sulphur ÷ 0.4) should be less than 2.5%. Over and above the requirements listed in Table 8.19, the suitability of the raw mix can be assessed using three ratios, viz:

The silica ratio (SR)

$$SiO_2/(Al_2O_3 + Fe_2O_3)$$

Preferred range 2.5–3.0

The alumina ratio (AR)

$$Al_2O_3/Fe_2O_3$$

Preferred range 2.0–2.5

The lime saturation factor (LSF)

$$LSF = CaO \times 100/[2.8(SiO_2) + 1.2(Al_2O_3) + 0.65(Fe_2O_3)]$$

Typical range 94–98%

Table 8.18 Estimated kiln feeds produced by blending

				Calculated kiln feed	
	Unit 4a (%)	Units 7 + 9 (%)	Unit 3 (%)	59.37% 4a 30.00% 7/9 10.63% 3 (%)	95.25% 4a 4.75% 3 (%)
SiO_2	12.20	6.00	47.20	14.06	13.87
Al_2O_3	3.90	0.80	13.30	3.97	4.35
Fe_2O_3	1.20	0.80	4.10	1.39	1.34
Mn_2O_3	0.02	0.06	0.05	0.04	0.02
P_2O_5	0.61	0.04	0.29	0.40	0.59
TiO_2	0.10	0.01	0.37	0.10	0.11
CaO	43.60	49.70	14.40	42.33	42.21
MgO	0.50	0.60	0.60	0.54	0.50
SO_3	0.26	0.01	0.18	0.18	0.26
S	0.47	0.11	0.84	0.40	0.49
LOI	35.30	41.70	15.30	35.09	34.35
K_2O	0.51	0.16	1.20	0.48	0.54
Na_2O	0.04	0.02	0.11	0.04	0.04
C	1.20	—	2.10	0.94	1.24
$CaCO_3$	77.80	88.70	25.70	75.60	75.30
S/(A + F)	2.39	3.75	2.71	2.62	2.44
A/F	3.25	1.00	3.24	2.86	3.25
LSF	—	—	—	0.94	0.94

Table 8.19 Composition of kiln feed for the production of Ordinary Portland Cement

SiO_2	Al_2O_3	Fe_2O_3	CaO	MgO	Na_2O	K_2O	Total S as SO_3	P_2O_5	SR*	AR*	LSF*
13.78	3.83	1.66	42.60	0.67	0.13	0.40	1.83	0.10	2.51	2.13	96.2

*Defined in text.

The raw mix is usually designed to a specific LSF which is related to the combinality temperatures of the raw materials within the kiln and the fineness of the raw feed grind. The silica ratio is important for silica imparts a good hardening quality to the cement. High silica levels in a limestone require a harder burn during the production of the clinker but this produces a better cement (at greater cost).

Contaminants P, total SO_3 and C should be below critical levels of 0.6%, 2.5% and 1.0% respectively in the raw feed. British Standard BS12 (1978) requires that the levels of minor constituents in Ordinary Portland Cement (i.e. not the raw feed) be as follows:

SO_3	below	2.5%
MgO	below	4.0%
P_2O_5	below	0.9%
Alkalis	below	0.6%
Carbon	below	1.5%

Although the influence that various elements have on the cement making process and on the quality of the cement is a complex matter, some general comments are worthwhile in the context of this case history. For example, high phosphorus levels will reduce the strength of the cement; high sulphur affects the quality of the cement and causes a build up on the lining of the kiln, reducing its efficiency; high carbon could lead to explosions during the production of the cement; high alkalis should be avoided as they result in a reaction between the cement and silica aggregates known as 'cement cancer'.

The two mixes of Los Cedros material proposed in Table 8.18 broadly conform to these qualitative parameters but, in order to optimize the silica ratio and the alumina ratio, the preferred mix is that which involves the addition of high grade limestone. As can be seen from Table 8.18, the incorporation of high grade limestone in the mix increases SR to 2.6 and decreases the level of SO_3, P_2O_5 and carbon in the raw mix to 1.2, 0.4 and 0.9 respectively. At these levels, the P_2O_5 and carbon would not be expected to adversely affect kiln operation or cement quality but possible periodic increases in SO_3 input will necessitate bleeding of a substantial proportion of the kiln gases. The kiln feed was designed to a LSF of 94% based upon the moderately easy combinality characteristics of the raw materials.

8.9.4 Reserve estimation

On each of the geological sections drawn across the deposit (e.g. Figure 8.51 – Section C–C′), the outline of a possible quarry was inserted using the following guidelines.

1. Ultimate batters of 45° in Unit 3, 60° in Unit 4a.
2. Minimization of the amount of Unit 3 mined as secondary feed or waste. (Note that additional limestone could be won, as indicated on the section, at the expense of the removal of more Unit 3 waste.)
3. Stripping ratios for cement limestone quarries rarely exceed 1 : 1.
4. Maximum working depth 530 m ASL.
5. Minimum quarry floor width of 70 m (in syncline).
6. Base of extraction at 4a/4b contact.
7. Average density of Unit 4a = 2.65 t/m^3.
   Average density of Unit 3 = 2.53 t/m^3.
   Average density of Unit 7+9 = 2.70 t/m^3.

Tonnages related to each section were then determined by planimetry of cross-section areas of the different rock types and by using the relevant tonnage factors. These values were then combined to produce global tonnages for the quarry area outlined on Figure 8.49. Reserves of approximately 40 m tonnes of Unit 4a were determined in this manner, together with 4.3 m tonnes of leached Unit 3 and 2.8 m tonnes of waste (fresh Unit 3). These reserves were checked by the grid superimposition method (section 3.10.1). The grades of Unit 4a in each borehole were determined by length-weighted assays and finally these were weighted by volume to produce the global grade values. Because of the wide spacing of borehole intersections, local block estimation techniques are not readily applicable.

The reserves of high grade limestone (Units 7 and 9), in a second quarry to the N, were calculated in the same way. This revealed that 1.64 million cubic metres of waste, consisting of Unit 6 and 8 material, sandstone interbeds in Unit 9 and 3–4 m of sandy limestone at the base of Unit 7, would have to be stripped to access 8 m tonnes of high grade limestone.

As the anticipated kiln output is 1800 TPD (tonnes per day), the amount of dry raw feed required from the quarry to meet this demand is 280 TPD, as 1.6 t of dry raw feed produces 1 t of clinker. As the moisture contents of Units 4a, 3 and 7+9 are estimated as 3%, 6% and 1% respectively, the relative tonnages of the various components of the two proposed mixes can be calculated. The results are presented below:

Mix	4a	7+9	3	Wet tonnes
1	1762	873	326	2961
2	2828	—	146	2974

On the basis of these calculations, the reserves of both the La Luna Limestone and Shale (Units 4a and 3) are sufficient for more than 50 years whereas those for the high grade limestones are only adequate for 30 years, if a mix requiring 30% of this material is used (Mix 1). The shortfall will have to be obtained from elsewhere.

8.9.5 Operational details

The quarry outlined on the basis of the geological information was adequate for the purposes of reserve estimation. However, the final quarry design uses 15 m high benches in the main quarry and 20 m benches in the northern high grade limestone quarry. Haul roads, 15 m wide, were also included in the designs with 10% gradients. It was proposed that the main quarry should use a multi-bench quarrying system with faces advancing to the SSW to allow blending of material from the vertically heterogeneous Units 3 and 4a. This will help ensure that the chemical composition of the feed remains within acceptable limits. Quarrying will be achieved by a combination of blast-hole drilling, blasting and loading with shovels, ripping and bulldozing. As the La Luna formation is generally friable, as 8 × 8 m grid of blast-holes is adequate as this will allow sufficient shaking of the rock to facilitate direct face loading. These holes are to be drilled by crawler-mounted airtrack down-the-hole drills fitted with dry dust collectors of the Ilmeg type. The cuttings are sucked into a pre-cyclone which removes the coarse chips into a plastic bag. The fine particles are then collected into another bag by the main cyclone and the superfine particles are trapped on filters from which they are removed by a vibrator. It is proposed that the cuttings be collected to form 4–5 m composites and that the analyses be used for grade control purposes.

The plan envisages that the northern quarry should work the limestone and sandstone horizons separately to obtain a clean split and reduce the amount of silica contamination. The benches will be cut against the dip of the formations (i.e. to the W), with the main quarry face advancing in a southerly direction.

BCT advise that a 3-day working stockpile of ex-quarry product should be maintained to feed an 800 t/hour crusher. A storage stockpile of approximately 35 000 t should also be established for kiln feed.

The quarry was successfully brought into operation in 1979/80 and is closely following the recommendations made by BCT. Figure 8.52 shows initial haul road development and quarrying activity in September 1982. The feed mix does, however, contain somewhat less than 30% of high grade limestone as dilutant.

Fig. 8.52 Los Cedros quarry at an early stage of development (1982). Photograph taken looking west.

8.10 CASE HISTORY – NAVAN Zn-Pb MINE, EIRE (TARA MINES LTD)

8.10.1 Introduction

The Navan orebody, located 1 km W of the town of Navan in County Meath, Eire, was discovered in 1970 by Tara Exploration and Development Ltd. Underground development began in 1973 and production commenced in 1977. The mine is now the largest zinc mine in Europe (100% owned by Outokumpu Oy) and is currently (1989) scheduled to produce 2.6 million tonnes of ore per annum grading close to 8.3% Zn and 2.00% Pb. Although initial reserves of 69.9 million tonnes at 10.09% Zn and 2.63% Pb were delineated, only 60.9 million tonnes are available for mining because of problems concerning the ownership of the mineral rights to the N of the Blackwater River. To date, a total of 25 million tonnes has been mined, requiring 200 km of underground development.

Mining methods employed depend on the thickness of the ore and have ranged from multiple-lift (15 m) cut-and-fill methods, to open-stoping in thick ore, to room and pillar where the ore is thinner. Blast-hole open-stoping is currently employed. A typical stope layout is shown in Figure 8.53. Hydraulic backfilling of stopes with cemented tailings allows the recovery of the intervening pillars. Approximately 60 stopes and pillars are extracted per annum and 20–25% of the ore is derived from development.

8.10.2 Geology

The zinc-lead mineralization at Navan is hosted by rocks of the Navan Group and, in particular, the central section of this Group which is referred to as the 'Pale Beds'. The Navan Group rocks are Lower Carboniferous in age (Courceyan to Chadian) and consist of a sequence of limestones, sandstones and dolomites which overlie variable thicknesses of red beds (sandstones and conglomerates) which in turn rest unconformably on a Lower Palaeozoic basement. Figure 8.54 contains a diagrammatic representation of the mine stratigraphy but a more detailed account of the mine geology can be obtained from Libby *et al.* (1985), Andrew and Ashton (1985) and Ashton *et al.* (1986).

The Pale Beds represent a sequence of shallow-water calcarenites overlying a basal micrite unit, immediately above the Muddy Limestone, and terminate at the first significant silty shale bed above the Upper Sandstone Marker (USM). This shale represents the base of the Shaley Pales, a deeper water offshore equivalent of the Pale Beds. The Pale Beds are approximately 200 m thick and include a series of marker horizons (Figure 8.54; base – LDM, LSM, NOD, UDM, USM – top) which vary in thickness from 2 m to 8 m. The intervals between the lowest four markers are potentially mineralized and are referred to as Lenses (1 Lens at the top; 5 Lens at the base, beneath the LDM). The relationship of the mineralization to the stratigraphic section is illustrated in Figures 8.55 and 8.56. The ABC Group above the Shaley Pales consists of argillaceous bioclastic limestones and pale crinoidal micrites, representing the local facies of the Waulsortian Reef. An erosional surface separates these rocks from the overlying Upper Dark Limestone (UDL). A debris flow boulder-conglomerate is locally present above this unconformity.

The mineralization (sphalerite, galena, ± marcasite ± pyrite, with minor carbonates and barytes as gangue) occurs as a series of vertically stacked stratiform to stratabound sulphide lenses which dip to the SW at 15–20°. The limits of these lenses are irregular and are truncated laterally by a series of ENE normal faults, thought to be related to reactivation of basements faults, and locally, by a Lower Carboniferous erosional surface (Figure 8.56). The 'A' and 'B' faults subdivide the ore-lenses into three zones (1, 2 and 3) and thus the lens number is usually prefixed by a zone number (e.g. 2–5: the lowest lens in Zone 2). Ninety-five per cent of the ore comes from the lowest 130 m of the Pale Beds and the thickest and highest grade ore is from the easterly updip portion of Zone 2 where a 80 m vertical section, from the base of the Pale Beds up to the 2–2 Lens, is mineable (Figure 8.56). Towards the SW, the main sulphide zone breaks up into isolated lenses separated by low-grade material (Figure 8.55). The 2–1 Lens is always an independently mineable unit, where it makes ore grade. The most extensive mineralization occurs in the 5 Lens and its lateral limits are still being defined to the SW. Most of the overlying lenses are confined to Zone 2.

The presence of both concordant and trangressive mineralization, as described by Andrew and Ashton (1985), can be best explained in terms of a long history of mineralization by hydrothermal fluids during diagenesis. This process was largely controlled by ENE fractures, by the permeability of the host rock and by the presence of pre-ore dolomite horizons (Anderson, 1990).

8.10.3 Exploration

The exploration of the Navan deposit commenced with BQ-size drilling on an approximately 30 m grid. However, this was subsequently increased so that currently an 80 m grid is used which is infilled by 40 m holes where necessary. Since 1980, this surface drilling has been augmented by AQ drilling from drives in the hangingwall of 5 Lens (Figure 8.57). Additional confidence in ore-zone outlines and grades was then gained by production drilling (AX or AQ) on 10 m section lines within stopes.

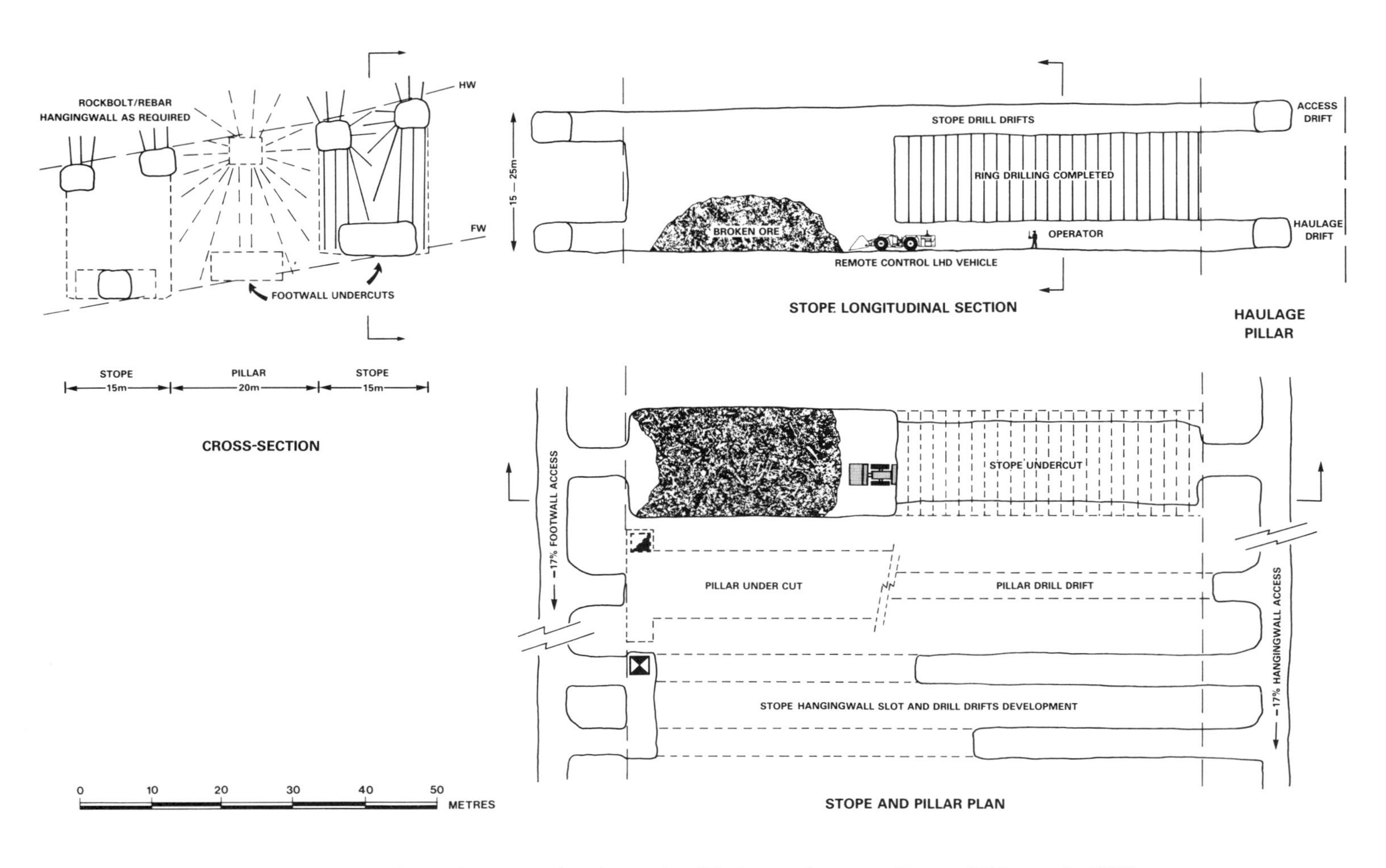

Fig. 8.53 Plan and sections showing a simplified stope layout at Navan (Libby *et al.*, 1985).

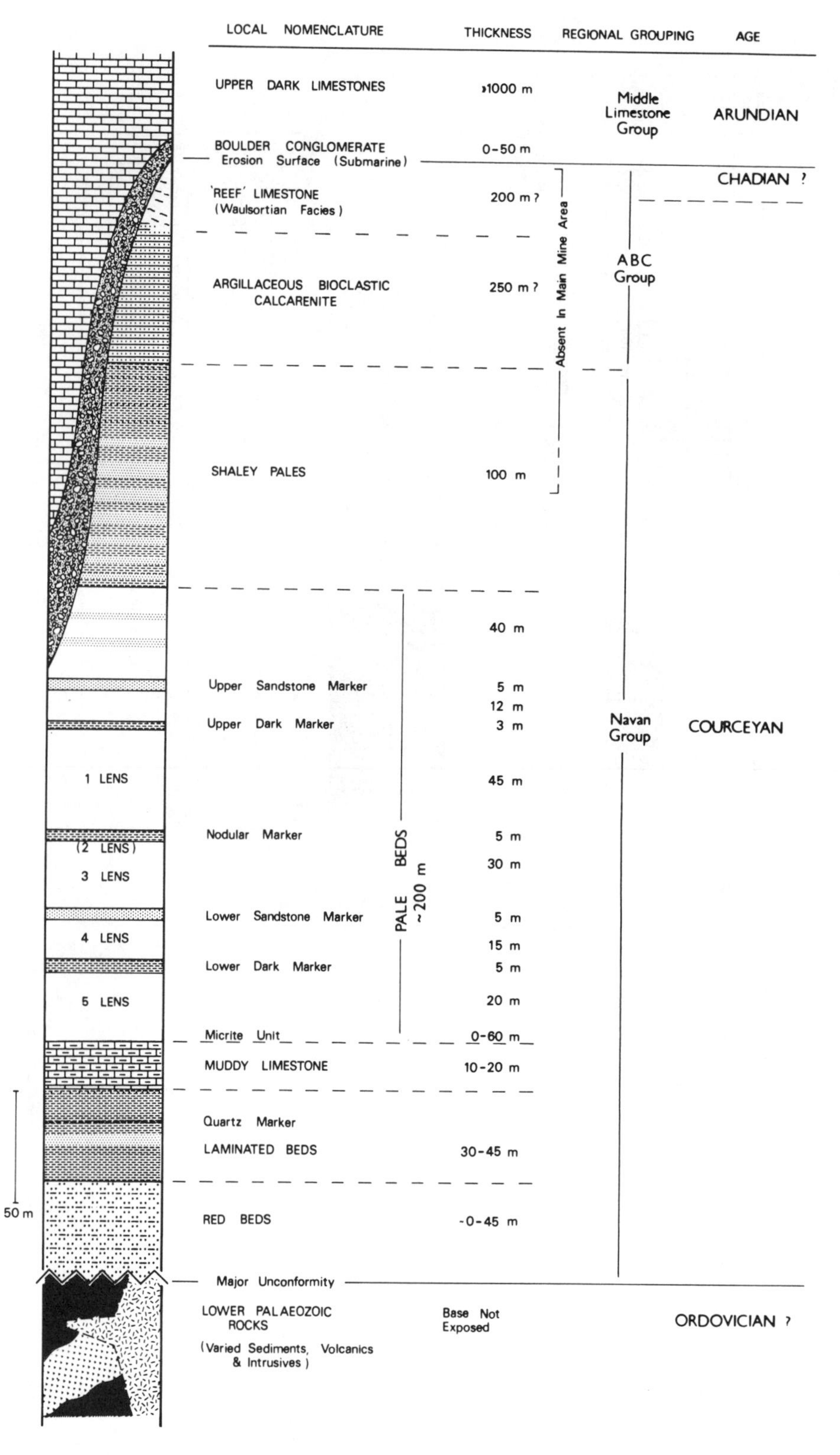

Fig. 8.54 Navan Mine stratigraphy showing the main marker horizons associated with the Pale Beds (Libby *et al.*, 1985).

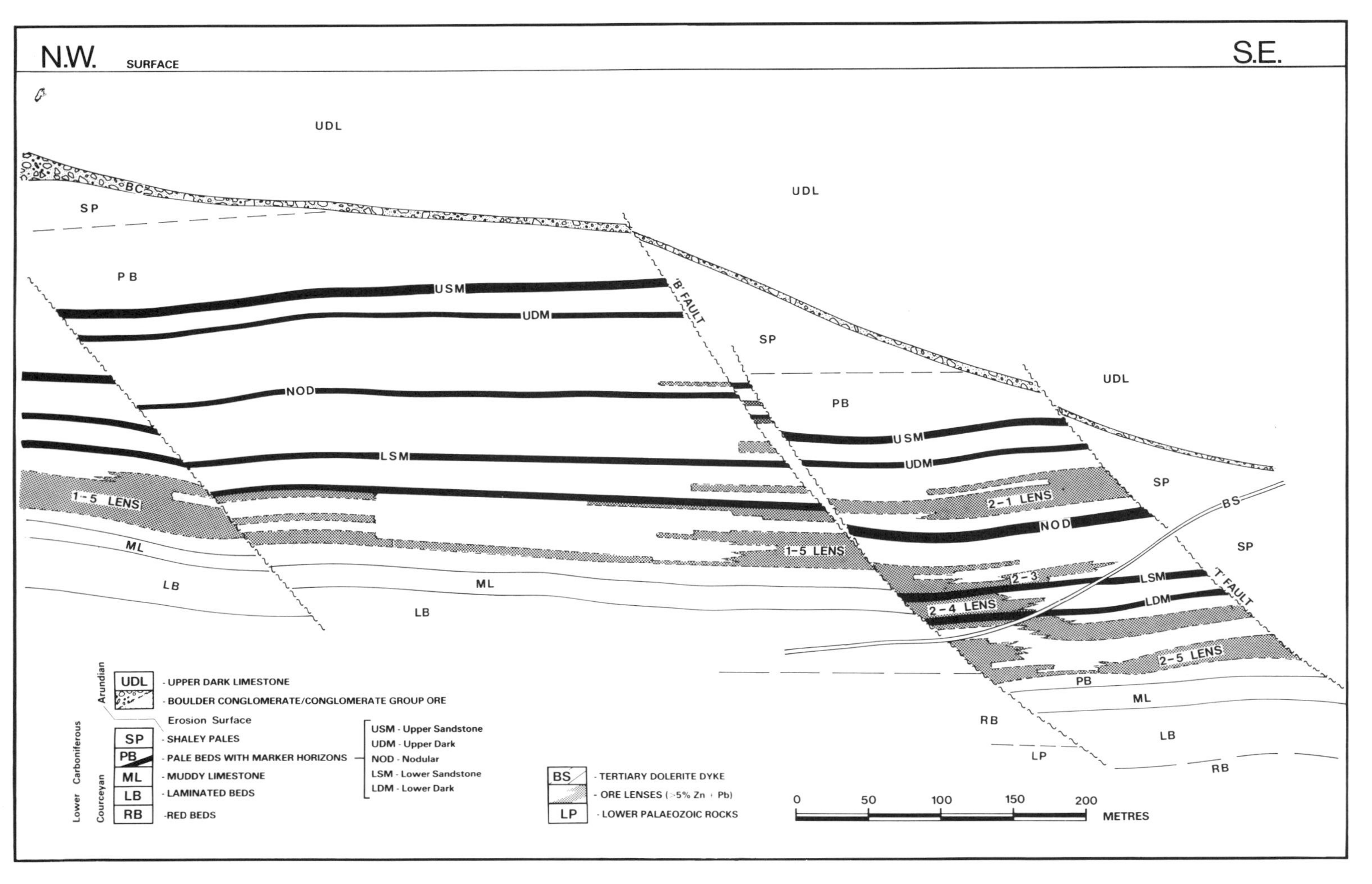

Fig. 8.55 Strike section showing the ore lenses in relation to the marker horizons plus the B and T faults. (Down-dip area.)

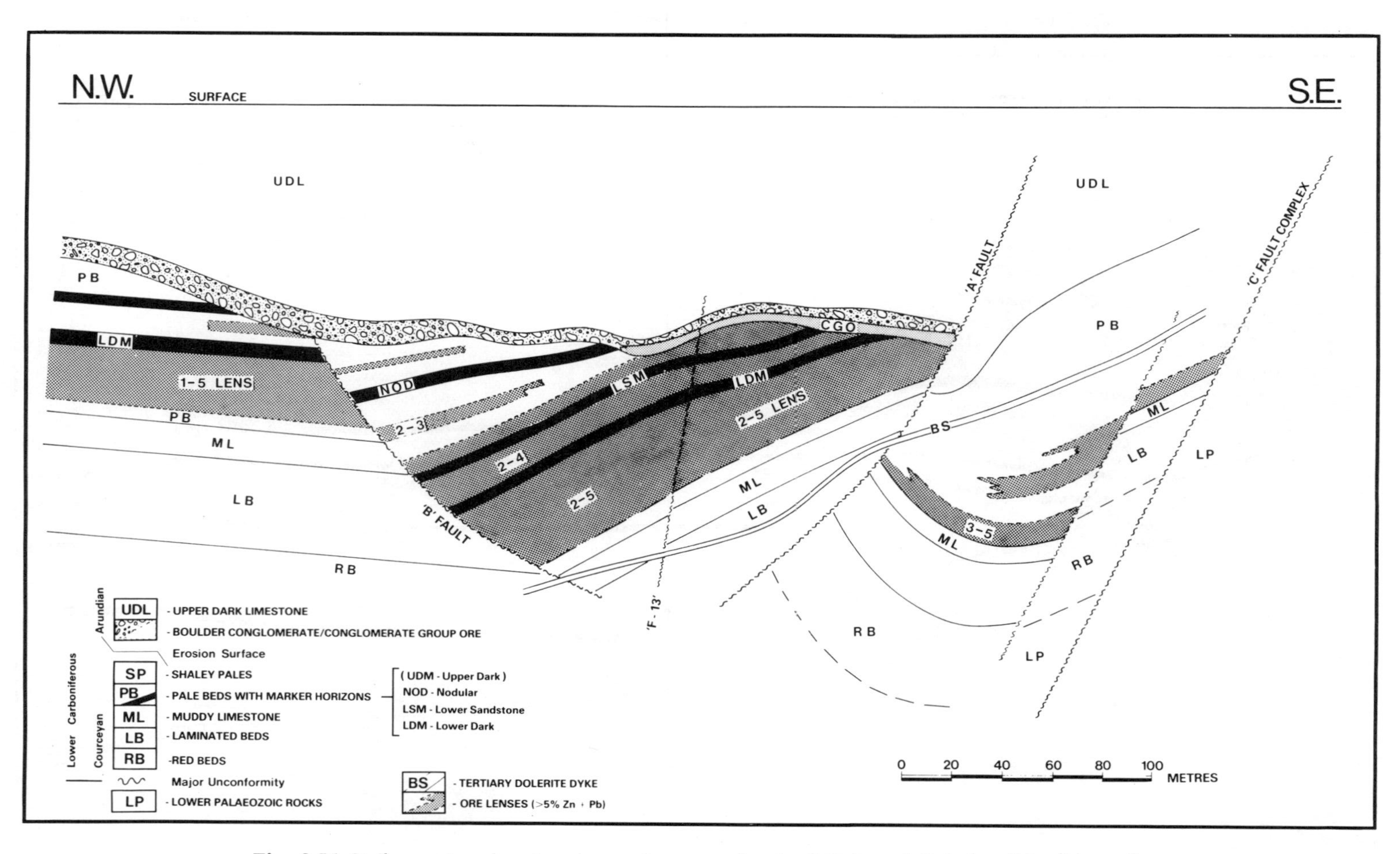

Fig. 8.56 Strike section showing the ore lenses cut by the B,F,A, and C faults. (Up-dip area.)

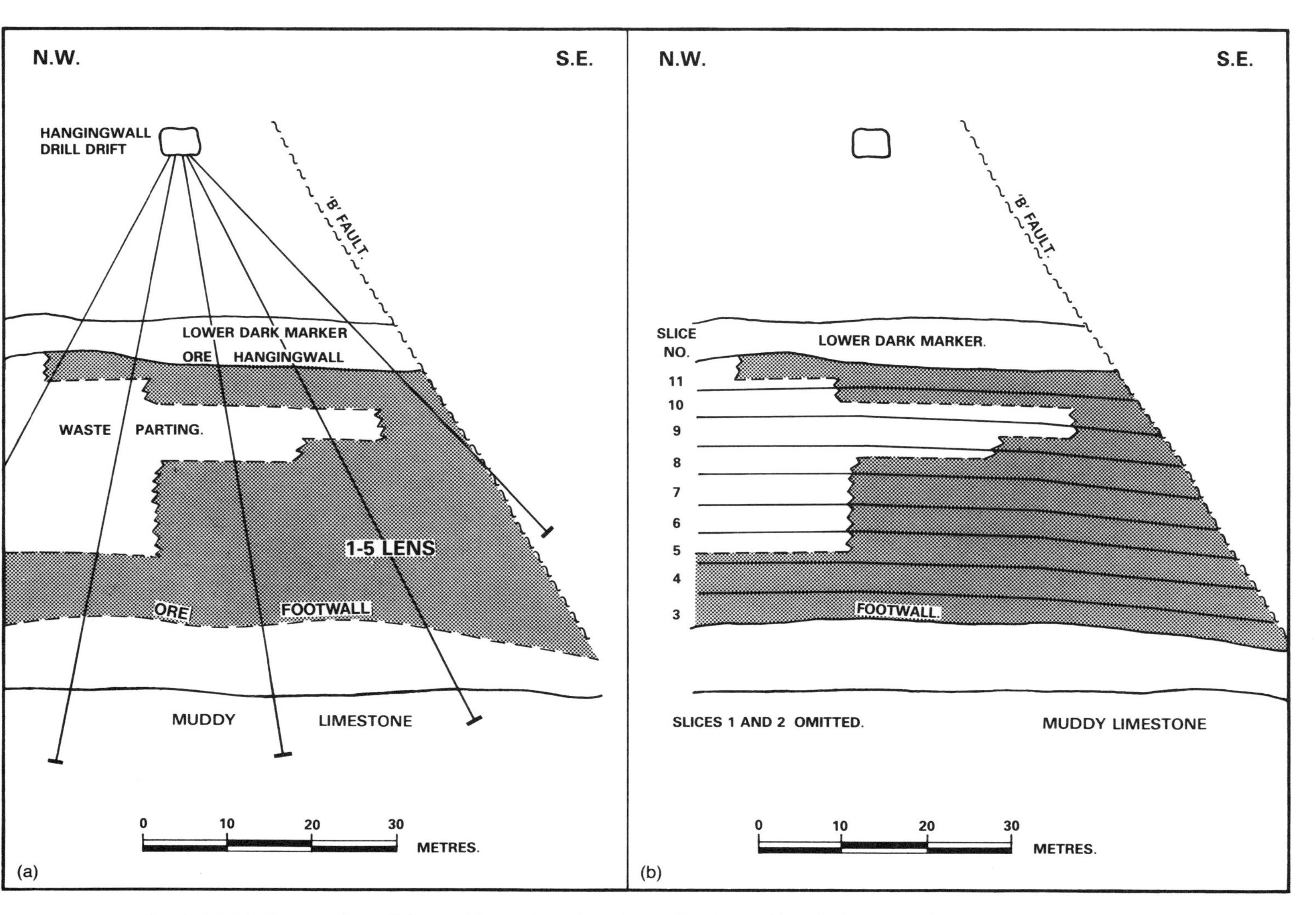

Fig. 8.57 (a) Exploration of the 1–5 Lens from hangingwall drives. (b) Sub-division of the 1–5 Lens slices for reserve estimation purposes (Libby *et al.* 1985).

The introduction of computer systems, and the use of accurate multishot surveys in drill-holes, has since led to a minor change in strategy in that production drilling tends to be used to augment the intersections gained by surface and development drilling and its density is very much dependent on the geologist's judgement. The final spacing of holes is thus usually in the range 10–15 m.

To date a total of 500 km of drilling has been completed in 7000 holes and this figure is being augmented at the rate of 500 holes per annum. This information is supplemented by underground geological mapping of all development drifts at a scale of 1 : 250. This geology is transferred on to level plans drawn on plastic film. Direct digitization of mapping sheets into a computer graphic data base is currently being implemented.

Drill cores are logged in detail to distinguish the various marker horizons and faults within the Pale Beds. A graphic logging system is used to simplify and standardize all hand-written logs. Drill sections are then produced by computer on which correlations between marker horizons are made.

Cores from surface BQ holes are split longitudinally using either a core splitter or a diamond saw, but cores from underground holes are sampled whole. Sampling is usually done over 1.5 m intervals and is controlled by the location of marker horizons and ore-lenses. These sampling intervals may be reduced to delineate clean ore/waste partings and may be increased to 3 m in more homogeneous areas. The samples are then routinely analysed for Zn, Pb and Fe (occasionally Ag also).

The information from each new hole is used to update a computerized drill-hole data base, which consists of the assay data, drill-hole coordinate and down-hole survey information, together with a series of three character codes representing different rock types, structural features, ore-lenses and hydrological data at specific depths, or over specific depth ranges. The assay data consist of sample numbers, depth ranges and the results of the above analyses. Where a drill-hole has made an incomplete intersection, chip sample assays are occasionally used to complete the intersection. All these data are validated before being used to calculate mine or stope reserves.

Experience has shown that lateral grade variations are gradual and that rapid vertical changes occur across the stratigraphy, necessitating its subdivision into slices.

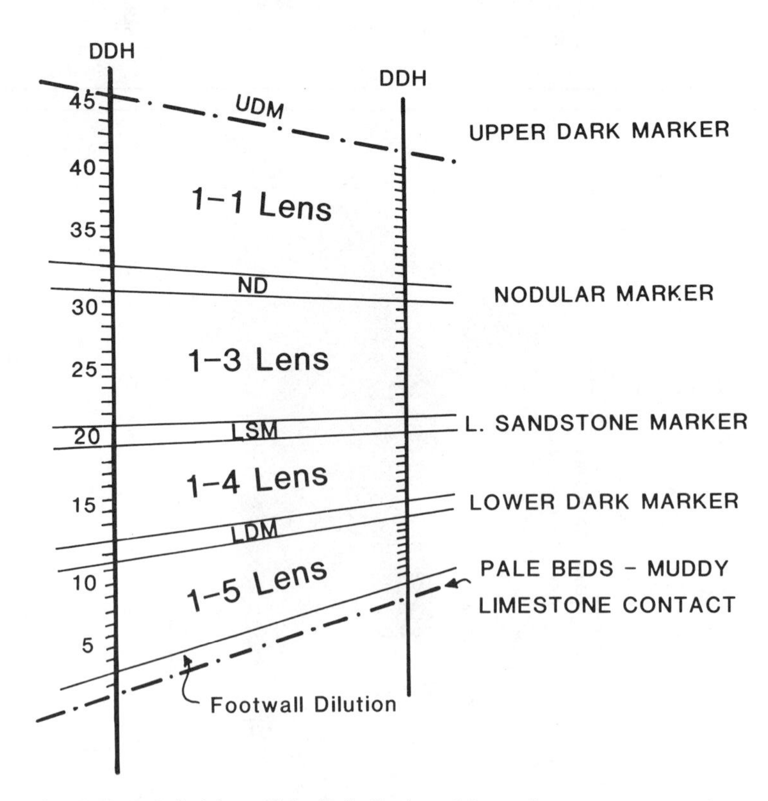

Fig. 8.58 Subdivision of the Pale Beds at Navan into ore-reserve slices.

Figure 8.58 shows how the stratigraphy of the Pale Beds is subdivided into up to 45 stratigraphic slices whose vertical thickness lies between 2 m and 4.5 m. The computer program produces a composite grade for each of the slices in each drill-hole.

Updated plans and sections including this drill-hole information are produced by the computer. Different types of information can be selected for use on these plans/sections by assigning graphic layer numbers to each set of data. Thus it is possible to superimpose drill-hole information, of various degrees of complexity, on plans of the mine development, together with the various elements of the geology as recorded underground or in previous drill-holes. The geologist then undertakes a manual geological interpretation of the orebody outline on sections which are then digitized for computer storage. Structure contours of fault planes, or the footwalls of mineralized zones or lithological units, can also be manually inserted on to computer-generated plans showing the location and elevation of intersection points. This additional information is then digitized. Proposed stope outlines can also be added to the geological plans and digitized stope outlines added to sections. It is the intention of the mine to produce 3D representations of these stopes for use in blast-hole design and layout.

8.10.4 Reserve evaluation

The reserve evaluation procedure used at Navan is based on a large data base containing the composite grades for each of the slices in each drill-hole. Densities are assigned by the computer to each of these slice composites on the basis of best-fit regression analysis of combined metal grade against density for a representative sample set.

Semi-variogram modelling of service variables derived from the slice composite information (using programs originally developed by A. G. Royle of Leeds University), allows the production of the slice semi-variogram parameters (Co, C and 'a'). The mine is subdivided into three areas for the purposes of semi-variogram modelling, two of which are in Zone 1 and the third in Zone 2. These parameters are cross-validated by point kriging and are produced for each slice in each area of the mine as significant differences exist between slices. These parameters are only revised on acquisition

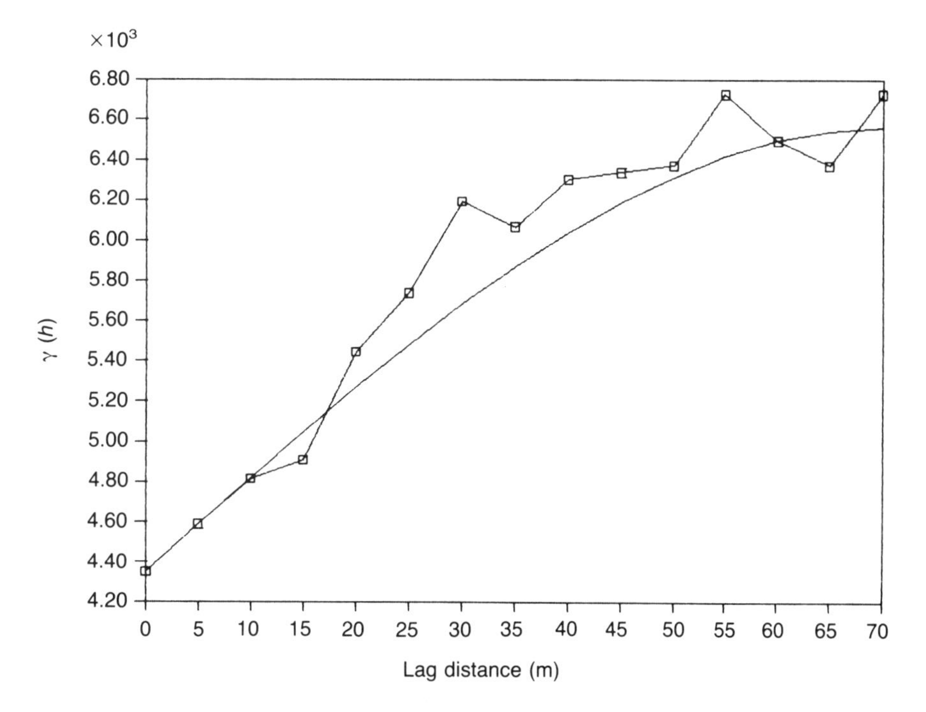

Fig. 8.59 Isotropic semi-variogram for the service variable M × SG × Zn, Slice 4, Lens 1–5, Navan. □, Experimental SV; ——, model SV.

of large amounts of additional data. The service variables involved include M × SG, M × SG × Zn, M × SG × Pb and M × SG × Fe, where M is slice vertical thickness and SG is specific gravity. Figure 8.59 shows a typical example of a semi-variogram for M × SG × Zn, to which a simple spherical model has been fitted with a range of 70 m, a nugget variance Co of 4350 and a regionalized variance C of 2212.

The ore-reserve blocks used in each slice during the local block kriging estimation procedure which follows, depend on the type of reserve that is being determined. Where long-term possible and probable reserve categories are required for annual reports etc., square blocks are used, which are either 20 × 20 m or 40 × 40 m and which are arranged parallel to the NW/SE mine section lines. However, when stopes or pillars are being evaluated, then the dimensions are more variable. Figure 8.60 shows part of a typical stope whose width may vary from 12.5 m up to 30 m and whose length may be up to 150 m. Because grades in such stopes may vary considerably along their lengths, the stope length is subdivided into 10 m intervals, thus forming blocks 10 × 12.5–30 m. This subdivision is done automatically by the computer and the block position can be verified using computer graphics. Each 10 m long block will contain a vertical stack of slices (each approximately 4 m high) and is numbered as part of a consecutive sequence along the length of the proposed stope. Because the walls of the planned stope may be inclined due to a fault, for example, the limiting coordinates for each slice are defined, thus enabling quite complex stope configurations to be considered. The tonnage of each slice in each block will

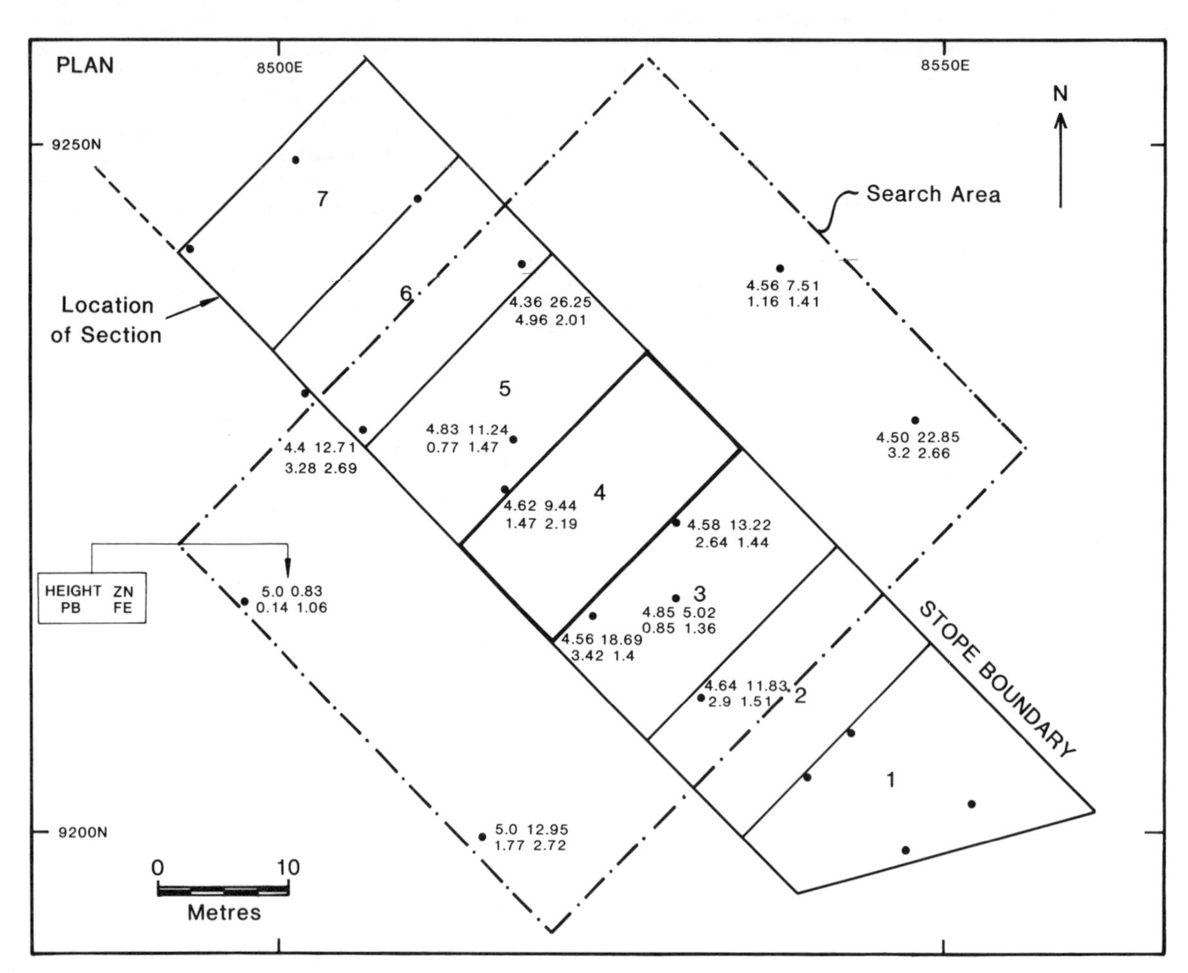

Fig. 8.60. Part of a stope at Navan in the 1–5 Lens showing the slice composite grades and thicknesses for slice 4, together with the kriging search area for block 4.

thus change as a result of this variation in width and also because of the variable vertical thickness of each slice between different pairs of marker horizons.

Once the blocks in each slice have been defined, the computer sets up a search area around each block centre, the radius of which is specified by the geologist. The size of the search area, and the minimum number of holes required to calculate a block, vary with the type/size of the block to be evaluated. They may be varied on a stope by stope basis at the geologist's discretion, and based on experience, but typically search areas would extend 10–15 m beyond the margins of each block and the minimum

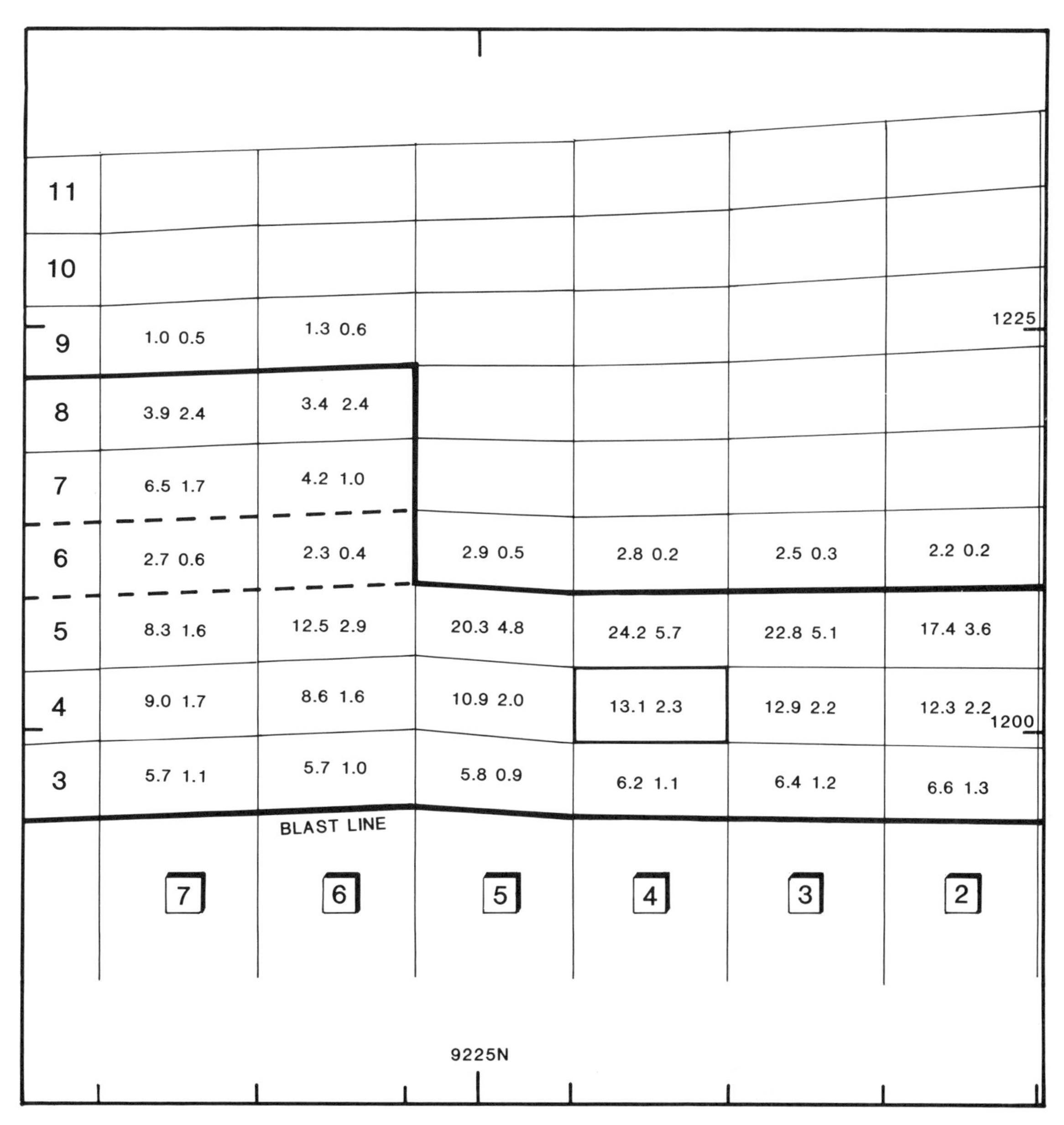

Fig. 8.61 Longitudinal block section for the south west wall of the 1–5 Lens stope shown in Fig. 8.60, showing kriged Pb and Zn grades and the proposed blast line.

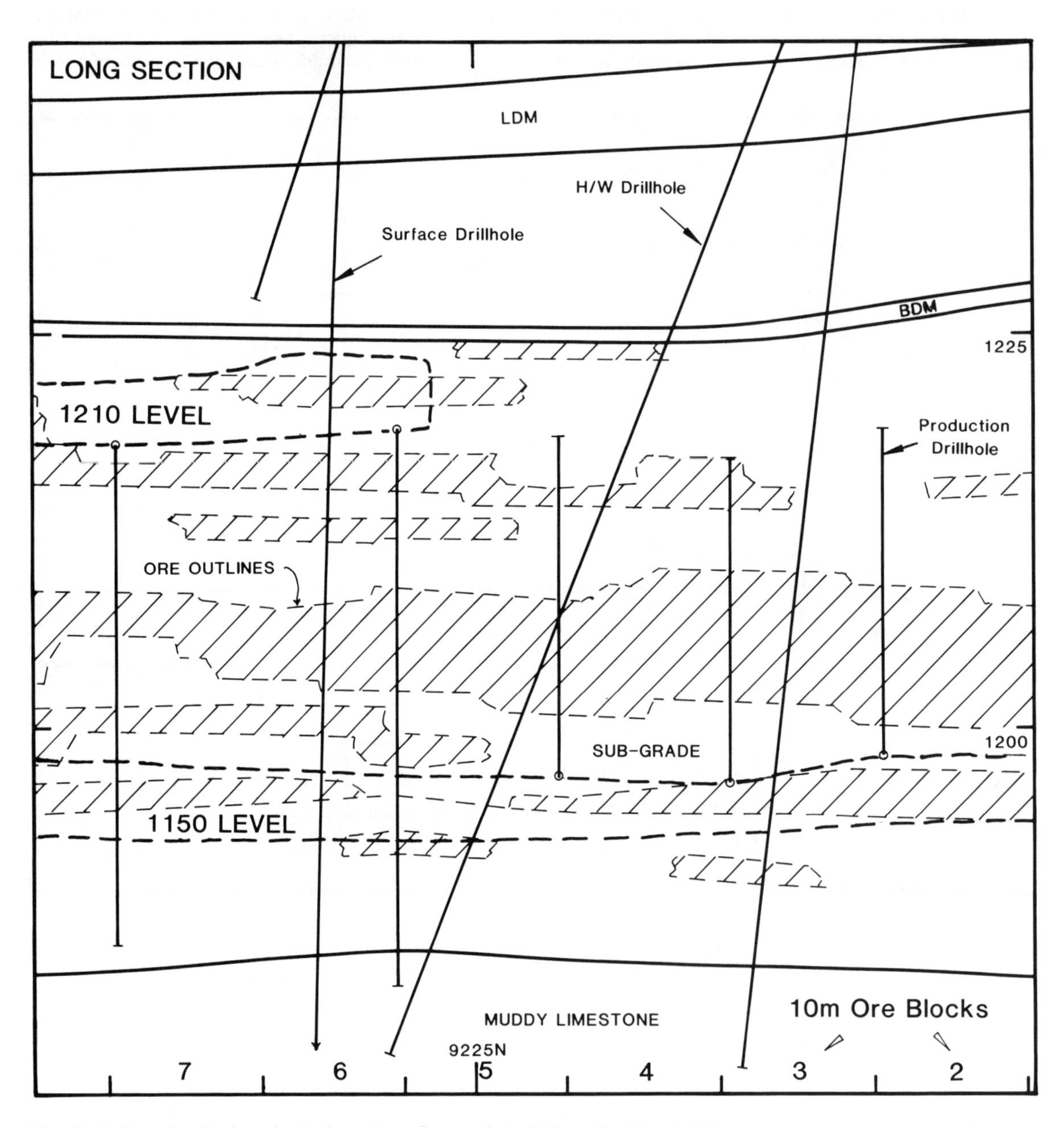

Fig. 8.62 Longitudinal geological section of part of the SW wall of the 1–5 Lens stope (Figs 8.56, 8.57) showing drill-hole traces and a geological interpretation.

number of holes that would be required for a block to be evaluated would be set at 4. In the case of slice 4 in block 4 (Figure 8.60), the search area outlined has captured 12 drill-holes. Also, if the geologist suspects that one hole is biased in any way it can be excluded from the calculations. For each block in each slice therefore, the program computes the kriged grades (Zn, Pb and Fe) and associated kriging variances and also the tonnage from the block area times the kriged slice thickness times specific gravity. This information is output as a printout and as sections (Figure 8.61). Figure 8.61 shows that the calculated zinc and lead grades for slice 4 block 4 are 13.1% and 2.3% respectively. At this stage, this slice data is compared with the geological interpretation by overlaying a plot of slice grades on a longitudinal section (Figure 8.62) on which a geological interpretation has been made and ore-zones outlined on the basis of a cut-off grade of 5% combined metals. Figure 8.63, based on a text figure in Ashton and Harte (1989), is a 3D representation of the information that is produced.

Blast-line delineation can now be undertaken in conjunction with the stope planners. This is an iterative procedure which allows the optimization of stope height and the location of the footwall and hangingwall blast-lines on the basis of current economic factors. In order to arrive at a local mining cut-off grade, the mine planning department investigates the options available in respect of mining method, considering in each case the cost of additional development, the mucking method to be used and the cost of haulage, production drilling and cement-fill. Stope design is based on many factors including mining practicality, the grade distribution, the degree of interfingering of ore and waste, the structural complexity, and the current metal prices. The hangingwall and footwall blast-lines may be constantly re-examined and manually modified in the light of these factors. This is the advantage of the slice modelling procedure as it allows this to be done rapidly without having to re-krige an ore-zone on the basis of a new cut-off grade. Once the blast-lines have been defined, overall pillar and stope tonnages,

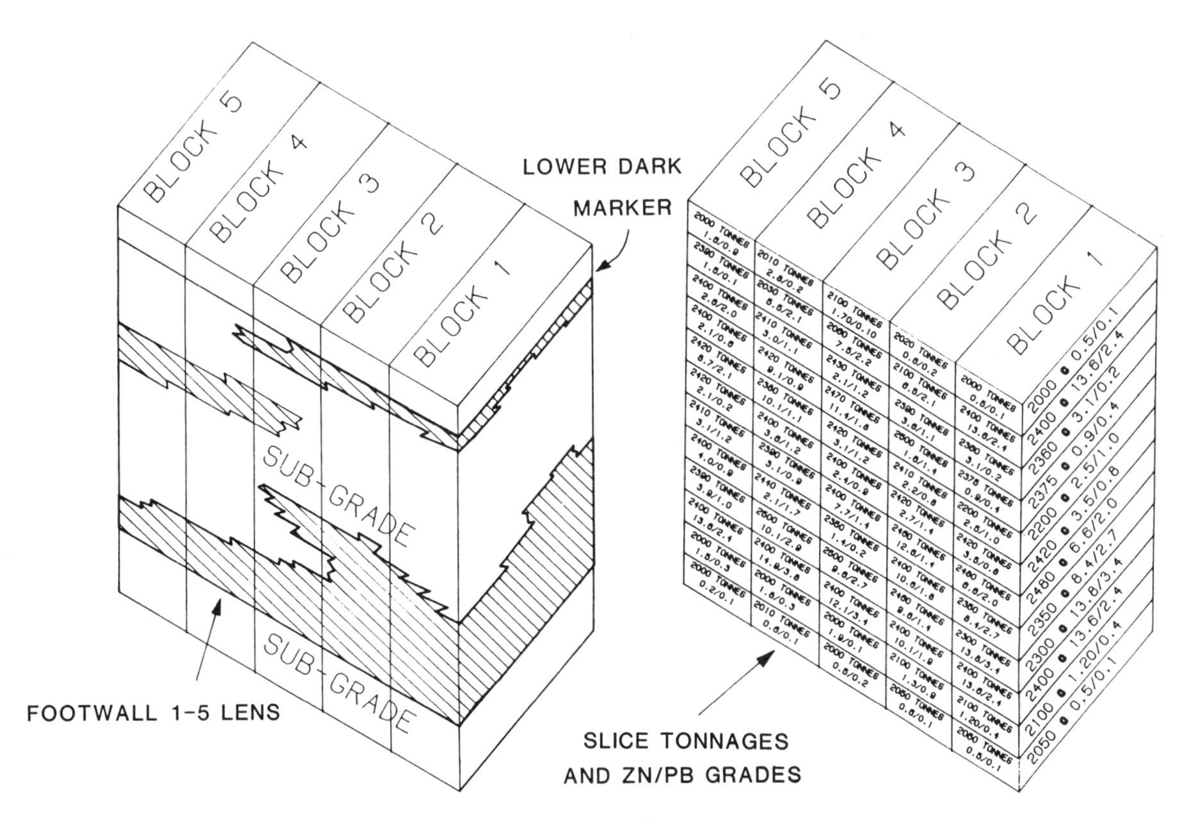

Fig. 8.63 3D block representation of the geology and reserve information from a portion of a 1–5 Lens stope. Based on a text figure from Ashton and Harte (1989).

and tonnage-weighted grades, can be produced. Allowance can be made for material which has to be left as support pillars and also for material already mined out in the form of drives and cross-cuts. At the present time, this is manually input to the computer via the keyboard. Calculation can also be made of the amount of internal waste in each stack of blocks. Hangingwall and footwall dilution can also be taken into account by including a proportion of each block above, or below, the blast-line at a grade equal to its kriged grade.

Although computerized geostatistical ore-reserve methods can be applied to most of the Navan orebody, ore occurring in some fault wedges and in Zone 3, is still calculated manually by inverse distance weighting methods. This is partly due to the complexity of the structure (abundant faulting and thickness fluctuations) and partly due to the small tonnages involved in these areas.

Much of the information produced can be colour coded on the basis of tonnage, zinc grade, zinc : lead ratio, etc., and displayed on a colour graphics terminal. Colour-coded plans can also be produced, for each or any slices, to display metal distribution patterns within the entire mine area.

The computer techniques developed at Navan (Ashton and Harte, 1989) have set a standard for the mining industry world-wide. They have produced a method which takes full regard of the geological controls of mineralization and of knowledge of its distribution stratigraphically. Also, it allows manual control at all stages in the process to ensure that the results at each stage are geologically meaningful. The use of advanced graphics packages has also allowed the mine to produce a wide range of different plans and sections and to update existing plans with the minimum of additional effort. The ore-evaluation process is clearly working well as there is a high degree of correlation between predicted grades and mill head grades. These predicted grades are based on stope estimates and development grades based on visual examination of clean faces.

REFERENCES

Anderson, I. K. (1990) Depositional Processes in the Genesis of the Navan Orebody, Unpublished PhD thesis, University of Strathclyde.

Andrew, C. J. and Ashton, J. H. (1985) Regional setting, geology and metal distribution patterns of the Navan orebody, Ireland. *Trans. IMM, B: Appl. Earth Sci.*, **94**, B66–93.

Ashton, J. H., Downing, D. T. and Finlay, S. (1986) The geology of the Navan orebody, in *Geology and Genesis of Mineral Deposits in Ireland* (eds C. J. Andrew, R. W. A. Crowe, S. Finlay *et al.*), Irish Association for Economic Geology, Dublin, pp. 243–280.

Ashton, J. H. and Harte, G. (1989) Technical computerization at Tara Mines Ltd, Navan. *Trans. A IMM*, **98**, 85–97.

Libby, D. J., Downing, D. T., Ashton, J. H. *et al.* (1985) The Tara Mines Story. *Trans. A IMM*, **94**, 1–41.

Index